A Brief Introduction to World Regional Geography

Edited by Steve Wolfe

A Brief Introduction to World Regional Geography

ISBN: 978-1-943536-38-2
Edition 1.0 Fall 2017
© 2017 Chemeketa Community College

Chemeketa Press

Publisher: Tim Rogers
Managing Editor: Steve Richardson
Production Editor: Brian Mosher
Design Editor: Ronald Cox IV
Layout: Emily Evans, Michael Ovens, Keyiah McClain, and Shaun Jaquez
Cover photo by National Aeronautics and Space Administration (NASA) is in the public domain.

Edited by Steve Wolfe, Chemeketa Community College, with additional contributions by Megan Cogswell and Michaela Daughters. Additional support has been provided by the Social Sciences Division at Chemeketa Community College, R. Taylor, dean.

Text and image acknowledgments appear on pages 373 to 384 and constitute an extension of the copyright page.

Printed in the United States of America.

Contents

Chapter 1

Introduction to the World

1.1 Geography Basics

Learning Objectives:

1. Understand the focus of geography and the two main branches of the discipline.
2. Learn about the tools geographers use to study the earth's surface.
3. Utilize the grid system of latitude and longitude to define locations and explain how it relates to seasons and time zones.
4. Distinguish between the different types of regional distinctions recognized in geography.
5. Understand the spatial nature of geography and how each place or region is examined, analyzed, and compared with other places or regions.
6. Identify the basic geographic realms and their locations.

A. What is Geography?

Geography (from the Greek geo, which means "Earth," and graphein, which means "to write") is the spatial study of the earth's surface. Geographers study the earth's physical characteristics, its inhabitants and cultures, phenomena such as climate, and the earth's place within the universe. Geographers examine the spatial relationships between all physical and cultural phenomena in the world, since what happens in one place affects other places and people. Geographers also look at how the earth, its climate, and its landscapes are changing due to cultural intervention.

Geography is a much broader field than many people realize. Most people think of area studies as the whole of geography. Some think it is simply memorizing lists of places and their locations. In reality, geography is the study of the entire earth, including how human activity has changed it. Physical geography involves all the planet's physical systems. Human geography incorporates studies of human culture, spatial relationships, interactions between humans and the environment, and many other areas of research that involve the different subspecialties of geography.

The discipline of geography bridges the social sciences with the physical sciences and can provide a framework for understanding our world. By studying geography, we can begin to understand the relationships and common factors that tie our human community together and link us to the natural environment. For instance, as a result of the rapid transfer of information and technology and the growth of modes of transportation and communication, the world is undergoing globalization on a massive scale. Globalization refers to the spread of ideas and information worldwide, making distant places become more and more similar. A common example is fast food restaurants, such as McDonald's, Pizza Hut, and Kentucky Fried Chicken, which are now found in countries around the world, creating a more homogeneous culinary landscape. Some are concerned that globalization is creating a uniform landscape around the world, often at the expense of local culture. The more we understand our world, the better prepared we will be to address issues such as this.

There are many approaches to studying world geography. A systematic approach focuses on studying one topic at a time, such as population, and looking at the worldwide patterns of that phenomenon before moving on to another topic. This textbook takes a regional approach, in which one region at a time is studied in its entirety, describing its landforms, climate, population, languages, religions, natural resources, political geography, and so on. This book also focuses on themes that illustrate the globalization process, in order to help you better understand your global community and its current affairs.

B. Thinking Geographically

Geography helps us make sense of the world through four historical traditions:

1. Spatial analysis
2. Earth science
3. Area (or regional) studies
4. Human-environment interaction

Spatial analysis involves analyzing where phenomena are located on the surface of the earth, as well as where they are located relative to other phenomena. For example, a geographer might use maps of climates and population to discover which types of climates are home to the most people. **Earth science** includes the study of landforms, climates, and the distribution of plants and animals. **Area or regional studies** focuses on a particular region to understand the dynamics of a specific interaction between human activity and the landscape. Researchers studying **human-environment interaction** examine the impact of humans on the natural landscape and find out how different cultures have used and changed their environments.

Geography provides the tools to integrate knowledge from many disciplines into a usable form by providing a sense of place to natural or human events. You will find that geography often explains why or how something occurs in a specific location. World geography utilizes the spatial approach to help understand the components of our global community.

When thinking geographically, geographers try to answer three fundamental questions:

1. Where are things located?
2. Why are they located where they are?
3. What difference does this make (how does it affect us)?

C. Branches of Geography

The discipline of geography can be broken down into two main branches: **physical geography** and **human geography**. These two main areas are similar in that they both use a spatial perspective, and they both include the study of place and the comparison of one place with another. Each branch, though, focuses on different aspects of the world.

Physical geography is the spatial study of natural phenomena that make up the environment, such as rivers, mountains, landforms, weather, climate, soils, plants, and any other physical aspects of the earth's surface. Physical geography focuses on geography as a form of earth science. It tends to emphasize the main physical parts of the earth—the **lithosphere** (surface of the earth's crust), the **atmosphere** (air), the **hydrosphere** (water), and the **biosphere** (living organisms)—and the relationships between these parts. Each of these four spheres is the focus of study of one of the major subfields within physical geography:

- **Geomorphology** (the study of the earth's surface features)
- **Climatology** (the study of climate and climate change)
- **Hydrography** (the study of water resources and their distribution)
- **Biogeography** (the study of the geographic patterns of species distribution)

Some physical geographers study the earth's place in the solar system. Others are environmental geographers, part of an emerging field that studies the spatial aspects and cultural perceptions of the natural environment. Environmental geography requires an understanding of both physical and human geography, as well as an understanding of how humans conceptualize their environment and the physical landscape.

Physical landscape is the term used to describe the natural terrain at any one place on the planet. The natural forces of erosion, weather, tectonic plate action, and water have formed the earth's physical features. Many U.S. state and national parks, such as Yellowstone, Yosemite, and the Grand Canyon, attempt to preserve unique physical landscapes for the public to enjoy.

Human geography is the study of human activity and its relationship to the earth's surface. Human geographers examine the distribution of human populations, religions, languages, ethnicities, political systems, economics, urban dynamics, and other components of human activity. They study patterns of interaction between human cultures and various environments and focus on the causes and consequences of human settlement and distribution over the landscape. While the economic and cultural aspects of humanity are primary focuses of human geography, these aspects cannot be understood without describing the physical landscape on which economic and cultural activities take place.

The **cultural landscape** is the term used to describe those parts of the earth's surface that have been altered or created by humans. For example, the urban cultural landscape of a city may include buildings, streets, signs, parking lots, or vehicles, while the rural cultural landscape may include fields, orchards, fences, barns, or farmsteads. Cultural forces unique to a given place—such as religion, language, ethnicity, customs, or heritage—influence the cultural landscape of that place at a given time.

D. Tools of Geography

Geospatial techniques are tools used by geographers to illustrate, manage, and manipulate spatial data. Because maps are powerful graphic tools that allow us to illustrate relationships and processes at work in the world, cartography and geographic information systems have become important in modern sciences. **Cartography** is the art and science of making maps, which illustrate data in a spatial form and are invaluable in understanding what is happening at a given place at a given time.

Making maps and verifying a location have become more exact with the development of the **global positioning system (GPS)**. A GPS unit can receive signals from orbiting satellites and calculate an exact location in latitude and longitude, which is helpful for determining where one is located on the earth or for verifying a point on a map. GPS units are standard equipment for many transportation systems and have found their way into products such as cell phones, handheld computers, fish finders, and other mobile equipment. GPS technology is widely implemented in the transport of people, goods, and services around the world.

Remote sensing technology acquires data about the earth's surface through aerial photographs taken from airplanes or images captured by satellites orbiting the earth. Remotely sensed images allow geographers to identify, understand, or explain a particular landscape or determine the land use of a place. These images can serve as important components in the cartographic (map-making) process. These technologies provide the means to examine and analyze changes on the earth's surface caused by natural or human forces. Google Earth is an excellent example of a computer tool that illustrates remotely sensed images of locations on the earth.

Geographic information systems (GIS) use a computer program to assimilate and manage many layers of map data, which then provide specific information about a given place. GIS data are usually in digital form and arranged in layers. The GIS computer program can sort or analyze layers of data to illustrate a specific feature or activity. GIS programs

Figure 1. Low elevation air photo of a cultural landscape in Hillsboro, Oregon.

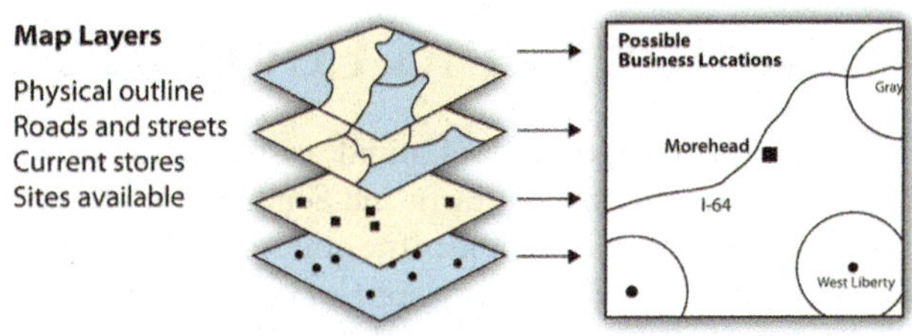

GIS programs can process layers of map data into one specific map of a location.

Figure 2. Illustration of layers in a GIS.

are used in a wide range of applications, from determining the habitat range of a particular species of bird to mapping the hometowns of university students.

GIS specialists often create and analyze geographical information for government agencies or private businesses. They use computer programs to take raw data to develop the information these organizations need for making vital decisions. For example, in business applications, GIS can be used to determine a favorable location for a retail store based on the analysis of spatial data layers such as population distribution, highway or street arrangements, and the locations of similar stores or competitive establishments. GIS can integrate a number of maps into one to help analysts understand a place in relation to their own specific needs.

GIS has revolutionized the field of cartography: nearly all cartography is now done with the assistance of GIS software. Additionally, analysis of various cultural and natural phenomena through the use of GIS software and specialized maps is an important part of urban planning and other social and physical sciences. Students interested in a career in geography are well served to learn geospatial techniques and gain skills and experience in GIS and remote sensing, as they are the areas within geography where employment opportunities have grown the most over the past few decades.

E. Absolute and Relative Location

When identifying a region or place on the earth, the first step is to understand its relative and absolute location. **Relative location** describes a location on the earth's surface with reference to other places, taking into consideration features such as transportation access or terrain. For example, Salem, Oregon is located about an hour south of Portland along the Willamette River, and is between the Coast Range to the west and the Cascade Range to the east.

Absolute location, on the other hand, describes a location on the earth's surface using some kind of coordinate system, such as a street address or latitude and longitude. Absolute location is vital to the cartographic process and to human activities that require an agreed-upon method of identifying a place or point.

Geographers and cartographers organize locations on the earth using a series of imaginary lines that encircle the globe. Lines of latitude and longitude allow any location on the earth to have an identifiable address of degrees north or south and east or west. For example, Salem, Oregon is located at 45°N 123°W. This means it is 45 degrees north of the equator, and 123 degrees west of the prime meridian.

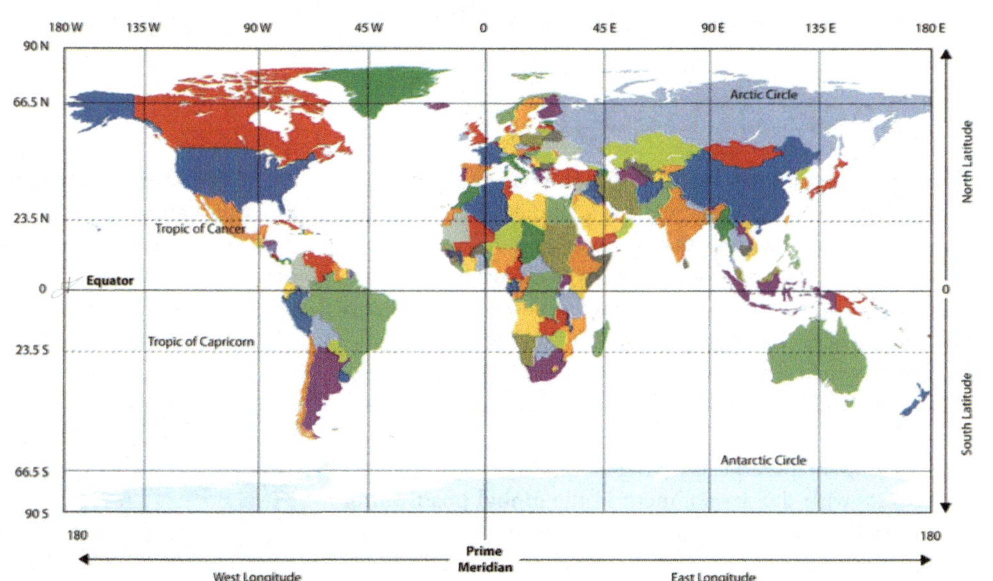

Figure 3. Basic lines of latitude and longitude.

F. Latitude and Longitude

The **equator** is the largest circle of latitude on Earth. The equator divides the earth into the Northern and Southern Hemispheres and is located at 0 degrees latitude. The other lines of latitude are numbered from 0 to 90 degrees going toward each of the poles. The lines north of the equator toward the North Pole are north latitude, and each of the numbers is followed by the letter "N." The lines south of the equator toward the South Pole are south latitude, and each of the numbers is followed by the letter "S." The equator (0° latitude) is the only line of latitude without any letter following the number. Notice that all lines of latitude are parallel to the equator (they are often called parallels) and that the North Pole equals 90 degrees N and the South Pole equals 90 degrees S. Noted parallels include both the Tropic of Cancer and the Tropic of Capricorn, which are 23.5 degrees from the equator. At 66.5 degrees from the equator are the Arctic Circle and the Antarctic Circle near the North and South Pole, respectively.

The **prime meridian** sits at 0 degrees longitude and divides the earth into the Eastern and Western Hemispheres. The prime meridian is defined as an imaginary line that runs through the Royal Observatory in Greenwich, England, a suburb of London, and in fact it is often referred to as the Greenwich meridian. The Eastern Hemisphere includes the continents of Europe, Asia, and Australia, while the Western Hemisphere includes North and South America. All meridians (lines of longitude) east of the prime meridian are numbered from 1 to 179 degrees east (E); the lines west of the prime meridian are numbered from 1 to 179 degrees west (W). The 0 and 180 lines do not have a letter attached to them. The **International Date Line** (located at or near 180 degrees longitude) is opposite the prime meridian. Each day officially starts at 12:00 a.m., at the International Date Line. The International Date Line does not follow the 180-degree meridian exactly. A number of alterations have been made to the International Date Line to accommodate political agreements to include an island or country on one side of the line or another.

G. Climate and Latitude

The earth is tilted on its axis 23.5 degrees from vertical. As it rotates around the sun, the tilt of the earth's axis provides different climatic seasons because of the variations in the angle of direct sunlight on the planet. Places receiving more direct sunlight, such as near the equator, experience a warmer climate. Near the poles, the lower angle of incoming solar radiation results in more reflected sunlight and thus a cooler climate. The Northern Hemisphere experiences summer when the Northern Hemisphere is tilted toward the sun, and winter when it is tilted away from the sun. Seasons in the Southern Hemisphere are the opposite of the Northern Hemisphere, since when the Northern Hemisphere is tilted toward the sun, the Southern Hemisphere is tilted away from the sun, and vice versa.

The **Tropic of Cancer** is the parallel at 23.5 degrees north of the equator, which is the most northerly place on Earth to receive direct vertical rays of sunlight (on or about June 21). The **Tropic of Capricorn** is the parallel at 23.5 degrees south of the equator and is the most southerly location on Earth to receive direct vertical rays of sunlight (on or about December 21).

The region between the Tropics of Cancer and Capricorn is known as the tropics. This area does not experience dramatic seasonal changes because the amount of direct sunlight received does not vary widely. The higher latitudes (north of the Tropic of Cancer and south of the Tropic of Capricorn) experience significant seasonal variation in climate.

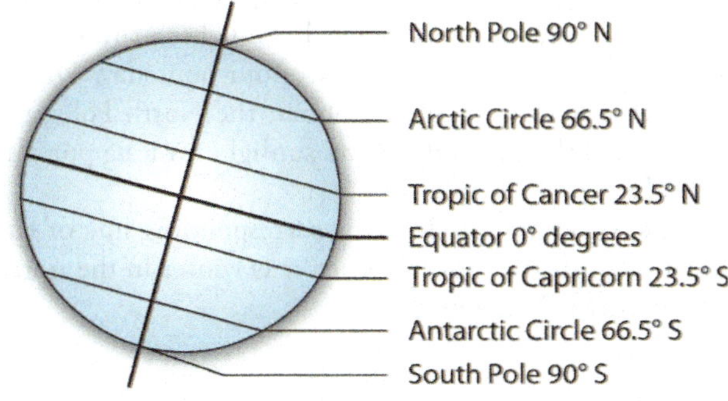

Figure 4. Important lines of latitude.

Figure 5. 45th parallel sign near Baker City, Oregon.

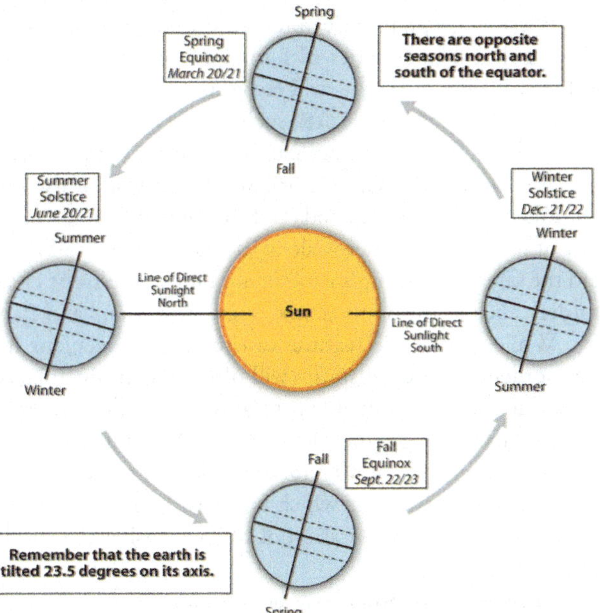

Figure 6. Graphic of the four seasons.

The **Arctic Circle** is the line of latitude at 66.5 degrees north. During winter, the North Pole is tilted away from the sun and does not receive much sunlight. At times, it is dark for all or most of the twenty-four-hour day. During the Northern Hemisphere's summer, the North Pole faces more toward the sun and receives sunlight for long portions of the day.

The **Antarctic Circle** is the corresponding line of latitude at 66.5 degrees south. When it is winter in the north, it is summer in the south.

The Arctic and Antarctic Circles mark the extremities (southern and northern, respectively) of the polar day (twenty-four-hour sunlit day) and the polar night (twenty-four-hour sunless night). North of the Arctic Circle, the sun is above the horizon for twenty-four continuous hours at least once per year and below the horizon for twenty-four continuous hours at least once per year. This is true also near the Antarctic Circle, but it occurs south of the Antarctic Circle, toward the South Pole. Equinoxes, when the line of direct sunlight hits the equator and days and nights are of equal length, occur in the spring and fall on or around March 21 and September 22.

H. Regions in Geography

A region is a basic unit of study in geography—a unit of space characterized by a feature such as a common government, language, political situation, or landform. A region can be a country governed by political boundaries, such as France or Canada; a region can be defined by a landform, such as the drainage basin of all the water that flows into the Mississippi River; and a region can even be defined by the area served by a shopping mall. Cultural regions can be defined by similarities in human activities, traditions, or cultural attributes such as language or religion.

There are three basic types of regions: formal, functional, and vernacular. **Formal regions** (also sometimes called uniform regions) are usually precisely defined and are rather objective, with easily recognized boundaries. Examples include political units, such as counties, states, provinces, or countries. Physical regions are often formal regions, including landform regions, such as the Rocky Mountains or Great Plains, climate regions, or vegetation regions. Cultural regions can also often have formal boundaries, including language regions, religious regions, or areas dominated by certain ethnic groups.

Physical geographic features have a huge influence on where political boundaries of formal regions are set. If you look at a world map, you will recognize that many political boundaries are natural features, such as rivers, mountain ranges, and large lakes. For example, between the United States and Mexico, the Rio Grande makes up a portion of the border. Likewise, between Canada and the United States, a major part of the eastern border is along the Saint Lawrence Seaway and the Great Lakes. Alpine mountain ranges in Europe create borders, such as the boundary between Switzerland and Italy.

While geographic features can serve as convenient formal borders, political disputes will often flare up in adjacent areas, particularly if valuable natural or cultural resources are found within the geographic features. Oil drilling near the coast of a sovereign country, for example, can cause a dispute between countries about which one has dominion over the oil resources. The exploitation of offshore fisheries can also be disputed.

Functional regions have boundaries related to a practical function within a given area. For example, a function-al region can be defined by a newspaper service or delivery area. If the newspaper goes bankrupt, the functional region no longer exists. Church parishes, shopping malls, a city and its suburbs, and business service areas are other examples of functional regions. They function to serve a region and may have established boundaries for limits of the area to which they will provide service. An example of a common service area—that is, a functional region—is the region to which a local pizza shop will deliver.

Vernacular regions have loosely defined boundaries based on people's perceptions or thoughts. Because of this, they are sometimes referred to as popular or perceptual regions. Vernacular regions are subjective—that is, different people may have different opinions about the limits of the regions. Vernacular regions include concepts such as the region called the "Middle East." Many people have a rough idea of the Middle East's location but do not know precisely which countries make up the Middle East. Also, in the United

Size in Area: Approximately 98,466 square miles
Population: About 4 million people (2015)
Population Density: About 40.6 people per square mile

Figure 7. Comparison information for state of Oregon.

States, the terms Pacific Northwest, Midwest, or South have many variations. Each individual might have a different idea about the location of the boundaries of these regions. Whether the state of Kentucky belongs in the Midwest or in the South might be a matter of individual perception. Similarly, various regions of the United States have been referred to as the Rust Belt, Sun Belt, or Bible Belt without a clear definition of their boundaries. The limit of a vernacular area is more a matter of perception than of any formally agreed-upon criteria. Nevertheless, most people would recognize the general area being discussed when using one of the vernacular terms in a conversation.

I. World Regional Geography

World regional geography studies various world regions as they compare with the rest of the world. Factors for comparison include both the physical and the cultural landscape. Major questions include: Who lives there? What are their lives like? What do they do for a living? Physical factors of significance can include location, climate type, vegetation, and terrain. Human factors include cultural traditions, ethnicity, language, religion, economics, and politics.

World regional geography focuses on regions of various sizes across the earth's landscape and aspires to understand the unique character of regions in terms of their natural and cultural attributes. This textbook takes a regional approach with a focus on themes that illustrate the globalization process, which in turn helps us better understand our global community.

The regions studied in world regional geography can be combined into larger portions called realms. **Realms** are large areas of the planet, usually with multiple regions, that share the same general geographic location. Regions are cohesive areas within each realm. This textbook divides the world into the following twelve realms:

1. Europe
2. Russia
3. North America
4. Australia and New Zealand
5. The Pacific and Antarctica
6. Middle America
7. South America
8. Sub-Saharan Africa
9. North Africa and Southwest Asia
10. South Asia
11. East Asia
12. Southeast Asia

Key Takeaways

1. Geography is the spatial study of the earth's surface. The discipline of geography bridges the social sciences with the physical sciences. The two main branches of geography include physical geography and human geography. GIS, GPS, and remote sensing are tools that geographers use to study the spatial nature of physical and human landscapes.

2. A grid system divides the earth by lines of latitude and longitude that allow for the identification of absolute location on the earth's surface through coordinates measured in degrees. There are twenty-four time zones that are each roughly 15 degrees wide and organize time intervals around the world.

3. The tilt of the earth's axis at 23.5 degrees helps create the earth's seasonal variations. For example, the northern hemisphere exper-iences summer when it is tilted toward the sun, and winter when it is tilted away from the sun. The line of direct sunlight always hits the earth between 23.5 degrees north (the Tropic of Cancer, about June 21) and 23.5 degrees south (the Tropic of Capricorn, about December 21).

4. A region (or realm) is the basic unit of study in geography. There are three main types of regions: formal, functional, and vernacular. World regional geography is the study of the world realm by realm.

1.2 The Environment and Human Activity

Learning Objectives

1. Explain how climate and human habitation are related and distinguish between the main climate types.
2. Explain the dynamics of tectonic plates and their relationship to earthquakes and volcanic activity.
3. Outline the main causes of and problems with deforestation. Explain the relationship between deforestation and climate change.
4. Point out where the rain shadow effect takes place and explain why it occurs in those places and how it may influence human activity.
5. Understand how climate change occurs and the relationship between greenhouse gasses such as carbon dioxide and the planet's temperature regulation.

A. Climate and Human Habitation

The earth's ability to receive and absorb sunlight is a primary factor in the earth's environment, and it also has a big impact on human populations. There are no large cities or human communities in Antarctica because it is so cold; most of the sunlight filtering down to Antarctica is reflected off the earth at that latitude because of the tilt of the earth's axis and the resulting low angle of incoming solar radiation. Answering the basic questions of where most humans live on Earth and why they live there depends on understanding climate. Temperate midlatitude climates usually provide the greatest opportunities for human habitation.

Since the region between the Tropic of Cancer and the Tropic of Capricorn receives the most direct sunlight throughout the year, it is favorable to plant and animal life, provided there is adequate moisture or precipitation. Humans have been living in the tropics for a long time, even when the ice sheets were covering parts of the midlatitudes. The problem with the tropics is that the soils are usually of poor quality and the nutrients have been leached out. **Leaching** is a process in which nutrients in the soil dissolve in water and are carried below the root zone of a crop as the water percolates downward.

Today, when we look at the earth and the distribution of human population, two main factors attract human habitation: moderate climates and access to water. More than 70 percent of the earth's surface is covered with water. The problem is that less than 3 percent of the water is fresh, and most of that fresh water is stored in ice caps near the North or South Pole. This leaves less than 1 percent of the world's fresh water for human use, usually in lakes, rivers, streams, or groundwater and underground aquifers. Climate plays an important role in where humans live because precipitation is necessary for growing crops, raising livestock, and supplying fresh water to urban communities.

Figure 8. Glaciers in Glacier National Park, Montana
Mountainous type H climates vary with elevation, with warmer temperatures at the base and colder temperatures at higher elevations.

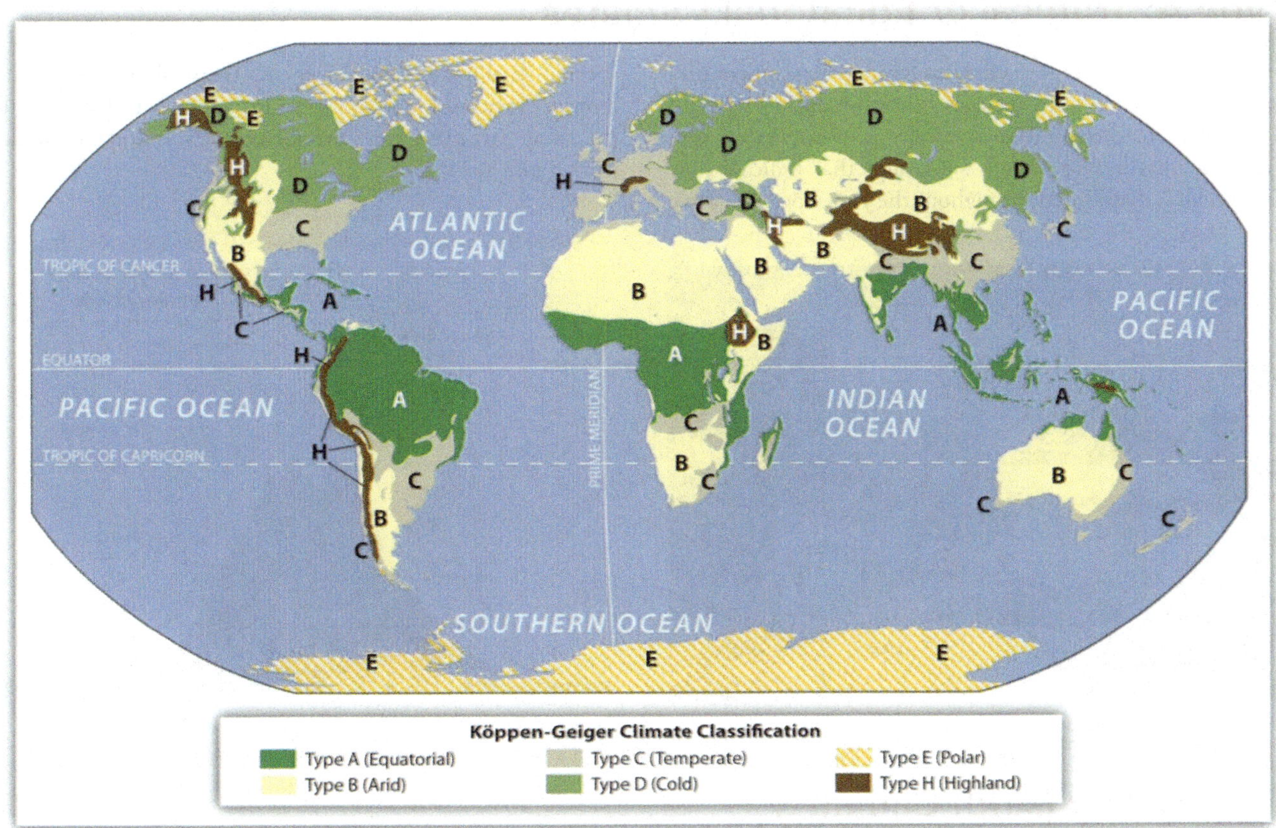

Figure 9. Basic climate regions based on the Köppen-Geiger Classification System (local conditions may vary widely)

Geographers have developed various classification systems to identify climate types. **Climate** refers to the long-term average weather conditions in a particular region of the world. **Weather** refers to the atmospheric conditions at a given time and place. In other words, "Climate is what you expect, but weather is what you get!" The two main elements in describing both weather and climate are temperature and precipitation. For the purposes of this overview of world geography, the various climate types have been broken down into six basic types—A, B, C, D, E, and H—after the Köppen-Geiger classification system.

- Type A: Tropical or equatorial climates

- Type B: Dry (arid) climates

- Type C: Moderate or temperate (mild midlatitude)climates

- Type D: Cold or continental (severe midlatitude)climates

- Type E: Polar or extreme climates

- Type H: (Unclassified) highland climates

Type A: Tropical or Equatorial Climates

The humid tropical type A climates have warm temperatures year round (every month averages at least 64 degrees F), with a high level of precipitation, typically in the form of rain. Type A climates have various subgroups that indicate how the rainfall is distributed throughout the year. There are three specific A climates. Two of them (**tropical monsoon** and **tropical savanna**) have a dry season and a wet season, while the other (**tropical rain forest**) receives abundant rainfall throughout the year.

Figure 10. Coastal Belize, a type A climate example.

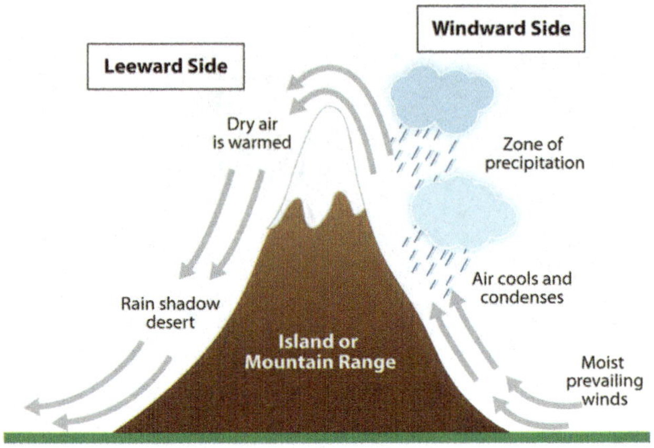

Figure 11. Rain shadow effect for a location in the tropics (windward and leeward sides are reversed in the midlatitudes).

Type B: Dry (Arid) Climates

There are two types of Type B climates: **desert** and **steppe (semi-desert)**. Deserts are defined by aridity and average less than about 10 to 15 inches of precipitation per year. Moisture losses by evaporation exceed precipitation. **Xerophytic plants** (which have adaptations such as deep roots, small leaves or needles, or thick, waxy leaves to reduce transpiration and conserve moisture) dominate the rocky or sandy soils which lack humus. Steppes average about twice as much precipitation as deserts and support more plant life. Short grasses are the dominant vegetation, creating fertile soils that are rich in **humus** (decaying plant material). Livestock are often grazed in these grasslands.

There is a direct relationship between highlands and type B climates in various places in the world. This condition, known as the **rain shadow effect**, occurs when one side of a mountain range (the **windward** side, or side facing the wind) receives abundant precipitation while the region on the other side of the mountain range (the **leeward**, or downwind side) is a desert or has more arid climate conditions. This phenomenon is evident wherever there is terrain with enough elevation to force air to rise, where it cools causing water vapor to condense and form clouds which may then drop precipitation.

The Hawaiian island of Kauai has an extreme example of the rain shadow effect. The island's windward (eastern) side receives more rain than almost any other place on Earth: as much as 460 inches (almost 40 feet!) a year. Only a part of the island, however, receives that amount of rain. The height of the mountains causes a rain shadow on the leeward (western) side, creating semi-desert conditions and type B climates.

Death Valley in California is also an example of the rain shadow effect. Little rain falls on Death Valley because most moisture in the prevailing westerly winds falls on the western side of the coastal mountains and Sierra Nevada. This leaves the leeward eastern side dry, including not only Death Valley, but the entire state of Nevada.

Note that since California and Nevada are in the midlatitudes, winds are generally moving from west to east, so the rain shadow is on the eastern side of the mountains. In Hawaii and other tropical regions, winds generally blow from east to west, so the rain shadow is on the western side. In both cases, however, the rain shadow is on the downwind, or leeward, side of the mountains.

Figure 12. Saguaro Cacti in Arizona, a type B climate example.

Type C: Moderate or Temperate (Mild Midlatitude) Climates

Often described as moderate in temperature and precipitation, type C climates are the most favorable to human habitation and therefore host the largest human population densities on the planet. Humans inhabit these areas because of the abundance of forests, farmland, and fresh water found in type C climate regions.

Type C climates are found mostly in the midlatitudes bordering the tropics. Seasonal changes are pronounced, with distinct winters and summers. Winters are cool to cold and summers are usually warm. Precipitation varies from low to high, depending on location.

Figure 13. Northern mixed forest in Appalachia, a type C climate example.

Type D: Cold or Continental (Severe Midlatitude) Climates

Type D climate regions are often found in the interiors of continents away from the moderating influence of large bodies of water. They are usually farther north than type C regions, resulting in colder winters. Seasonal variations exist, with cool to hot summers and cold winters. Precipitation is usually in the form of rain in summer and snow in winter. Regions with type D climates can be found in the Great Lakes region of the United States, much of Canada, and a large portion of Russia. Due to the virtual absence of land at these latitudes in the southern hemisphere, type D climates are only found in the northern hemisphere.

Figure 14. Snow covered neighborhood in Fargo, North Dakota, a Type D Climate example.

Type E: Polar Climates

Type E is an extreme climate type found in the polar regions near or to the north of the Arctic Circle and near or to the south of the Antarctic Circle. Regions with type E climates are cold with permanent ice or **permafrost** (permanently frozen ground). Vegetation is minimal, and there are no trees. Temperatures may warm slightly during the short summer months but rarely rise above 50 degrees. The two polar climates are **tundra** and **ice cap**. Tundra is dominated by low-growing vegetation (dwarf shrubs, grasses, mosses, and lichens) and is mostly found in extreme northern locations bordering the Arctic Ocean, such as northern Alaska, Canada, and Russia. Ice cap mostly occurs in the interior of Greenland and Antarctica, and has no vegetation.

Type H: Undifferentiated Highland Climates

Type H highland climates are not actually a climate type. Instead, they reflect the fact that in mountainous areas, the climate may change dramatically over short distances, resulting in numerous **microclimates** (small areas of the same climate). Mountain ranges can create a variety of climate types because of the change in elevation from the base of the range to the summit. Different climate types can be found on the same mountain at different elevations. Climates at the base of mountains will vary depending on whether the mountains are found in the tropics or in the higher latitudes. For example, high mountains near the equator may have a type A climate at their base and a type E climate at their summit with various type C and type D climates between them. Moving up in elevation, wind speed and precipitation tend to increase, while atmospheric pressure and temperature typically decrease.

B. Environmental Concerns

Deforestation

The planet's growing population has increased demands on natural resources, including forest products. Humans have been using trees for firewood, building homes, and making tools for millennia. Trees are a renewable resource, but deforestation occurs when they are removed faster than they can be replenished. Most people in rural areas in developing countries rely on firewood to cook their food. Many of these areas are experiencing a fast decline in the number of trees available. People living in mainly type B climates may not have access to a lot of trees to start with; therefore, when trees are cut down for firewood or for building materials, deforestation occurs. In the tropical areas, it is common for hardwood trees to be cut down for lumber to gain income or to clear the land for other agricultural purposes, such as cattle ranching. Countries that lack opportunities and advantages look to exploit their natural resources—in this case, trees—for either subsistence agriculture or economic gain. Deforestation has increased across the globe with the rapid rise in worldwide population.

Countries that are better off economically no longer have to cut down their own trees but can afford to substitute other resources or import lumber from other places. Developing regions of the world in Latin America, Africa, and parts of Asia are experiencing serious problems with deforestation. Deforestation is widespread: Residents of Haiti have cut down about 99 percent of the country's forests; most of the wood has been used as fuel to cook food. People in Afghanistan have cut down about 70 percent of their forests. Nigeria has lost about 80 percent of its old-growth forests since 1990. Ethiopia has lost up to 98 percent of its forested acreage, and the Philippines has lost about 80 percent of its forests.

Tropical rain forests only make up about 5 percent of the earth's surface but contain up to 50 percent of the earth's biodiversity. These forests are cut down for a variety of reasons. Norman Meyers, a British environmentalist, estimated that about 5 percent of deforestation in tropical regions is caused by the push for cattle production. Nineteen percent of these forests are cut down by the timber industry, 22 percent are cut down for the expansion of plantation agriculture, and 54 percent are removed due to slash-and-burn farming. Most tropical rain forests are located in the Amazon basin of South America, in central Africa, and in Southeast Asia. All these areas are looking for advantages and opportunities to boost their economies; unfortunately, they often target their tropical rain forests as a revenue source.

Forest ecosystems provide for a diverse community of organisms. Tropical rain forests are one of the most vibrant ecosystems on the planet. Their abundant biodiversity can provide insight into untapped solutions for the future. Plants and organisms in these habitats may hold the key to medical or biological breakthroughs, but wildlife and vegetation will be lost as deforestation eliminates their habitat and accelerates the extinction of endangered species.

Trees and plants remove carbon dioxide from the atmosphere and store it in the plant structure through the process of photosynthesis. Carbon dioxide is a major greenhouse gas that is a part of the climate change process. Carbon dioxide and other similar gases reduce the amount of long-wave radiation (heat) that escapes from the earth's atmosphere, resulting in increased temperatures on the planet. As more carbon dioxide is emitted into the atmosphere, more warming can occur. The removal of trees through deforestation results in less carbon dioxide being removed from the atmosphere. Slash-and-burn farming methods that burn forests release the carbon in the plant life directly into the atmosphere, further increasing the climate change effect.

Figure 15. Forest removal example. Forests are removed for timber, and burning the excess then clears the land for other purposes.

Figure 16. A man in Malawi carries firewood for cooking and heating purposes.

Figure 17. Lumber mill processing hardwood timber.

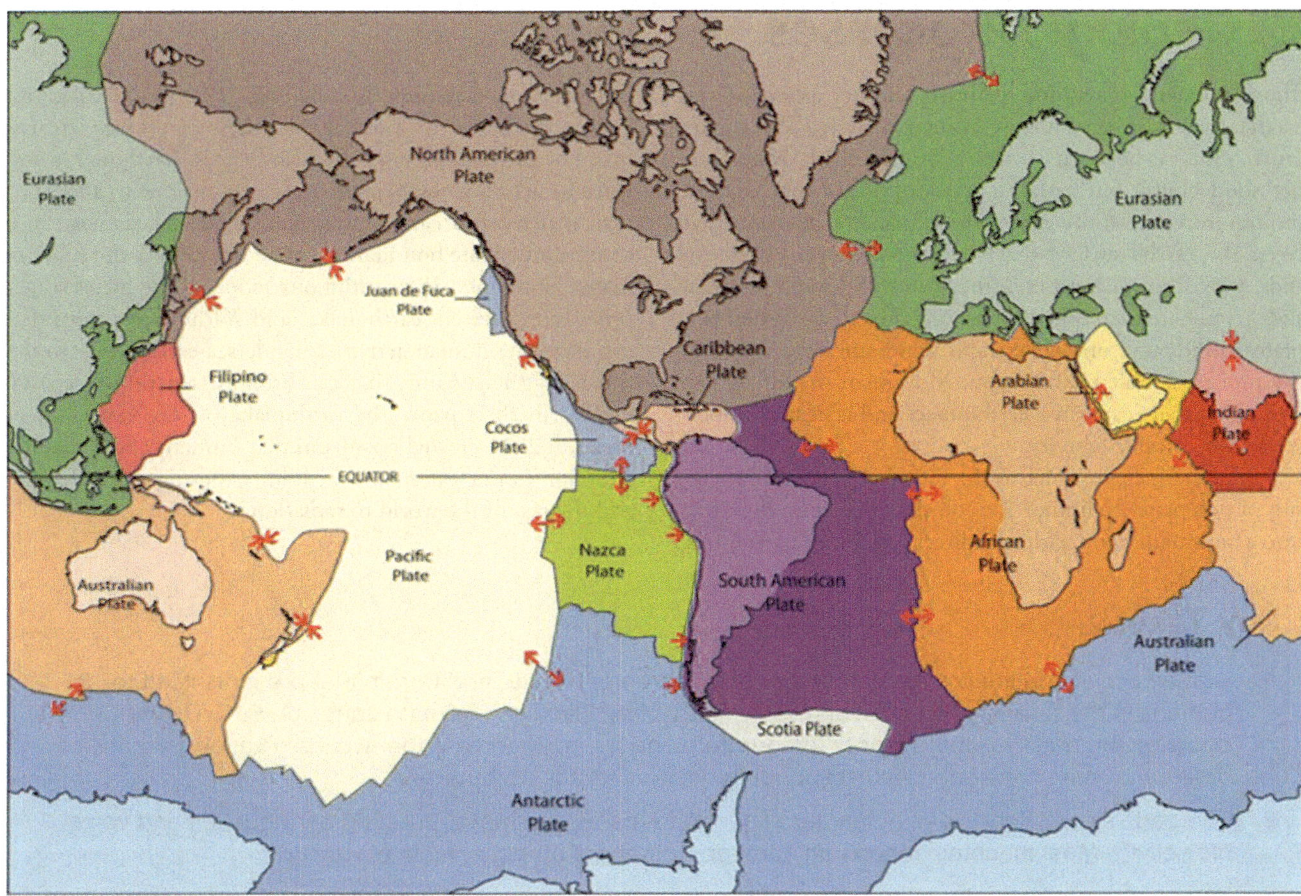

Figure 18. General pattern of tectonic plates.

Climate Change

The increase in temperature in our environment has gained much attention in recent years. Questions have been raised about the rate and extent of climate change around the world. Understanding the dynamics of the temperature increase can assist in understanding how it is related to human activity.

Since the 1960s, scientists have been concerned about the concentrations of carbon dioxide, methane, nitrous oxide, and chlorofluorocarbons in the atmosphere. These so-called greenhouse gases can trap heat energy emitted from the earth's surface and may increase global temperatures. Since the Industrial Revolution, human activity—the burning of fossil fuels and large-scale deforestation—has increased the amount of heat-trapping greenhouse gases in the atmosphere. Carbon dioxide and similar gases act like the glass panels of a greenhouse that allow shortwave radiation from the sun to enter but do not allow the long-wave radiation of heat to escape into space.

An increase in carbon dioxide and greenhouse gases in the atmosphere will normally cause an increase in the temperature, which in turn may cause changes in weather conditions in various places on Earth. Temperature changes may affect precipitation patterns and alter weather patterns, which may affect agricultural outputs and influence energy needs that can create increasing economic instability. Changes in climate also impact environmental conditions for organisms adapted to specific habitat ranges. When climates change, an organism's habitable zone may also change, which in turn can impact entire ecosystems.

Deforestation and the burning of fossil fuels can contribute to climate change. Fossil fuels such as coal, oil, and natural gas are created when dead plant and animal life are under pressure, decay for long periods, and retain their carbon component. Burning fossil fuels releases the carbon back into the atmosphere. The increasing need for energy and lumber by human activity may continue to contribute to climate change unless alternatives can be found. The increase in temperatures may result in melting of the ice caps, which in turn may raise sea levels, impacting human activity around the world.

C. Geologic Processes

The movement of tectonic plates is another aspect of the earth's dynamics that affects human activity. The earth's crust, which is between 10 and 125 miles thick, is not one big solid chunk but rather a series of plates that cover a molten iron core at the center of the planet. The plates that cover the earth's surface slowly shift and move. Plates can slide away from each other or they can collide, and they can slide parallel to each other in opposite directions. When two plates collide and one plate slides under an adjacent plate, the process is called subduction. Movement or shift where two plates meet can cause earthquakes and is usually associated with volcanic activity.

Mountain chains, such as the Himalayas, are a direct result of two plates colliding. The collision pushes up the earth into a mountain chain, either by direct pressure or by volcanic activity. Plates can move up to an inch a year in active regions. Driven by the earth's internal heat, these plates have created many of the planet's mountain landscapes. Earthquakes and volcanic action along plate boundaries continue to affect human activity and can cause serious economic damage to a community. Plate boundaries can be found near the edges of many continents. The continuous action of the plates sometimes causes serious earthquakes and volcanic eruptions that can devastate human activities. Undersea earthquakes sometimes trigger tsunamis that can bring destruction to coastal regions in their path. The earthquake off the east coast of Japan in 2011 created a tsunami that brought additional destruction to nuclear energy facilities, exposing parts of Japan and the rest of the world to radiation.

Key Takeaways

1. Human activity on the planet correlates with the type of climate and terrain that presents itself to humans in the form of natural resources or habitability. Five basic climate zones (A, B, C, D, and E) describe the earth's climate types. Temperature and precipitation are the two main variables that define a climate zone and its corresponding environmental attributes.

2. The earth's crust consists of a number of separate plates that move, creating earthquakes and volcanic activity. Most mountain ranges on Earth are a product of tectonic plate activity.

3. Removing trees faster than they can grow back is called deforestation. Humans are cutting down the forests in many areas at an unsustainable rate. Deforestation can result in soil erosion, changes in weather patterns, and the loss of habitats. Trees are being cut down for firewood, building materials, or to clear land for other activities.

4. Mountains can force air to rise and cause precipitation to fall on the windward side. The leeward side of the mountain range receives much less precipitation, creating an arid rain shadow.

5. Climate change is a phenomenon whereby gases such as carbon dioxide and methane increase in the atmosphere and restrict long-wave radiation from escaping the planet, which can result in warmer temperatures on Earth. Trees remove carbon dioxide from the atmosphere, which may reduce climate change.

1.3 Population and Culture

Learning Objectives

1. Explain the demographic transition process. Understand the concept of carrying capacity as it relates to the planet's human population.

2. Outline the relationship between urbanization and family size. Show how rural-to-urban shift relates to industrialization and the change in rural populations.

3. Interpret a population pyramid and determine if the population is increasing or declining and if the pace of growth is intensifying or slowing.

4. Distinguish between the concepts of culture and ethnicity as these terms are used in this textbook.

5. Understand the difficulty in determining the number of languages and religions existing on Earth. Name the main language families and the world's major religions.

A. Demographic Transition

Demography is the study of how human populations change over time and space. It is a branch of human geography related to population geography, which is the examination of the spatial distribution of human populations. Geographers study how populations grow and migrate, how people are distributed around the world, and how these distributions change over time.

For most of human history, relatively few people lived on Earth, and world population grew slowly. Only about five hundred million people lived on the entire planet in 1650 (that's less than half India's population in 2000). Things changed dramatically during Europe's Industrial Revolution in the late 1700s and into the 1800s, when declining death rates due to improved nutrition and sanitation allowed more people to survive to adulthood and reproduce. The population of Europe grew rapidly. However, by the middle of the twentieth century, birth rates in developed countries declined, as children had become an economic liability rather than an economic asset to families. Fewer families worked in agriculture, more families lived in urban areas, and women delayed the age of marriage to pursue education, resulting in a decline in family size and a slowing of population growth. In some countries today (e.g., Russia and Japan), population is actually in decline, and the average age in developed countries has been rising for decades. The process just described is called the demographic transition.

At the beginning of the twentieth century, the world's population was about 1.6 billion. One hundred years later, there were roughly six billion people in the world, and as of 2017, the estimate is now over 7.4 billion. This rapid growth occurred as the demographic transition spread from developed countries to the rest of the world. During the twentieth century, death rates due to disease and malnutrition decreased in nearly every corner of the globe. In developing countries with agricultural societies, however, birth rates remained high. Low death rates and high birth rates resulted in rapid population growth. Meanwhile, birth rates—and family size—have also been declining in most developing countries as people leave agricultural professions and move to urban areas. This means that population growth rates—while still higher in the developing world than in the developed world—are declining. Although the exact figures are unknown, demographers expect the world's population to stabilize by around 2100 and then decline somewhat.

In 2017, the world's population is growing at the rate of about 1.2 percent per year, or about 90 million people per year, with that growth found almost exclusively in developing countries, as populations are stable or in decline

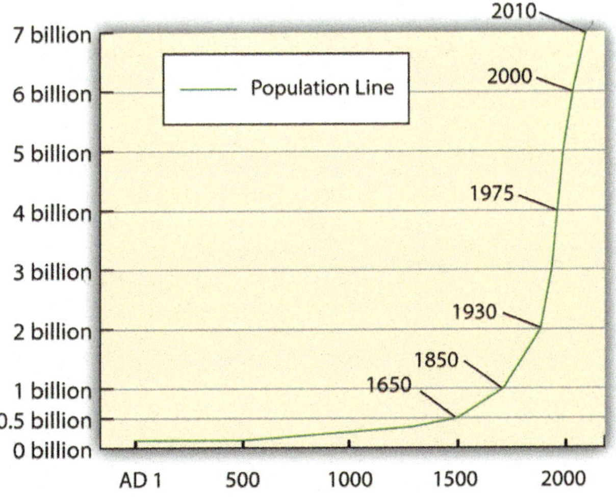

Figure 19. Population growth from year 1 to year 2010 AD.

in places such as Europe and North America. World population increase is pronounced on the continent of Asia: China and India are the most populous countries in the world, each with more than a billion people, and Pakistan is an emerging population giant with a high rate of population growth. The three largest population clusters in the world currently are the regions of east Asia, south Asia, and Europe, followed closely by southeast Asia. The continent of Africa, however, has the highest fertility rates in the world, with countries such as Nigeria—Africa's most populous and the world's eighth most populous country—growing rapidly each year.

The most striking paradox within population studies is that while there has been a marked decline in fertility (declining family size) in developing countries, the world's population will grow substantially by 2030 because of the compounding effect of the large number of people already in the world—that is, even though population growth rates are in decline in many countries, the population is still growing. A small growth rate on a large base population still results in the birth of many millions of people.

Providing food, energy, and materials for these additional humans will tax many countries of the world, and poverty, malnutrition, and disease are expected to increase in regions with poor sanitation, limited clean water, and lack of economic resources. In 2010, more than two billion people (one-third of the planet's population) lived in abject poverty and earned less than the equivalent of two US dollars per day. The carrying capacity of the planet (the number of people it can sustainably support) is not and cannot be known, but there is the possibility that we have already reached the threshold of the earth's carrying capacity.

Population growth exacts a toll on the earth as more people use more environmental resources. The resources most immediately affected by increased populations include forests (a fuel resource and a source of building material), fresh water supplies, and agricultural soils. These systems get overtaxed, and their depletion has serious consequences. Type C climates, which are moderate and temperate, are usually the most productive and are already vulnerable to serious deforestation, water pollution, and soil erosion. Maintaining adequate food supplies will be critical to supporting a sustainable population.

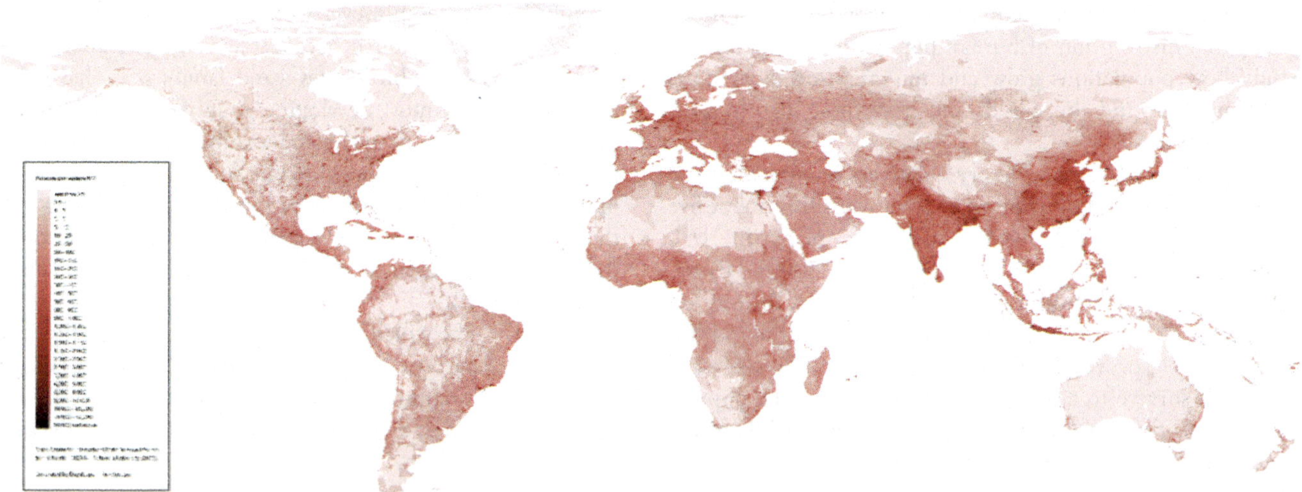

Figure 20. Human population clusters on the planet. The three main human population clusters on the planet are eastern Asia, southern Asia, and Europe. Most of these regions with high population densities are in type C climates.

B. Urbanization and Family Size

As countries move from an agricultural to an industrial economy, there is a major shift in population from rural to urban settings. The Industrial Revolution of the nineteenth century ushered in major technological developments and changes in labor practices, which encouraged migration from the farm to the city. Because of increased mechanization, fewer farm workers were needed to produce larger agricultural yields. At the same time, factories in urban areas had a great need for industrial workers. This shift continued into the information age of the late twentieth century and continues in many parts of the developing world in the current century.

A basic principle of population growth that addresses this rural-to-urban shift states that as countries industrialize and urbanize, family size typically decreases and incomes traditionally increase. Though this may not be true in all cases, it is a general principle that is consistent across cultural lines. Agricultural regions generally have a larger average family size than that of their city counterparts. Fertility rate is the average number of children a woman in a particular country has in her lifetime, whether or not they all live to adulthood. If a fertility rate for a given country is less than 2.1—the replacement level in developed countries—the population of that country is in decline, unless there is significant immigration. A fertility rate greater than 2.1 indicates that the country's population is increasing. Some children will never reach reproductive age nor have children of their own, so the replacement rate has to be slightly greater than 2. In developing countries with high infant mortality rates, the replacement level may be as high as 2.5 or more. The concept of fertility rate is slightly different from the term family size, which indicates the number of living children raised by a parent or parents in the same household. In this textbook, family size is used to illustrate the concept of population growth and decline.

C. Population Demands

A country's demographic statistics can be illustrated graphically by a population pyramid (also called an age-sex diagram). A population pyramid is essentially two bar graphs that depict male and female age cohorts either in absolute size or as a percentage of the total population. Male cohorts are typically shown on the left side of the pyramid, and females are on the right side.

The shape of a country's population pyramid tells a story about the history of its population growth. For example, a high-growth-rate country has a pyramid that is narrow at the top and wide at the bottom, showing that every year more children have been born than the year before. As family size decreases and women in a society have fewer children, the shape of the pyramid changes. A population pyramid for a postindustrial country that has negative growth would be narrower at the bottom than in the middle, indicating that there are fewer children than middle-aged people. Four basic shapes indicate the general trends in population growth:

1. Rapidly expanding
2. Expanding
3. Stationary
4. Contracting

These shapes also illustrate the percentage of a population under the age of fifteen or over the age of sixty-five, which are standard indicators of population growth. Many postindustrial countries have a negative population growth rate. Their population pyramids are narrow at the bottom, indicating an urbanized population with small family sizes.

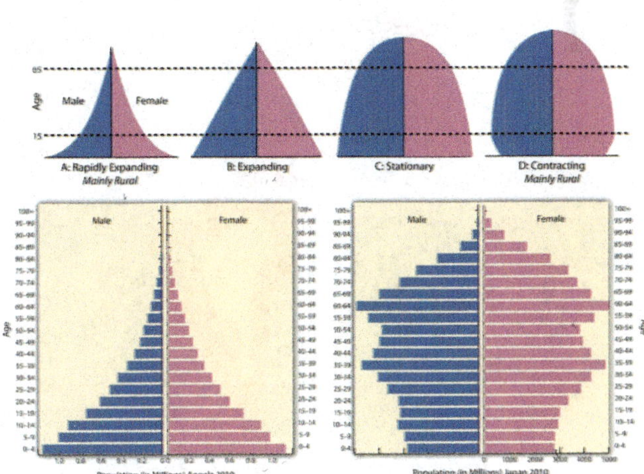

Figure 21. Population Pyramids. Angola had a fertility rate of 5.6 in 2011. Japan had a fertility rate of 1.4 in 2011.

D. Culture and Ethnicity

The term culture is often difficult to differentiate from the term ethnicity. In this textbook, ethnicity indicates traits people are born with, including genetic backgrounds, physical features, or birthplaces. People have little choice in matters of ethnicity. The term culture indicates what people learn after they are born, including language, religion, and customs or traditions. Individuals can change matters of culture by individual choice after they are born. These two terms help us identify human patterns and understand a country's driving forces.

The terms culture and ethnicity might also be confused in the issue of **ethnic cleansing**, which refers to the killing or forced removal of a people from their homeland by a stronger force of a different people. Ethnic cleansing might truly indicate two distinct ethnic groups: one driving the other out of their homeland and taking it over. On the other hand, ethnic cleansing might also be technically cultural cleansing if both the aggressor and the group driven out are of the same ethnic stock but hold different cultural values, such as religion or language. The term ethnic cleansing has been used to describe either case.

E. Languages of the World

Of the more than 6,000 languages, about a dozen are spoken by more than one hundred million people each. These are the world's main languages used in the most populous countries. However, the vast majority of the world's languages are spoken by a relatively small number of people. In fact, many languages have no written form and are spoken by declining numbers of people. Language experts estimate that up to half the world's living languages could be lost by the end

of the twenty-first century as a result of globalization. New languages form when populations live in isolation, and in the current era, as the world's populations are increasingly interacting with each other, languages are being abandoned and their speakers are switching to more useful tongues.

There are nine dominant language families in the world. Each of the languages within a language family shares a common ancestral language. An example of a language family is

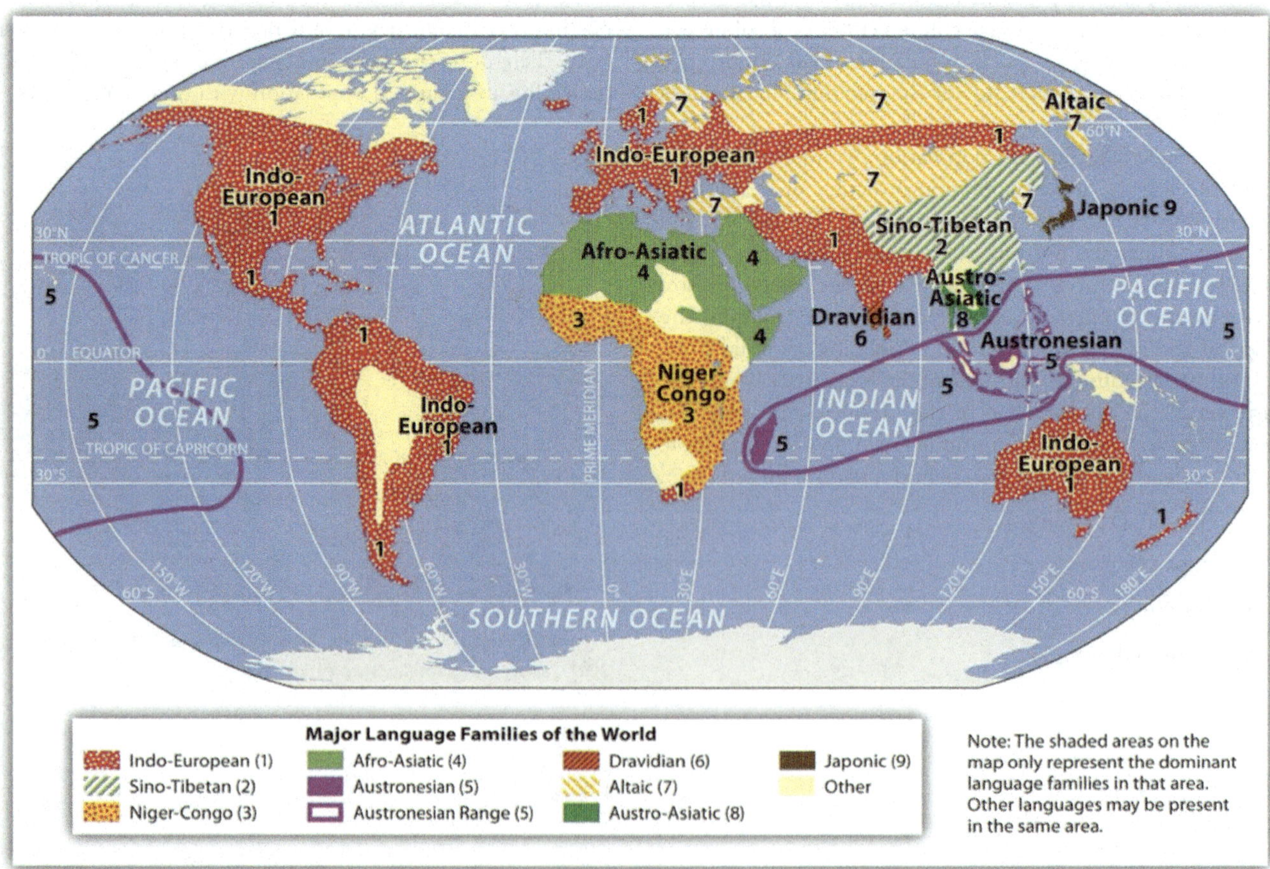

Figure 22. Major language families of the world.

the Indo-European family, which has a number of branches of language groups that come from the same base: a language called Proto-Indo-European that was probably spoken about six thousand years ago. As populations migrated away from the ancestral homeland, their language evolved and separated into many new languages. The three largest language groups of the Indo-European family used in Europe are the Germanic, Romance, and Slavic groups. Other Indo-European languages include Hindi (spoken in India) and Persian (spoken in Iran).

Languages with more than one hundred million speakers (speakers given in millions)			
Language	**First Language**	**Second Language**	**Total Speakers**
Mandarin	845	180	1,025
Hindi/Urdu	242	224	466
Arabic (All)	206	246	452
English	340	110	450
Spanish	329	53	382
Russian	144	106	250
Bengali	181	69	250
Portuguese	178	42	220
Indonesian	23	140	163
German	95	28	123
Japanese	122	1	123
French	65	55	120
Punjabi	109	6	115

Table 1. These thirteen languages are spoken by more than four billion people, or about 60 percent of the current world population in 2009.

Language Characteristics

The following terms are used to describe language characteristics:

- **Creole**. Similar to pidgin, a Creole language arises from contact between two other languages and has features of both. However, Creole is a pidgin that becomes a primary language spoken by people at home. Creole languages are often developed in colonial settings as a dialect of the colonial language (usually French or English). For example, in the former French colony of Haiti, a French-based Creole language was developed that is spoken by people at home, while French is typically used for professional purposes.

- **Dialect**. A dialect is a regional variety of a language that uses different grammar or pronunciation. Examples include American English versus British English. Linguists suggest that there are three main dialects of the English language in the United States: a Southern dialect, a midland dialect, and a Northern dialect. Television and public media communication have brought a focus on more uniform speech patterns that have diminished the differences between these three dialects.

- **Isolated language**. An isolated language is one not connected to any other language on Earth. For example, Basque is not connected to any other language and is only spoken in the region of the Pyrenees between Spain and France.

- **Lingua franca**. A lingua franca is a second language used for commercial purposes with others outside a language group but not used in personal lives. For example, Swahili is used by millions in Africa for doing business with people outside their own group but is not used to communicate within local communities.

- **Official language**. The official language is the language that is on record by a country to be used for all its official government purposes. For example, in India the official language is Hindi, though in many places the lingua franca is English and several local languages may be spoken.

- **Pidgin**. A pidgin is a simplified, created language used to communicate between two or more groups that do not have a language in common. For example, Residents of New Guinea mix English words with their own language to create a new language that can bridge speakers of different local language groups. Though the words are in English, the grammar and sentence structure is mixed up according to local vocabulary. There are many English-based pidgin languages around the world.

F. Religions of the World

Religious geography is the study of the distribution of religions and their relationship to their place of origin. Cultural geographers recognize three main types of religions: universal (or universalizing), ethnic (or cultural), and tribal (or traditional) religions. Universal religions include Christianity, Islam, and various forms of Buddhism. These religions attempt to gain worldwide acceptance and appeal to all types of people, and they actively look for new members, or converts. Ethnic religions appeal to a single ethnic group or culture. These religions do not actively seek out converts. Broader ethnic religions include Judaism, Shintoism, Hinduism, and Chinese religions that embrace Confucianism and Taoism. Finally, traditional religions involve the belief in some form of supernatural power that people can appeal to for help, including ancestor worship and the belief in spirits that live in various aspects of nature, such as trees, mountaintops, and streams (this is often called **animism**). Sub-Saharan Africa is home to many traditional religions.

Although the world's primary religions are listed here, many other religions are practiced around the world, as well as many variations of the religions outlined here. The top four religions by number of adherents are Christianity, Islam, Hinduism, and Buddhism. Because the official doctrine of Communism was nonreligious or atheist, there are actually many more followers of Buddhism in China than demographic listings indicate. The percentage of the world's population that follows Buddhism is probably much higher than the 6 percent often listed for this religion.

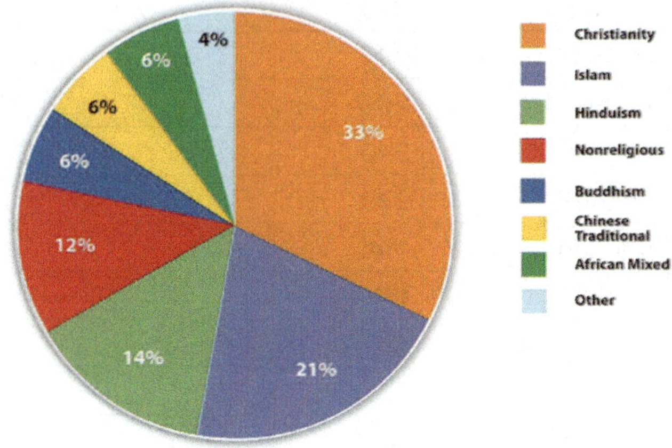

Figure 23. Major religions of the world and their respective percentage of the world population.

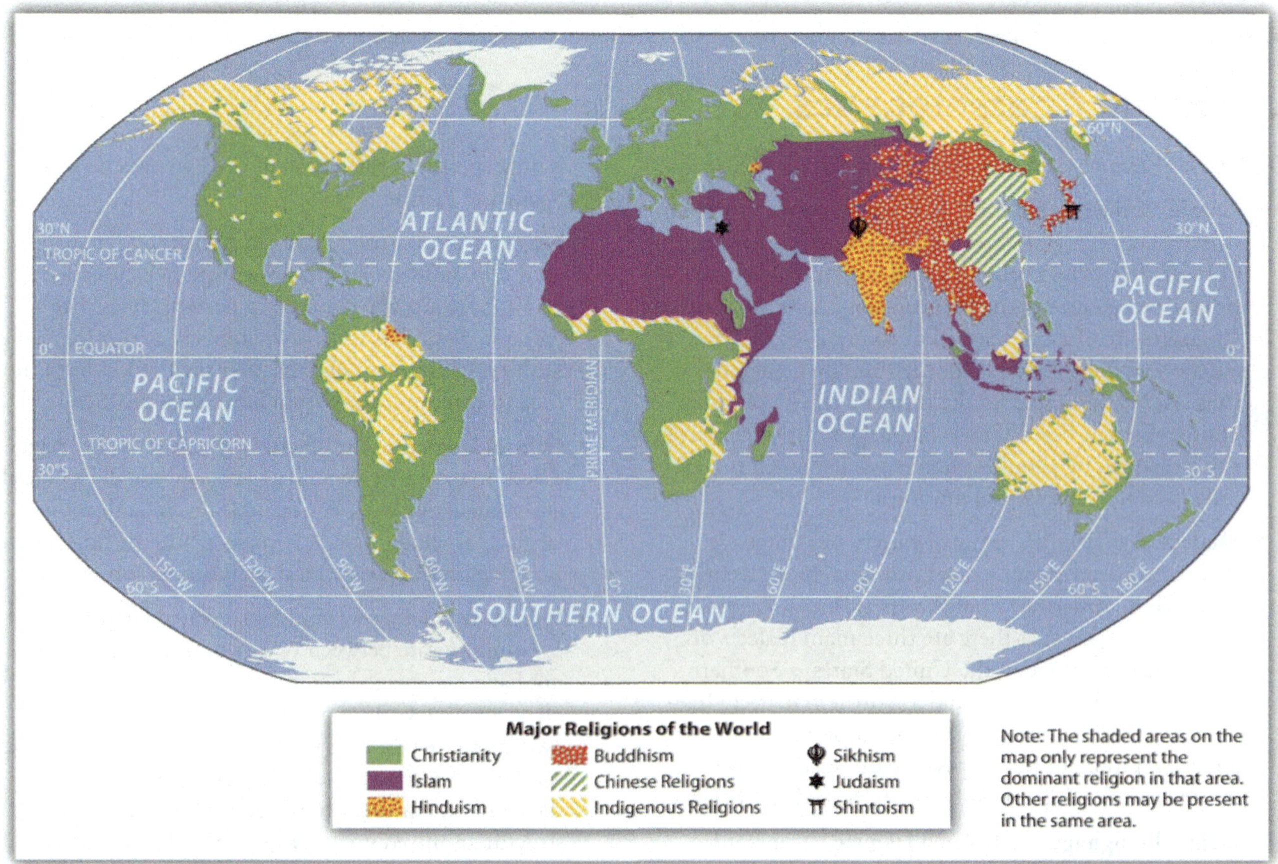

Figure 24. Major religions of the world.

- Christianity and Islam originated out of Judaism in the eastern Mediterranean and the Arabian Peninsula. Both are monotheistic religions that look to the Jewish patriarch Abraham as a founding personage. Christianity, based on the life and teachings of Jesus Christ, who lived in Palestine in the first century CE, spread rapidly through the Roman Empire. Islam is based on the teachings of Muhammad, a seventh-century religious and political figure who lived on the Arabian Peninsula. Islam spread rapidly across North Africa, east across southern Asia, and north to Europe in the centuries after Muhammad's death.

- Buddhism is a religion or way of life based on the teachings and life of Siddhartha Gautama, who lived in what is now India/Nepal around the fifth century BCE. There are three main branches of Buddhism: southern or Theravada Buddhism, eastern or Mahayana Buddhism, and northern or Vajrayana (Tibetan) Buddhism.

- Hinduism is a religious tradition that originated on the Indian subcontinent. Unlike other world religions, Hinduism has no single founder and is a conglomerate of diverse beliefs and traditions. Hinduism has a large body of scripture, including the Vedas, the Upanishads, and epic tales such as the Mahabharata and the Ramayana.

- Sikhism, a religion founded in the Punjab region of southern Asia, is a monotheistic religion centered on justice and faith. High importance is placed on the principle of equality between all people. The writings of former gurus are the basis for the religion.

- Judaism is the religion of the Jewish people, whose traditions and ethics are embodied in the Jewish religious texts, the Tanakh, and the Talmud. According to Jewish tradition, Judaism began with the covenant between God and Abraham around 2000 BCE.

- Shintoism is a major ethnic religion of Japan focused on the worship of kami, which are spirits of places, things, and processes.

- Confucianism and Taoism are ethnic Chinese religions based on morality and the teachings of religious scholars such as Confucius.

Key Takeaways

1. The human population is estimated at over seven billion in 2016 and is increasing rapidly, mainly in developing regions of Asia and Africa. No one can agree on the earth's carrying capacity for our human population, but unless the growth rate changes, the human population will double in about sixty years.

2. Since the Industrial Revolution, humans have been moving from rural areas to urban areas. Workers were needed in the factories and fewer workers were needed on the farms because of improved technology. This trend is still happening in many developing countries. Population pyramids are one method of illustrating demographic data for a country to show if the population is declining or increasing and at what rate.

3. Though often interchangeable in general terms, for the purpose of geography in this textbook, ethnicity is what you are born with and culture is what you learn after you are born.

4. There are over six thousand individual languages in the world today, with about thirteen of them spoken by one hundred million people or more. Of the main language families, nine include at least 1 percent or more of the human population.

5. There are thousands of religions or variants of them in the world. Cultural geographers recognize three main types of religions: universal, ethnic, and traditional. The four largest religions of the world are Christianity, Islam, Hinduism, and Buddhism.

1.4 Globalization and Development

Learning Objectives

1. Recognize that globalization has been a human activity since the era of European colonialism and that more recent globalization has been evident in the post-Cold War era through an increase in global activ-ities by multinational corporations.

2. Explain how the concepts of opportunity and advantage create a stronger rural-to-urban shift and fuel migration in various regions of the world.

3. Understand the dynamics of the core-periphery spatial relationship and determine whether a country is a part of the core or periphery by its respective attributes.

4. Determine how countries gain national income and which activities are renewable or have val-ue-added profits. Understand the vital roles that labor and resources play in the economic situation for each country.

5. Comprehend the patterns illustrated in the index of economic development—especially in terms of how it illustrates a country's development status in regard to family size and economic indicators. Learn the relationship between the concepts of rural-to-urban shift, core-periphery spatial relation-ships, opportunity and advantage, and haves and have-nots.

A. Globalization

Geographers and professionals in other disciplines understand that the world is not static. Cultural forces continue to act on human activities as globalization creates new alliances and global networks. The goal is to understand globalization and to make sense of what is happening. The better we understand the world and human dynamics, the better we will be prepared to address the changes that are occurring. Geography provides a means to spatially examine these changes.

Globalization is a process with a long history. People have been exploring, migrating, and trading with each other throughout human history, and these activities have created interactive networks connecting the different parts of the planet and producing dependent economic relationships. In modern times, globalization can be recognized by noting iconic global corporations, such as Wal-Mart, McDonald's, or Toyota, that trade across international borders and inte-grate labor and resources from different countries to sell a product or service in the global marketplace. In a number of countries, people have protested against the building of a new Wal-Mart or McDonald's, and such protests exemplify concerns about globalization and the growing expansion of dominant global economic units into local communities. These ubiquitous corporations represent corporate interests that are primarily concerned with company profits. Global corporations tend to view countries or communities as either markets for their products or sources of labor or raw materials. Globalization can seriously impact local communities for better or for worse, depending on local circumstances. The main force that encourages globalization is economic activity based on technological advancements. Cultural and societal changes often occur as a consequence and are no less significant.

European colonialism was an early wave of globalization that changed the planet and shaped most of the world's current political borders. This early wave of global conquest was fueled by the Industrial Revolution. Colonialism transferred technology, food products, and ideas around the globe in merchant ships that centered on the European power bases of the colonial empires of Europe—mainly Britain, Spain, France, Portugal, and the Netherlands. When the United States became independent of these European colonial powers, it began to extend its power and influence around the world. Thus the first major wave of globalization was a result of European colonialism.

Figure 25. Cultural landscapes representing the urban core region of Los Angeles and the peripheral region of rural Montana.

The space race and the information age of the latter portion of the twentieth century initiated a second major wave of globalization. The space race was a competition between the United States and the Soviet Union to develop space-related technologies, including satellites, and to land on the moon. The end of the Cold War, with the collapse of the Soviet Union in 1991, coincided with advancements in computer technology that fueled the second major wave in modern globalization. Technology and corporate activity have stimulated a wave of globalization that is impacting the economies of countries around the world. In European colonialism, the land and people were physically conquered by the mother country and became colonies ruled by the European colonizer's government. Great Britain was the most avid colonizer and amassed great fortunes through its colonial possessions. One difference between European colonialism and globalization today is that globally active multinational corporations do not wish to own the country or run the government directly. Corporations are not concerned with what government type is in power or who is running the country as long as they can operate and make a profit. This **neocolonialism** (new colonialism or **corporate colonialism**), like European colonialism, continues to exploit natural resources, labor, and markets for economic profits. Its critics claim that corporate colonialism is nothing more than a legal method of pillaging and plundering, and its supporters claim it is the most efficient use of labor and resources to supply the world with the lowest-priced products.

Examples of corporate colonialism can be seen in the trade relationships between the United States and places such as Mexico and China. US corporations move their manufacturing plants to Mexico to earn more profits by exploiting cheap labor. The corporations do not take over Mexico politically; they exploit it economically. The many US corporations that have started manufacturing their products in China do not attempt to overthrow the Communist Chinese govern-

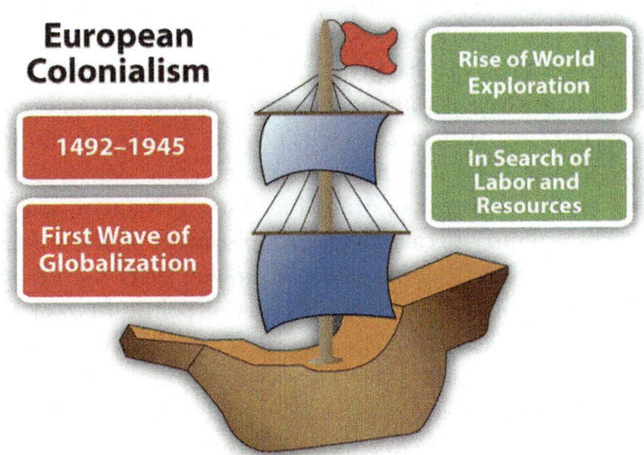

Figure 26. European Colonialism changed many things. The expansion of European empires was driven by the search for profits from resources and labor in the acquisition of new lands. European colonialism had a significant impact on people and cultures.

Figure 27. After the turn of the twenty-first century, Wal-Mart became the world's largest corporation. Its global expansion and economic relationships are good examples of the activities of corporate colonialism.

ment; they want to exploit the cheap Chinese labor pool and open up markets to sell products to Chinese consumers. Desire for profits drives corporate colonialism.

B. Core-Periphery Spatial Relationship

Economic conditions vary across the globe. There are wealthy countries (the so-called "haves") and poor countries ("have-nots"), and the determination of which countries are wealthy and which countries are poor has generally been determined by the availability of economic opportunities and advantages. There are three core areas of wealthy industrialized countries, all of which are in the Northern Hemisphere: North America, Western Europe, and eastern Asia. The main market centers of these regions are New York City, London, and Tokyo. These three core areas and their prosperous neighbors make up the centers of economic activity that drive the global economy. Other wealthy countries can be found dispersed in regions with large amounts of natural resources, such as the Middle East, or places of strategic location, such as Singa-

pore. The world's poorer countries make up the **peripheral** countries. A few countries share qualities of both and may be called semi-peripheries.

The periphery countries and the core countries each have unique characteristics. Peripheral locations are providers of raw materials and agricultural products. In the periphery, more people earn their living in occupations related to securing resources: farming, mining, or harvesting forest products. For the workers in these occupations, the profits tend to be marginal with fewer opportunities to advance. In the periphery, there is a condition known as **brain drain**, which describes a loss of educated or professional individuals. Young people leave the peripheral areas for the cities to earn an education or to find more advantageous employment. Few of these individuals return to the periphery to share their knowledge or success with their former community.

Brain drain also happens on an international level—that is, students from periphery countries might go to college in core countries, such as the United States or countries in Europe. Many international college graduates do not return to their poorer countries of origin but instead choose to stay in the core country because of the employment opportunities. This is especially true in the medical field. There is little political power in the periphery; centers of political power are almost always located in the core areas or at least dominated by the core cities. The core areas pull in people, skills, and wealth from the periphery. Lack of opportunities in the periphery pushes people to relocate to the core.

Power, wealth, and opportunity have traditionally been centered in the core areas of the world. These locations are urbanized and industrialized and hold immense economic and political power. Ideas, technology, and cultural activity thrive in these core areas. Political power is held in the hands of movers and shakers who inhabit the core. The core depends on the periphery for raw materials, food, and cheap labor, and the periphery depends on the core for manufactured goods, services, and governmental support.

The core-periphery spatial relationship can be viewed on various levels. On a local level, consider the state of Oregon. Portland is by far the largest city, followed by Salem and Eugene to the south in the fertile Willamette Valley. This area is the core of Oregon, with about 75% of the state's population, its largest cities, state capital, and the bulk of the state's manufacturing and high tech industries. The rest of Oregon can be considered the periphery, and has an economy based

more on farming, ranching, and timber, which is typical of a peripheral region.

If we move up a level, we can understand that entire regions of the United States can be identified as peripheral areas: the agricultural Midwest, rural Appalachia, and the mountain ranges and basins of the western United States. The large metropolitan areas of the East and West Coasts and the Industrial Belt act as the core areas. Los Angeles and New York City anchor each coast, and cities such as Chicago, St. Louis, Denver, and Indianapolis represent the heartland. All the other large cities in the United States act as core areas for their surrounding peripheral hinterlands. Southern cities such as Atlanta, Memphis, Dallas, or Phoenix act as core centers of commerce for the South in a region known as the Sun Belt.

On a global scale, we can understand why North America, Western Europe, and eastern Asia represent the three main

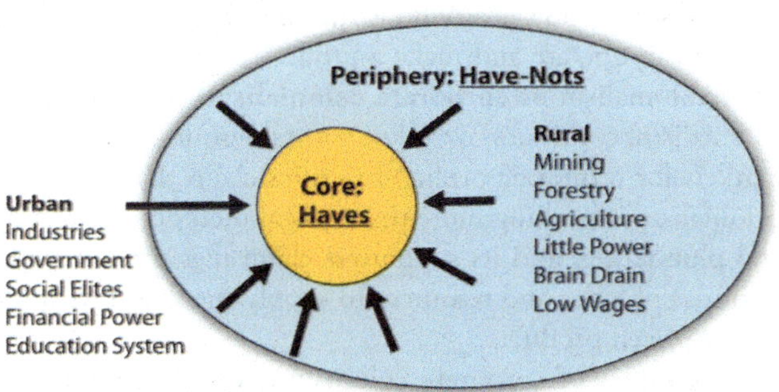

Figure 28. The Core-Periphery spatial relationship.

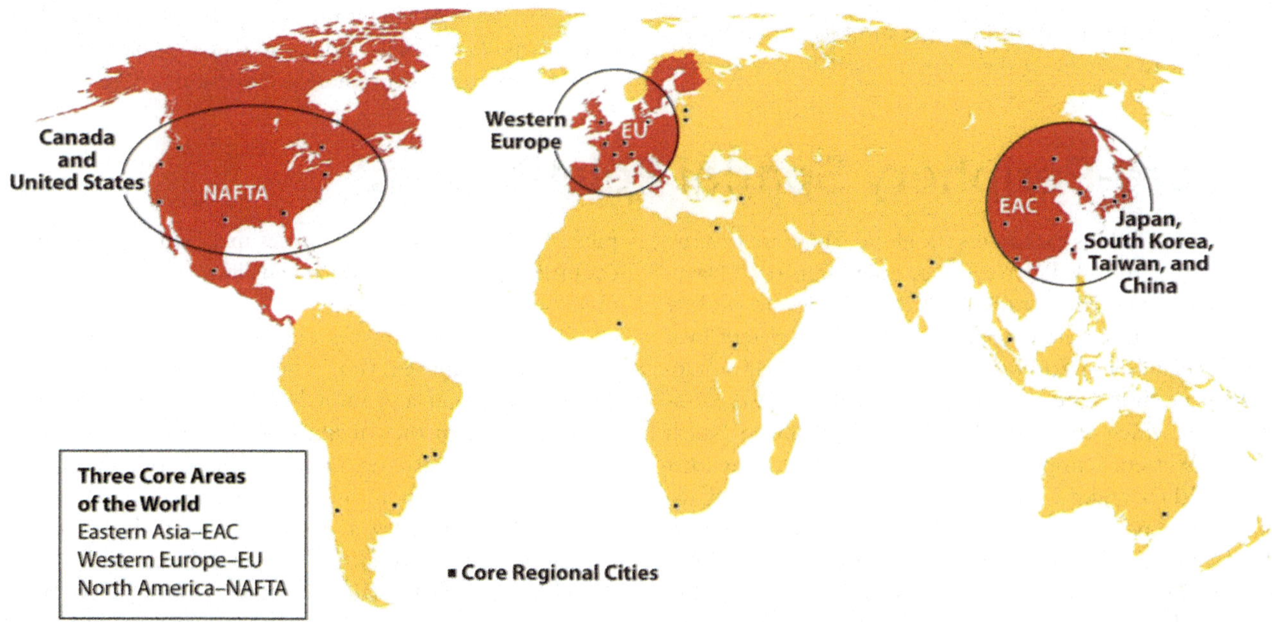

Figure 29. Three core economic areas of the world: North American Free Trade Agreement (NAFTA); the European Union (EU); and the East Asian Community (EAC). EAC is not an "official" organization but is a recognized economic group.

economic core areas of the world. They all possess the most advanced technology and the greatest economic resources. Core regions control the corporate markets that energize and fuel global activity. Peripheral regions include portions of Africa, Asia, Latin America, and all the other places that primarily make their living from local resources and support the economic core. These peripheral regions may include key port cities. A semi-periphery would be a transitional area between the core and the periphery, which could include countries such as Russia, India, or Brazil that are not exactly in the core and not really in the periphery but might have qualities of both. World migration patterns follow the core-periphery spatial relationship in that people and wealth usually shift from the peripheral rural regions to the urban core regions. The "have" countries of the world are in the core regions, while the "have-not" countries are most likely in the peripheral regions.

C. Economic Sectors

For the purpose of understanding economic geography, all economic activities can be grouped into one of four categories, each with its respective terms, depending on the nature of what is being produced:

Primary economic sector activities include everything that pertains to the collection of raw materials, such as agriculture, forestry, fishing, and mining—in other words, growing and extracting activities.

Secondary economic sector activities involve the processing of those raw materials through manufacturing, which has been the mainstay of economic growth for most developed countries, as well as construction.

Tertiary economic sector activities are those that produce services, not physical products, such as wholesale and retail sales, personal services, and business services.

Quaternary economic sector activities are those that deal with information collecting and processing, as well as management and decision-making.

The tertiary and quaternary economic sectors are often thought of together as the **service sector**. Only primary and secondary activities produce actual physical products, and manufacturing traditionally earns the highest value-added profits. Tertiary activities are selective in gaining national wealth. For example, service activities such as tourism can bring in national wealth if the visitors are from outside the country. Tourism within a country can also influence economic conditions by increasing the amount of consumer spending.

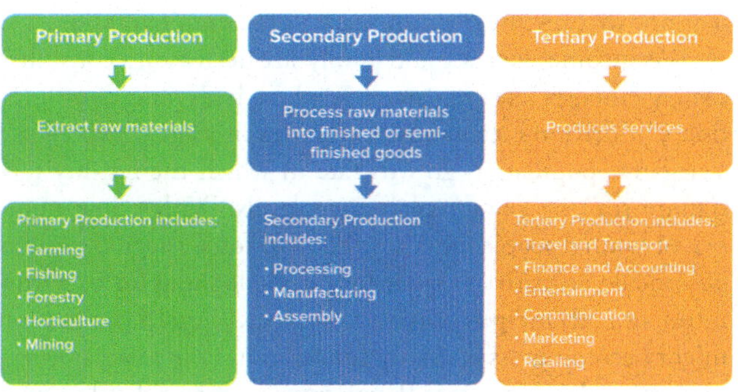

Figure 30. The three main sectors of the economy.

D. Development and Demographics

The Industrial Revolution, which prompted the shift in population from rural to urban, also encouraged market economies, which have evolved into modern consumer societies. Various theories and models have been developed over the years to help explain these changes. For example, in 1929, the American demographer Warren Thompson developed the **demographic transition model** (DTM) to explain population growth based on an interpretation of demographic history. A revised version of Thomson's model outlines five stages of demographic transition, from traditional rural societies to modern urban societies:

- **Stage 1:** High birth and death rates; rural pre-industrial society

- **Stage 2:** Declining death rate; high birth rate; developing country

- **Stage 3:** Declining birth rate; low death rate; high urbanization rate; industrial country

- **Stage 4:** Low birth and death rates; stabilized population; mostly urban population

- **Stage 5:** Declining population; urban post-industrial society

As a general trend, when a country experiences increasing levels of industrial activity and urbanization, the outcome is usually a higher standard of living for its people. One way

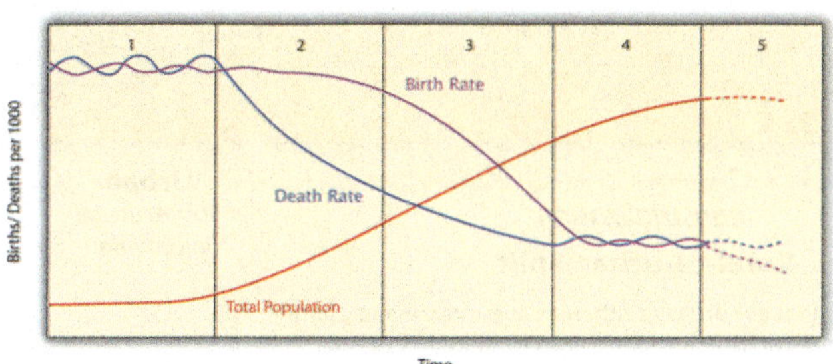

Figure 31. The five stages of the demographic transition model.

to define development, in fact, is simply as a process of improving the standard of living of people.

The **human development index** (HDI) was developed in 1990 and used by the United Nations Development Program to measure a standard of human development, which refers to the widening opportunities available to individuals for education, health care, income, and employment. The HDI incorporates variables such as standards of living, literacy rate, and life expectancy to indicate a measure of well-being or the quality of life for a specific country. The HDI is used as an indicator of a country's economic and technological development.

Additionally, rural-to-urban shift takes place, driven by the pull of opportunities and advantages in the industrializing and urbanizing areas. Though there are exceptions, a decrease in family size usually coincides with a higher level of urbanization. There are cases in which only core regions within a country transition through the five development stages without the peripheral regions experiencing the coinciding levels of economic benefit. The five stages of the DTM assist in illustrating these general patterns.

Stage 1 of the DTM indicates traditional rural societies, which are usually based on agriculture and not as dependent on the outside world. Stage 1 families are larger, their income levels are low, and their advantages and economic development opportunities are low. Health care, education, and social services are in short supply or nonexistent. High birth and death rates maintain a high fertility rate/family size and a low population-growth status. Populations in stage 1 development have a stationary population pyramid. Though there may be small, remote regions of the world that exhibit stage 1 development patterns, no countries fall into this category any longer.

Stage 2 countries experience high population growth rates because family size remains high but modern medicine or improved nutrition allows people to live longer, which lowers the death rate. Population is exploding in countries in stage 2. During this stage, young people from rural areas often migrate to the cities looking for employment. Rural stage 2 regions are starting to urbanize and integrate their economic activities with the outside world. Regions in stage 2 often have a surplus of cheap labor. Income levels remain low and family size continues to be large. Countries in this stage often have a rapidly expanding population pyramid with a very wide base.

Societies that have made business connections

that provide for manufacturing of products, industrial activities, or an increased service sector might progress to **stage 3**, the rural-to-urban shift stage. These regions are experiencing a high rate of rural-to-urban shift in their populations. These regions are often targeted by multinational corporations for their labor supply, and as people migrate from rural areas to the cities looking for employment, urban populations grow and core or central cities experience high rates of self-constructed housing (slums). Income levels start to increase and family size starts to drop significantly. Stage 3 countries still have an expanding population, but the rate of population growth is slowing as birth rates drop.

Societies that have urbanized and industrialized and are members of the global marketplace might enter **stage 4**. Members of an urban workforce assist in building a networked economy. Family size is lower as urban women enter the workforce and have fewer children. Health care, education, and social services become increasingly available, and income levels continue to rise. In stage 4, there is typically a high level of growth in the industrial and service sectors with a great need for infrastructure in the form of transportation, housing, and human services. Countries in stage 4 have populations that resemble a stationary population pyramid.

As incomes increase and family size decreases, a consumer society emerges, creating **stage 5**, where high mass consumption can drive the economy. Countries in stage 5 experience a negative population growth rate in which the fertility rate (family size) is below replacement levels. With a low number of young people entering the workforce, stage 5 regions become an attractive magnet for people looking for opportunities and advantages in the job market. Illegal immigration might become an issue. Many countries in Europe are now experiencing this condition. Populations in stage 5 development have a contracting population pyramid with a narrow base.

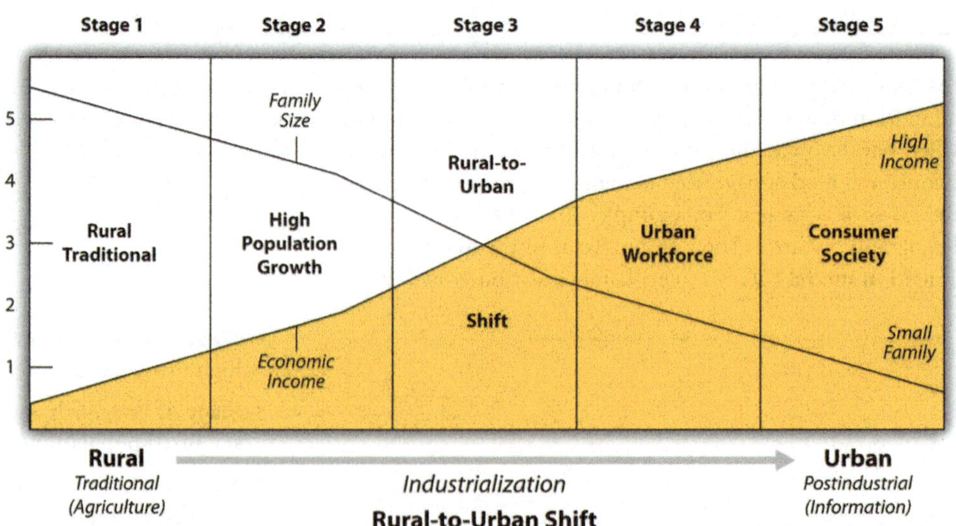

Figure 32. Index of Economic Development. The stages correlate with the stages of the DTM.

E. Geography of Opportunity

Highly industrialized countries in Western Europe, eastern Asia, or North America can offer more economic opportunities compared with developing countries. The **push-pull factors** that push people out of poorer countries and pull them to an industrialized country are strong. Portions of the population of countries in the earlier stages of the demographic transition often migrate to countries in the latter stages looking for work and other opportunities. People who do not have opportunities often want to move to places that do have them so they can work to attain greater economic security for themselves and their families.

Countries in stage 4 or 5 of the DTM are often attractive places for those seeking greater economic opportunities. Populations in these stages generally have fewer children, so the demand for entry-level workers is often higher. Immigrants with fewer skills take the entry-level jobs to enter the economic workforce. An established country with a long-standing history and culture does not always welcome an influx of new immigrants, however. The arrival of an immigrant labor pool often includes individuals who hold different cultural traditions or customs than those of the mainstream society. Social tensions may arise if different ethnic groups are vying for the same cultural spaces and opportunities. One example is the large number of people entering the United States across its southern border. Europe is experiencing a similar immigration issue, with immigrants from North Africa, the Middle East, and former colonies. Japan, on the other hand, has taken pride in holding on to its homogeneous Japanese culture, but it is facing the same employment situation.

Key Takeaways

1. The search for cheap labor and resources drives the need for profits by multinational corporations that fuel the global economy. This activity is creating a second major wave of globalization, often referred to as neocolonialism or corporate colonialism.

2. The concepts of opportunity and advantage provide a means to understand the attractiveness or unattractiveness of a place to immigrants or economic activities. Opportunities and advantages drive rural-to-urban shift, migration, and movement of corporate activity.

3. Core areas are usually urban with high levels of industrial and economic development. Peripheral areas are typically suppliers of food and raw materials used in the core. Political and economic power is held in the core, while the periphery suffers from lower incomes and brain drain.

4. Economic activities can be grouped into one of four economic sectors: primary (extractive); secondary (processing and manufacturing); tertiary (sales and personal and business services); and quaternary (data collecting and processing and decision-making). In more developed countries, most workers are employed in services (the tertiary and quaternary sectors), although the primary and secondary sectors may contribute more to national income.

5. Development and population models can help one understand a country's socioeconomic dynamics. Family size and economic income are two indicators that can be tracked to assist in understanding the industrialization or urbanization levels of a country or demographic region.

6. Globalization has prompted a greater understanding of how opportunities and advantages relate to haves and have-nots. Whether it is individuals or countries, some have greater levels of opportunity and advantage than others. Human migration patterns usually coincide with the push-pull forces of opportunity or advantage levels.

7. Many of the concepts used to understand the dynamics of a region or place are related. The four main types of population pyramids can help illustrate the various stages of a country's socioeconomic situation as illustrated in the demographic transition model (DTM).

End-of-Chapter Summary

- Geography is the spatial study of the earth's surface. The dynamic discipline of geography bridges the social sciences and the physical sciences. Geography's spatial nature can be illustrated by the creation of maps as an important means of communicating information. Human geography and physical geography are the two main fields of the discipline. GPS, GIS, and remote sensing are tools geographers use to spatially study a location or the physical or cultural landscape.

- Geographers divide the earth into a grid to provide location references to places on its surface. Human activity is orientated around the grid system to provide time zones, navigation, and the organization of communication systems. Climate and seasonal changes can be tracked using this grid system of latitude and longitude.

- Regions are the basic units of geographic study. World geography divides the world into sets of regions called realms that are used as comparison studies regarding human and physical landscapes and activities. Climate regions or zones are helpful in understanding the earth's environmental conditions. Historically, human activity has been strongly affected by the variation in climates. Type C climates have generally attracted large human populations.

- The relationship between the environment and human activity is an important component of geography. The movement of tectonic plates sometimes causes earthquakes and volcanic activity, which affects human activity. The rain shadow effect can also impact where and how humans live. Human activity contributes to environmental problems such as deforestation, which impacts the environment through the loss of habitats, soil erosion, and possibly climate change.

- The human population continues to increase. The earth's carrying capacity for humans is debatable, but the impact that humans have on the planet is undeniably extensive. Some factors and conditions encourage high population growth, while others discourage population growth. Migration, rural-to-urban shift, and urbanization are related to changes in population. As population increases, the number of languages continues to decrease—an example of the impact of globalization.

- European colonialism created a major wave of globalization that extended up until about the time of World War II. After the Cold War ended, a second wave of globalization was fueled by the information age and the introduction of new communication and transportation technologies. Concepts such as labor and resources, opportunity and advantage, or haves and have-nots can help explain the dynamics of globalization.

- The core-periphery spatial relationship helps explain how human economic activity is organized around either an urban core or a rural periphery. The methods countries use to gain national income are a means to understand the ways in which economic activity is categorized and explained. The demographic transition model (DTM) is a model relating to the stages a country may transition through to reach a postindustrial development level. All these concepts, models, and theories are tools used to understand the human activities that are elements of the globalization process.

Chapter 2

Europe

2.1 Introducing the Realm

Learning Objectives

1. Describe the various climate types and physical landforms of the European continent.
2. Explain how Europe's physical geography has supported its development.
3. List Europe's various natural resources.
4. Summarize the environmental concerns Europe faces.

Europe is a continent of peninsulas, islands, and varied landforms. The traditional boundaries of the European continent include the North Atlantic Ocean to the west and Russia as far as the Ural Mountains to the east. Since the Soviet Union's collapse in 1991, Russia has been given its own identification and, in this text, is not included in the study of Europe. Russia will be discussed in Chapter 3. Greenland is located next to the North American country of Canada but has traditionally been considered a part of Europe because of Denmark's colonial acquisition of the island. The Arctic Ocean creates a natural boundary to the north. The southern boundary of Europe is the Mediterranean Sea and includes the islands of Malta and Cyprus as independent countries. A portion of Turkey is in Europe, but Turkey is considered a part of Asia Minor and is usually included in the study of the Middle East region. The waterway in Turkey between the Black Sea and the Aegean Sea is the Bosporus, or the Istanbul Strait, which creates a natural border between Asia and Europe. Europe is also close to North Africa, and Morocco's coast can be seen across the Strait of Gibraltar from Spain.

From the Roman Empire to the European Union (EU), Europe's historical pattern of development is a model study in regional geography. From historic empires to diverse nation-states to a multi-country union, the continent struggles to confront the cultural forces that unite and divide it. The powerful impact European colonialism has had on the world since the Industrial Revolution is still felt today. The rural-to-urban shift prompted by the Industrial Revolution first impacted Europe and continues to impact developing countries. Understanding the geographic region of Europe is essential to understanding our world. This short summary of the basic concepts will provide a valuable lesson in globalization, which affects every human being on the planet. The concepts and principles that apply to Europe can also apply to other countries and regions.

EUROPE

Figure 1. Political map of Europe.

A. Location and Climate

Europe is a northern continent. All the British Isles, for example, fall above the fiftieth parallel. Much of Europe lies north of the United States. Paris, France, is at about the same latitude as Fargo, North Dakota. Athens, Greece, is at about the same latitude as St. Louis, Missouri. Europe's northern position affects its growing seasons and people's moods, and should be taken into consideration as an important influence in the evolution of the European character. Europe is also surrounded by water: the Atlantic Ocean borders Europe on the west, the Arctic Ocean borders it to the north, and many seas surround the peninsulas and coastal regions.

The oceans exert significant influence on the world's climates. The oceans collect and store vast amounts of solar energy, particularly around the equator, and transport that heat with their currents. Ocean currents can move water for thousands of miles from one temperature zone to another. Because oceans can absorb so much heat, **maritime** climates are milder than **continental** ones, with smaller temperature variations from day to night as well as from winter to summer. Since water heats and cools more slowly than land, coastal communities have climates that tend to be more moderate than one might imagine for places so far north. Interior Europe does not benefit from coastal waters and is thus more continental and can have winters as cold as those found within the upper Midwestern United States.

The **Gulf Stream** is perhaps the most important current for Western Europe's climate and is responsible for producing a temperate climate for a northern latitude location. Most of Western Europe has a moderate type C climate. The Gulf Stream originates in the Gulf of Mexico, where the waters are warmed. This powerful current follows the Eastern Seaboard of the United States before crossing the Atlantic Ocean for Europe, where it is called the **North Atlantic Drift** or North Atlantic Current. The North Atlantic Current's most dramatic effect can be found in the western coastal islands of Scotland, which have a mild enough climate to support some forms of tropical flora, even though they are about as far north as Hudson Bay, Canada. The coast of Norway provides another example. While most of Norway's coastal area lies within the Arctic region, it remains free of ice and snow throughout the winter.

People living farther inland and closer to Eastern Europe and Russia encounter the colder type D climates. Colder air sweeps down from the Arctic north or from eastern Siberia and provides colder winters in this eastern region. The Mediterranean Sea moderates the temperature to the south, providing a type C climate around its shores. Type C climates meet up with type E climates at or near the Arctic Circle in Norway and in Iceland.

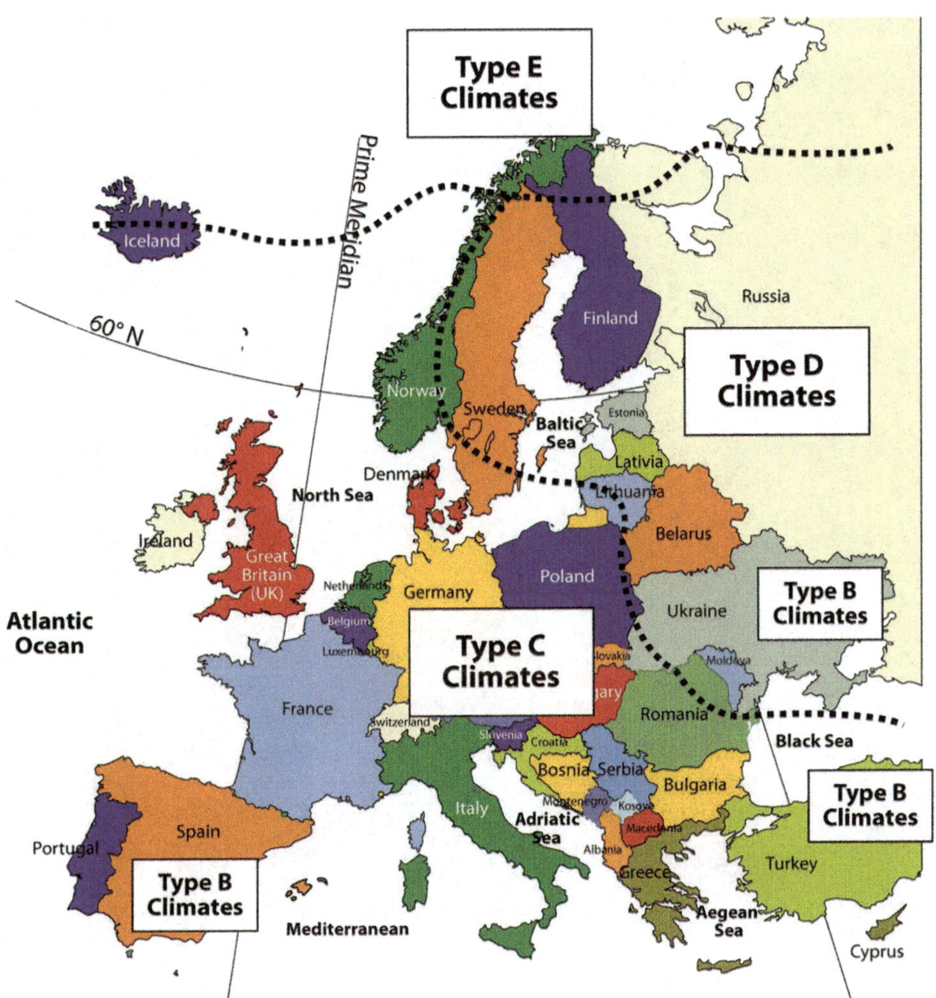

Figure 2. The dominant climate types of Europe.

B. Four Main European Landforms

Europe has four main landforms, many islands and peninsulas, and various climate types. The four main landforms include the **Alpine region, Central Uplands, Northern Lowlands, and Western Highlands**. Each represents a different physical part of Europe. The wide-ranging physical environment has provided Europe with an abundance of biodiversity. Biodiversity refers to the diversity of the number of species in an ecosystem and the quantity of members in each species. The physical environment also provides natural resources and raw materials for human activities. Europe's moderate climates and favorable relative location are supported by its access to the many rivers and seas.

These advantageous developmental factors supported the development of the Industrial Revolution in Europe, which gave rise to highly technical and urban societies. Europe has emerged as one of the core economic centers of the global economy. Associated with the urbanization of Europe are high human population densities that have placed a strain on the natural environment. As result, there has been significant deforestation and the loss of natural habitat, which has in turn has decreased the realm's level of biodiversity.

Rivers are abundant in Europe and have provided adequate transportation for travel and trade throughout its history. Most of Europe is accessible by water transport either via the many rivers or along the extensive coastlines of the peninsulas and islands. Two main rivers divide Europe: the **Danube** and the **Rhine**. Both have their origins in the region of southern Germany on or near the border with Switzerland. The Rhine River flows north and empties into the North Sea in Rotterdam, The Netherlands, one of the world's busiest ports. The Danube flows east through various major European cities, such as Vienna, Budapest, and Belgrade before emptying into the Black Sea.

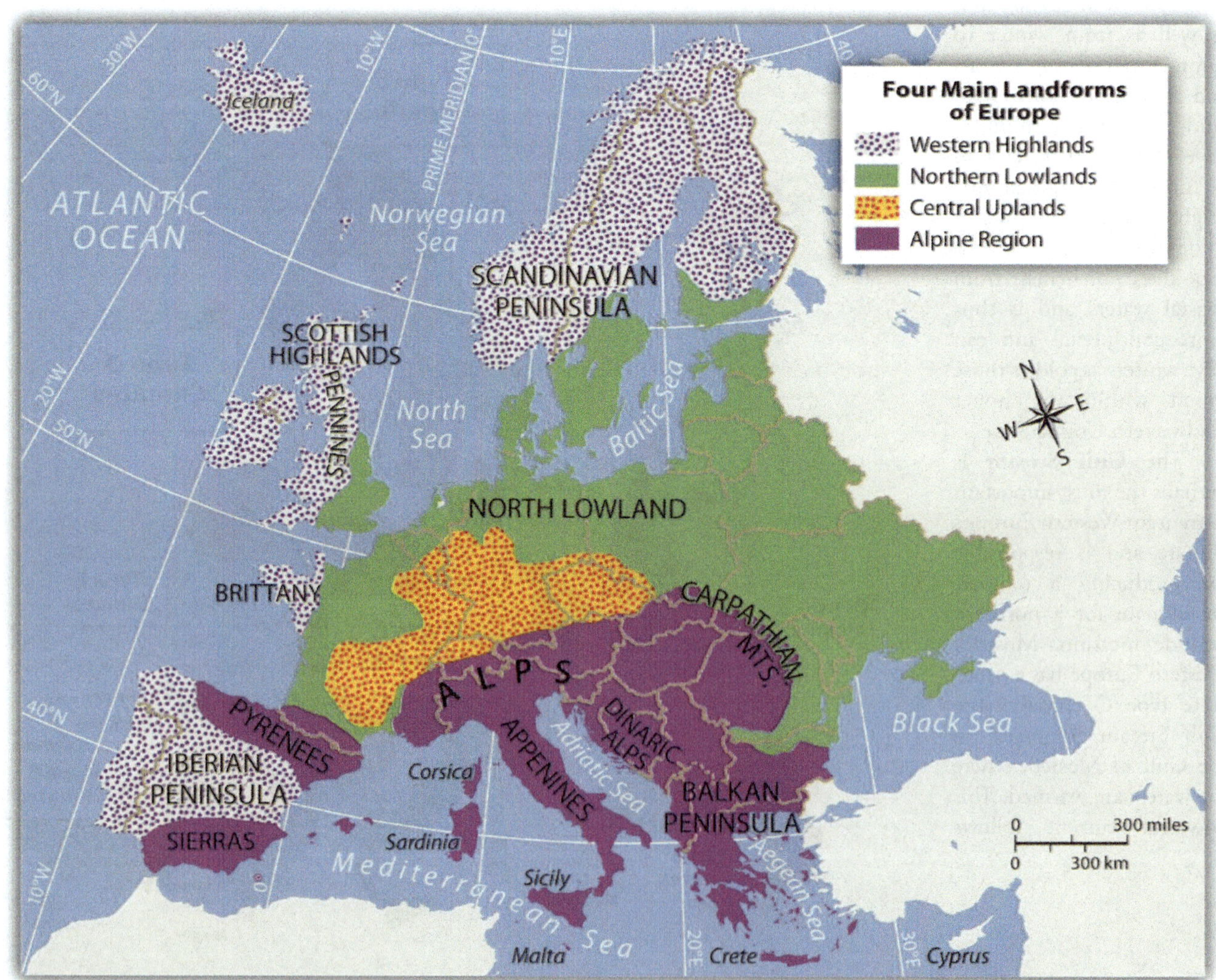

Figure 3. Four main landforms of Europe: Western Highlands, Northern Lowlands, Central Uplands, and the Alpine Region.

C. Alpine Region

The **High Alps**, which range from eastern France to Slovenia, are central to the Alpine region. Included in the **Alpine Range** are the **Pyrenees**, located on the border between France and Spain; the Apennines, running the length of Italy; the **Carpathians**, looping around Romania from Slovakia; and finally, the shorter **Dinaric Alps** in former Yugoslavia. Mountains usually provide minerals and ores that were placed there when the earth's internal processes created the mountains. Mountains also isolate people by acting as a dividing range that can separate people into cultural groups.

The Alpine region encircles the Mediterranean coastlines, which have more temperate type C climates that are particularly warm with hot, dry summers and cool, wet winters. This climate type allows for the cultivation of food products such as olives, citrus fruit, figs, apricots, and grapes.

Evergreen scrub oaks and other drought-resistant shrubs are common in the Mediterranean region.

Figure 4. The Alpine Region: Eiger, Mönch, and Jungfrau from Männlichen—Swiss Alps.

D. The Central Uplands

The region bordering the main Alps to the north, which includes a large portion of southern Germany extending eastward, is known as the **Central Uplands**. These foothills to the Alps are excellent sources of raw materials such as forest products and coal, which are valuable resources for industrial activities. The Central Uplands are also good locations for dairy farming and cattle raising. This middle portion of the continent has a mixed deciduous-coniferous forest, and the vegetation includes oak, elm, and maple trees intermingled with pine and fir trees. There are four distinct seasons in this region with moderate amounts of precipitation year round.

E. Northern Lowlands

Similar to the breadbasket of the Midwestern United States, Europe's **Northern Lowlands** possess excellent farmland. Major agricultural operations here provide for a large European population. The land is flat to rolling with relatively good soils. The Northern Lowlands are a great plain that extends across northern Europe from southern France, north through Germany, and then all the way east to the Ural Mountains of Russia. This area is typified by prairies and areas of tall grasses and is mostly used as farmland. The lowlands area also contains bogs, heaths, and lakes. The eastern part of this great plain around Ukraine is characterized by a steppe biome. It is a flat and relatively dry region with short grasses and is generally an agricultural region. This eastern area has great swings in temperature, both from day to night and from summer to winter. Winter temperatures in the eastern steppe can drop to below −40 °F, with summer temperatures reaching as high as 105 °F. This is similar to the steppes of eastern Montana or western North Dakota in the United States.

Figure 5. The Western Highlands meet the lowlands in central Scotland.

F. Western Highlands

On the western edges of the European continent arise short rugged mountains called highlands that extend throughout Norway, parts of Britain, and portions of the **Iberian Peninsula** of Portugal and Spain. These Western Highlands hold sparser populations and are less attractive to large farming operations. Agriculture is usually limited to grazing livestock or farming in the valleys and meadows. The **Scottish Highlands** are noted for their wool products and Highland cattle. In England, the central chain of highlands called the **Pennines** proved valuable during the Industrial Revolution because they enabled hydropower and, later, coal mining. Coal mining was prominent in the highland regions of Wales. In the far northern regions of Scandinavia, tundra environments prevail. In this coldest and driest biome, permafrost dominates the landscape, and the land becomes soggy for brief periods during the few weeks of summer. The flora consists primarily of lichens, mosses, low shrubs, and wildflowers.

G. Natural Resources in Europe

The physical landforms of Europe provide diverse geographic opportunities that have catapulted Europe through the Industrial Revolution and into the information age. With an abundance of natural resources, European countries have gained wealth from the land and leveraged their geographic location to develop a powerhouse of economic activity for the global marketplace. Europe has placed a strong focus on manufacturing activity to take advantage of its natural resources. The highly urbanized society has struggled to find a balance between modernization and environmental concerns. Industrial activities have contributed to the degradation of the environment and the demise of a number of species.

Different regions of Europe are blessed with fresh water supplies, good soils, and various minerals. Chief among the mineral deposits in Europe is iron ore, which can be found in Sweden, France, and Ukraine. Other minerals exist in smaller quantities, including copper, lead, bauxite, manganese, nickel, gold, silver, potash, clay, gypsum, dolomite, and salt. Extraction activities have supported the continent's industrialization.

The ready access to vast areas of the Atlantic Ocean and a number of major seas, lakes, and rivers has elevated fish to an important natural resource in Europe. The seas around Europe provide about 10 percent of the world's fish catches. Europeans are becoming increasingly aware of the effects of overfishing. Stocks of Atlantic cod and Atlantic mackerel are considered to be at risk because of the twin threats of overfishing and changes in the environment that are affecting natural mortality and slowing spawning.

Forest covers more than 40 percent of the continent's land area, with the majority on the Russian side. Forests exist primarily in the less populous Nordic and Baltic countries and in Central Europe. About half the forest land in Europe is privately owned. The percentage of forested land in Europe is rebounding because of an extensive tree-planting initiative since 2000.

Coal, now substantially depleted, is abundant in several areas of Great Britain as well as in the industrial centers of Germany and in Ukraine. Other coal deposits are found in Belgium, France, Spain, the Czech Republic, Poland, Slovakia, and Russia. The burning of coal has produced high levels of air pollution. Acid rain has been a major concern in the northern countries, where wind currents carry pollutants north into Scandinavia from the industrial regions of Central Europe. In Scandinavia, acid rain has diminished fish populations in many of the lakes. Forest health is also being challenged, which is diminishing the economic conditions of regions that depend on forests for their economic survival.

Petroleum and natural gas deposits exist underneath the North Sea and were first tapped in the 1970s. Five European countries have rights to these resources, including Norway, the United Kingdom, Denmark, the Netherlands, and Germany, with Norway holding the bulk of the rights. The governments of these five nations agree that, although tapped only decades ago, half the North Sea oil reserves have been consumed.

Before the extraction of petroleum products from the North Sea, Russia and the former Soviet Union's other republics supplied petroleum to Europe. These areas still have a number of active extraction operations. Hydroelectric power has been important in Europe as well. With both coal and oil resources largely depleted and the desire to avoid the environmental damage caused by dams, the European Energy Commission is devoting substantial energy and resources to encouraging use of renewable resources such as wind and solar energy. In March 2007, European leaders agreed that a binding target of 20 percent of all energy must be from alternative sources by 2020. Also, 10 percent of the transportation fuels used by EU members must be sustainable biofuels.

Key Takeaways

1. The Gulf Stream provides a moderate type C climate for much of Western Europe. Eastern Europe can experience colder type D climates.

2. Europe has four main physical landforms that provide a diversity of natural resources. The North European Lowland holds the majority of its agricultural potential.

3. An increase in population has also increased the demand on the environment. Various environmental concerns are becoming more evident. Acid rain from industrialization has caused extensive damage to forests and fish populations in northern Europe. Atlantic fisheries are also experiencing a decline in production.

2.2 Historical Development Patterns

Learning Objectives

1. Outline how the Roman Empire and the Viking era contributed to European development.
2. Describe how European colonialism changed or influenced other countries.
3. Explain the major developments that prompted the Industrial Revolution.
4. Summarize the impact of the rural-to-urban shift and its impact on urbanization specifically.
5. Outline the concept of a nation-state and explain how this applies to Europe.
6. Explain how cultural forces can positively or negatively influence political units.
7. Identify the three main language groups and the three main religious denominations of Europe.

Europe didn't become a center for world economics with high standards of living by accident. Historical events in global development have favored this realm because of its physical geography and cultural factors. In southern Europe, the Greeks provided ideas, philosophy, and organization. Greek thinkers promoted the concept of democracy. The Romans carried the concept of empire to new levels. From about 150 BCE to 475 CE, the Romans brought many ideas together and controlled a large portion of Europe and North Africa. The Roman Empire introduced a common infrastructure to Europe. The Romans connected their world by building roads, bridges, aqueducts, and port facilities. They understood how to rule an empire. By taking advantage of the best opportunities of each region they controlled, they encouraged the best and most-skilled artisans to focus on what they did best. This created the specialization of goods and a market economy. No longer did everyone have to make everything for themselves. They could sell in the market what they produced and purchase products made by others, which would be of higher quality than what they could make at home. Regions that specialized in certain goods due to local resources or specialty skills could transport those goods to markets long distances away. The Roman Empire connected southern Europe and North Africa.

The Vikings of Scandinavia (Norway, Sweden, and Denmark; 900–1200 CE) are often inaccurately referred to as rogue bands of armed warriors who pillaged and plundered northern Europe. Though they were fierce warriors in battle, they were actually farmers, skilled craftsmen, and active traders. They developed trade routes throughout the north. Using their seafaring knowledge and skills, the Vikings used Europe's waterways for transportation. They were the early developers of the northern world from Russia to Iceland and even to North America. They developed colonies in Iceland, Greenland, and what is present-day Canada. Their longships were renowned for versatility and provided an advantage on the sea. The Vikings made advances deep into Europe—all the way to Constantinople. History indicates that the Byzantine Empire employed Scandinavian Vikings as mercenaries.

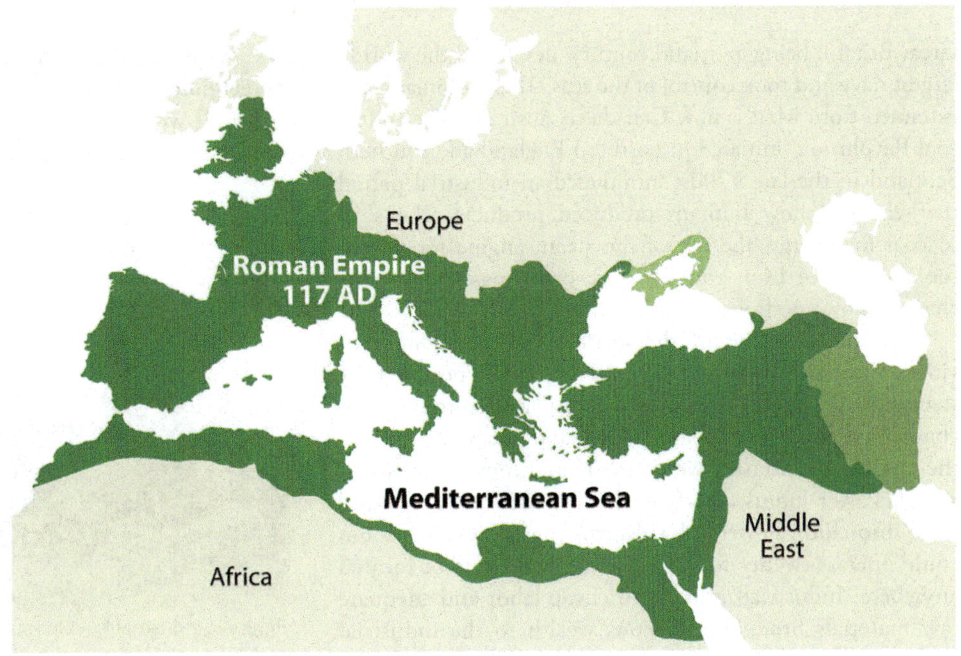

Figure 6 Extent of the Roman Empire, 117 CE.

A. Colonialism

It wasn't until after the Dark Ages of Europe ended that a rebirth of ideas, technology, and progress took hold. The Renaissance of the late fifteenth century prompted activity in Europe that changed the world. In 1492, Columbus and his three ships crossed the Atlantic to land on the shores of the Americas, in what are now known as the Bahamas. This event symbolized the beginning of the era of **European colonialism**, which only diminished after World War II. Colonialism's effects remain in the colonies or protectorates that European countries still possess. Colonialism was fueled by the economic concept of mercantilism that included the drive of governments to control trade, promoting the acquisition of wealth by the quick gain of gold or silver from their colonies.

Colonialism included the development of colonies outside the home country, usually for the expansion of imperial power and the exploitation of material gain. The building of larger ships and an understanding of sea travel allowed an exchange of new goods and ideas between continents. North and South America were opened up to the European explorers for colonial expansion. European colonialism brought newfound wealth from the colonies back to Europe. All the regions of the world outside Europe were targeted for colonialism. Africa was divided up, "Latin" America was created, and Asia became a target for resources and trade. The few powerful countries along the Atlantic coast of Europe began the drive to dominate their world. If you live in the Western Hemisphere, consider the language you speak and the borders of your country: both were most likely products of European colonialism. Most of the current political geographic boundaries were drawn up or shaped through colonial conflict or agreement. For better or worse, understanding much of the world's current cultural and political geography requires an understanding of European colonialism.

B. The Agrarian Revolution

The post-Renaissance era introduced a number of agricultural changes that impacted European food production. Before this time, most agricultural methods were primitive and labor intensive, but new technologies were introduced that greatly enhanced agricultural production. Plows, seeders, and harvesting technologies were introduced, and land reform and land ownership transitioned to adapt to the changing times. These innovations supported the expanding port cities that created urban markets for agricultural surpluses. Colonial ships returned from the colonies with new crops such as the potato that revolutionized crop production.

This era's progress in agricultural advancements is often referred to as the agrarian revolution. The agrarian revolution led to the industrial developments such as the steam tractor and steel implements that further advanced agricultural production worldwide.

C. The Industrial Revolution

Great Britain, being an island country, developed the world's largest navy and took control of the seas. Their colonial reach extended from what is now Canada to Australia. The Industrial Revolution, initiated in northern England and southern Scotland in the late 1700s, introduced an industrial period that changed how humans produced products. The shift to coal for energy, the use of the steam engine for power, the smelting of iron, and the concept of mass production changed how goods were produced.

The development of the steam-powered engine provided a mobile power source. Waterwheels powered by steep-flowing rivers or streams were an early source of mechanical power. With coal for fuel and steam for power, the engines of industry were mobile and moved full speed ahead. Power looms converted textiles such as cotton and wool into cloth. Powered by a steam engine, a power loom could operate twenty-four hours a day and could be located anywhere. Industrialization with cheap labor and adequate raw materials brought enormous wealth to the industrial leaders and their home countries.

With the mass production of goods and advancements in technology, there was a major shift in human labor. Fewer people were needed on the farms, and more workers were needed in factories. There was a large rural-to-urban shift

Figure 7. Industrial Revolution. Mass production during the Industrial Revolution prompted a rural-to-urban shift in the population. Coal for energy and steam for power energized industrial activities.

in the human population. Europe experienced the development of the major cities of its realm during this period. In Britain, for example, in 1800 only nine percent of the population lived in urban areas. By 1900, some 62 percent were urban dwellers. By 2010, it was more than 90 percent. Europe, as a whole, is about 75 percent urban. As a comparison, the US population is about 80 percent urban.

As discussed in earlier sections, the rural-to-urban shift that began with the Industrial Revolution in Europe continues today in developing countries. The Industrial Revolution, which started in northern England's Pennine mountain chain, rippled through Europe and across the Atlantic to the United States. The majority of countries in Europe are currently in stage 5 of the demographic transition. The five stages of the demographic transition illustrate a pattern of development and population dynamics for a country or region. The model outlines how rural societies with an agrarian economy in stage 1 can make the transition to stage 5: the stage that indicates an urban society with a consumer economy. As a general trend, when a country's levels of industrial activity and urban growth rise, the outcome is usually a higher standard of living and smaller family sizes. Additionally, rural-to-urban shift takes place, driven by the pull of opportunities and advantages in the industrializing and urbanizing areas. Countries in stage 5 have small families with a fertility rate below the replacement level. Their incomes, based on a consumer economy, are generally at high levels.

As Europe industrialized and progressed through the stages of the demographic transition, certain core regions reached the postindustrial stages earlier than others. Western Europe established a core industrial region with an extended periphery. The postindustrial activity in this core area contin-

Figure 8. The core region of Europe and the four main centers of industrial activity.
Attractive "pull" forces draw immigrants to the core economic area of Europe, seeking opportunity and advantage. "Push" forces cause people to leave other areas due to negative cultural or environmental forces, and/or the lack of opportunity and advantage.

ues today in four main centers of innovation: (1) Stuttgart in southern Germany, (2) Lyon in southeastern France, (3) Milan in northern Italy, and (4) Barcelona in northeastern Spain. These four industrial centers have been referred to by some as the **Four Motors of Europe** because they promote business and industry for the European community. The European core region extends as far as Stockholm in the north to Barcelona in the south.

D. Central Business Districts and Primate Cities

European urban development centered on port cities that had industrial activity. Ships could import raw materials, the factories could manufacture the goods, and the ships could export the products. **A central business district (CBD)** developed around these activities. Since walking was the main transportation mode, all business activities had to be located in the same vicinity. Banks, retail shops, food markets, and residential dwellings had to be close to the factories and port facilities. Modern cities emerged from this industrialization process, and Europe is one of the more urban realms on the planet.

Many European countries have what geographers call a primate city. A primate city is a city that is much more than twice as large as the country's second-largest city and is especially important, usually dominant economically, culturally, and/or politically (although not all primate cities are national capitals). Primate cities represent a country's persona and often are symbolic of the country's heritage and character. Though common, not all countries of the world have a primate city. Financial and business centers in cities such as London, Rome, or Paris support the industrial activity that led to their development as primate cities. Most primate cities are ports or are located on a major river.

E. Rural-to-Urban Shift and Population Growth

Consider the trend that the Industrial Revolution brought to Europe. As countries industrialize and urbanize, family size naturally goes down and incomes traditionally go up. Integrate this with the rural-to-urban shift that occurs when countries progress through the five stages of the index of economic development. By understanding these basic trends, one can determine the average family size in Europe and why it is declining.

Because Europe is an urbanized realm, family size in Europe is small. As a matter of fact, various countries in Europe have negative population growth rates. Family size (or, more precisely, the **fertility rate**) in Spain and Italy is around 1.2, with the average family size in all of Europe at 1.4 children. The replacement rate to maintain an even population-growth pattern would be a fertility rate of about 2.1 children. Small families do not provide enough young people to cover the available entry-level service jobs. Most countries in Europe are in stage 4 or 5 of the demographic transition and are facing low or negative population growth and a deficit in their labor supply. As a core economic global power, Europe has experienced an increase in immigration.

With a lower fertility rate and an increase in postindustrial activity, Europe is a magnet for people from poorer peripheral countries and even peripheral regions within Europe who are looking for opportunities.

With the planet's human population increasing overall, one might think that a smaller family size is a positive trend. It may be, but there are challenges as well. If there are fewer young people in a community—fewer children and fewer people of employment age—consider how this affects the economic situation. With a declining European population, who will apply for the entry-level jobs? Who will support the growing numbers of retirees?

Economic core areas attract immigrants seeking opportunities and advantages. Europe follows this pattern. There have been increasing tensions between the long-standing European cultural groups and immigrants from developing countries who often speak non-European languages or follow religions other than Christianity. The main religion of immigrants from North Africa or the Middle East is Islam, which is the fastest-growing religion in Europe.

F. Nation-States and Devolution

The agrarian revolution and the Industrial Revolution were powerful movements that altered human activity in many ways. Innovations in food production and the manufacturing of products transformed Europe, which in turn impacted the rest of the world. Even before the agrarian revolution was under way, other transitions in European political currents were undermining the established empire mentality fueled by warfare and territorial disputes. The **political revolution** that transformed Europe was a result of diverse actions that focused on ending continual warfare for the control of territory and introducing peaceful agreements that recognized sovereignty of territory ruled by representative government structures. Various treaties and revolutions continued to shift the power from dictators and monarchs to the general populace.

The political revolutions laid the groundwork for a sense of **nationalism** that transformed Europe into nation-states. The term **nation** refers to a homogeneous group of people with a common heritage, language, religion, or political ambition. The term **state** refers to an independent, self-governing political unit; for example, the United States has a State Department with a secretary of state who represents the entire country. In international usage, and in this textbook, the term state can generally be thought of as synonymous with the word country, unless specifically referring to a smaller political unit within a country, such as states in the US or Mexico. When nations and states come together, there is a true **nation-state**, wherein most citizens share a common heritage and a united government.

European countries have progressed to the point where the concept of forming or remaining a nation-state is a driving force in many political sectors. To state it plainly, most Europeans, and to an extent every human, want to be a member of a nation-state, where everyone is alike and shares the same culture, heritage, and government. The result of the drive for nation-states in Europe is an Italy for Italians, a united Germany for Germans, and a France for the French, for example. The truth is that this ideal goal is difficult to come by. Though the political borders of many European countries resemble nation-states, there is too much diversity within the countries to consider the ideal of creating a perfect nation-state a reality.

Various ethnic populations in Europe desire their own nation-states. They want to devolve or separate from the larger state. The term devolution refers to the process whereby regions or people within a state demand independence and autonomy at the central government's expense. There are now a number of cases where devolution is occurring in Europe. For example, Scotland and Wales seek to devolve from the United Kingdom. The Basque region between Spain and France would like to have its own nation-state. Former Yugoslavia broke up into seven smaller nation-states. Various other minority groups in Europe seek similar arrangements. Thus both cohesive cultural forces and divisive cultural forces are active in the European community.

G. Centrifugal and Centripetal Forces

Cultural forces continually apply pressure on a country. Some of these cultural forces pull the nation together (**centripetal forces**) and others pull it apart (**centrifugal forces**). Primary sources of these cultural forces include religion, language, ethnicity, politics, and economic conditions.

When there is division, conflict, or confrontation, the centrifugal forces are at play. When unification, agreement, or nationalism are being exercised, centripetal forces are evident. The sources that tie a country together can also be the sources that divide a country. Ethnic unity can be a positive force, while ethnic division and conflict can be a divisive force. If centrifugal forces become strong, the result may be outright civil war, as was seen in the United States in the 1860s. Unity can also be evident through national struggles, such as the nationalism displayed immediately after the 9/11 terrorist attacks in the United States. After the attacks, an outpouring of goodwill and agreement strengthened the bonds within the United States.

To understand our world, it is helpful to understand the cultural forces that are active in any one location. Disagreement, inequity, or injustice related to the cultural factors of ethnicity, religion, language, and economics of a region or country is the cause of most conflicts.

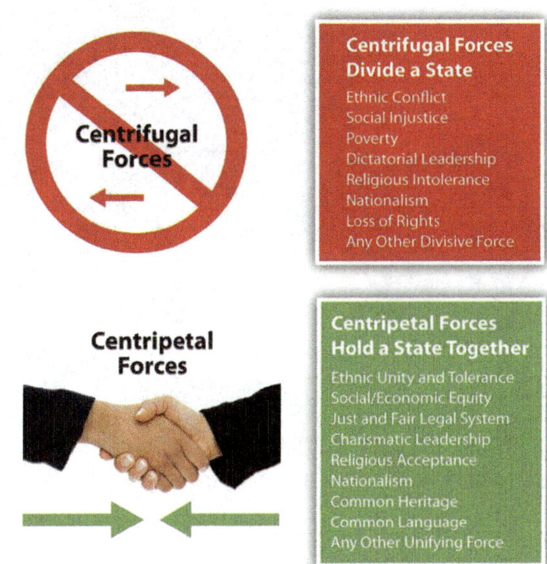

Figure 9. Centrifugal and centripetal forces in a state. Centrifugal forces divide a state and centripetal forces hold a state together power energized industrial activities.

H. Religion and Language in Europe

Europe has historically been considered a Christian realm. The three main branches of Christianity in Europe are Roman Catholic, Protestant, and Eastern Orthodox. Rome has been the geographical base for the Roman Catholic Church since the Roman Empire. Operating on the Romance language of Latin, the Catholic Church has provided southern Europe with a common religion for over 1,500 years.

The Roman Catholic Church split when Constantinople, now called Istanbul, gained preeminence. The Eastern Orthodox Church launched itself as the primary organization in the Slavic lands of Eastern Europe and Russia. The reformation of the fourteenth century, led by people such as Martin Luther, brought about the Protestant Reformation and a break with the Roman Catholic Church. Protestant churches have dominated northern Europe to this day.

Three main Indo-European language groups dominate Europe. Though there are additional language groups, the dominant three coincide with the three main religious divisions. In the east, where the Eastern Orthodox Church is dominant, the Slavic language group prevails. In the north, along with Protestant Christianity, one finds the Germanic language group. In southern Europe, where Roman Catholicism is dominant, the Romance languages are more commonly spoken.

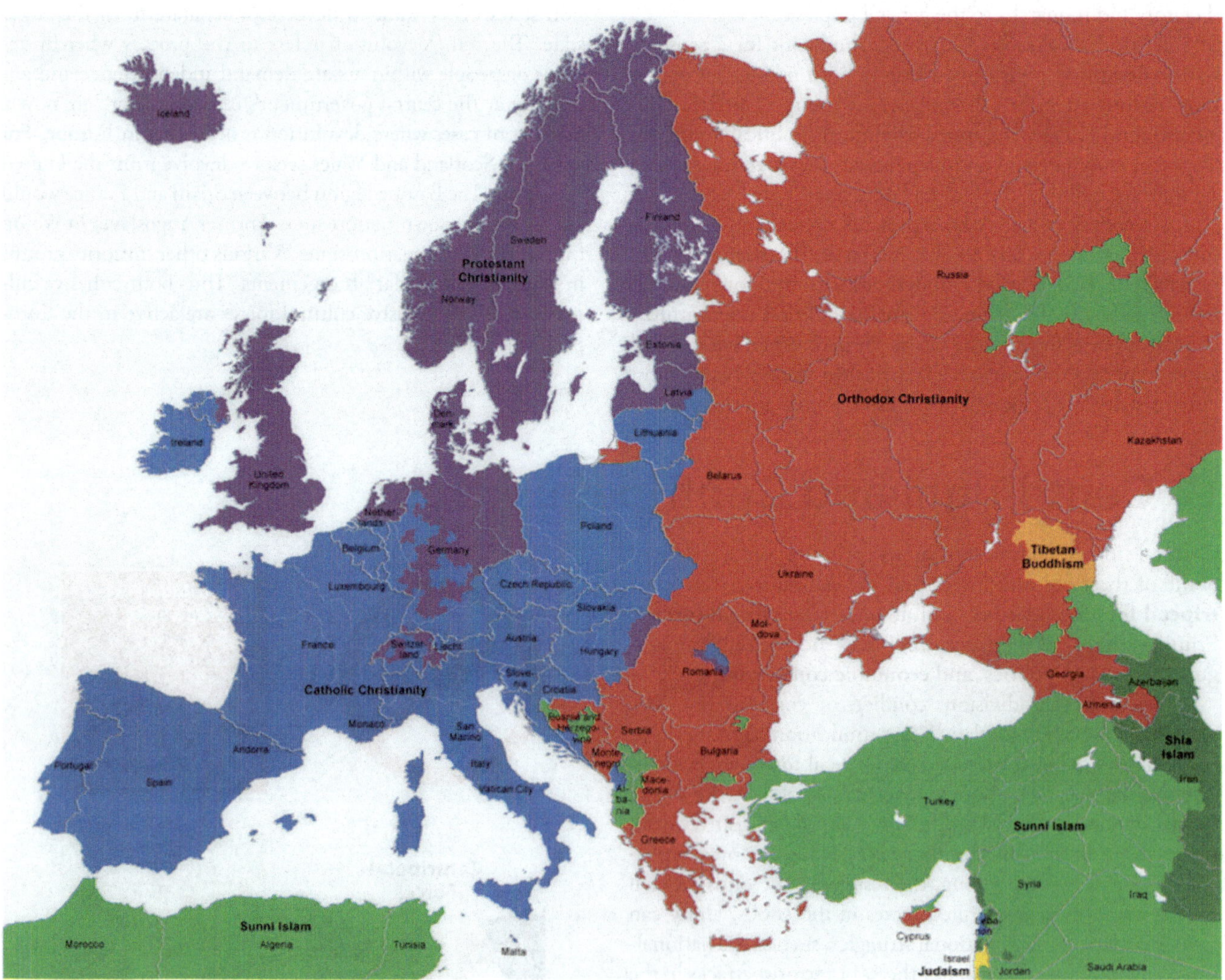

Figure 10. The three dominant branches of Christianity in Europe (Protestant, Catholic, and Eastern Orthodox).

Key Takeaways

1. The Roman Empire connected southern Europe and created an infrastructure to help promote trade and intercultural connections. The Vikings connected northern Europe through trade and exploitation.

2. Technological advancements helped European colonialism dominate other countries and exploit their labor and resources. Coastal European countries created colonies and external sources of wealth.

3. The Industrial Revolution was promoted by the development of steam power with coal as a fuel source. The mass production of goods gave the European countries an advantage in the world marketplace.

4. The Industrial Revolution prompted a shift in population from the rural agricultural regions to the urban centers. More people were needed in the factories and fewer workers were required on the farms because of improved agricultural methods. This shift resulted in smaller families and more women entering the workplace.

5. The early empires of Europe gave way to the concept of a similar people (the nation) unifying under a common government (the state) to create nation-states.

6. Divisive centrifugal cultural forces tend to divide and separate people in a state, whereas cohesive centripetal cultural forces tend to unify a state.

7. The Indo-European language family has three dominant groups in Europe: Germanic in the north, Romance in the south, and Slavic in the east. The Christian religion has three main divisions in Europe: a Protestant north, Catholic south, and an Orthodox east.

2.3 The European Union and Supranationalism

Learning Objectives

1. Outline how Europe has been divided during the twentieth century.
2. Describe the measures or methods that have been implemented to help unify Europe.
3. Explain the dynamics of supranationalism and its advantages and disadvantages.
4. Summarize how globalization has increased with the advent of the European Union.

The world economy has a competitive marketplace. Each independent country has to compete economically to earn national income, but not all countries are equal in natural or human resources. The smaller countries of Europe may have difficulty competing with the world's core economic powers such as Japan or the United States. Some of the countries of Europe are small in terms of physical area. In fact, the state of Texas is larger in square miles than any individual European country, and Oregon is larger in size than more than half the European countries. The total physical area of all the European countries together equals only 60 percent of the total physical area of the United States. However, in regard to Europe, physical size may not an indicator of economic ability. A number of European countries possess a high level of economic output and are major forces in the global economy.

The economic forces of globalization have motivated the nation-states of Europe to work together rather than compete with each other. Western Europe as a region is highly industrialized and has a high standard of living. Unified, the countries could be a major economic power in the world. Separately and independently, they may not be able to compete at the same level as other globally recognized trading blocs. To become unified after a century of centrifugal forces dividing them has not been easy. Consider the cultural forces that have been active in Europe. Centripetal forces unifying the realm include a common Christian religion, Indo-European language groups, and a Caucasian ethnic background. These forces, however, have not resulted in a unified Europe. The closest resemblance to a unified Europe was the Roman Empire, which was held together by military force.

During the twentieth century, there were three major divisions in Europe, all products of centrifugal forces. First, World War I, which was supposed to be the war to end all wars, divided Europe and the industrialized world. Second, World War II pitted the Axis powers (led by Germany and Italy with Adolph Hitler leading the German contingency) against the Allied powers (led by Great Britain and France with the United States entering later). Third, after World War II, the confrontation between communism and capitalism developed into the Cold War. The Iron Curtain, built of concrete, barbed wire, and land mines, separated Communist Eastern Europe, which was dominated by the government of the Soviet Union, from the capitalist democracies of Western Europe, which were allied with the United States.

A. Unification Efforts in Europe

Despite many problems, since World War II, steady efforts have been made toward European unification. After World War II ended in Europe, the three small countries of Belgium, the Netherlands, and Luxembourg realized that together they would be much stronger and recover more quickly from the war than if they remained separate. Belgium had banking and business; the Netherlands had industry, farming, and the world-class port of Rotterdam on the Rhine River; and Luxembourg had agricultural and mineral resources. To help recover from World War II, in 1944 the three countries signed an economic pact called the Benelux Agreement (after the first letters of each country's name), which provided a successful example of cooperation.

Implemented from 1948 to 1952, US Secretary of State George Marshall's Marshall Plan helped rebuild war-torn Europe with American aid and business connections. US business-
es and corporations benefited from the increased international trade with Europe. But to deter the European nations from going to war again, there needed to be an economic trade policy that encouraged a strong business climate. In 1957, the more prominent countries of France, West Germany, the Netherlands, Luxembourg, Italy, and Belgium all signed the **Treaty of Rome**, which created the Common Market. This agreement provided the structure necessary to unify Europe under a **European Union** (EU) in 1992. The EU was the structure for a common economic system with an agreed-upon governing body, and it was designed as an economic trading bloc that could compete with the United States and Japan. A mechanism was finally in place for a supportive, unified Europe, but it would be up to the independent countries that joined this union to make it work.

B. Supranationalism

Supranationalism is defined as the voluntary association of three or more independent states willing to yield some measure of sovereignty for mutual benefit. The Benelux Agreement of 1944 was a model for European supranationalism. Nations are often hesitant to give up any sense of independence or autonomy, especially with the strong drive to-

Figure 11. The 27 EU members as of 2010. In 2013, Croatia was admitted to the EU as a 28th member.

ward nation-state status. It is different for the United States: though the United States is not classified as a nation-state because of its ethnic diversity, English is the only major language and the dollar is the national currency. One can drive for three thousand miles across the United States and still experience a similar cultural landscape, complete with the same English-language road signs, identical franchised restaurants, common chain big-box stores, and similar advertising icons. This is not the case in Europe, where each country might have its own language, currency, traffic laws, and legal system. Supranationalism and unification can be a painful process. With common standards and a common currency comes the reality that a common cultural landscape might develop. Supranationalism could erode the uniqueness of each state as Europe becomes more of a "United States of Europe." To address the differences in the many currencies used in Europe, the EU introduced a common currency called the euro. This solution encountered resistance but has been accepted by most EU members. Most of the old currencies of EU members were phased out and the single currency of the euro has become the standard. Even some countries not currently in the EU have adopted the euro as their national currency.

Unification has created economic problems between the wealthy industrialized countries and the poorer regions of southern Europe. There is disagreement over how taxes or funds will be allocated. The poorer countries would like economic assistance in developing their industries. The wealthier nations are at times resistant to sharing their wealth. Italy has experienced this problem within its borders: Italy's wealthy northern regions have hinted at separating from their poorer southern regions. This is an age-old problem that confronts governments of most countries.

Travel within the EU has also changed. Before the EU, people traveling between European countries encountered border checks at which their passports were checked and stamped, and different traffic laws existed for each country. This is changing. Once inside the EU, there are no more border stops, and traffic and travel are becoming streamlined with common laws, making travel between member nations similar to travel between US states.

European countries have to confront both the centrifugal forces that rally their nations to remain uniquely independent and the centripetal forces that call for integration into the EU, which results in the loss of some of their autonomy. European cultures have a history of struggling to retain their heritage and traditions. The strong devolutionary forces that advocate for nation-state status are challenged by the need to belong to a larger union for economic survival. Europeans are caught between holding on to cultural heritage and moving forward economically in a competitive global economy. Similar forces are also felt in other regions of the world.

Germany, which united its western and eastern regions after the Iron Curtain came down in the early 1990s, has climbed to the top of the economic ladder in Europe. Germany is also the largest country in Europe by population. Other EU nations are concerned that Germany might once again dominate their realm or bring about massive division in Europe. The Cold War, World War I, and World War II are not easily forgotten; they are vital events in European history that affected the whole world. Many Europeans are suspect of what the EU may evolve into. Others welcome a more open European community.

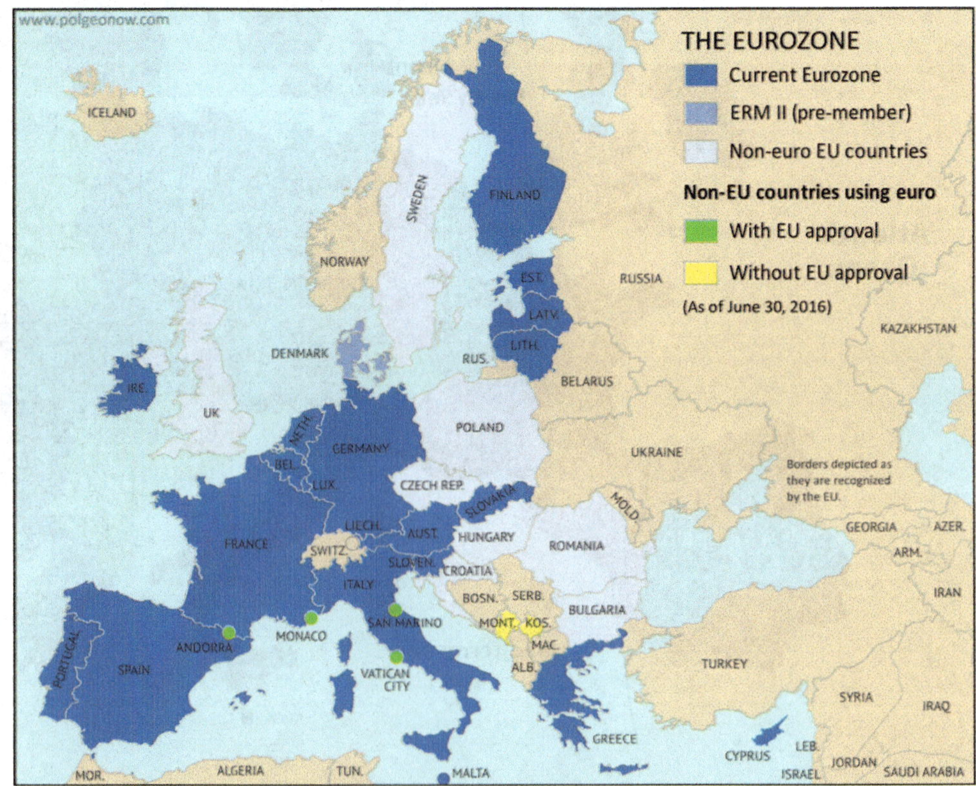

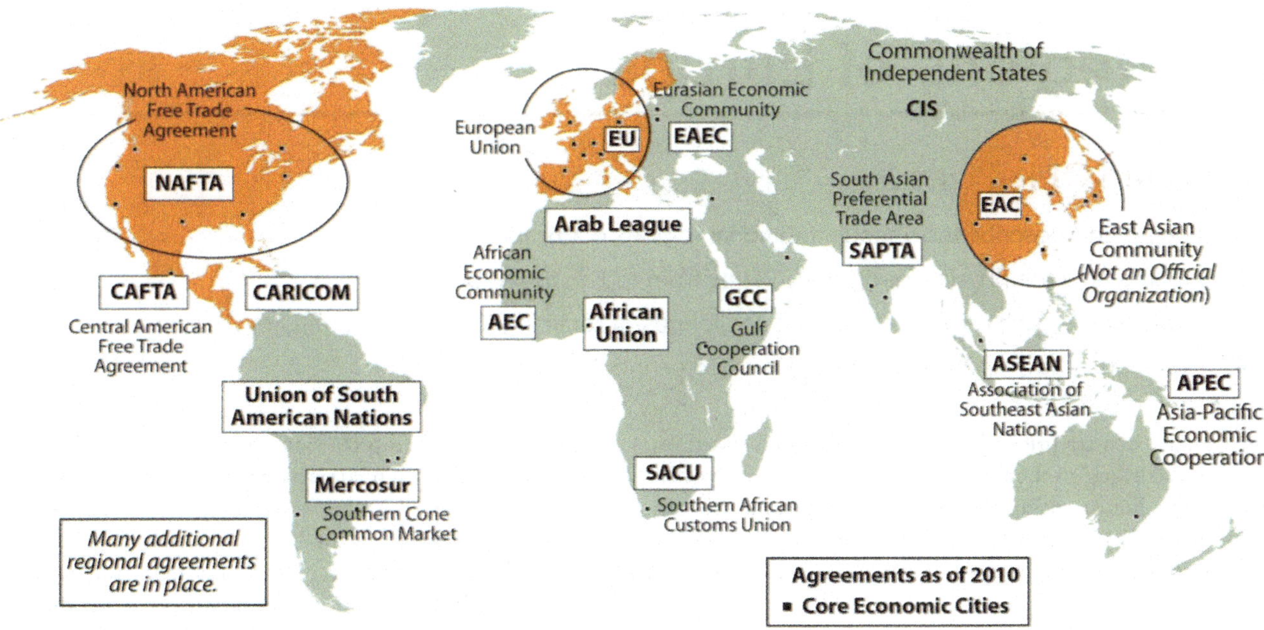

Figure 13. Global groups with the big three core areas of North America, Western Europe, and Eastern Asia.

C. About the European Union

According to the 1993 Copenhagen European Council, "a country has to meet certain requirements to join the EU. These requirements include a stable democracy which respects human rights and the rule of law; a functioning market economy capable of competition within the EU; and the acceptance of the obligations of membership, including EU law. Evaluation of a country's fulfillment of the criteria is the responsibility of the European Council."

As of 2016, the EU had 28 member states representing just over 500 million people. A number of additional states are applying for membership, so the number of EU member states continues to change. In 2016, there were 24 official EU languages. Each EU state has its own military for national defense. A majority of EU members also hold membership in the North Atlantic Treaty Organization (NATO), which is a political alliance between Western European countries, Canada, and the United States.

The EU is one of many supranational entities. Such economic and political associations have increased across the globe. Competition for labor and resources drives the need for smaller countries to join to compete economically in the global marketplace.

The EU is in the driver seat of one of the three main core economic areas of the world: North America, Western Europe, and Eastern Asia. The EU is an example of what supranationalism can produce. To compete in a global economy, the nation-states of Europe must cooperate and coordinate their industrial activities to support their high standard of living. The EU member states are a part of the elite "have" nations of the world. They face many questions about their future, and they will be watched closely by the rest of the world.

Key Takeaways

1. Although World War I, World War II, and the Cold War divided Europe during the twentieth century, the EU emerged as a unifying force for the European people.

2. The Benelux Agreement, the Marshall Plan, and the Treaty of Rome all helped set the stage for the European unification that evolved into the EU.

3. Supranationalism has provided European countries with the ability to compete economically in the global marketplace. Difficulties have been in the areas of cultural and historical differences that have influenced the continuing economic and political challenges.

4. The EU represents a core economic region for the planet. North America and Eastern Asia each have worked to create competitive trading relationships to compete with the EU economically.

2.4 Regions of Europe

Learning Objectives

1. Locate and describe the various traditional regions of Europe.
2. Outline how the physical geography varies from region to region.
3. Explain how each region has met the challenges of retaining its cultural identity or uniqueness.
4. Summarize how each region has developed an industrialized economy.

Europe has been traditionally divided into regions based on location according to the four points of the compass: Northern Europe, Southern Europe, Western Europe, and Eastern Europe. The British Isles are often considered a separate region (as they are in this book), but sometimes are included as a part of Western Europe.

The traditional regions of Europe are not as relevant today as they have been historically with the creation of the European Union (EU). Economic and political relationships are more integrated than they were in past eras when nation-states and empires were more significant. Economic conditions have often superseded cultural factors and have intensified the need for increased integration. Cultural forces have traditionally supported nationalistic movements that work to preserve the culture, heritage, and traditions of a people. Regional cultural differences remain the social fabric of local communities that support the retention of their identity. Modern transportation and communication technology has brought this cornucopia of European identities into one single sphere of global recognition.

A. Northern Europe

Europe has many different cultural identities within its continent. Northern Europe has traditionally included Iceland, Finland, and the three Scandinavian countries of Norway, Sweden, and Denmark. These countries are often referred to as the Nordic countries. All these countries were influenced by Viking heritage and expansion. Their capital cities are also major ports, and the largest cities of each country are their primate cities. The languages of the three Scandinavian countries are from the Germanic language group and are mutually intelligible. Finnish is not an Indo-European language but is instead from the Uralic language family. Most of Iceland's inhabitants are descendants of Scandinavian Vikings. Protestant Christianity has prevailed in northern Europe since about 1000 CE. The Lutheran Church has traditionally been the state church until recent years. These countries were kingdoms, and their royal families remain highly regarded members of society. The colder northern climate has helped shape the cultural activities and the winter sports that are part of the region's heritage. Peripheral isolation from the rest of Europe because of their northern location and dividing bodies of water have allowed the northern culture to be preserved for centuries and shape the societies that now exist in northern Europe.

Human rights, education, and social concerns are high priorities of the governments of northern Europe, and the quality of these elements rank highly by global comparisons. Standards of living are among the highest in Europe. Isolation in northern Europe does create an element of economic cost, and transportation technology has been leveraged to address this. A modern bridge has been constructed across the Baltic Sea from Denmark to Sweden to increase the flow of people, goods, and materials between the Scandinavian Peninsula and mainland Europe. Iceland is the most remote of the Nordic countries. Its small population—less than a half million people—is connected to Europe by sea and air transportation and communication technologies. Almost all elementary school children in the Nordic countries are taught English as a second language. Fish, meat, and potatoes are traditional dietary staples; fish in particular has been an important food source. The cuisine of the region is not noted for being spicy. Northern Europe has worked diligently to integrate itself with the global community and yet maintain its cultural identity.

As a standard practice, the northern European countries have exploited the opportunities and advantages of their natural resources to expand their economies. Sweden, northern Europe's largest country, has used its natural iron ore supply to develop its manufacturing sector. Sweden was the production base of Saab and Volvo vehicles as well as other high-tech products; however, GM purchased the Saab auto division in 2000 and some of its automobiles were manufactured in

Mexico. In 2010, Saab was sold back to European investors and production resumed in Sweden. Volvo Car Corporation was purchased by Ford Motor Corporation in 1999 and then acquired by a Chinese automaker in 2010.

Finland has vast timber resources and is one of Europe's major sources of processed lumber. It was the original manufacturer of Nokia cell phones, an example of its technological advancements. Nokia is the largest manufacturer of mobile phones in the world and has production facilities in eight different countries.

Norway has been benefiting from the enormous oil and natural gas reserves under the North Sea. Fishing and modest agricultural activities had been Norway's traditional means of gaining wealth, but now it is the export of the much-in-demand energy resources. Because of its economic and energy independence, Norway has opted not to join the EU.

Figure 14. Traditional Regions of Europe member.

Vikings were masters of the seas and colonized Greenland, which is located next to Canada and is considered to be the world's largest island. Danish colonization in the eighteenth century included Greenland and the Faeroe Islands, which are located between Scotland and Iceland. Both are now under the government of Denmark but retain a high level of self-rule and autonomy, which has aided them in holding on to their cultural identity. Greenland has also opted not to become a part of the EU even though Denmark is a member. Greenland only had a population of about 57,670 in 2011, and 80 percent of its surface is covered with ice. Fish is Greenland's main export, but minerals, diamonds, and gold are also present in viable amounts.

Denmark has a consumer economy with a high standard of living. This Scandinavian country is often ranked as the least corrupt country in the world and has the happiest people. The country has supported a positive environment and implemented strong measures to protect its natural areas. Denmark's main exports are food products and energy. The country has sizeable oil resources in the North Sea but also receives over 15 percent of its electricity from wind turbines.

The Baltic states of Lithuania, Latvia, and Estonia have often been included in the northern European designation because of their northern location. Estonia has the strongest similarities in religion, traditions, and culture, and geographic literature often has included it as a part of northern Europe. The Baltic states have been associated with Eastern Europe through the Soviet era but, like their neighbors to the north, are becoming more economically integrated with Western Europe.

Northern Europe is a peripheral region. Southern Sweden has an advanced industrial base and resembles a core area. Sweden's northern portion and the main parts of the other Nordic countries act as sources of raw materials for Europe's urban core industrial region. In the core-peripheral spatial relationship, northern Europe most resembles a semiperipheral region that has attributes of both the urban core and the rural periphery.

Norway, Sweden, and Finland are quite large in area but

Figure 15. Port in Stockholm, Sweden.

are not densely populated compared with other European nations. Sweden ranks as the fourth-largest European country in physical area. Sweden is larger than the US state of California, but in the 2010 census, it had less than ten million people. In 2010, Finland, Norway, and Denmark all had less than six million people each.

The cultures and societies of northern Europe have progressed along similar paths; that is, they have advanced from once Viking-dominated lands into modern democratic and socially mature nations. Northern Europe is known for its concern for the social welfare of its citizens. Their strong egalitarian ideals have contributed to extensive advancements in free medical care, free education, and free social services for all, regardless of nationality or minority status. Civil rights for minorities, women, and other groups is assured and protected. Denmark doesn't have a legal age for consumption of alcoholic beverages, though tradition sets the age at about fourteen. Culture and the arts are well developed; examples include everything from the Nobel Peace Prize to Hans Christian Anderson to the 1970s chart-topping pop group ABBA. Sweden has become a major exporter of music worldwide. Rock, hip-hop, and pop music are common genres. With English as a widely spoken language in Scandinavia, music and cultural trends have a larger export market in places such as the United States and Britain.

B. Southern Europe

Southern Europe includes three large peninsulas that extend into the Mediterranean Sea and the Atlantic Ocean. The Iberian Peninsula consists of Spain and Portugal. The Pyrenees mountain range separates the Iberian Peninsula from France. Greece, the most southern country on the Balkan Peninsula, includes hundreds of surrounding islands and the large island of Crete. The Italian Peninsula is the shape of a boot with the Apennine Mountains running down its center. Italy also includes the islands of Sicily and Sardinia. Technically, the island country of Cyprus is also included in southern Europe. There are five microstates in this region. The small island of

Figure 16. Street view in Naples, Italy.

Malta is located to the south of Sicily and is an independent country. Monaco, San Marino, Andorra, and Vatican City are also independent states located within the region. Southern Europe's type C climate, moderated by the water that surrounds it, is often referred to as a Mediterranean climate, which has mild, wet winters and hot, dry summers.

Rural-to-urban shift in southern Europe has not been as strong as that of Western Europe. Only about 50 percent of the people in Portugal are urban; in Spain and Greece, about 60 percent are urban.

Italy is more representative of Europe, with about 68 percent of the population urbanized. Italy is also divided, with northern Italy being more industrialized than southern Italy. The southern regions of Italy, including the island of Sicily, are more rural with fewer industries. Northern Italy has the metropolitan city of Milan as an anchor for its global industrial and financial sector in the Lombardy region, which includes the city of Turin and the port of Genoa. This northern region of Italy has the economic muscle to be one of Europe's leading manufacturing centers. The so-called Ancona Line can be drawn across the middle of Italy from Ancona on the east coast to Rome on the west coast to separate the industrial north with the more agrarian south. The north also has the noted cultural cities of Venice, Florence, and Pisa.

A similar situation exists in Spain. The urbanized Catalonia region around Barcelona in the northeast has high-tech industries and a high standard of living. Southern Spain has large rural areas with economies heavily based on agricultural production. Portugal and Greece are not as industrialized and do not have the same economic opportunities. Historically, southern Europe, Portugal, and Greece in particular each have had a much lower gross domestic product (GDP) per capita than northern or Central Europe. Their economies have been much more aligned with the economic periphery than with the industrial core region of Europe. Greece has had serious economic difficulty in the past few years.

Southern European countries have much larger populations than their northern European counterparts. Italy has about sixty million people in an area smaller than Norway, which has less than five million. Spain has about forty million; both Portugal and Greece have more than ten million. Cultural factors are also different here than in northern Europe. The culture of southern Europe has been built around agriculture. Traditional cuisine is based on locally grown fresh food and wine. Olive oil and wine have been major agricultural exports. The main languages of Iberia and Italy are based on the Romance language group, and Greek is an independent branch of the Indo-European language family. The most dominant religious affiliation in the south is Roman Catholicism, except in Greece, where the Eastern (Greek) Orthodox Church is prominent.

Spain is the most diverse nation in southern Europe with a number of distinct ethnic groups. The Basques in the north along the French border would like to separate and create their own nation-state. The region of Galicia in northwest Spain is an autonomous region and was once a kingdom unto itself. There are many other autonomous communities in Spain, each with its own distinct heritage and culture. Farther east in the Mediterranean is the island state of Cyprus, which is divided by Greek and Turkish ethnic groups. The southern part of the island is dominated by Greek heritage and culture, and the northern part of the island is dominated by Turkish culture and traditions. Islam is the main religion of the Turkish north. The people of southern Europe are diverse and hold to many different traditions but are tied together by the sea and the land, which create similar lifestyles and economic activities.

Figure 17. Spain's autonomous communities. Because of its diversity, Spain is not categorized as a nation-state.

C. Western (or Central) Europe

When discussing the political geography of the European continent, the states located in the western part of the European mainland are often referred to as Central Europe. Germany and France are the two dominant states, with Belgium, the Netherlands, and Luxembourg making up the Benelux countries. Switzerland and Austria border the Alpine region. The microstate of Liechtenstein is located on the border between Switzerland and Austria. France is the only country with coastlines on both the Atlantic Ocean and the Mediterranean. These countries are located in the core economic region of Europe and have stable democratic governments and a relatively high standard of living by world comparisons.

Central Europe is a powerhouse of global economics. The Rhine River is a pathway for industrial activity from southern Germany to Europe's busiest port of Rotterdam in the Netherlands. Western France has the political capital of the EU along the Rhine at Strasbourg. To the south is France's second-largest city, Lyon, which is a major industrial center for modern technology. Germany had the historical Ruhr industrial complex along the Rhine that supported the high-tech industries in southern Germany in the cities of Stuttgart, Mannheim, and Munich. Germany is the most populous country in Europe, with over eighty-two million people in 2010. Germany is also Europe's largest economy and has the largest GDP overall as a country. Belgium has major business centers in Brussels and Antwerp. Switzerland is noted for its banking and financial markets. Luxembourg has one of the highest GDP per capita in all of Europe. Austria is noted for its high level of cultural activities in Vienna and Salzburg. All these countries complement each other in creating one of the dominant economic core areas in the world.

In the first half of the twentieth century, the political geography of Central Europe was not conducive to the high level of economic cooperation that now exists. In World War I and World War II, Germany and France were on opposite sides, and the Benelux countries were caught in the middle. The cultural differences between the Germans and the French start with the differences in language and religious affiliation. Germany was divided after World War II into East Germany and West Germany, separated by the so-called Iron Curtain. East Germany was under a Communist government, and West Germany was a capitalist democracy. They were reunited in 1990 when the Iron Curtain and the Berlin Wall came down. The two countries merged under one government. Europe is gradually being united economically, but each country or region still retains its cultural uniqueness.

The Benelux Countries

The Benelux countries have a great deal in common historically. Before the economic union that created the term Benelux, these countries were collectively referred to as the Low Countries, so called because of their relative position to sea level. The Benelux countries are some of the most densely populated countries. They have managed to work together toward a common economic objective in spite of their cultural differences.

The capital and largest city in Belgium is Brussels, with the other urban areas being the ports of Antwerp and Ghent. Belgium is split into three large geographic areas. The dominant language in the northern region of Flanders is Dutch (Flemish), and the people are known as Flemings. In the southern region of Wallonia, most people speak French and are known as Walloons. German is the third official language and is spoken along the eastern border.

Belgium is a major exporter of manufactured products, including finished diamonds, food products, nonferrous metals, technology, petroleum products, and plastics. In general, Belgium imports the raw materials to manufacture these goods for export. Belgium also has a significant service sector, including real estate, hotels, restaurants, and entertainment, which thrives in part because Brussels is the headquarters of the North Atlantic Treaty Organization (NATO) and components of the EU. Many countries and organizations maintain offices in Brussels to have easy access to these headquarters; therefore, Brussels is the temporary home to many diplomats and foreign business people.

The Netherlands is sometimes called Holland, which is actually the name of two provinces (North Holland and South Holland) in the northwest part of the country. The largest city is Amsterdam. The Hague is the seat of government and is home to the United Nations International Court of Justice. Rotterdam is located at the mouth of the Rhine

Figure 18. Brussels high-speed train. Brussels is connected to other European cities through high-speed rail networks.

River and is one of the busiest ports in the world. The country is famous for its Zuider Zee, which is the large inland region below sea level that has been drained of water and surrounded with an extensive dike system to protect it from the North Sea. Reclaimed land from the sea in areas called polders has provided this densely populated country with more land area for its people to expand their activities.

For a small country with few natural resources, the Netherlands has an impressive GDP. The Dutch have made good use of their location on the North Sea and of the location of several large navigable rivers. This has facilitated voluminous exports to the inland parts of Europe. The major industries include food processing, chemicals, petroleum refining, and electrical machinery. The Netherlands is a top exporter of agricultural products, which contribute substantially to its economy. Dutch agricultural exports consist of fresh-cut plants, flowers, and bulbs as well as tomatoes, peppers, and cucumbers.

Luxembourg's one major city is Luxembourg City. Luxemburg has an enviable economic situation with a stable and prosperous economy, low unemployment, and low inflation. Thanks to rich iron-ore deposits, this country was able to develop a very robust steel industry, which was the cornerstone of the nation's prosperity until the 1970s. As steel declined, Luxemburg remade itself as an important world financial center. Luxembourg leads Europe as the center for private banking and insurance industries and is second only to the United States in terms of being an investment fund center.

France

France covers 211,209 square miles and is the second-largest European country in area; only Ukraine is slightly larger. The physical landscapes of France vary widely from the northern low-lying coastal plains to the Alpine ranges of the east. Mont Blanc, the highest mountain in the Alpine range at 15,782 feet, is located in France near the Italian border. In the far south, the Pyrenees run along the border with Spain. The south-central region of the country is home to the Massif Central, which is a plateau and highland region made up of a large stretch of extinct volcanoes.

During the colonial era, France was a major naval power and held colonies around the world. The French Empire was the second largest at the time. The French language is still used for diplomacy in many countries. Though the French Empire no longer exists, France has progressed into a postindustrial country with one of the most developed economies in the world. It is a major player in European affairs, the EU, and the United Nations (UN). France is a democratic republic that boasts a high-quality public education system and long life expectancies.

In 2010, France's population was about sixty-five million, with about ten million living in France's primate city of Paris. The city of Paris is on an excellent site, favorably situated with regard to its surrounding area. It is the core area of France serving a large peripheral region of the country. The next largest city of Lyon, which is a major high-tech industrial center for Europe's economy, only boasts a population of about 1.4 million. Even with a large population, the country is able to produce enough food for its domestic needs and for export profits.

France enjoys a robust economy and is one of the world's leading industrial producers. Its industrial pursuits are diverse, including the manufacture of planes, trains, and automobiles, as well as textiles, telecommunications, food products, pharmaceuticals, construction and civil engineering, chemicals, and mechanical equipment and machine tools. Additionally, defense-related industries make up a significant sector of the economy. France's production of military weapons is recognized worldwide. The country has been a leader in the use of nuclear energy to produce electricity. Nuclear energy supplies about 80 percent of the country's electricity, which reduces the need for fossil fuels and imported oil.

Agriculture is an important sector of the French economy, as it has been for centuries, and is tied to industry through food processing. Food processing industries employ more people than any other part of the French manufacturing sector. If you think of cheese and wine when you think of

Figure 19. The Eiffel Tower, a symbol of the primate city of Paris, was built in 1889. The city of Paris started as a citadel on an island in the middle of the Seine River.

France, you have identified two of its largest food processing endeavors, along with sugar beets, meats, and confectionaries. World-renowned wines are produced in abundance, sometimes in areas that bear their names, such as in Burgundy, around the city of Bordeaux, and in Champagne in the Loire Valley. French cuisine and fashion have long been held in highest esteem worldwide and are a source of national pride.

Thanks to the climate and favorable soil conditions, agriculture is highly productive and lucrative for France. France is second only to the United States in terms of agricultural exports. Exports mainly go to other EU countries, to the United States, and to some countries in Africa. France leads Europe in agricultural production.

The plains of northern France are excellent for wheat. Dairy products are a specialty in the western regions of France, which also produce pork, poultry, and apples. Beef cattle are raised in the central portion, where a cooler, wetter climate provides ample tracts of grasslands for grazing. Fruit, including wine grapes, is grown in the central and southern regions, as are vegetables. The region around the Mediterranean is blessed with hot, dry weather ideal for growing grapes and other fruits and vegetables.

Germany

Germany's location in Central Europe has meant that throughout history many peoples—all with their own cultures, ideas, languages and traditions—have traversed Germany at one time or another. Thus Germany's culture has received many influences over the centuries.

Germany's present geopolitical configuration is quite young, as it reunified the eastern and western portions into a single entity in 1990. Germany was formed in 1871 during the leadership of Otto von Bismarck in an attempt to create a Germanic power base. World War I was fought during the last years of the German Empire. Germany, as part of the Central powers (Germany, Austria-Hungary, and Bulgaria), was defeated by the Allies with much loss of life. The German Republic was created in 1918 when, having been defeated in World War I, Germany was forced to sign the Treaty of Versailles.

In 1933, with an environment of poverty, disenfranchisement of the people, and great instability in the government, Germany gave way to the appointment of Adolf Hitler as chancellor of Germany. Within a month of taking office, Hitler suspended normal rights and freedoms and assumed absolute power. A centralized totalitarian state quickly resulted. In a move to expand Germany, Hitler started to expand its borders. Germany's invasion of Poland in 1939 kicked off what would become World War II. In 1941, Germany invaded the Soviet Union and declared war on the United States. After Germany's defeat, the country was divided into East Germany, controlled by the Soviet Union, and West Germany, controlled by the Allied powers. The Iron Curtain divided the two Germanys, with the Berlin Wall dividing the city of Berlin. The Iron Curtain and the Berlin Wall were major symbols of the Cold War. In 1989, the Berlin Wall came down, and the two Germanys were reunited in 1990. Today, Germany is a vibrant country and an active EU member.

Germany is Europe's largest economy, with strong exports of manufactured goods. Most exports are in automobiles, machinery, metals, and chemical goods. Germany has positioned itself strategically to take economic advantage of the growing global awareness of environmental issues and problems by focusing on improvements and manufacturing of wind turbines and solar power technology. The service sector also contributes heavily to the economy. Deutsche Bank holds the enviable position of being one of the most profitable companies on the Fortune 500 list. Germany is also a major tourist destination. The Black Forest, Bavaria, the Alpine south, a variety of medieval castles, national parks, and a vibrant assortment of festivals such as Oktoberfest attract millions of tourists to Germany every year.

Figure 20. President Ronald Reagan speaking at the Brandenburg Gate, Germany, June 12, 1987. From this speech comes his famous quote, "Mr. Gorbachev, tear down this wall."

The Alpine Center

Landlocked in the center of Europe are the two main states of Switzerland and Austria. Sandwiched on the border of these two states is the microstate of Liechtenstein. This region is dominated by the Alpine ranges. Switzerland, officially known as the Swiss Confederation, is divided into twenty-six cantons (states). Because of its location and close ties with neighboring countries, four official languages are spoken in Switzerland: German, French, Italian, and Romansh. Typically, one language predominates in any given canton. Berne is the country's capital, and Geneva, Zurich, and Basel are the other major cities. As of 2010, Switzerland's population was about 7.8 million.

Internationally, Switzerland is known for its political neutrality. The UN European offices are located there. The Red Cross and the main offices of many international organizations are located in Switzerland. Switzerland joined the UN in 2002 and has applied for EU membership. Swiss culture is thought to have benefited from Switzerland's neutrality. During times of war and political turmoil, creative people found refuge within the Swiss borders. Swiss banking practices and policies are known throughout the world, and Swiss banks have benefited greatly from the country's politically neutral status. Banking is one of the country's top employers and sources of income. The Swiss people enjoy a high standard of living.

Austria is larger than Switzerland and is similar in area to the US state of South Carolina. In 2010, the population was estimated at 8.4 million. Austria has various Alpine ranges, with the highest peak at 12,457 feet in elevation. Only about a fourth of the land area is considered low lying for habitation. The Danube River flows through the country, including the capital city of Vienna. Austria has a well-developed social market economy and a high standard of living.

Austria is a German-speaking country, and nearly the entire population self-identifies as ethnic Austrian. It is also predominantly Roman Catholic and was home to many monasteries in the Middle Ages, influencing a strong Austrian literary tradition. Austria's best-known cities are its capital of Vienna and Salzburg and Innsbruck. Vienna was the center of the Habsburg and Austrian Empires and earned a place as one of the world's great cities. It is famed for its baroque architecture; its music, particularly waltzes; and theater.

Figure 21. Swiss Cottage Café, Illfracombe, Switzerland.

D. The British Isles

The British Isles are an archipelago (group of islands) separated from the European mainland by the English Channel. The British Isles are often included in the region of Western Europe when discussing political geography; however, the fact that they are separated from the mainland of Europe by water provides them with a separate identity. The British Isles consist of two separate, independent countries: the Republic of Ireland and the United Kingdom. The United Kingdom (UK) of Great Britain and Northern Ireland consists of the regions of England, Scotland, Wales, and Northern Ireland. All four regions are now under the UK government. The Republic of Ireland, consisting of the southern portion of the island of Ireland, is independent of the United Kingdom and does not include Northern Ireland. The primate city and UK capital is London, which is a financial center for Europe. The capital city of the Republic of Ireland is Dublin.

Influenced by the Gulf Stream, the climate of the British Isles is moderate, in spite of its northern latitude location. The UK and Ireland are located above the fiftieth degree of latitude, which is farther north than the US-Canadian border. The northern latitude would normally place this region into the type D climates, with harsher winters and more extreme seasonal temperatures. However, the surrounding water moderates temperature, creating the moderate type C climate that covers most of the British Isles. The Gulf Stream pulls warm

Figure 22. The cliffs of Moher, Ireland.

water from the tropics and circulates it north, off the coast of Europe, to moderate the temperature of Western Europe.

The Western Highlands and the Northern Lowlands dominate the islands. Scotland, Wales, and parts of England have highland regions with short mountains and rugged terrain. The lowlands of southern England, Ireland, and central Scotland offer agricultural opportunities. The Pennines mountain chain runs through northern England and was the source of the coal, ores, and waterpower that fueled the Industrial Revolution. To the east of Britain is the North Sea, which provided an abundance of petroleum resources (oil) for energy and wealth.

Though the heritage of the British Isles is unique to this region, the geographic dynamics are similar to Central Europe—that is, smaller families, urbanization, industrialization, high incomes, and involvement with economic

Figure 23. Eastern Europe.

globalization. The EU has had an enormous influence on the British Isles. Ireland has embraced EU's economic connections, but the British people have been hesitant to relinquish full autonomy to the EU. This reluctance can be noted in the fact that the United Kingdom kept the British pound sterling as their currency standard after the euro currency was implemented. However, the Republic of Ireland converted to the euro currency.

The regions of the British Isles follow similar dynamics to those of other countries in Western Europe. Though some regions are not as wealthy as others, they all demonstrate a high level of industrialization, urbanization, and technology. These urban societies have smaller families and higher incomes and are heavy consumers of energy, goods, and services. Just as the Industrial Revolution attracted cheap labor, the aging workforce has enticed people from former British colonies to migrate to the United Kingdom in search of increased employment opportunities. The mix of immigrants with the local heritage creates a diverse community. London has diverse communities with many ethnic businesses and business owners.

Devolutionary forces are active in the United Kingdom. Scotland and Wales are already governing with their own local parliaments. Devolutionary cultural differences can be noted by studying the different heritages found in each region. Just as the Welsh language is lingering in Wales, Gaelic continues in Ireland and Scotland. Each region has made efforts to retain local heritage and rally support for its own nation-state. Until recently, this was all done under the umbrella of the EU. However, in June 2016, voters in the UK as a whole voted for a British exit, or "Brexit," from the EU. In Scotland, though, the majority voted to remain in the EU, prompting calls for another Scottish referendum on Scottish independence. It may take years for these issues to be resolved.

E. Eastern Europe

After World War II ended in 1945, Europe was divided into Western Europe and Eastern Europe by the Iron Curtain. Eastern Europe fell under the influence of the Soviet Union, and the region was separated from the West. When the Soviet Union collapsed in 1991, all the Soviet Republics bordering Eastern Europe declared independence from Russia and united with the rest of Europe. The transition Eastern Europe has experienced in the last few decades has not been easy; however, most of the countries are now looking to Western Europe for trade and economic development. Cooperation continues between Eastern and Western Europe, and the European Union (EU) has emerged as the primary economic and political entity of Europe.

The collapse of Communism and the Soviet Union led to upheaval and transition in the region of Eastern Europe in the 1990s. Each country in the region was under Communist rule. The countries bordering Russia were once part of the Soviet Union, and those countries not part of the Soviet Union were heavily influenced by its dominant position in the region. When the Soviet Union collapsed in 1991, the bordering countries declared independence and began the process of integration into the European community. Moldavia changed its name to Moldova. The countries of Czechoslovakia and Yugoslavia each broke into multiple countries and, because of the diverse ethnic populations, organized around the concept of nation-states. Czechoslovakia peacefully agreed to separate into two states: the Czech Republic and the Republic of Slovakia. Yugoslavia was not so fortunate. After a series of wars that raged throughout the 1990s (collectively called the **Yugoslav Wars**), the country eventually devolved into seven independent countries: Slovenia, Croatia, Bosnia and Herzegovina, Serbia, Montenegro, Kosovo, and Macedonia.

Most Eastern European political borders coincide with ethnic boundaries. Each of the regions once resembled nation-states. In principle, Romania is set apart for Romanians, Hungary for Hungarians, and so on. Few are true nation-states because of ethnic minorities located within their borders, but the countries held on to their common heritage throughout the Communist era. In most Eastern European countries, cultural forces have brought people together to publicly support the move to unite and hold onto a heritage that is as old as Europe itself.

Governments that were controlled by Communist dictators or authoritarian leaders before 1991 were opened up to democratic processes with public elections. With the fall of Communism came economic reforms that shifted countries from central planning to open markets. Under central planning, the governments dictated which products were produced and how many of each were to be produced. The open markets invited private capitalism and western corporate businesses.

The power of the state was transferred from the Communist elite to the private citizen. People could vote for their public officials and could choose businesses and work individually. With the EU looming over the realm, the now-independent countries of Eastern Europe shifted their economic direction away from Moscow and the collapsing Communist state and toward the core industrial countries of Western Europe and the EU.

Figure 24. Reforms in Eastern Europe in the 1990s.

Key Takeaways

1. Europe can be divided into a number of smaller geographic regions, including Northern Europe, Southern Europe, Western Europe, the British Isles, and Eastern Europe.

2. Differences in climate, terrain, and resources provide for diverse economic activities that influence cultural development within Europe.

3. The countries of Northern and Western Europe and the British Isles have high standards of living by world standards. The southern countries of Portugal, Greece, and southern Italy are more agrarian and have been struggling economically since the global recession began in 2007.

4. The British Isles comprise the two independent countries of the United Kingdom and the Republic of Ireland. The UK consists of four administrative units (England, Scotland, Wales, and Northern Ireland), with Wales and Scotland having more autonomy to govern local affairs.

5. Since World War II, the many states of Europe have been evolving into a more integrated realm. Each nation has worked to develop its economy to take advantage of its physical geography and its cultural history and heritage. Devolutionary forces remain strong in many areas to counterbalance the forces of supranationalism.

6. After World War II ended, Europe was divided into Western Europe and Eastern Europe by the Iron Curtain. Western Europe promoted capitalist democracies, and Eastern Europe came under the Communist influence of the Soviet Union. After the collapse of the Soviet Union in 1991, Eastern Europe began to transition toward Western European ideals, including shifting toward democratic governments, open market economies, private ownership, and the EU.

7. Countries with stable governments and industrial potential have been accepted into the EU and have expanding economies. Other countries that have not reached that level of economic development or political reforms have not been admitted into the EU.

End-of-Chapter Summary

- The European continent extends from the North Atlantic Ocean to the Ural Mountains in Russia. The Russian segment of the European continent is usually studied with Russia as a whole. Europe is bordered by the Arctic Ocean to the north, the Atlantic Ocean to the west, and the Mediterranean Sea to the south.

- The Gulf Stream helps create a type C climate for much of Western Europe. Type D climates dominate the north and eastern portions of Eastern Europe. Europe has four main physical landforms that provide a diversity of resources for human activity.

- European colonialism brought increased wealth and economic activity to Western Europe. The Industrial Revolution also began here. Europe developed into an industrialized realm with powerful economic forces that continue to drive the postindustrial engine of globalization. Rural-to-urban shift and urbanization were products of industrialization. The result for Europe has been smaller families and higher incomes.

- The Roman Empire created early networks of infrastructure for southern Europe, while the Vikings connected northern Europe through trade and warfare. The latest attempt to unify the European nations is through supranationalism in the formation of the European Union (EU). Countries that have strong economies and stable governments are allowed to join the EU.

- Europe can be divided into various geographic regions based on the points of the compass. The distinctions often relate to the type of economic activity the people are engaged in or are based on cultural traits such as the variations of Christianity or the branches of the Indo-European language.

- Since the Soviet Union's collapse in 1991, Eastern Europe has been transitioning from Communist governments to democratic governments with capitalist-style economies. Some countries have made this transition more easily than others. Many of the more progressive countries have been accepted into the EU. Other countries continue to struggle to establish stable democratic governments and a growing economy.

Chapter 3

Russia

Introduction

Russia is the world's largest country in physical area—almost twice the size of the United States. The country extends from its European core, where most of the population live, across the Ural Mountains into Siberia and the Russian Far East, where residents have more economic and social connections with China than with Europe. A train journey from St. Petersburg to Vladivostok, the western and eastern termini of the Trans-Siberian Railway, takes about one week of constant travel. No paved highways cross the entire country. The main part of Russia is so big that it requires eight time zones, with an additional time zone for the European enclave of Kaliningrad, near Poland. Russia includes world-class cities such as Moscow, with its many billionaires and famous Red Square; vast territories of the Arctic north; immense forests of Siberia; grain farms rivaling those in Kansas; and mountain communities in the Caucasus. Russia has a complicated history of monarchy and totalitarianism, rich natural resources, extremes of wealth and poverty, and a slowly declining population. It is a dynamic country transitioning from a Communist state to part of the global economy.

Russia is located in both Europe and Asia. The Ural Mountains are considered the separation boundary for the two continents. The Asian side of Russia is physically bordered to the south by Kazakhstan, Mongolia, and China, with an extremely short border shared with the tip of North Korea. The Amur River creates a portion of the boundary with China. The Pacific Ocean is to the east with the Bering Strait separating Russia from North America. The Arctic Ocean creates the entire northern boundary of Russia stretching all the way from Norway to Alaska. The Arctic Ocean can be ice covered for much of the long winter season. Russia is a northern country with the majority of its physical area above the latitude of 50 degrees north. The Arctic Circle runs the entire length through the middle of the northern half of the country.

The boundaries of European Russian include its southern border in the Caucasus Mountains with Georgia and Azerbaijan. This portion of Russia protrudes south of 50 degrees latitude. The Caucasus Mountains are the tallest mountain chain in Russia and Europe. The Black Sea and the Caspian Sea create natural boundaries on either side of the Caucasus Mountains. The main borders with Eastern Europe include the large countries of Ukraine and Belarus. Farther north, Russia borders Latvia, Estonia, Finland, and Norway. European Russia is much smaller than its Asian counterpart but is the dominant core area for the country anchored by the capital city of Moscow.

Figure 1. Map of Russia.

3.1 Introducing the Realm

Learning Objectives

1. Identify Russia's climatic influences and physical regions.

2. Determine how the czars expanded their territorial power to create the Russian Empire.

3. Contrast the ways that the governments of the Russian Empire and the Soviet Union dealt with the issue of diverse nationalities within their countries.

4. Describe some of the environmental problems facing the Russian republics today.

The massive expanse of Russia exhibits a variety of physical environments, such as tundra, steppe, mountains, and birch forests. Type D (continental) climates dominate most of the country and characterize large landmasses such as Eurasia and North America. Land in the center of a large continent, far from the moderating effects of oceans, tends to heat up rapidly in the summer and cool down rapidly in the winter. These areas are known for hot summers and cold, harsh winters. Northern Russia borders the Arctic Ocean, and frigid air masses from the Arctic swoop south across Russia each winter. Moreover, Russia's northerly latitude means that it experiences a short growing season and has never been an agricultural superpower; the country usually has to import grain to feed its people. Mountain ranges to the south block summer rains and warm air masses that would otherwise come from South and Central Asia, thus creating deserts and steppes in southern Russia.

Most of Russia's population lives in the European part of the country on the Eastern European Plain, also known as the Western Russian Plain, or simply the **Russian Plain**, the most agriculturally productive land in Russia. The eastern edge of the plain is marked by the **Ural Mountains**, a low-lying mountain chain (about 6,000 feet) that crosses Russia from the Arctic Ocean to Kazakhstan. The mountains contain deposits of coal, iron ore, and precious and semiprecious stones and are considered the boundary between Europe and Asia. To the south of the Russian Plain is another mountain range, the **Caucasus Mountains**, which bridges the gap between the **Caspian** and **Black Seas.** East of the Urals are the **West Siberian Plain**, the **Central Siberian Plateau**, the **Yakutsk Basin**, the **Eastern Highlands**, and the **Central Asian Ranges**. Russia has rich natural resources, such as petroleum, natural gas, and forest products.

A. Expansion of the Empire

The territory that makes up the **Russian Federation** was gradually conquered by the **Russian Empire** as the country expanded from its political core around Moscow/St. Petersburg during the sixteenth through the nineteenth centuries. By the end of the eighteenth century, Czarina Catherine the Great had expanded Russia to include the area that is now Ukraine (the north side of the Black Sea), the northern Caucasus Mountains, and Alaska (which Russia later sold to the United States). During the next century, the Russian Empire expanded eastward into Central Asia (what is now Kazakhstan, Uzbekistan, and the other Central Asian republics), southward into the rest of the Caucasus region, and westward into Poland and Finland. In the twentieth century, when the Russian Empire disintegrated and was replaced by the **Soviet Union** (the Union of Soviet Socialist Republics or **USSR**), the central government continued to expand and strengthen its control of the vast area from Eastern Europe to the Pacific Ocean.

Both the Russian Empire and the Soviet Union were **imperial powers**. In other words, these governments ruled a large variety of ethnic groups in distant places: people who spoke many languages, people who worshiped different gods in different ways, people who had various skin and hair colors, and people who did not consider themselves to be Russian. Although the British and some other European powers had an arguably more difficult task of ruling empires that were widely scattered around the world, Russia had the largest empire in terms of territory. Ruling this diverse, immense empire was an incredible challenge.

The czars ruled this empire with **Russification** and the sword. Russification refers to the attempt to minimize cultural differences and turn all Russian subjects into Russians, or at least to make them as Russian as possible. People were taught the Russian language and were encouraged to convert to Russian Orthodoxy. Russification was not very successful, and the farther people were from Moscow the less likely they were to be Russified. When the Soviets took over the Russian Empire, millions of Muslims still lived in Central Asia, on the Crimean peninsula of southern Ukraine, in the Caucasus Mountains, and elsewhere.

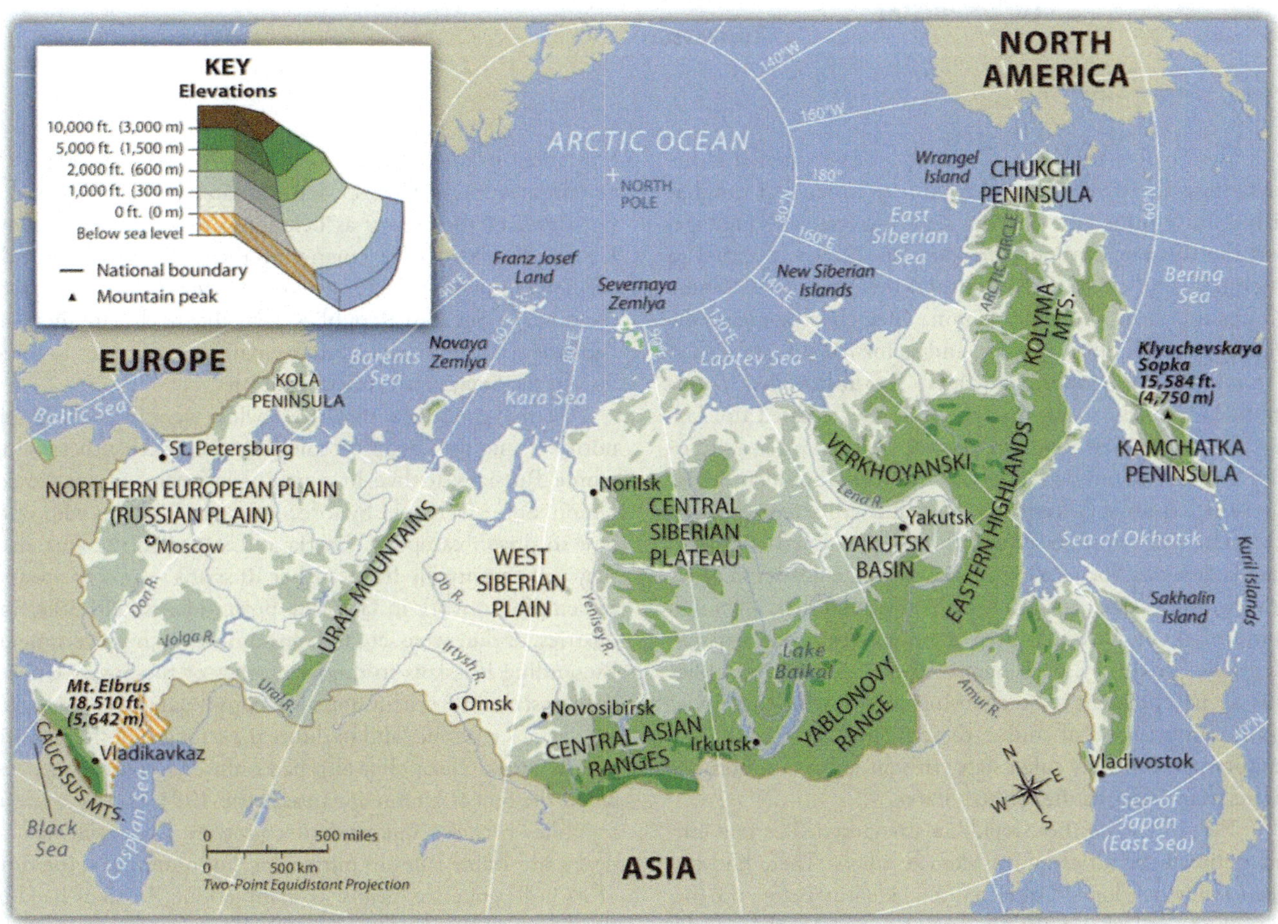

Figure 2. Physical regions of Russia.

Figure 3. Former Soviet Union.

The Soviets took a different approach when it came to taming the diversity of the empire. Instead of emphasizing unity under the Russian czar, the Russian language, and the Russian Orthodox religion, the Soviets decided to organize—and thus try to control—the diversity of ethnic groups found in the Soviet Union. They chose some of the major groups (Uzbek, Kazakh, and so forth) and established a total of fifteen **Soviet Socialist Republics** that corresponded to these major groups. Thus they created the Uzbek Soviet Socialist Republic (Uzbek SSR), the Kazakh Soviet Socialist Republic (Kazakh SSR), the Ukrainian Soviet Socialist Republic (Ukrainian SSR), and a different republic for each of eleven other ethnic groups, plus the Russian Soviet Federative Socialist Republic. About eighty-five other ethnic groups were not given their own republics, although some of them were allocated regions within the Russian Republic. In theory, each of the Soviet Socialist Republics was an independent state choosing to ally with the Soviet Union. In practice, of course, these republics were part of a totalitarian, centrally ruled state with far fewer autonomous rights than states in the United States.

The creation of these republics strengthened certain ethnic/national identities and weakened others. There had not been a fully developed Uzbek national identity before the formation of the Uzbek SSR. The same was true for the Kazakh SSR, the Turkmen SSR, and others. Although people in a certain area might have spoken the same language, they did not think of themselves as belonging to a nation of fellow Uzbeks, Kazakhs, or Turkmen until they were put into one by the Soviet rulers. In 1991, when the Soviet Union collapsed, these Soviet Socialist Republics were able to declare their independence from Russia, and the national identities fostered during the Soviet era came to fruition.

At the same time that the Soviets were organizing minority ethnic groups into republics, they were also sending ethnic Russians to live in non-Russian parts of the Soviet Union. Some were sent by force—such as Russians who were sent to prison camps in Siberia and stayed in the area after they were eventually freed. Other Russians were sent around the empire to work in factories, power plants, and other industries, or they were sent to help administer the government. By sending Russians to the far reaches of the Soviet Union, the Russian government hoped to consolidate its control over the various republics and to dilute the strength of the minority ethnicities. This policy also had unintended consequences: when the Soviet Union collapsed after 1991 and the various republics became independent countries, they each had to deal with sizable Russian minorities. For example, at the time of its independence, nearly as many ethnic Russians lived in Kazakhstan (38 percent) as ethnic Kazakhs (40 percent). In the years since then, many Russians moved to Russia from the former Soviet republics. In 2010, Kazakhstan's population was only about 24 percent ethnic Russian.

The Russian federation was created with eighty-three federal subjects: two autonomous federal cities; forty-six provinces (oblasts) and nine territories (krais) that function in the same way and are the most common type of federal unit; twenty-one republics; four autonomous districts (okrugs); and one autonomous oblast. Moscow and St. Petersburg are the two federal cities that function as their own units. The oblasts and krais each have a governor appointed by the central government and a locally elected legislature. The governorship was an elected position in the 1990s, but President Vladimir Putin changed the structure to strengthen the power of the central state. The republics, designed to be home to certain ethnic minorities, are allowed to have their own constitutions and governments and to select an official language that will be used besides Russian, but they are not considered independent countries with the right to secession. The autonomous districts were also formed for ethnic minorities and are administered either by the central state or by the province or territory in which they are located. The only autonomous oblast was created in the 1930s to be a home for Jews in the Russian Far East, but only about 1 percent of the population remains Jewish today.

Figure 4. New Russian federal districts.

B. Regional Environmental Problems

Each region of the Russian republic has its own environmental issues. The core region surrounding Moscow, with all its industrial activity and large urban expanses, introduces sewage and chemicals to the country's waterways, contributing to serious water pollution. The same water pollution is found east of the Ural Mountains—and in the waterways in that region—because of the industrial cities found there. Moscow and the ring of industrial cities surrounding it have seen a dramatic increase in automobile use since 1991, contributing to air pollution. Russia is blessed with abundant natural resources, but significant environmental damage has been the price of exploiting and extracting those resources. Massive oil spills have occurred in the **taiga** and tundra areas, where the lack of safety management has increased environmental damage during oil exploration and development. The taiga is the large expanse of evergreen or boreal forests in the north just south of the tundra in North America, Europe, and Asia. The taiga is most common in type D climates and is one of the largest biomes on earth. The taiga is the largest biome in Russia. Mining and smelting processes in Siberian cities have added to the region's air and water pollution. These ecosystems are rather fragile and will take years to recover from such damage.

Water pollution from the rivers extends into the Black Sea, the Caspian Sea, and other bodies of water. **Lake Baikal**, the largest and deepest freshwater lake in the world, was at one time pristine, but pollutants have entered its waters from nearby industrial activity. Increased pollution in the Black and Caspian Seas, as well as overfishing, the lack of fishing regulations, and the lack of law enforcement, has resulted in the devastation of fish populations such as the caviar-producing sturgeon. The Arctic waters of the **Barents Sea** off the northern coast of Russia have been a dumping ground for nuclear waste products and expended nuclear reactors from naval vessels. The consequences of this nuclear pollution are not widely known or studied. Many additional aging nuclear reactors from the Soviet era dot the landscape, and they will need to be decommissioned at some point, adding to the nuclear waste issues. Various regions in Siberia were used for nuclear testing and are also contaminated with nuclear radiation.

Key Takeaways

1. Russia's climate is characterized by long, cold winters and short summers, and because of a short growing season, the country has such poor agriculture that usually it needs to import grain.

2. The Russian Empire gradually expanded its territory to the east, west, and south of Moscow and by the end of the nineteenth century had accumulated a vast area of land and a great variety of people of many ethnicities, languages, and religions.

3. The Soviets sent ethnic Russians across the Union of Soviet Socialist Republics (USSR) to better control and govern the territory and its people.

4. When the Soviet Union collapsed, its internal nationality-oriented republics declared independence, and the Russians who had been sent to live around the USSR were now living in non-Russian countries. While most of the ethnic Russians stayed in their current countries, many others returned to Russia after 1991.

3.2 The USSR and the Russian Federation

Learning Objectives

1. Define the main tenets of a socialist economy.
2. Describe some of the conditions of life in the Soviet Union.
3. Briefly explain why the Union of Soviet Socialist Republics (USSR) collapsed.
4. Describe the post-Soviet economic and political situation.

The Russian Empire was built by the czars over the course of a few hundred years. However, the economic and political systems of the Russian Empire were not sustainable in the modern era. The vast majority of the population was poor, and most were landless peasant farmers—and in a place with short growing seasons, farming was not an easy path to riches. Political decisions were made by a very small elite group. At the dawn of the twentieth century, one hundred years after the Industrial Revolution swept through Great Britain and Western Europe, Russia remained an agricultural country and had not yet begun large-scale industrialization. Outside of the aristocracy, few supported the status quo in Russia, and there was widespread desire for a new political system and government.

However, no one could agree on what a new government would look like. In the aftermath of the First World War, a civil war erupted in Russia. During these chaotic times, the last czar, Nicholas II, was forced from office, and he and his family were executed. The most powerful group battling for control of Russia was a Communist group called the **Bolsheviks**, which literally meant the "larger group." Other groups, including the "smaller group," the **Mensheviks**, lost the civil war. The Bolshevik leader was **Vladimir Lenin**, and in 1917 he and his supporters embarked on a quest to turn Russia into a Communist state.

The capital city was moved back to **Moscow** from **St. Petersburg**, where it had been since the time of Czar Peter the Great in the eighteenth century. St. Petersburg's name was changed to Petrograd and then Leningrad in honor of Vladimir Lenin, as the atheist Soviets did not want any references to Christian saints. The entire territory of the Russian Empire was turned into the Union of Soviet Socialist Republics (USSR). The Russian people traded a monarchy for a Marxist totalitarian state.

The Soviet Union lasted from 1922 to 1991. **Josef Stalin**, the Soviet dictator who took over after Lenin, was incapacitated in 1922 (and died in 1924). He was a ruthless leader who murdered his way to power and killed or exiled anyone who got in his way. Stalin is famous for initiating economic plans that helped move Russia from a poor, agrarian state to a large, industrial superpower. He pushed for rapid industrialization, the eradication of family farms in lieu of large communal farms, the end of personal ownership of land or businesses, and the dramatic weakening of organized religion. All these changes came at a great price. During his reign of terror, an estimated thirty million people lost their lives. The forced collectivization of agriculture brought about a devastating famine in 1932–33, in which between six and eight million people starved to death or were killed outright, many of them in Ukraine. Stalin led periodic purges of his perceived political enemies. The largest of these is known as the Great Purge. At that time (1936–38), about one million so-called enemies of the state were executed. More people lost their lives under Stalin than in all the concentration camps of Adolf Hitler's Nazi regime. The full extent of Stalin's purges of his people may never be fully known. Stalin's rule ended in 1953, when he reportedly died of natural causes. However, some historians believe he was poisoned by his close associates.

Figure 5. Vladimir Lenin and Josef Stalin, 1922.

Soviet Union Leaders	Time as Leader	Life Span
Vladimir Ilich Ulyanov (Lenin)*	1917–1924	1870–1924
Josef Vissarionovich Djugashvili (Stalin)*	1924–1953	1878–1953
Nikita Sergeyevich Khrushchev	1953–1964	1894–1971
Leonid Ilyich Brezhnev*	1964–1982	1906–1982
Yuri Vladimirovich Andropov*	1982–1984	1914–1984
Konstantin Ustinovich Chernenko*	1984–1985	1911–1985
Mikhail Sergeyevich Gorbachev	1985–1991	1931–
Russian Federation presidents (1991+)		
Boris Nikolayevich Yeltsin	1991–1999	1931–2007
Vladimir Vladimirovich Putin	2000–2008	1952–
Dmitry Anatolyevich Medvedev	2008–2012	1965–
Vladimir Vladimirovich Putin	2012-	1952–
*died while in office		

Table 1. Soviet leaders and Russian presidents.

A. Marxist-Leninist Central Planning

The Soviet Union espoused the philosophies of **Karl Marx,** a nineteenth-century German theorist. Marx wrote that all political and economic life can be understood as a struggle between the various classes in society. People who adhere to Marx's philosophy are called Marxists, and the Soviet version of **Marxism** is called Marxism-Leninism. In Marxist thought, capitalism is an oppressive economic system in which the working class (the **proletariat**) is oppressed by the **bourgeoisie** (the wealthy middle class). Marxists believe that the proletariat should revolt, rise up against the bourgeoisie, take the property away from the rich, and give it to the government to control it for the benefit of the common people. Ultimately, a pure Communist system would result, with no social or economic classes, no private property, no rich people, and no poor people. In real life, governments that adopt these ideas practice socialism and are said to be socialist.

As a socialist state, the Soviet Union did not include open markets. The Soviet Union was a command economy, in which economic decisions were made by the state and not left to the market to decide. During the Soviet era, for example, industrial production was planned by the central government. The government would decide what would be produced, where it would be produced, the quantity produced, the number of workers who would produce it, where the raw materials would come from, and how the final product would be distributed. By mobilizing the entire country to work toward common goals, the USSR was able to achieve the rapid industrialization that it so desired. However, the Soviets underestimated the power and efficiency of free-enterprise capitalism, and their socialist system was undermined by waste, fraud, and corruption.

Another main economic feature of the Soviet Union was collectivized agriculture. The Soviet leaders did not want individual, capitalist farmers to become rich and threaten their economic system. Nor did they want thousands of small, inefficient farms when the country was perpetually unable to feed itself. Instead, they decided to streamline agricultural production into large farm factories. All the farmland in each area was consolidated into a government-owned collective operation. Some collective farms were run by the state, while others were run by private cooperatives. During the transition period to collective farming, individual farmers were forced to give up their land, animals, farm equipment, and farm buildings and donate them to the collective farm in their area. The state also demanded a high percentage of the crops produced. At times, the government collected the entire harvest, not even allowing seed crops to be held for the following season. This brought about widespread famine in 1932–33. Collectivized agriculture remained the norm in the Soviet Union until the country's dissolution in 1991 and even afterward in some areas.

B. The Cold War

From the end of World War II in 1945 until the collapse of the USSR in 1991, the Soviet Union and the United States competed in the global community for the control of labor, resources, and world power. Each side attracted allies, and most countries were on the side of either the United States or the Soviet Union; very few remained neutral. This era, known as the **Cold War**, did not involve direct military armed conflict between the United States and the Soviet Union, but it transformed the world into a political chessboard, with each side wanting to block the other side from gaining ground. Whenever the Soviets would enter into an alliance with a certain country, the United States was right there to try to counter the move. Wars, armed conflicts, sabotage, spying, and covert activities were the methods of the Cold War. Both sides stockpiled as much deadly weaponry as possible, including nuclear warheads and missiles. They also competed in the race to put people in outer space.

The Cold War led to wars fought in Vietnam, Korea, Grenada, Afghanistan, Angola, and the Middle East, with the Soviet Union funding one side and the United States supplying the other. Covert wars or guerilla wars with secret agents and political assassinations were fought in Cuba, Nicaragua, Chile, Guatemala, Mozambique, Laos, Cambodia, and a host of other third-world countries. The Cold War divided the world into two main camps, each with a high number of nuclear weapons. Eastern Europe was sectioned off by the Iron Curtain, and the Berlin Wall divided the city of Berlin, Germany. These physical barriers divided the communist countries of Eastern Europe with the capitalist democracies of Western Europe. Germany itself was divided into two separate countries, as explained in Chapter 2 on Europe.

Various Soviet dictators came to power and died in office before the end of the Cold War. The last Soviet leader was **Mikhail Gorbachev**, who assumed power in 1985. The US president at the time was Ronald Reagan. During the 1980s the United States was outspending the Soviets mil-

itarily, and its economy was growing at a much faster rate than that of the USSR. At the same time, the Soviets were engaged in a costly war in Afghanistan, and their economy was faltering and in danger of collapse. Gorbachev realized that reforms had to be implemented to modernize the Soviet system: political life needed to be more open so that people would feel ownership of the country, and the economy needed to be restructured. Gorbachev implemented perestroika (restructuring of the economy with market-like reforms) and glasnost (openness and transparency of all government activities). The restructuring exposed fundamental problems in the economy, and by 1990 the Soviet economy was in worse shape than ever before.

The end came in 1991: the Soviet Union collapsed when fourteen of the Soviet republics broke away and declared their independence. At this point, the Soviet state was too weak to prevent it. All the republics, including Russia itself (now called the **Russian Federation**), became independent countries. The only territories that did not achieve independence were the smaller republics and autonomous regions that existed within the Russian Federation's boundaries. The **Iron Curtain** melted away seemingly overnight, and people were free to travel to and from the former Communist countries. The old Russian flag flew over the Kremlin—the seat of the Russian government—for the first time since Czar Nicholas II had been in power seventy-three years earlier. The Communist era of the Soviet Union and the Cold War were over.

The post-Soviet transition was filled with political, economic, and social turmoil. Boris Yeltsin, the first president of the new Russian Federation, ushered in a series of economic reforms that privatized state-owned enterprises. Russian leaders tried to reverse socialism rapidly through what they called "shock therapy," which they knew would be painful but hoped would be brief. These reforms created a new class of capitalist entrepreneurs.

Figure 6. October revolution celebration 1983, Moscow, during the Cold War.

Figure 7. The first McDonald's restaurant in the former Soviet Union.

Wealth, once controlled by the political elite, was now being shifted to the business elite, a pattern found in most capitalist countries. Western goods and consumer products became more widely available than they were during the Soviet era. Bread lines and empty store shelves became distant memories. Many ordinary workers faced unemployment for the first time, however, as the new owners of companies trimmed unnecessary staff. Private ownership forced housing costs to skyrocket, and purchasing houses or condominiums became out of reach for many people. The value of the ruble, Russia's currency, declined rapidly, and older people watched as their life savings evaporated overnight. Yeltsin became increasingly less popular, as citizens became dissatisfied with corruption and the high social costs of the post-Soviet transition. He resigned in 1999 and was replaced by **Vladimir Putin**.

Figure 8. Red Square in Moscow. Kremlin is a Russian word for a walled city. Inside Moscow's Kremlin is the famous Red Square.

Outline of Russia's Historical Geography

1. **The Region's Early Heritage**
   - Vikings created fortified trading towns called gorods
   - Genghis Khan's Mongol Empire invaded (1240 CE)
   - Feudal states arose around dominant trading centers

2. **Czarist Russia, 1547–1917**
   - Czars unified empire by internal colonialism
   - Forward capital of St. Petersburg created
   - Pioneers pushed eastward to Siberia and North America

3. **Bolshevik Revolution, 1917–22**
   - Czar Nicholas II and his family executed
   - Russian Civil War fought
   - Vladimir Lenin created Communist imperial state
   - Capital moved to Moscow; republics broke away

4. **The Soviet Union (USSR), 1922–91**
   - Czarist Empire became the Soviet Union
   - Central planning, collectivization, and the Cold War began
   - Republics kept together by military force
   - External interaction of glasnost initiated (1980s)
   - Economic restructuring and reforms of perestroika introduced (1980s)

5. **The Russian Federation, 1991–Present**
   - Independent republics lost with internal unrest
   - Economics privatized
   - Democracy introduced (1990s)
   - Central state strengthened (twenty-first century)

Figure 9. Moscow and the Moskva river. Behind the Borodinsky Bridge on the right are the government buildings of the Russian Federation.

C. Twenty-First-Century Russia

A Russian style of capitalism replaced the social, political, and economic system of the Communist era with a growing market economy. The export of Russia's vast quantities of natural resources, such as oil, natural gas, and timber, to Europe and the rest of the world helped the country rebound from the economic collapse of the 1990s. Russia has benefited from the recent increase in energy prices, and oil, natural gas, metals, and timber account for more than 80 percent of exports and 30 percent of government revenues. However, Russia still needs to modernize its dilapidated manufacturing base if it is to economically compete against the European Union, North America, or eastern Asia. During Vladimir Putin's presidency (2000–2008), Russia witnessed substantial economic growth that inspired foreign investors to pump money into the Russian economy and catapulted Moscow into an investment haven and one of the richest cities in the world. In 2008, Moscow claimed to have more billionaires than any other city in the world. Russia has reestablished itself as a major player in the global economy, although much of its population still suffers from poverty and social problems.

Russia's population grew steadily during the Soviet era, except during periods of famine or warfare, and the country underwent a rural-to-urban shift as farm workers moved to cities to labor in factories. However, when the USSR collapsed, Russia's population began a steep decline, falling from a peak of 149 million in 1991 to about 143 million in 2005. The trend of low birth rates actually began during the Soviet period. The population decline occurred because birth rates always decline during periods of economic and social crisis as people delay or decide against having children, and the country experienced particularly high death rates because of alcoholism, heart disease, and the collapse of the social safety net. Any time death rates surpass birth rates, a country's population will decline unless the difference is offset by immigration. Russia's birth rate was also impacted by very high rates of abortion: in 1992, for example, there were 221 abortions in Russia for every 100 live births. Although Russia's population has seemingly bottomed out, the only reason it is not continuing to decline is immigration from the former Soviet republics. Russia still has a negative rate of natural increase.

To put Russia's demographic profile in context, its fertility rate was only about 1.5 in 2010, meaning that the average woman would have 1.5 children in her lifetime. This is below the 2.1 children each woman would need to have for the population to remain stable. The fertility rate in Russia is similar to that of countries in Europe (1.5) but lower than that of the United States (2.1). Life expectancy for Russian men is variously reported as sixty years up to sixty-three years, while

women can expect to live seventy-three to seventy-five years. In Western Europe, life expectancies are about eighty years.

About 80 percent of Russia's population is ethnically Russian. The next largest group is Tatar (3.8 percent), a group that traditionally has spoken the Tatar language and practiced Islam. More than 150 ethnic groups are represented in Russia, including indigenous people of the Arctic who herd reindeer for a living. Each of the nationalities of the former Soviet republics has a presence in Russia, and because of its relatively strong economy, Russia (especially Moscow) is an immigration magnet for residents of those countries. Most of these groups have their own language and cultural traditions.

The Russian Orthodox Church is the dominant religious denomination in Russia. For generations, it was the country's official religion, and Russian people were automatically considered to be Orthodox, no matter what their personal beliefs. During the Soviet era, the government did much to weaken the church, including killing tens of thousands of priests, monks, and nuns and closing most churches. The much smaller church that survived was largely controlled by the state. Because of state-sanctioned atheism during the Soviet era, only 15 to 20 percent of Russia's population today actively practices Orthodoxy, although a much greater number claim to be Russian Orthodox Christians. Another 15 percent of the country's population practices Islam, especially in places such as the Caucasus region and the southern Ural Mountains, and about 2 percent practice other forms of Christianity, such as Catholicism and Protestantism. Even though Orthodoxy is practiced by a minority of Russia's population, the church has increased its influence since 1991 and often acts as an official church.

Figure 10. The Cathedral of Intercession of the Virgin on the Moat, also known as the Cathedral of St. Basil the Blessed, on the Red Square, Moscow.

Key Takeaways

1. The Communist state was created in 1917 as the result of a civil war, which evolved into the creation of the Soviet Union in 1922. During this era, the Russian people traded a monarchy for a Marxist totalitarian state.

2. One of the primary goals of the new USSR was rapid industrialization, and this goal was achieved through central planning and the collectivization of agriculture. Inefficiencies in the system persisted throughout the Soviet era.

3. The Soviet Union's Communist system was maintained at great cost: millions were killed by purges and in government-produced famines.

4. A weakened USSR tried to reform in the 1980s through Mikhail Gorbachev's policies of perestroika and glasnost but instead collapsed in 1991.

5. Early post-Soviet years were ones of democratization, rapid privatization, and the unraveling of the social safety net. The shift to capitalism restructured the Russian economy.

6. The Russian economy strengthened after 2000, and Russian president Vladimir Putin strengthened the power of the central state.

7. Russia has a low fertility rate and a negative rate of natural increase. High rates of abortions and alcoholism have been contributing factors.

8. The Russian Orthodox Church is the dominant religious denomination in Russia.

3.3 Regions of Russia

Learning Objectives

1. Name the major cities, rivers, and economic base of Russia's core region.

2. Identify the economic base of cities in the Eastern Frontier and the identity of the world's most voluminous freshwater lake.

3. Describe the physical attributes of Siberia and the Far East.

4. Explain why Chechnya has been at war with Russia twice since 1994.

5. Learn why Russia invaded Georgia in 2008 and discover the role of fossil fuels in the economies of Armenia and Azerbaijan.

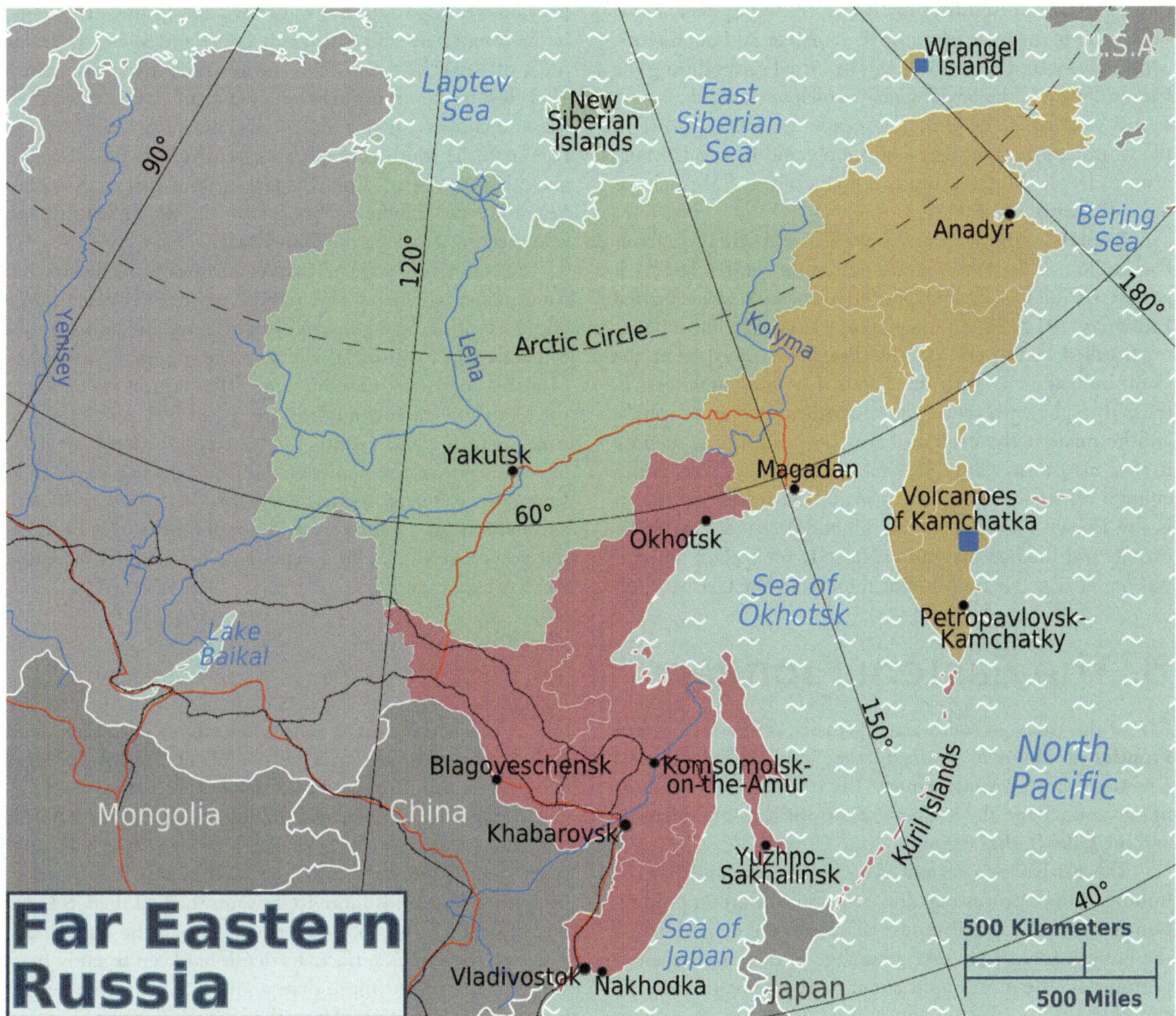

Figure 11. Russia's Eastern frontier, the far east, and Siberia.

A. The Core Region

Most of Russia's population and its major industries are located west of the Ural Mountains on the Russian Plain. Known as Russia's geographic core, this includes the Moscow region, the Volga region, and the Ural Mountain region. Moscow, Russia's capital city, anchors a central industrial area that is home to more than fifty million people. Moscow alone has more than ten million residents, with about thirteen million in its metropolitan area, making it slightly smaller than the Los Angeles, California, metro area. A ring of industrial cities surrounding Moscow contains vital production centers of Russian manufacturing. During the Communist era, Moscow expanded from its nineteenth-century core (although the city dates from at least the twelfth century) and became an industrial city with planned neighborhoods. This world-class city has an extensive subway and freeway system that is expanding to meet current growth demands. Although rents, commodities, and domestic goods had fixed prices during the Communist era, the Soviet Union's collapse changed all that. Today Moscow is one of the most expensive places to live in the world, with prices based on supply and demand. Many want to live in Moscow, but it is financially out of reach for many Russians.

Russia's second-largest city, with a population of about five million, is St. Petersburg. Located on the Baltic Sea, it is western Russia's leading port city. The city was renamed Petrograd (1914–24) and Leningrad (1924–91) but today is often called Petersburg, or just "Peter" for short. Peter the Great built the city with the help of European architects in the early eighteenth century to rival other European capitals, and he made it the capital of the Russian Empire, naming it after his patron saint, St. Peter in the Bible. Today it is a cultural center for Russia and a major tourist destination. It is also known for shipbuilding, oil and gas trade, manufacturing, and finance. Its greatest tragedy took place when it was under siege for twenty-nine months by the German military during World War II. About one million civilians died of starvation or during the bombardment, and hundreds of thousands fled the city, leaving the city nearly empty by the end of the siege.

To the far north of St. Petersburg on the Barents Sea are the cities of **Murmansk** and **Arkhangelsk**. Murmansk is a major military port for Russia's navy and nuclear submarine fleet. Relatively warm water from the North Atlantic drift circles around Norway to keep this northern port city fairly free of ice. Arkhangelsk (which literally means "archangel"), used as a port for lumber exports, has a much shorter ice-free season than Murmansk. Both of these cities are in Russia's far north, with long winters and exceedingly brief summers.

The **Volga River** flows through the core region of Russia, providing transportation, fresh water, and fishing. The Volga is the longest river in Europe at 2,293 miles, and it drains most of Russia's western core region. This river has been a vital link in the transportation system of Russia for centuries and connects major industrial centers from the Moscow region to the south through an extensive network of canals and other waterways. The Volga River flows into the **Caspian Sea**, and a canal links the Volga with the **Black Sea** through a connecting canal via the **Don River**.

At the eastern edge of Russia's European core lie the Ural Mountains, which act as a natural divide between Europe and Asia. These low-lying mountains have an abundance of minerals and fossil fuels, which make the Ural Mountains ideal for industrial development. The natural resources of the Urals and the surrounding area provide raw materials for manufacturing and export. The eastern location kept these resources out of the hands of the Nazis during World War II, and the resources themselves helped in the war effort. Oil and natural gas exploration and development have been extensive across Russia's core region and have greatly increased Russia's export profits.

B. The Eastern Frontier

East of the Urals, in south-central Russia, is Russia's **Eastern Frontier**, a region of planned cities, industrial plants, and raw-material processing centers. The population is centered in two zones here: the Kuznetsk Basin (or Kuzbas, for short) and the **Lake Baikal** region.

The Kuzbas is a region of coal, iron ore, and bauxite mining; timber processing; and steel and aluminum production industries. Central industrial cities were created across the Eastern Frontier to take advantage of these resource opportunities. The most important of these is **Novosibirsk**, the third-largest city in Russia after Moscow and St. Petersburg and home to about 1.4 million people. The city is not only noted for its industries but it is also the region's center for the arts, music, and theater. It is host to a music conservatory and a philharmonic orchestra, a division of the Russian Academy of Sciences, and three major universities.

Agriculture, timber, and mining are the main economic activities in the eastern Lake Baikal region, which is more sparsely settled than the Kuzbas. Lake Baikal (400 miles long, 50 miles wide) holds more fresh water than all the US Great Lakes together and about 20 percent of all the liquid fresh water on the earth's surface. Its depth has been recently measured at 5,370 feet (more than a mile). Some of the longest river systems in the world flow through the **Eastern Frontier**.

The Irtysh, Ob, Yenisey, and Lena are the main rivers that flow north through the region into Siberia and on to the

Arctic Ocean. To the east, the Amur River creates the border between Russia and China until it flows north into the Sea of Okhotsk. In addition to waterways, the Trans-Siberian Railway is the major transportation link through the Eastern Frontier, connecting Moscow with the port city of **Vladivostok** in the Far East.

C. Siberia

Siberia, as a place name, actually refers to all of Asian Russia east of the Ural Mountains, including the Eastern Frontier and the Russian Far East. However, in this and some other geography textbooks, the term *Siberia* more specifically describes only the region north of the Eastern Frontier that extends to the **Kamchatka Peninsula**. The word *Siberia* conjures up visions of a cold and isolated place, which is true. Stretching from the northern Ural Mountains to the Bering Strait, Siberia is larger than the entire United States but is home to only about fifteen million people. Its cities are located on strategic rivers with few overland highways connecting them.

Type D (continental) climates dominate the southern portion of this region, and the territory consists mainly of coniferous forests in a biome called the taiga. This is one of the world's largest taiga regions. Type E (polar) climates can be found north of the taiga along the coast of the Arctic Sea, where the tundra is the main physical landscape. No trees grow in the tundra because of the semifrozen ground. **Permafrost** may thaw near the surface during the short summer season but is permanently frozen beneath the surface. On the eastern edge of the continent, the mountainous Kamchatka Peninsula has twenty active volcanoes and more than one hundred inactive volcanoes. It is one of the most active geological regions on the Pacific Rim.

The vast northern region of Russia is sparsely inhabited but holds enormous quantities of natural resources such as oil, timber, diamonds, natural gas, gold, and silver. There are vast resources in Siberia waiting to be extracted, and this treasure trove will play an important role in Russia's economic future.

Figure 12. Mount Koryasky, an active volcano, and Petropavlovsk-Kamchatsky, a town on Russia's Kamchatka Peninsula.

D. The Far East

Across the strait from Japan is Russia's **Far East** region, with the port of **Vladivostok** (population about 578,000) as its primary city. Bordering North Korea and China, this Far East region is linked to Moscow by the Trans-Siberian Railway. Before 1991, Vladivostok was closed to outsiders and was an important army and naval base for the Russian military. Goods and raw materials from Siberia and nearby **Sakhalin Island** were processed here and shipped west by train. Sakhalin Island and its coastal waters have oil and mineral resources. Industrial and business enterprises declined with the collapse of the Soviet Union in 1991. Today, the Far East is finding itself on the periphery of Russia's hierarchy of productivity. However, it has the potential to emerge again as an important link to the Pacific Rim markets.

Figure 13. Vladivostok tram opposite the Admiral's Club on its way to the railway station, Vladivostok.

E. Southern Russia

In the southern portion of the Russian core lies a land bridge between Europe and Southwest Asia: a region dominated by the **Caucasus Mountains**. To the west is the Black Sea, and to the east is the landlocked Caspian Sea. The Caucasus Mountains, higher than the European Alps, were formed by the Arabian tectonic plate moving northward into the Eurasian plate. The highest peak is **Mt. Elbrus** at 18,510 feet. Located on the border between Georgia and Russia, Mt. Elbrus is the highest peak on the European continent as well as the highest peak in Russia.

Most of this region was conquered by the Russian Empire during the nineteenth century and held as part of the Soviet Union in the twentieth. However, only a minority of its population is ethnic Russian, and its people consist of a constellation of at least fifty ethnic groups speaking a variety of languages.

Since the collapse of the Soviet Union, the Caucasus region has been the main location of unrest within Russia. Wars between Russia and groups in the Caucasus have claimed thousands of lives. Some of the non-Russian territories of the

Figure 14. Republics of Southern Russia, including Chechnya.

Caucasus would like to become independent, but Russia fears an unraveling of its country if their secession is allowed to proceed. To understand why the Russians have fought the independence of places such as Chechnya but did not fight against the independence of other former Soviet states in the Caucasus such as Armenia, it is necessary to study the administrative structure of Russia itself.

Of the twenty-one republics, eight are located in southern Russia in the Caucasus region. One of these, the **Chechen Republic** (or Chechnya), has never signed the Federation Treaty to join the Russian Federation; in fact, Chechnya proposed independence after the breakup of the Union of Soviet Socialist Republics (USSR). Although other territories to the south of Chechnya, such as Georgia, Armenia, and Azerbaijan, also declared their independence from Russia after 1991, they were never administratively part of Russia. During the Soviet era, those countries were classified as Soviet Socialist Republics, so it was easy for them to become independent countries when all the other republics (e.g., Ukraine, Belarus, and Kazakhstan) did so after 1991. However, Chechnya was administratively part of the USSR with no right to secession. After 1991, Russia decided that it would not allow territories that had been administratively governed by Russia to secede and has fought wars to prevent that from happening. It feared the consequences if all twenty-one republics within the Russian Federation were declared independent countries.

Chechnya has fought against Russia for independence twice since the USSR's collapse. The First Chechen War (1994–96) ended in a stalemate, and Russia allowed the Chechens to have de facto independence for several years. But in 1999, Russia resumed military action, and by 2009 the war was essentially over and Chechnya was once more under Russia's control. Between twenty-five thousand and fifty thousand Chechens were killed in the war, and between five thousand and eleven thousand Russian soldiers were also killed.

F. Transcaucasia

The independent countries of **Georgia, Armenia**, and **Azerbaijan** make up the region of **Transcaucasia**. Although they are independent countries, they are included in this chapter because they have more ties to Russia than to the region of Southwest Asia to their south. They have been inextricably connected to Russia ever since they were annexed by the Russian Empire in the late eighteenth and early nineteenth centuries, and they were all former republics within the Soviet Union. When the Soviet Union collapsed in 1991, these three small republics declared independence and separated from the rest of what became Russia.

Geographically, these three countries are located on the border between the European and Asian continents. The Caucasus Mountain range is considered the dividing line. The region known as Transcaucasia is generally designated as the southern portion of the Caucasus Mountain area.

The country of Georgia has a long history of ancient kingdoms and a golden age including invasions by the Mongols, Ottomans, Persians, and Russians. For a brief three years—from 1918 to 1921—Georgia was independent. After fighting an unsuccessful war to remain free after the Russian Revolution, Georgia was absorbed into the Soviet Union. Since it declared independence in 1991, the country has struggled to gain a stable footing within the world community. Unrest in the regions of **South Ossetia**, **Abkhazia**, and **Adjara** (where the populations are generally not ethnic Georgian) has destabilized the country, making it more difficult to engage in the global economy.

A democratic-style central government has emerged in Georgia, and economic support has been provided by international aid and foreign investments. The country has made the switch from the old Soviet command economy to a free-market economy. Agricultural products and tourism have been Georgia's main economic activities.

Figure 15. Oil fields of Azerbaijan.

In 2010, Armenia, to the south of Georgia, had a population of only about three million in a physical area smaller in size than the US state of Maryland. It is a country with its own distinctive alphabet and language and was the first country in the world to adopt Christianity as a state religion, an event traditionally dated to 301 CE. The Armenian Apostolic Church remains the country's central religious institution, and the Old City of Jerusalem in Israel has an Armenian Quarter, an indication of Armenia's early connection with Christianity.

Like the other former Soviet republics, Armenia has shifted from a centrally planned economy to a market economy. Before independence in 1991, Armenia's economy had a manufacturing sector that provided other Soviet republics with industrial goods in exchange for raw materials and energy. Since then, its manufacturing sector has declined and Armenia has fallen back on agriculture and financial remittances from the approximately eight million Armenians living abroad to support its economy. These remittances, along with international aid and direct foreign investments, have helped stabilize Armenia's economic situation.

Azerbaijan is an independent country to the east of Armenia bordering the Caspian Sea. It is about the same size in area as the US state of Maine. This former Soviet republic has a population of more than eight million in which more than 90 percent follow Islam. Azerbaijan shares a border with the northern province of Iran, which is also called Azerbaijan. Part of Azerbaijan is located on the western side of Armenia and is separated from the rest of the country.

Located on the shores of the Caspian Sea, **Baku** is the capital of Azerbaijan and is the largest city in the region, with a population approaching two million. During the Cold War era, it was one of the top five largest cities in the Soviet Union. The long history of this vibrant city and the infusion of oil revenues have given rise to a metropolitan center of activity that has attracted global business interests. Wealth has not been evenly distributed in the country, and at least one-fourth of the population still lives below the poverty line.

Azerbaijan is rich with oil reserves. Oil and natural gas are the country's main export products and have been a central focus of its economy. Large oil reserves are located beneath the Caspian Sea, and offshore wells with pipelines to shore have expanded throughout the Caspian Basin. As much as the export of oil and natural gas has been an economic support for the country, it has not been without costs to the environment. According to US government sources, local scientists consider parts of Azerbaijan to be some of the most devastated environmental areas in the world. Serious air, soil, and water pollution exist due to uncontrolled oil spills and the heavy use of chemicals in the agricultural sector.

Key Takeaways

1. The vast majority of Russia's population lives in the western core area of the country, the region around the capital city of Moscow.

2. Most of the cities on the Barents Sea and in the Eastern Frontier were established for manufacturing or for the exploitation of raw materials.

3. The Volga River and its tributaries have been an important trans-portation network for centuries. The Volga is the longest river in Europe.

4. Very few people live in Siberia, but the region is rich with natural resources.

5. The most contentious region in Russia is the Caucasus Mountain region, especially the area of Chechnya. The Caucasus is characterized by ethnic and religious diversity and by a desire for independence from Russia.

6. South of Russia in the Caucasus is the region of Transcaucasia. It is ethnically, religiously, and linguistically diverse. Countries there are independent of Russia, although they have a long history of being part of the Russian and Soviet Empires. Some of the countries are rich in petroleum reserves.

End-of-Chapter Summary

- Russia is a large country that crosses the boundary between Europe and Asia. It has abundant natural resources, continental and arctic climates, mountains, plains, and massive river systems.

- Russia's vast size has made it challenging to govern, both for the Russian Empire and the Soviet Union. Each government dealt with the size and cultural diversity in different ways.

- Both the Russian Empire and Soviet Union were empires—large countries in which Russian political control dominated peoples of various cultures and ethnicities within its boundaries.

- The Union of Soviet Socialist Republics (USSR) was founded through a violent rebellion and civil war. It was ruled by the Bolshevik party, a socialist group led by Vladimir Lenin. The second leader of the USSR, Josef Stalin, was renowned for the millions of people that he killed as he consolidated his power and sought economic growth for his country.

- The USSR was a command economy, in which economic decisions were made by the central state. Economic objectives of the early leaders included rapid industrialization and agricultural collectivization.

- The Soviet economy was ultimately corrupt and inefficient—two factors that, along with other problems, led to the unraveling of the Soviet Union in 1991. The reforms of the final Soviet leader, Mikhail Gorbachev, were not enough to prevent its collapse.

- The early post-Soviet years were ones of rapid privatization, immense wealth for a small few, economic hardship for most, and the disappearance of the social safety net.

- Since 1999, Russian presidents Vladimir Putin and Dmitry Medvedev have strengthened Russia's economy and consolidated the power of the central state.

- Most Russians live in the western part of the country near Moscow, and other large population centers are also located in the country's European core. There are a few industrial cities in the Eastern Frontier region, but most of Russia east of the Urals is a vast wilderness.

- Southern Russia—the Caucasus Mountain region—is the portion of the country that has caused the most unrest for Moscow. Non-Russian groups such as the Chechens would like to be independent, but Russia has engaged in warfare to prevent them from seceding.

- The countries of Georgia, Armenia, and Azerbaijan were once part of the USSR but are now independent states. They are south of the Russian border, in the southern Caucasus Mountains. These countries are not without their challenges, as they are influenced both by Russia and by the Middle East to their south. They are home to large components of Muslims and Christians and a variety of ethnic groups. While Armenia is the poorest of the three countries, Georgia and Azerbaijan have some wealth from petroleum exports.

Chapter 4

North America

Introduction

North America as a continent extends from the polar regions of the Arctic in northern Canada and Alaska all the way south through Mexico and the countries of Central America. Geographers usually study the continent by dividing it into two separate realms based on differences in physical and cultural geography: North America (the US and Canada) and Middle America (Mexico and Central America, as well as the Caribbean). Both the United States and Canada share similar physical geography characteristics as well as a common development history with either a British or French colonial legacy. Mexico and Central America are dominated by more tropical climates and were colonized mainly by the Spanish. The United States and Canada—the second- and third-largest countries in the world in physical area, respectively—make up more than 13 percent of the world's total landmass. The Atlantic Ocean borders their eastern edge, and the Pacific Ocean creates their western boundary. To the north is the Arctic Ocean. The North American region is highly urbanized—about 80 percent of the population lives in cities—but vast areas, especially in Canada, are sparsely populated. Although a small percent is native, most of North America's diverse population consists of immigrants or descendants of immigrants from other world regions. The United States is the world's largest economy, and both countries enjoy high standards of living as technologically developed countries.

Figure 1. The geographic center of North America located near Rugby, North Dakota.

4.1 Introducing the Realm

Learning Objectives

1. Define the physiographic regions of North America.

2. Explain the two dominant climate patterns in North America.

3. Find out which three European countries had the most significant early influence on North America, what parts of the region they dominated, and what their long-term impacts have been.

4. Determine the population distribution of the United States and Canada.

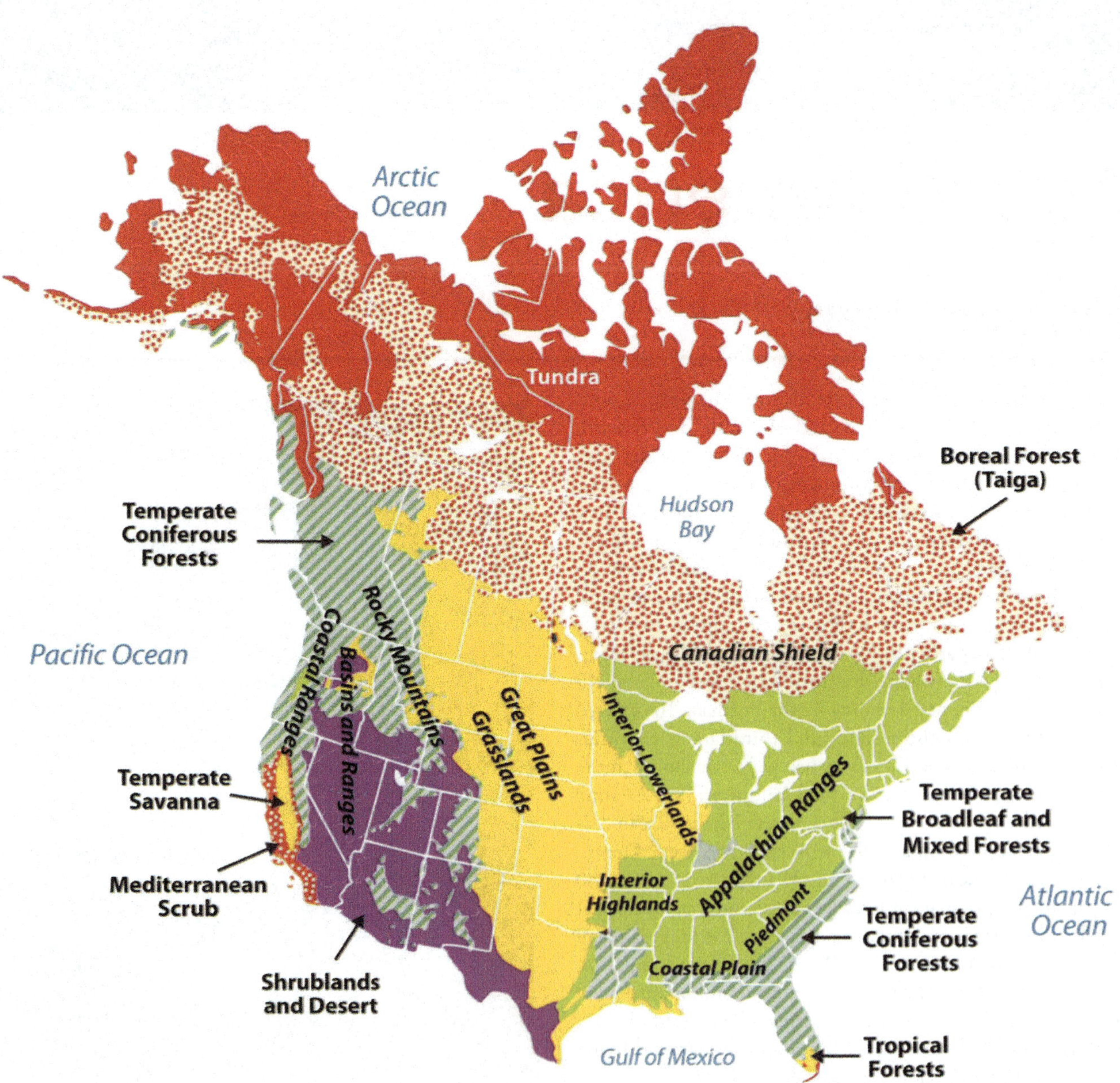

Figure 2. Physiographic regions of the US and Canada.

North America is divided into a number of physical regions with distinct landforms. The western part of the continent is marked by north-south mountain ranges in the Rocky Mountains and Pacific Mountains and Valleys physiographic provinces, with the Intermontane Basins and Plateaus in between. The eastern portion of North America is defined by the ancient Appalachian Highlands, a mountain range that is much less rugged than the Rockies but with no less influence on the history and development of the United States. The interior of the continent is characterized by plains—the Interior Lowlands and the Great Plains. To the north is the Canadian Shield, geologically the oldest part of North America, and a sparsely populated area with poor soils. At the southern and eastern edge of the continent is the Gulf-Atlantic Coastal Plain, a relatively flat zone that extends from New York to Texas.

The climates of the United States and Canada include the frigid type E climate of the tundra of northern Canada and Alaska, the tropical type A climate of southern Florida and Hawaii, the type C climates of the humid eastern United States, the seasonal type D climates of the northern United States and most of Canada, and the arid type B climates of the Southwest and Great Plains. In general, there are two different climate patterns common in North America. The first pattern is that temperatures get warmer as you travel from north to south and get closer to the equator. The second pattern is that there is a decrease in precipitation as you move from east to west across the continent until you reach the Pacific Coast, where rainfall is abundant again.

The second climate pattern is created by the **rain shadow** effect of the western mountain ranges. As wet air masses move from the Pacific Ocean over the North American continent, they run into the Cascades and the Sierra Nevada. The Cascade ranges of Washington and Oregon cut off moisture from falling on the leeward side of the mountains; thus eastern Washington and eastern Oregon are semiarid. The western United States experiences a strong rain shadow effect. As the air rises to pass the mountains, water vapor condenses and is released as rain and snow. This means that west of these mountain ranges there is much more precipitation than to their east, resulting in arid and semiarid lands. The entire Great Plains of the western United States are affected by the rain shadow effect and have a semiarid type B climate.

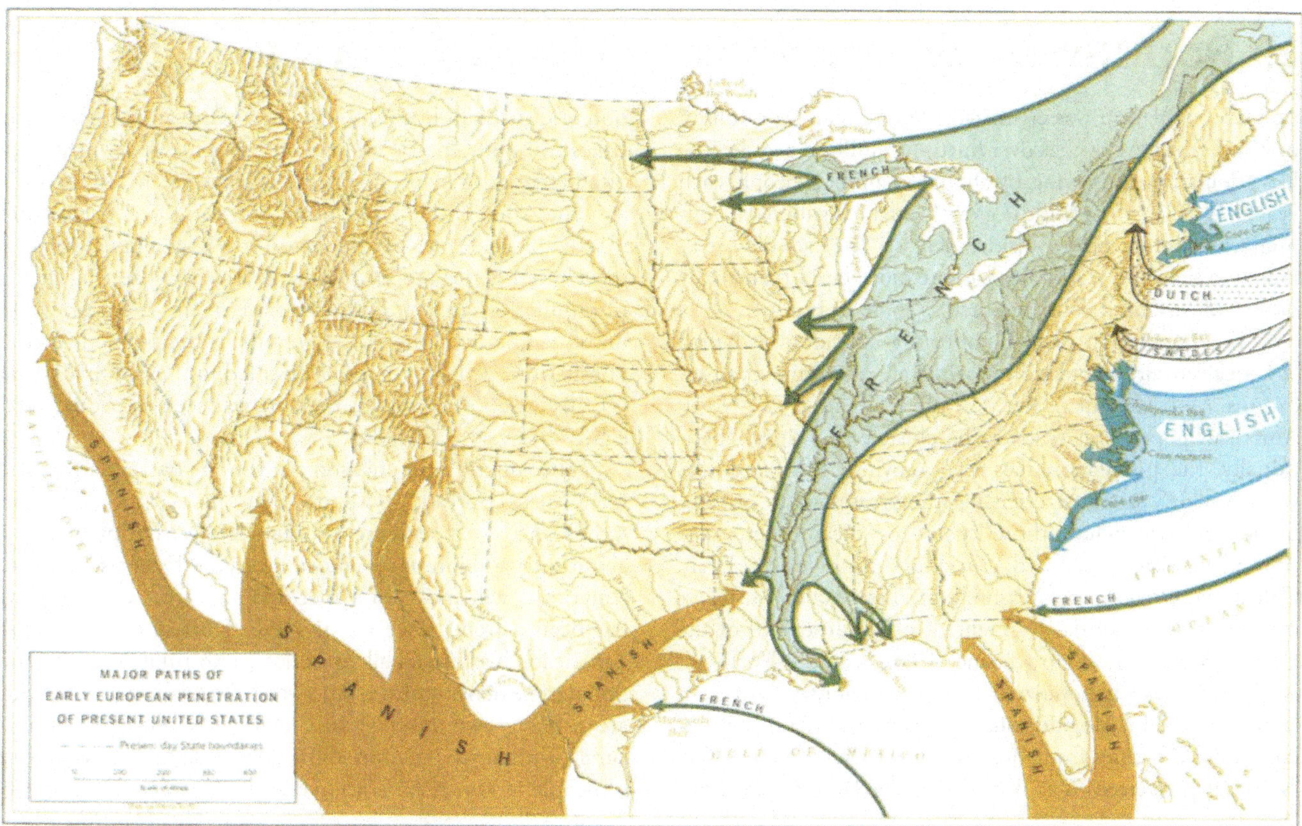

Figure 3. European influence in the Colonial United States

A. European Realms in North America

Both the United States and Canada are products of European colonialism. North America was inhabited by many Native American groups before the Europeans arrived. Complex native societies, federations, and traditional local groups faced the European invasion. While the indigenous population of North America was robust at the time of the European encounter, within a few generations, these native peoples were overwhelmed by the diseases, weapons, and sheer numbers of the European arrivals.

The Europeans—mainly the Spanish, French, and British—left a strong imprint on their North American colonies. The oldest colonial city in North America is St. Augustine, Florida (1565), founded by Spain when Florida was a remote portion of the Spanish Americas. Spain also had outposts in what are now California, Arizona, New Mexico, and Texas. The forms of settlement characteristic of those areas were similar to the Spanish colonies of Central America. While Spain governed what is now the southern United States, France ruled Canada and much of the interior of the North American continent. The French first came to Canada in the late 1500s to engage in fishing in the North Atlantic and soon expanded their reach by creating a fur trade in the area surrounding the **Great Lakes** and throughout the **Mississippi River** system.

Although there were fewer settlers from France than from other European countries—especially in what became the United States—this French era left behind place names (Baton Rouge and Detroit), patterns of land use, and a French-speaking population in Canada. Despite the early influence of Spain and France in North America, most North Americans speak English as their native language as a result of Britain's colonial dominance in the United States and Canada. The earliest permanent British colony, Jamestown, was founded in 1607 in what became Virginia. The British built up a successful empire in the New World. Their thirteen American colonies became populous, economically robust, and militarily strong enough to gain independence in 1776. Canada functions as an independent country but remains part of the British Commonwealth.

B. Population Distribution in North America

The US population was estimated at more than 325 million in 2017. Canada now has over thirty-five million people. The US population is growing by about 2.5 million people each year. A little less than half the growth can be attributed to immigration and the rest to natural increase. The pace of growth is slower than the world average but more rapid than many other industrialized countries such as those in Europe.

The population is not uniformly spread over North America, nor are population growth rates the same in all locations. Most Canadians live near the US border. The North American population tends to be clustered in cities. Additionally, the population has been moving southward and westward. US states experiencing the greatest rates of population growth include those located on the southern portion of the eastern seaboard, as well as Texas, Nevada, Utah, California, Oregon, and Washington. Three states—California, Texas, and Florida—accounted for about a third of the entire US population growth since 1990. Still, the Northeast is the most densely populated area of the country thanks, in large part, to the **Megalopolis** ("mega-metropolis") that encompasses the cities from Boston south to Washington, DC (sometimes referred to as Bos-Wash). The largest concentration of Canadians lives in southern Ontario and Quebec near Lakes Erie and Ontario and the St. Lawrence River in an area often referred to by geographers as **Canada's "Main Street."**

In general, the population of minorities is growing most rapidly. Some of the fastest-growing populations in the United States are Hispanics. Another interesting factor in population growth is the increase in life expectancy. As more people live longer, the growth of the segment of the population aged sixty-five has doubled in the last fifty years. Of this group, the greatest increase was seen in people aged eighty-five years and older.

The American population tends to be on the move. The US Census Bureau data show that the average American moves once every seven years; these data further predict that about forty million people move each year. Data also indicate Americans will move to a metropolitan area. Urbanization has been a trend since about 1950. Until that time, most Americans lived in small towns or more rural settings. The population density of the cities, and especially the suburban areas, has grown steadily since that time, bringing about a rural-to-urban population shift. Now a significant majority of people in North America live in suburban areas.

Key Takeaways

1. The United States and Canada have mountain ranges along their eastern and western portions, with lowlands in the middle.

2. In general, temperatures get cooler as you move from south to north, and the climate gets more arid as you move from east to west across the continent.

3. The Spanish were the earliest Europeans to establish a permanent settlement in the United States or Canada. They controlled the territory in the southern edge of what is now the United States, and their influence is still felt today through the Mexican American culture in that region.

4. The French colonized eastern Canada, the Great Lakes, and the Mississippi River valley. Although the number of settlers was small outside Quebec, French place names and French land-use patterns are still evident.

5. The British colonized the eastern coast of what became the United States. The number of English-speaking settlers was so high that the English culture dominated the region and left a strong long-term impact in terms of language, religion, and many other cultural aspects.

6. The more than 325 million people who live in the United States and the more than 35 million people who live in Canada are not evenly distributed across North America. The realm continues to urbanize, and minority groups are the fastest-growing segment of the population.

4.2 United States: Early Development and Globalization

Learning Objectives

1. Explain how the United States acquired its geographic boundaries.
2. Examine patterns of immigration to and migration within the United States through the period of westward settlement.
3. Examine urban growth and its connection to development of new forms of transportation.
4. Explain which economic patterns helped the United States become the world's largest economy.
5. Consider how the concept of the American Dream has been exported globally.

A. Early Development Patterns

With abundant resources and opportunity, the original thirteen colonies prospered and expanded into what became the fifty US states. The political geography of this nation was a product of various treaties and acquisitions that eventually resulted in the country extending from the Atlantic to the Pacific Ocean. Fueling the expansion was the concept of **Manifest Destiny**, the belief of some Americans that the new nation was divinely predestined to expand across the continent. The United States negotiated with France for the **Louisiana Purchase** in 1803, acquiring millions of acres in the central United States. Florida was acquired from Spain in 1819, and Texas was annexed in 1845. The British sold portions of the Pacific Northwest to the United States, and the 49th parallel was established as the boundary between the northwestern United States and Canada in 1846. Through conflicts with Mexico, large portions of the West were ceded to the United States in the mid-nineteenth century. Alaska was purchased from the Russians in 1867 for only $7.2 million. Alaska and Hawaii were the last two possessions to enter into statehood, which they did in 1959.

B. Westward Settlement Patterns and European Immigration

The thirteen original colonies are often grouped into three regions, each with its own economic and cultural patterns. These three areas—New England, the Mid-Atlantic, and the South—are considered **culture hearths**, or places where culture formed and from which it spread. The three regions were source areas for westward migration, and migrants from these regions carried with them the cultural traditions of their culture hearths. New England was characterized by poor soils, subsistence agriculture, and fishing communities and was the birthplace of North America's Industrial Revolution. Its largest city was Boston. Settlers from New England traveled west across New York State and into the upper Midwest and the Great Lakes region. The Mid-Atlantic region, focused on Philadelphia, Pennsylvania, was known for its fertile soils, prosperous small-scale agriculture, and multinational population. Prosperous farming led to a vibrant economy and a robust network of towns and cities. People who wanted to migrate west from this region traveled down the Great Valley into the Appalachian Mountains and across the Cumberland Gap into Kentucky, or they crossed Pennsylvania and traveled west via the Ohio River valley. The heart of the South was Virginia, a region oriented around plantation agriculture. The South was overwhelmingly rural, and in time the bulk of its agricultural workforce consisted of slaves brought to the United States from West Africa.

Westward migration was spurred along by the gold boom in California (1849) and by the completion of the transcontinental railroad (1869). The settlement frontier pushed westward during the course of the nineteenth century and was declared "closed" by the Bureau of the Census in 1890. This did not mean that settlers were spread uniformly across the continent by 1890; indeed, vast areas of the Great Plains and the mountain west remained sparsely populated by Europeans at that time. **The Homestead Act of 1862** also encouraged westward migration by offering 160 acres of free land to households willing to move west. The con-

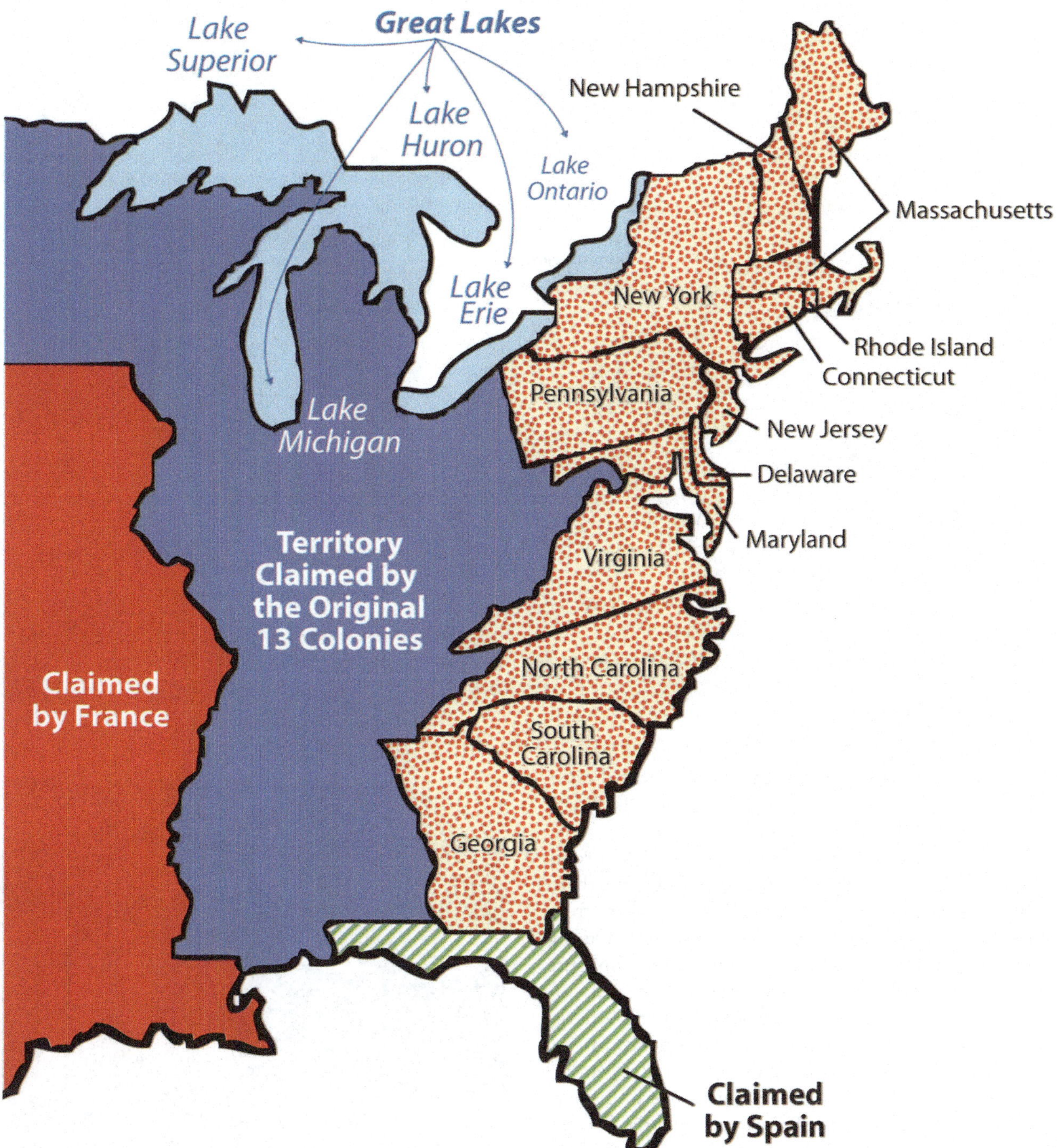

Figure 4. The thirteen original colonies.

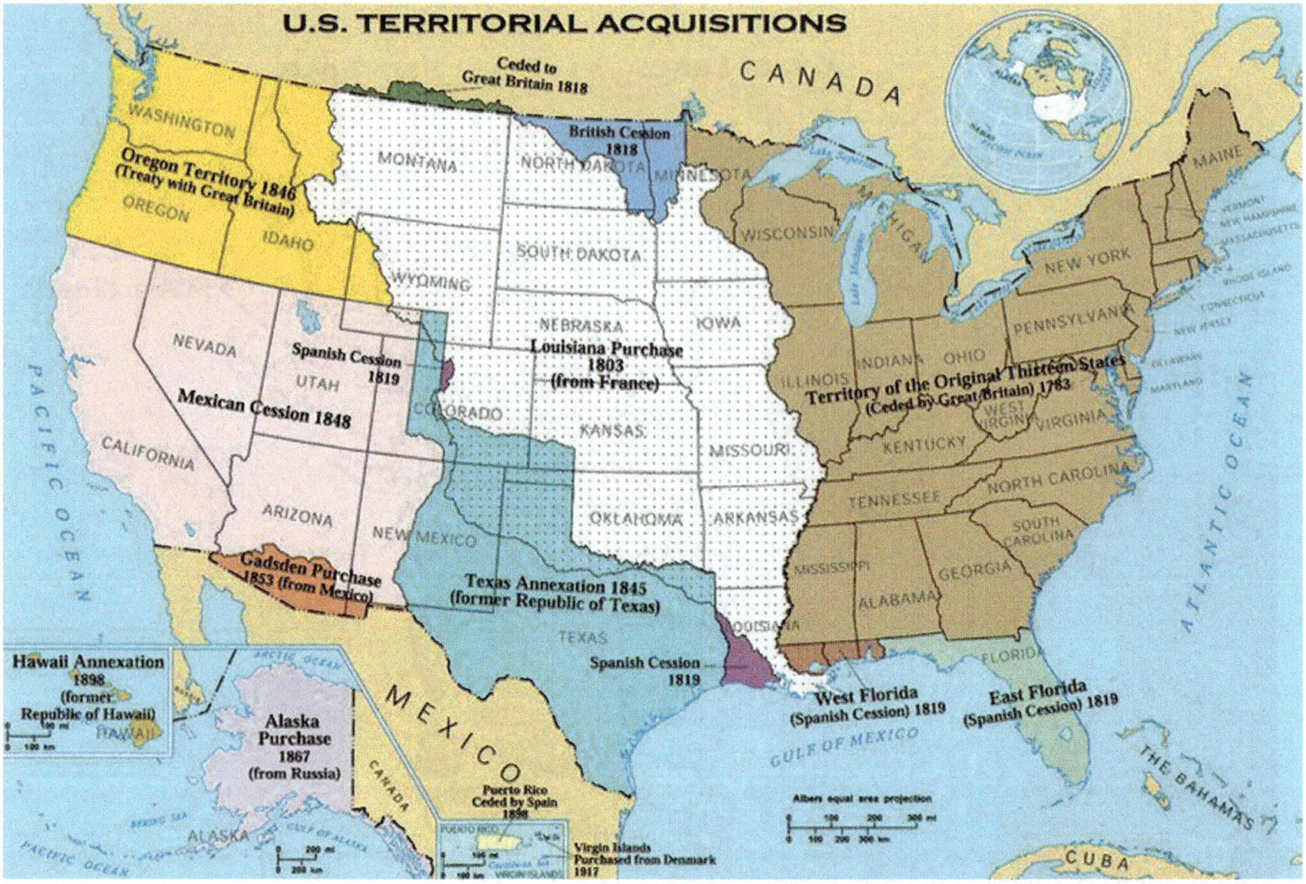

U.S. TERRITORIAL ACQUISITIONS

Figure 5. U.S. territorial acquisitions.

tiguous United States had been organized into official states by the end of the nineteenth century, except for Oklahoma (1907), Arizona (1912), and New Mexico (1912).

Most US residents at its founding in 1776 had roots in Great Britain, with large numbers from other northern and western European countries and many others from Africa (most of whom were slaves in the South). During the nineteenth century, migrants continued to immigrate to the United States as its economy grew, especially after the 1830s. Germans and Irish began arriving in large numbers, joining others from Britain and other countries, predominantly those in western Europe. As the century progressed, others from southern and eastern Europe, from countries such as Italy, Russia, and Austria, became the most significant stream of immigrants to the United States. The new arrivals were different from the early British immigrants: they practiced Roman Catholic or Eastern Orthodox Christianity (not Protestantism), they primarily moved to urban areas, and they found work in the new manufacturing sector growing rapidly in the Northeast and around the Great Lakes. Very few immigrants came from Latin America or Asia at that time.

Figure 6. The Holy Trinity Serb Orthodox Church in Butte, Montana This church gives testimony to the impact that European immigration has had on the North American continent and the freedom of religion that the US Constitution provides.

C. Industrial Development and Urbanization

The **Industrial Revolution** that began in Great Britain in the late eighteenth century eventually moved across the Atlantic and took hold in the United States. Rapid industrial growth emerged in the nineteenth century and was focused in the northeastern United States around the Great Lakes in an area called the **Manufacturing Belt**. Mechanized manufacturing began with textiles (New England), moved to steel and other metals (Pennsylvania and Indiana), and later was dominated by the manufacture of automobiles (Michigan). Of course, manufacturing would not have been possible without an abundant supply of power. Coal mining became an important industry in western Pennsylvania and in Appalachia.

Manufacturing took place in the cities and towns of the Manufacturing Belt. Not until the second half of the twentieth century did manufacturing move to rural areas; until then, it was almost entirely an urban activity. Since the 1980s, many manufacturing jobs have shifted away from the northeastern US to other regions of the country, especially in the southern and western states. This has resulted in the Manufacturing Belt being referred to as the **Rust Belt**.

As the United States went through its Industrial Revolution, its population shifted from being almost entirely rural to being mostly urban. In 1790, only about five percent of the US population lived in urban areas; by 1920, about 50 percent lived in cities. As the rural to urban shift took place, the function and form of US cities also changed.

From the colonial era until the late nineteenth century, US cities were walking cities. Because most Americans lived on farms, cities were small, compact, and centrally oriented: everything was located within walking distance. Only wealthy people had access to transportation by horse, and city dwellers needed to live within a short distance of where they worked, shopped, and carried out all their activities. The invention of the electric streetcar (1888) allowed cities to increase in size. People could live farther from their place of employment as long as they lived within walking distance of a streetcar line. Streetcar suburbs grew up along streetcar lines, and these neighborhoods were often segregated by ethnicity. Fewer people lived in downtowns, which became dedicated to retail

and manufacturing. Cities remained oriented around a **central business district** (CBD), which was often located near the railway station. Factories needed to be near transportation for both shipping in parts and shipping out completed products and so that workers could easily get to work.

Large numbers of middle class Americans began acquiring automobiles after about 1920; this eventually led to a complete rethinking of the spatial layout of the city. Automobile suburbs sprang up outside the traditional city limits as people were able to buy homes far from streetcar lines or railway stations. Cities became increasingly decentralized: people could go shopping in suburban malls instead of downtown department stores, factories could spring up at highway interchanges and not only near rivers and the railroad, and people could live in one suburb and work in another instead of living in the suburbs and working downtown. Neighborhoods became even more racially and economically segregated than they had in the past as middle-class whites moved into the new automobile suburbs and left the poorer African Americans behind in the cities.

By the late twentieth century, the automobile had led to a new urban form: the **edge city**. Edge cities are areas of dense urban development outside the boundaries of the traditional city. They often form at the intersection of major interstate highways and contain shopping malls, office complexes, high-rise apartment buildings, industrial parks, restaurants, and hotels. Sometimes edge cities are called **suburban downtowns**. Edge cities have supplanted the CBD as the destination of choice for Americans, whether they are heading to work or to play.

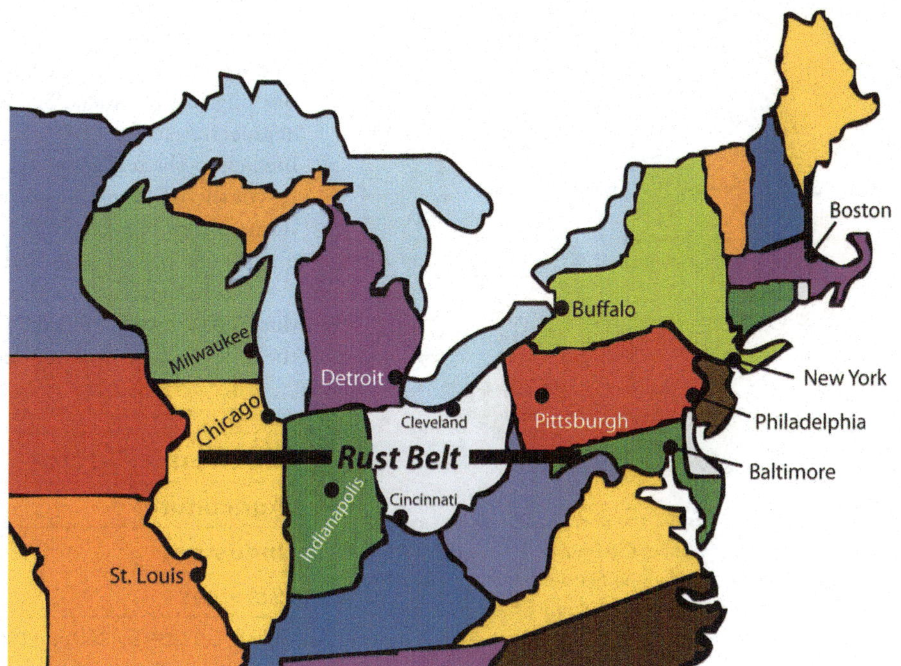

Figure 7. Manufacturing belt turned rust belt.

D. Economic Changes

During the colonial era and into the nineteenth century, when the majority of Americans lived on farms and worked in agriculture, most economic activity in the United States took place within the primary economic sector. Today, the primary sector is still an important component of the US economy, but far fewer people are employed in it. For example, less than 1 percent of Americans make their living by farming, but agricultural output has continued to grow because of advancements in mechanization and the development of high-tech seeds, fertilizers, and pesticides. The United States has been able to export surplus agricultural output to other parts of the world. Fewer people work in coal mines than in the past, but because of new mining technologies and methods such as mountaintop removal, coal production remains high.

The geographic distribution of primary activities depends both on the location of natural features such as physical geography and climate and on the location of the market for a particular crop or resource. The nineteenth-century German economist Johann von Thünen created a model that predicted land use around a central market. In his theory, land closest to the market would be used to produce crops that were expensive to transport, such as dairy. Land far from the market would be used for the production of crops that were less expensive to transport and less perishable, such as grain. **The von Thünen model** predicts a series of concentric rings surrounding a central market, with each ring producing a different kind of crop. If the von Thünen model is applied at a much larger scale to the United States as a whole, with the densely populated urban zone from Boston to Washington, DC (Megalopolis), used as the central market, the model does a fairly good job predicting the United States' agricultural land use. Dairy farms are found close to the market, grain farms are farther away, and ranch lands used for livestock production are even farther away.

Anything that involves the processing of raw materials—manufacturing—is a secondary activity. As the United States moved into the Industrial Revolution and into the mid-twentieth century, the percentage of the US workforce involved in manufacturing grew from almost nothing until it peaked in the late 1970s. It was the main area of economic growth for decades. Although manufacturing was present in most areas of the country, it was focused in the northeastern United States and along the Great Lakes. Factories were close both to the reserves of labor and to the markets for manufactured products found in the densely populated Northeast. The steel industry was located in Pittsburgh and its environs because of the area's access to iron ore (mined in Minnesota and transported via the Great Lakes) and to coal (mined in Pennsylvania, West Virginia, and other parts of Appalachia).

As manufacturing has grown in other parts of the world, the secondary economic sector has declined in the United States. US labor statistics indicate that the United States lost about five million manufacturing jobs between 2000 and 2010. Many of these jobs were lost to countries with lower labor costs, such as Mexico or China.

The third group of economic activities takes place in the tertiary and quaternary sectors, commonly known as the service sector. Tertiary and quaternary activities create services, not physical products. Service jobs include everything from engineering to finance, restaurants to sports, and childcare to medicine. The tertiary sector makes up more than three quarters of the US economy, as measured by its share of the **gross domestic product (GDP)**, which is the total value of all goods and services produced in a country in a given year.

These figures show that the US has shifted to a **postindustrial service economy**. The rise of the **information age** in the latter part of the twentieth century shifted the workforce

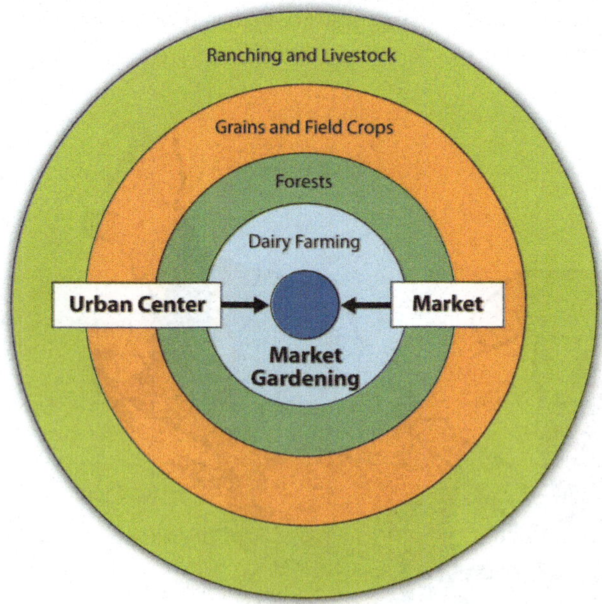

Figure 8. The von Thünen Model. The von Thünen model as it relates to the development west of the East Coast Megalopolis. For example, New Jersey is called the Garden State, Appalachia has its forests, the Midwest has its agricultural production, and the Great Plains has its large cattle ranching operations. This model was more applicable in past centuries for local communities when modern transportation technology was not available.

Economic Sector	GDP (%)
Agriculture	1.1
Industry	19.4
Services	79.5

Table 1. US GDP by Sector (2016 Data)
Source: "The World Factbook," Central Intelligence Agency, https://www.cia.gov/library/publications/the-world-factbook/geos/us.html.

into the information sector. By 2016, only about one percent of the US workforce was employed in agriculture, 20 percent in industry, and the rest in services (80 percent).

The locations of service-sector jobs are much more flexible than are jobs in the primary or secondary sectors. They are called **footloose jobs**: an accountant can live in New York or in Denver, whereas it is much more difficult for factories to move from one place to another and it is impossible for farms to relocate. Many of the information-technology jobs are emerging in the southern regions of the United States called the **Sun Belt**. Southern cities such as Atlanta, Dallas, and Phoenix are centers of innovation and population growth. The warmer climate, combined with a lower cost of living and less congestion, makes the Sun Belt an attractive location for emerging information-based companies. Note that the popularity of the South and West for service-sector jobs only came about after the invention and adoption of air-conditioning. Air-conditioning was not widespread until after the Second World War in the 1950s.

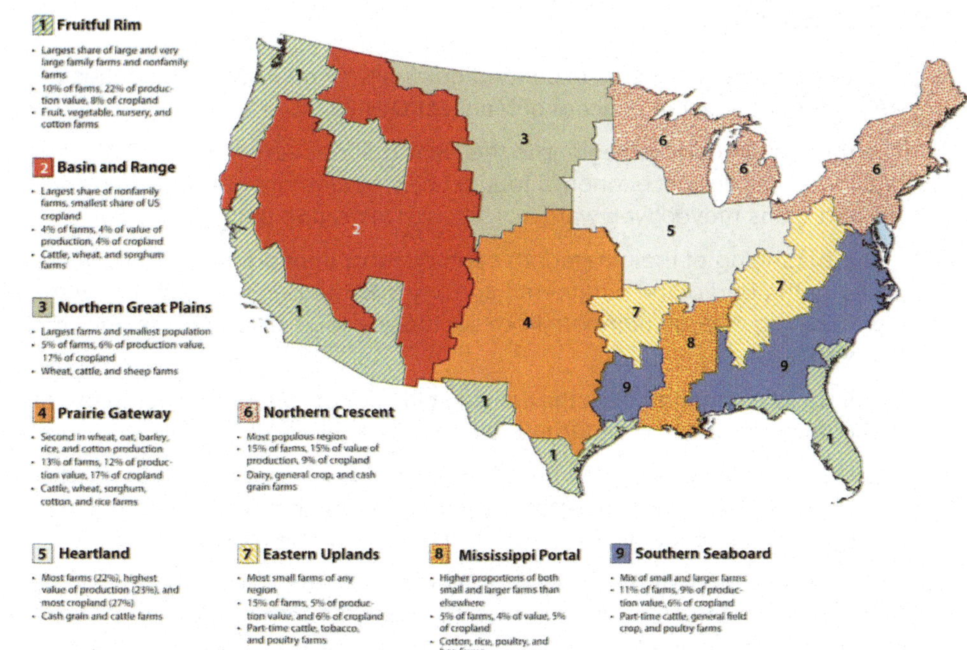

Figure 9. Farm resource regions.

E. Americanism and Globalization

The freedom of personal expression in the United States has supported individual ingenuity and creative ambition to create the largest economy in the world. US citizens have pushed American corporations to become a major force in the world markets. Products and franchises from the United States are being distributed throughout the world. Items such as fast food, computers, news networks, and Hollywood movies have become the products of choice in countries across the globe. The English language dominates the Internet, which has been heavily influenced by US corporations. The power of **the American Dream**—the idea that through hard work anyone can achieve upward mobility and financial success—as it is portrayed in the US media holds sway in the minds of people both in the United States and abroad.

US news networks, such as CNN, are so dominant that small countries, having no resources to create networks, rely on the US networks to deliver their world news. US fast food franchises, of which McDonald's is the largest, exist in over one hundred countries. Despite humble beginnings in Arkansas, Walmart grew to become the world's largest corporation. It has become the buyer and seller of retail trade that shapes and molds cultural attitudes and fashions internationally.

The size of the US population and the country's vast resource base have allowed it to become a world superpower. After the fall of the Soviet Union, the United States became the most powerful military force in the world. The United States has also dominated the world's economy and its communications networks. The advancements of multinational corporations have in essence enabled the sale of America to the rest of the world. The selling of American products and the large consumer market in the United States have provided profits that have fueled global economic markets.

The United States has become a worldwide franchise of its own. **Corporate colonialism** has advanced the American brand to a level that is now synonymous with consumerism, success, and power. Media advancements have promoted the concept of the American Dream worldwide. The reaction of the global community includes both admiration and disdain. Many view Americanism as interchangeable with globalization. Some welcome it; others reject it. The country of Iran is an example of this dichotomy. Young people in Iran wearing blue jeans gather in secret to watch American television programming from a hidden illegal satellite dish, while at the same time anti-American forces in their government condemn America as decadent, immoral, and imperialistic.

Corporate colonialism has become a dominant force impacting the global cultural fabric. Supporters appreciate access to American goods and services, while opponents claim that the English language and the American corporate franchise system are destroying the culture and heritage of millions who see their unique traditional ways of life being overshadowed and destroyed.

Key Takeaways

1. The United States' territory expanded gradually through various treaties and land acquisitions and was influenced by the concept of Manifest Destiny.

2. The three main colonial regions in the United States—New England, the Mid-Atlantic, and the South—had their own distinct economic foundations, settlement patterns, and social structures. People from these regions moved westward in particular migration patterns.

3. In the beginning of the nineteenth century, most immigrants were from western and northern Europe. By the end of the nineteenth century and the beginning of the twentieth century, immigrants were coming in large numbers from southern and Eastern Europe and moving to industrial cities of the Northeast.

4. City structure changed from the walking cities of colonial America, to the railroad and streetcar cities of the late nineteenth century, to the automobile cities of the mid- and late-twentieth century.

5. The US economy was initially based in the primary economic sector (particularly farming), then was based in the secondary economic sector (manufacturing), and is now oriented around the tertiary and quaternary economic sectors (services and information).

6. The diverse immigrants who have created American society have been unified by common aspirations and common ideals that created the concept of the American Dream. The concept indicates that regardless of one's station in life, by working hard, applying oneself, and following the rules, one can obtain upward economic mobility.

4.3 United States: Population and Religion

Learning Objectives

1. Explain the concepts of the cultural melting pot and the American Dream and how they have contributed to American society and culture.

2. Describe the current demographic profile of the United States.

3. Identify the size, distribution, and other characteristics of the Hispanic population in the United States. Explain the two most significant processes that led to the spatial distribution of African Americans in the United States.

4. Describe the distribution of the dominant religious affiliations in the United States.

A. The American "Melting Pot"

Early immigration to America was dominated by people from the British Isles, resulting in an American population for whom speaking English and practicing Protestant Christianity was the norm. There were some regional exceptions to this, such as Catholicism in Maryland and the widespread speaking of German in Pennsylvania, but by and large English and Protestantism were standard in the American colonies. As migrants arrived in the United States from non-English-speaking countries, within a generation they learned English and assimilated into American society, giving rise to the idea of the United States as a cultural melting pot. People were drawn to the United States by the hope of economic opportunities; most immigrants were poor and came to the United States to make a living and improve their financial well-being. They viewed assimilation into mainstream society as a necessity for success. They believed in the American Dream—that through hard work, you could achieve upward mobility and financial success no matter your background. The dream came true for millions of Americans but remains out of reach for many who live in poverty.

As of 2017, the United States was home to approximately 326 million people and was the third-most populated country in the world after China and India. Among developed countries, the US population is one of the fastest growing, at about 1 percent each year. This is thanks to a fertility rate of about 2.1 that is higher than the 1.5 for that of most European countries, as well as positive net migration (more people immigrating to the United States than emigrating from it). In terms of human well-being, life expectancy is more than seventy-eight years for men, and the average woman can expect to live more than eighty years. While this may seem high, especially when compared with a century ago, life expectancy in the United States is lower than in forty-nine other countries.

Although English has remained the dominant language, as a country of immigrants, the United States is home to people from all corners of the world and home to many cultural or ethnic minority groups. According to the 2010 census, the ethnic minority groups in the United States included 16.3 percent Hispanic (who can be of any race); 12.6 percent black or African American; 5.0 percent Asian and Pacific Islander; and 1.0 percent Native American (American Indians, Eskimos, and Aleuts).

Figure 10. The Statue of Liberty. has long been a symbol of the American ideals that welcomes immigrants to America.

B. The US Hispanic Population

One of the most striking demographic shifts in the past few decades has been the dramatic increase in the Hispanic/Latino population in the United States. In 1970, Hispanics made up less than 5 percent of the US population, but by 2010, forty-eight million Hispanics made up about 16 percent of the population. For the first time, Hispanics were the largest ethnic minority in the United States, surpassing blacks as the largest minority starting with the 2000 US census (12.5 percent Hispanic compared with 12.3 percent black). The US Hispanic population doubled between the 1990 and 2000 censuses. Between 2000 and 2006, Hispanic population growth accounted for about half the nation's growth and grew about four times faster than the country's population as a whole.

The growth of the US Hispanic population is a direct result of increased immigration from Latin America to the United States in the late twentieth and early twenty-first centuries and the Hispanic population having higher fertility

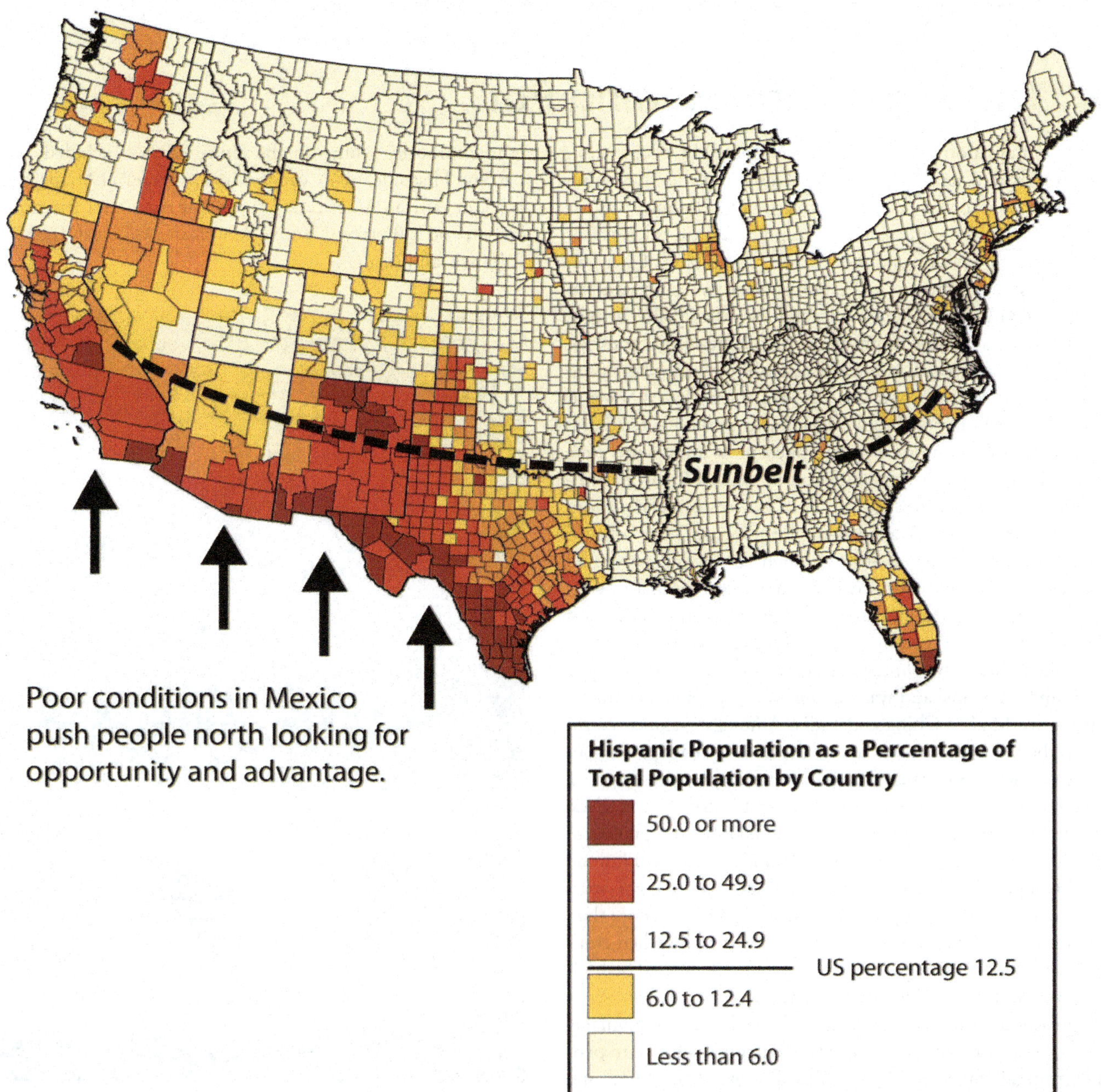

Figure 11. Hispanic population in the United States and the US Sun Belt.

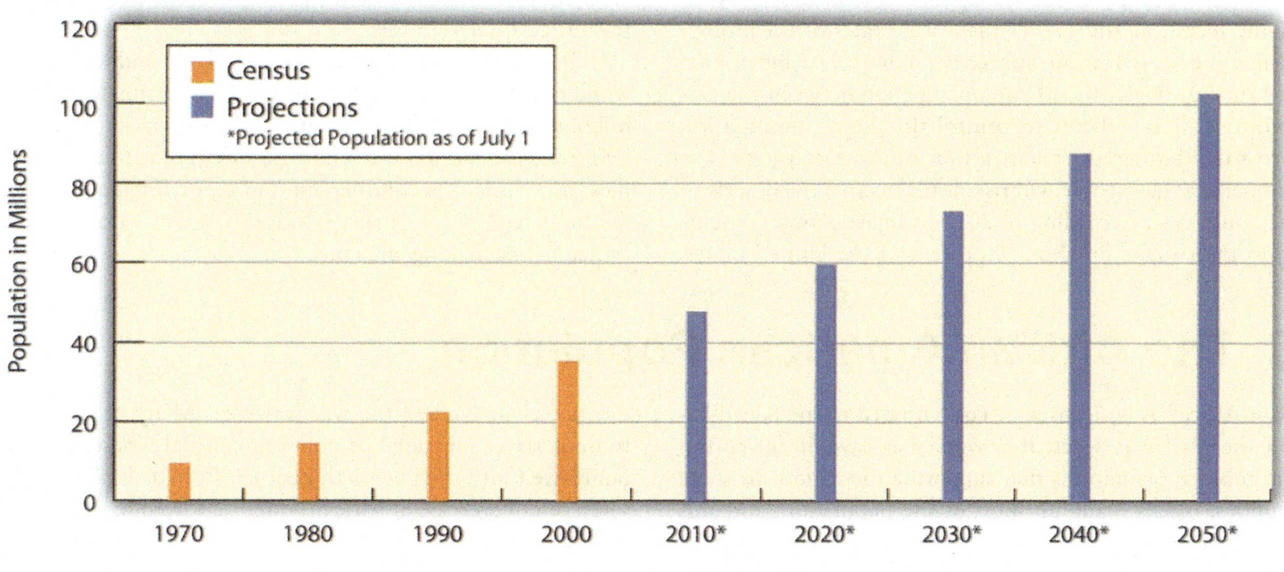

Figure 12. Hispanic population in the United States, 1970–2050.

Figure 13. US / Mexico border wall.
Barriers (walls, fences, or vehicular barriers) span about one-third of the nearly 2,000 mile border between the two countries.

rates than the non-Hispanic US population. Nearly half the Hispanics in the United States live in California or Texas, although there has been a large increase in the Hispanic population outside those states since 2000, especially in the South. For example, Arkansas, Georgia, Tennessee, and North and South Carolina all experienced Hispanic population growth rates between 55 and 61 percent from 2000 to 2006. All regions of the country saw double-digit growth rates of their Hispanic populations during that time. In places such as California, the large Hispanic population has an especially significant impact on the economy, politics, and every aspect of social life: more than one-third of Californians are Hispanic (37 percent), while 42 percent are non-Hispanic white, and a much smaller minority are African American (7 percent).

Who are the Hispanics living in the United States? Most were born in the United States (60 percent), while the rest are immigrants. Two thirds are either from Mexico or of Mexican descent, while others hail from the US territory of Puerto Rico, Cuba, or the Dominican Republic. Note that all Puerto Ricans are US citizens and can move to and from the US mainland without any special documentation requirements. More Hispanics come from Central America than from South America. Hispanics work in all professions but are found in professions such as agriculture, construction, and food service at higher rates than the country's non-Hispanic population.

The draw of opportunities and advantages has always pulled people toward the United States. While many Latin American immigrants enter the United States legally, according to the US Department of Homeland Security's Office of Immigration Statistics, there were about 10.8 million undocumented immigrants residing in the United States as of 2010; 62 percent were from Mexico. This number is lower than it was in previous years, possibly because of the economic recession and higher-than-usual rates of unemployment. The US-Mexican border is about 1,970 miles long and runs through an arid and open region between the two countries. It is difficult to control the illegal immigration across this border, as the attraction to American jobs is so compelling that people will risk death to cross the deserts of the Southwest. According to 2010 estimates, undocumented workers make up about 5 percent of the United States'

civilian workforce, including approximately 24 percent of the agricultural workforce.

The amount of **remittances** sent from undocumented workers in the United States to Mexico is estimated in the billions of dollars. The remittances from Mexican nationals living outside Mexico and sending money home to their families are Mexico's second-largest source of foreign income. Without remittances, many Mexican families would have a difficult time making ends meet.

C. The African American Population

Most African Americans were concentrated in the South before the Civil War, where they worked as slaves in the cotton and tobacco plantations that supported the region. In some counties, blacks made up most of the population, and this

did not change when the war was over. Many of the newly freed slaves remained as poor agricultural workers in the South well into the twentieth century. Even as late as 1910, seven out of every eight African Americans lived in the South.

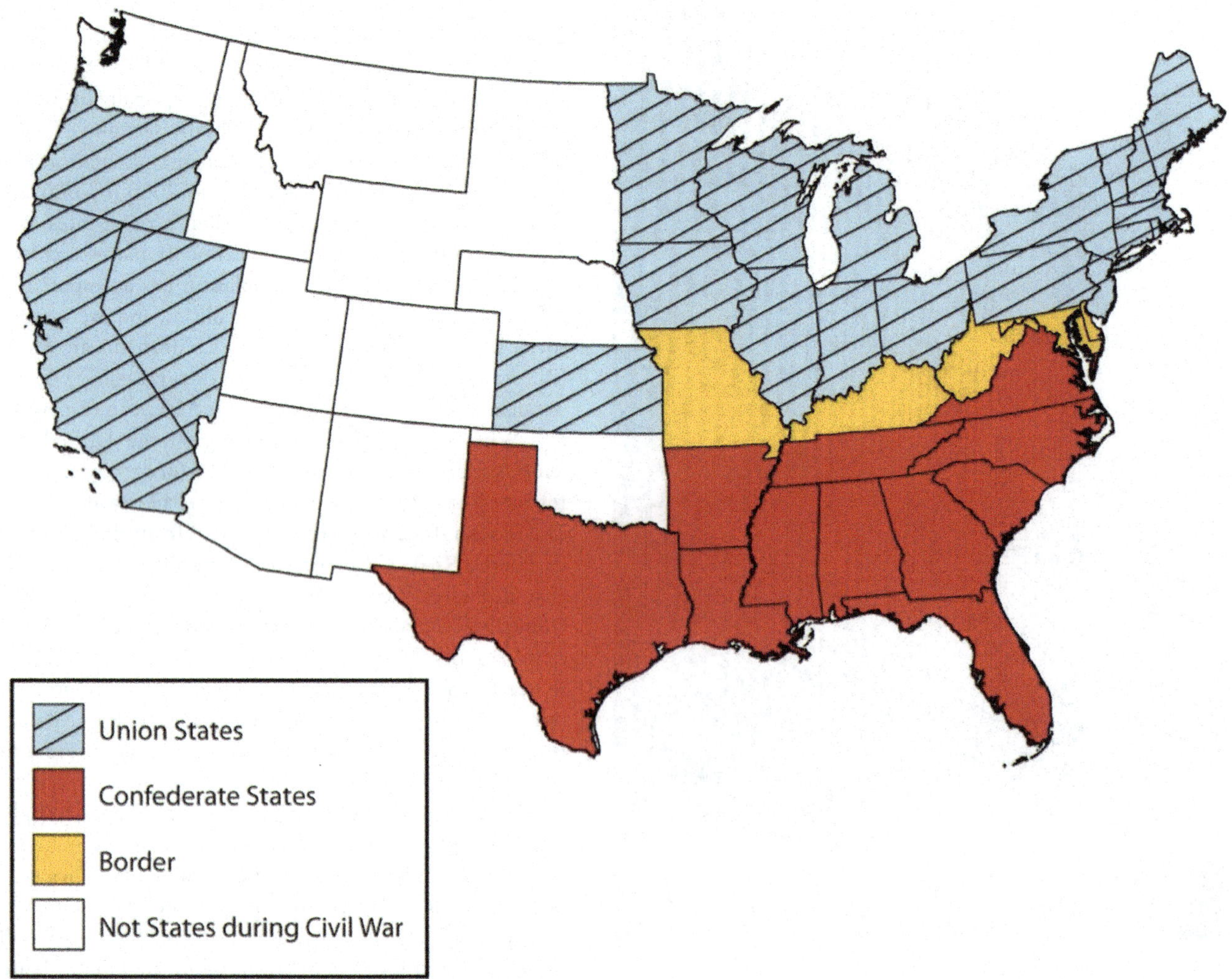

Figure 14. Civil War division in the United States, 1861–1865.

In the late nineteenth and early twentieth centuries, as the industrialization of northern cities was accelerating, the increased need for factory workers was largely met by immigration from Europe, especially from southern and eastern Europe. However, when the First World War began (1914), European immigration began to slow down. European immigration then nearly ground to a halt in the 1920s as Americans set quotas in place to reduce the number of Eastern European immigrants. At that time, the factories in the Manufacturing Belt continued to need workers, but instead of European workers, they recruited African Americans from the South.

This led to a massive migration of blacks from the South to cities of the North and West. This migration was so significant to African Americans in the United States that it is called the **Great Migration**. Between 1910 and 1925, more than 10 percent of African Americans made the journey north, and even more followed. Examining a map of the distribution of African Americans today shows the legacy of the Great Migration, as blacks live in many parts of the United States, both in the South and in postindustrial cities of the north and the Midwest. Blacks also now live in Sun Belt cities, as people of all races look for jobs related to the new information technology and service industries.

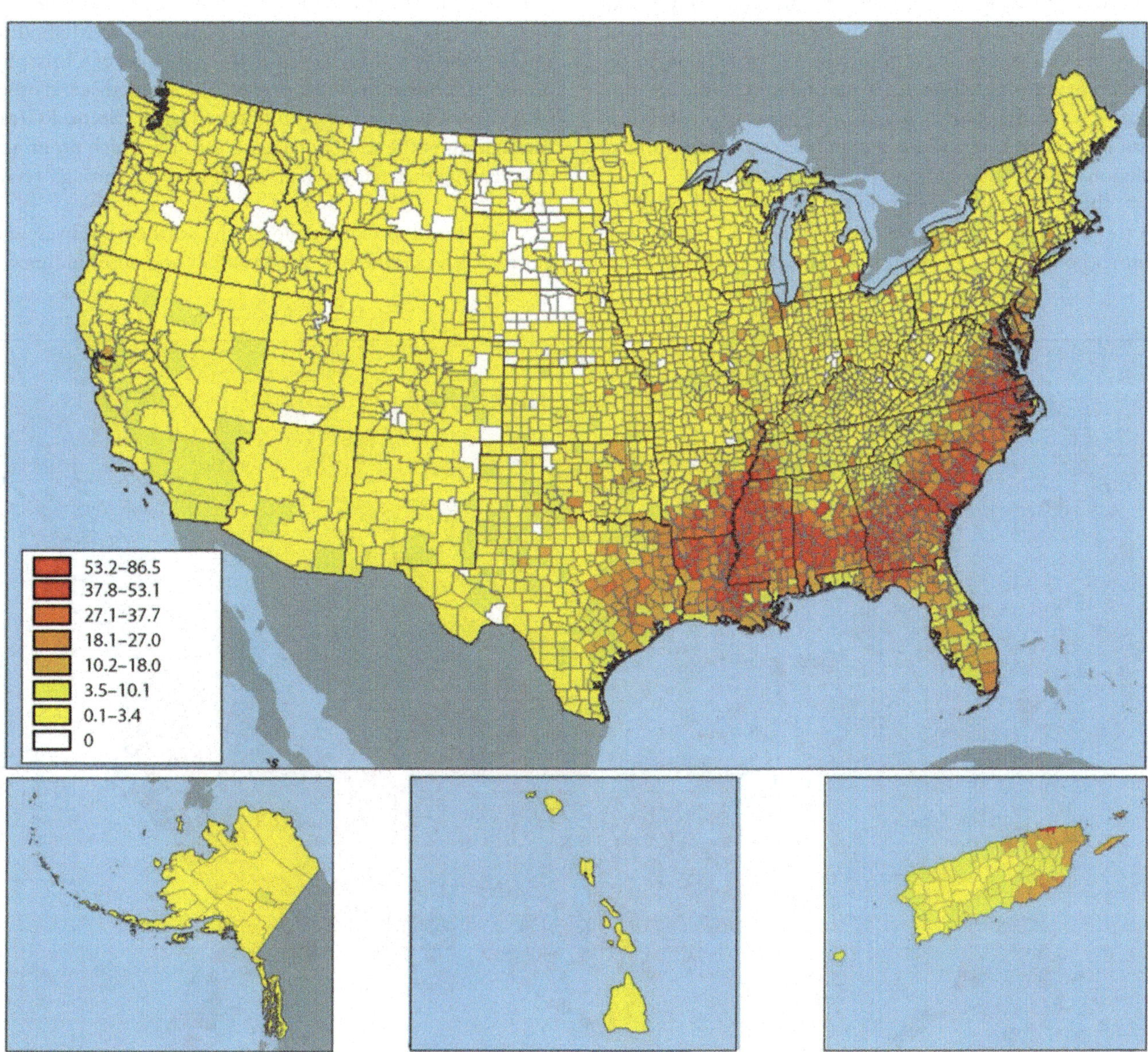

Figure 15. Black/African American population by county, 2000.

D. Geography of Religion in the United States

Most early settlers to the United States were Christians: Puritans lived in New England and Anglicans (later called Episcopalians) lived in Virginia. Roman Catholic immigrants settled in Maryland, and members of the Society of Friends (Quakers) founded Pennsylvania. Even within that overall picture, there was a great deal of religious diversity in the United States, and that diversity increased as new arrivals came from different countries with different religious backgrounds. The current pattern of religious affiliation in the United States remains quite complex, and one can find observers of nearly every major religion, and many minor ones, in virtually every area of the country. That being said, there are clear patterns to the geography of religion in the United States that tell stories of immigration and migration history, as well as stories about other aspects of American history. The map of leading church bodies shows regions of religious observance that are worth examining.

The most striking feature of the map is the block of red in the Southeast in which Baptist churches are the leading church body. Although Baptist churches are the leading religious body in about 45 percent of all counties in the United States, most of those counties are in the South. This region is considered the nation's **Bible Belt**, and it is a region in which churches are more likely than in other parts of the country to teach a literal interpretation of the Bible. Baptist churches grew in popularity in the South after the Civil War as more liturgical denominations such as Methodists went into decline. Baptist churches are popular among both African American and white residents.

Another Protestant region is northern Appalachia and the lower Midwest from Ohio to Iowa and Kansas. Some of these counties are mostly Baptist, some mostly Methodist, and in others it is Christian churches (Disciples of Christ and similar denominations) that prevail. The Methodist and other Christian areas were heavily influenced by the **Second Great Awakening** of the early nineteenth century, which promoted the theology that every person could be saved through revivals.

The third Protestant region is the northern Midwest and Great Plains: Minnesota, the Dakotas, and surrounding ar-

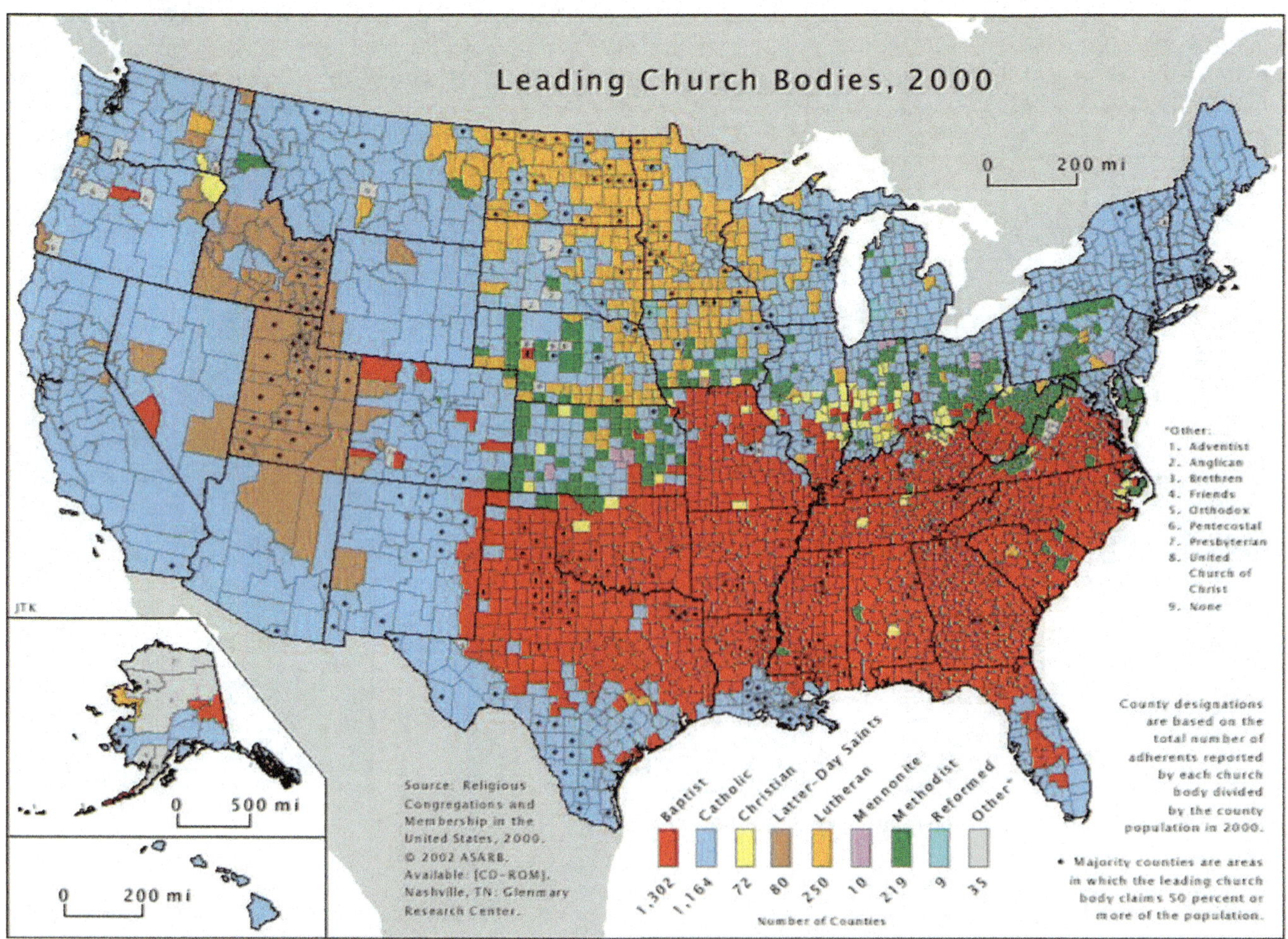

Figure 16. Religious congregations in the United States, 2000.

eas. This was the destination of German and Scandinavian Lutheran settlers during the late nineteenth and early twentieth centuries, and the leading denominations today in much of that area remain Lutheran.

The Roman Catholic Church, which is the leading religious body in 40 percent of US counties, is well represented in the Northeast, West, and Southwest. In the Northeast and Midwest, the Catholic dominance points to nineteenth- and early twentieth-century immigration from Roman Catholic countries in Europe such as Italy, Ireland, and Poland. Those earlier Catholics have been joined more recently by large numbers of Hispanic immigrants. The dominance of Roman Catholics in the western United States, the Southwest, and even Florida in the Southeast are a reflection of the strong Hispanic presence in those parts of the country.

Figure 17. Dehart's Bible and Tire, Eastern Kentucky.

In the western United States, the Church of Jesus Christ of Latter-Day Saints (LDS or Mormon church) dominates a region including Utah and surrounding states. Utah was the destination of Mormons as they migrated westward during the mid-nineteenth century.

Key Takeaways

1. US immigrants learned English and assimilated into American society, giving rise to the idea of the United States as a cultural melting pot.

2. Immigrants created the American Dream—the idea that by working hard and following the rules, one could achieve upward mobility and financial success regardless of one's background or heritage.

3. The Hispanic population is the largest minority group in the United States, and its population is growing.

4. Hispanics have an especially strong presence in California and Texas but are found all over the South as well as in rural and urban areas around the country.

5. One-third of Hispanics are immigrants to the United States, and most undocumented workers in the United States are Hispanic (58 percent are from Mexico).

6. Mexican nationals are an important component of the US workforce, and they send home billions of dollars each year in the form of remittances.

7. African Americans are heavily represented in former plantation agricultural states of the South, as well as in cities of the Rust Belt and far West.

8. Patterns of religious observance in the United States reflect immigration patterns.

4.4 Canada

Learning Objectives

1. Identify where in the country most Canadians live and why.

2. Identify and locate the dominant physical features of Canada.

3. Distinguish between the French-speaking and English-speaking areas of Canada and discuss the key activities in the effort to promote French culture in Canada.

4. Name the characteristics of some of the minority ethnic groups in Canada.

5. Determine which economic sectors are at the core of the Canadian economy and how the Canadian and US economies are connected.

Figure 18. Map of Canada.

A. Canadian Provinces and Territories

Canada's democratic state shares a similar developmental history and economic status with the United States. As of 2016, Canada had a population of just over thirty-five million, which is less than the population of California. Canada is slightly larger in area than the United States, making it the second-largest country in the world. However, despite this vast territory for a relatively small population, more than 90 percent of Canadians live within 150 miles of the US border. Northern Canada is not considered part of Canada's **ecumene**, or habitable zone, for permanent human settlement. Only a narrow band of territory in southern and eastern Canada has the climate and physical geography suitable for agricultural production and widespread settlement.

Moreover, Canada's economy is so closely tied to that of the United States that it makes sense for people to live close to the US border.

Canada consists of ten provinces and three territories. Ottawa is the nation's capital, and Toronto is its largest city. Ontario has by far the largest population of any of Canada's provinces, with nearly 40 percent of the total. Quebec, the dominantly French-speaking province, is home to about 23 percent. Almost everyone else lives in British Columbia, west of the Rocky Mountains (13 percent); in the prairies (18 percent); or along the Atlantic coast (7 percent). The wide-open areas of Canada's far north are home to only one-third of 1 percent of the population.

Population		Province/Territory	Capital
Maritime Provinces of the East	2.3 million	Newfoundland and Labrador	St. Johns
		Nova Scotia	Halifax
		New Brunswick	Fredericton
		Prince Edward Island	Charlottetown
French Canada	8.2 million	Quebec	Quebec City
South Canada	13.4 million	Ontario	Toronto
Prairie Provinces	6.4 million	Manitoba	Winnipeg
		Saskatchewan	Regina
		Alberta	Edmonton
Pacific Canada	4.6 million	British Columbia	Victoria
Territories	About 114,000	Yukon	Whitehorse
		Northwest Territories	Yellowknife
		Nunavut	Iqaluit

Table 2. Canadian Provinces, Territories, and Capitals. Source: 2016 Canada Census Data.

B. Physical Geography of Canada

Type D (continental) climates dominate most of central Canada, with their characteristically warm summers and cold winters, although the farther north you go, the cooler the summers are. Canada's west coast receives the most rainfall—between eighty and one hundred inches a year—while coastal areas in the Maritime Provinces can receive up to sixty inches per year. The northern territory of Nunavut barely receives ten inches per year, usually in the form of snow. Of course, far northern Canada has an arctic type E climate, and conditions there are so harsh that only a very few people inhabit it. The cultural influence of the colder climates and the long winters on the people is evident by the sports that are enjoyed by most Canadian citizens. Ice hockey is Canada's most prevalent sport and its most popular spectator sport. Other sports such as curling are also common in Canada.

Canada has abundant natural resources for its population. The **Canadian Shield** is an area of rock and forest that covers much of central Canada around Hudson Bay. This region, and the area to the east and west of it, provides timber and minerals for Canadian industries and for export. In the Maritime Provinces of the east, the main economic activities include fishing and agriculture. Some of Canada's best farmland is located along the **St. Lawrence River** and in the southern Prairie Provinces of Manitoba, Saskatchewan, and Alberta. The St. Lawrence River region includes dairy farms and agriculture, which provide food for the larger cities of the region. The province of Ontario has fertile farmland on the north and east sides of the Great Lakes. The farmland in the Prairie Provinces has much larger grain and beef operations.

Centered in the province of Alberta is a large region of fossil fuel exploration. Coal, oil, and natural gas are found there in abundance, and much of it is exported to the United States. Oil is found absorbed in surface soil called tar sands and is being extracted for energy. When tar sands are heated, the oil is separated and refined for fuel. Projections are that there is more oil in the tar sands of Canada than in the underground reserves of Saudi Arabia. Natural resources have even filtered into the cultural arena: Edmonton's professional hockey team is called the Edmonton Oilers.

The Rocky Mountains and the coastal ranges located in western Canada provide for mining and lumber mills. Vancouver, on the coast in British Columbia, has become a major port for Canada to export and import goods to and from the Pacific

Figure 19. The Canadian curling team during the 2006 Winter Olympics in Torino.

Rim. The Yukon Territory, also located in the mountains, has experienced a gold rush in years past. Canada has abundant natural resources to provide for its people and gain wealth.

C. British versus French Canada

French fishermen and fur traders initially colonized Canada, the British later took it from the French, and immigrants from various other countries moved there to farm and otherwise make a living. Although none of the provinces retain French names, countless cities—especially in Quebec—have French names, among them Montreal, Trois-Rivières, Charlesbourg, and Beauport.

The names of several provinces indicate the British connection: Nova Scotia means "New Scotland," and it was so named by the British when they took over the island from the French. Prince Edward Island was named for the father of the famous nineteenth-century British queen, Victoria. You can see remnants of British colonialism in the way Canadian government is organized. Canada, like many countries of Europe, including Great Britain, is a parliamentary democracy. The monarch of the United Kingdom is still the top-ranking government official in Canada, but only as a figurehead. The queen (or king) appoints a governor general to be her (or his) representative in the Canadian federal government. Again, this is a symbolic position. There are two chambers, a House of Commons and a Senate. Members of the House of Commons are elected and are called members of Parliament (MPs). Senators are appointed to a lifelong term by the prime minister.

As of 2006, about 58 percent of Canadians spoke English as their primary language, French was the mother tongue of 22 percent, and another one-fifth of the population (20 percent) spoke a mother tongue other than English or French. For most of Canada the **lingua franca** remains English. The French-speaking portion of Cana-

da is a reminder of Canada's history as a French colony. Many of today's Francophones (French speakers) are descendants of those earlier French settlers. The proportion of French speakers in Canada is declining as more and more immigrants (who also have higher fertility rates than

Figure 20. Bilingual stop sign in Ottawa, Ontario.

Francophones) arrive from other parts of the world and as more Francophones begin using English instead of French as their primary language. The new immigrants (along with native peoples) make up the 20 percent of Canada's population who speak neither English nor French as their native language. About 90 percent of Canadian Francophones live in Quebec, which is a center of French culture in Canada.

The separation between French Canada and British Canada goes back to colonial times. Beginning in the 1530s, the French were the first to develop fur-trading activities in the region and colonize what is present-day Canada, calling it *New France*. The French claimed much of the St. Lawrence River valley and the Great Lakes region, including the region that is now Ontario. When Britain began to dominate the eastern coast of North America in the 1680s, they entered into a series of wars with France. As a result of these wars, New France was eventually turned over to Britain.

Not wanting continued war with France, Britain allowed the French-dominated region to retain its land ownership system, civil laws, and Catholic faith. The Revolutionary War in 1776, which granted the United States independence from Great Britain, also pushed many people of English descent—especially those who had sided with the British during the Revolutionary War—from the United States into Canada. British North America no longer included the United States; Canada became the main British colony in North America.

In an attempt to keep the peace between French and English settlers, in 1791 the British Parliament divided Quebec into Upper and Lower Canada, which later evolved into the provinces of English-speaking Ontario and French-speaking Quebec. The Maritime Provinces of the east were then separated into individual provinces.

The cultural differences between Francophone Canada and the rest of Canada have since erupted into serious political conflicts. The Francophone areas, mainly southern Quebec, argue that they are treated unfairly, since they have to learn English but the rest of the country is not required to learn French.

During the second half of the twentieth century, many people in Quebec supported a **separatist movement** that sought to break Quebec off from the rest of Canada into an independent country. In 1995, the separatist initiative lost in a public vote, but only by a small margin. The issue of Quebec's sovereignty continued to be raised in the public arena. In 1998, Canada's Supreme Court ruled that Quebec could not separate from the rest of Canada under international law but that the matter would have to be negotiated between Quebec and the rest of Canada if secession was to proceed.

French and English are the two official languages of the Canadian government as a whole, but the French people in Quebec, fearing that English was dominating the media, the Internet, and industry to such an extent that it was endangering their French culture, have declared French as the only official language of the province. To combat the encroachment of English, laws were enacted in Quebec requiring all public advertising to be in French, or if other languages are used, they must be half the size of the French letters. All businesses employing more than fifty employees were required to conduct all business in French. Immigrants who wish to be citizens of Quebec must learn French. All primary and secondary education takes place in French unless the child's parents were educated in English elsewhere in Canada. Civil servants dubbed the "language police" monitor and enforce the French language laws. A business found to be out of compliance with the language laws could be fined or shut down. Even though the official language of Quebec is French, since the national government takes place in both English and French, some services are still available in Quebec in English.

When secession was being considered, it was found that businesses that employed more than fifty employees did not want to switch over and conduct all their business in French; they were conducting all their business in English to work with the global economic community. In 1994, the **North American Free Trade Agreement (NAFTA)** was established between Canada, the United States, and Mexico. The goal was to open up new lines of business operations between the three countries and to increase economic opportunities to better compete with the European Union and the Pacific Rim nations. What would Quebec do if it separated from Canada? Would Quebec be able to join NAFTA? What if the other NAFTA partners cut off Quebec? Quebec would have faced serious economic consequences if they had separated from Canada. A number of English-speaking businesses have already moved to Ontario, Canada's most populous province, to avoid changing to French. It appears Quebec will remain with the rest of Canada and work out any internal cultural issues.

Figure 21. Skyline of Toronto, one of the most diverse cities in North America.

D. Other Ethnic Groups in Canada

Of course, we cannot forget the native groups who were displaced when the Europeans arrived. About 1.2 million people who identify themselves as Aboriginal live in Canada, or about 3.8 percent of the total population. They include North American Indians (also called **First Nations**), **Métis** (descendants of both Europeans and American Indians), and Inuit (inhabitants of the far Arctic north). Of those three groups, Inuit are the smallest, with only about fifty thousand remaining. These native people represent more than six hundred recognized groups and sixty-five language dialects, although only a handful of these languages are still spoken by a large enough core of people to remain viable languages for the long term.

Quebec is not the only place where devolutionary forces have been dividing cultural groups in Canada. In 1999, Nunavut officially broke from the Northwest Territory to become its own territory. Nunavut has only about thirty-five thousand people in an area larger than any other province or territory in Canada. It comprises about one-fifth of Canada's land area. Most of the people who had claimed the land before the Europeans arrived are Inuit. Iqaluit, the capital city of Nunavut, is on Baffin Island near Canada's east coast.

Canada has a great deal of ethnic diversity. One measure of this is the number of languages spoken there. One source estimates that there are about 145 languages spoken in Canada, including English and French. This reflects both the rich native heritage and the history of immigration from around the world. As of 2006, the foreign-born population was 6.2 million, or nearly 20 percent of Canada's population. There

Figure 22. Evidence of Eastern European immigration. This Ukrainian Greek Orthodox Church is located in the southern region of the Prairie Province of Manitoba.

are few countries that match this level of immigration. Even the United States had only about a 12.5 percent foreign-born population in 2006.

The current surge of immigrants to Canada does not include many Europeans. Instead, these immigrants come from Asian countries, especially China and countries in South Asia such as India and Pakistan. About one in five people in Canada belong to a nonwhite minority group. According to the 2010 Canadian census, more than a million Chinese and more than a half million South Asians lived in Canada.

E. The Canadian Economy

Not surprisingly, Canada and the Unites States are each other's largest trading partner. More than 80 percent of Canadian exports go to the United States and 70 percent of imports to Canada come from the United States. Except for some natural-resource industries, most businesses are centered in Canadian cities to take advantage of the available labor force. Canada is rapidly moving toward a knowledge-based economy built on innovation and technology. Knowledge-intensive industries, such as biotechnology and information technology, are on the rise, and these are typically located in cities to facilitate partnering with universities and other researchers.

Although Canada is developing into a knowledge-based economy, the foundations of the Canadian economy have always been its abundant natural resources. Canada's primary industries have traditionally been agriculture, fishing, mining, fuel/energy, and logging/forestry. Success in tapping these natural resources for their economic benefit allowed the country to double in population since 1960 while the econ-

omy has increased sevenfold. The primary industries now make up less than 10 percent of the gross domestic product (GDP). Just as in the United States, the most dramatic structural change in the economy has been the rise of the service sector, which now employs about three-quarters of all Canadians and generates over 60 percent of the GDP. Canadian manufacturing has been a strong sector of the economy with close ties to United States and multinational corporations.

Canada's economy is tightly tied to that of the United States. One of the best examples is how the Canadian economy fluctuates depending on whether the Canadian dollar is weak or strong compared with the US dollar. For example, in 2002, one American dollar was worth about $1.60 in Canada. For many years, the American dollar was much stronger on the world market than the Canadian dollar; therefore, Canadian goods and labor were less expensive for Americans than comparable US labor and goods. During the time of the weak Canadian dollar, many film and television industries moved to Canada to film television shows and movies, as it was less expensive to do so in Canada. Many popular television shows and movies have been shot in Canada, particularly in Toronto and Vancouver.

More recently, the value of the American dollar declined against other major currencies. The Canadian dollar remained strong, which meant that goods produced in Canada became much less affordable in the United States, causing the television and film industries to move back south of the border and Canadian exports to the United States to decline. In 2007, the Canadian dollar and the US dollar reached parity for the first time in thirty years. The two currencies continue to fluctuate with market values. As of mid-2017, one US dollar was worth about $1.30 Canadian.

As mentioned earlier, **NAFTA**, the 1994 trilateral agreement between Canada, the United States, and Mexico, was one of the most significant economic events in North American history. For Canada, the agreement has meant more secure, stable access to US and Mexican markets. The agreement eliminated many tariffs; opened previously protected sectors in agriculture, energy, textiles, and automotive trade; and set specific rules for trade in services such as finance, transportation, and telecommunications. Perhaps most importantly for Canada, the agreement set rules for settling trade disputes.

The United States continues to exert its powerful influence on many countries in the world, but perhaps on none so strongly as Canada. Because of the geographical proximity of the two countries and the fact that the vast majority of Canadians live very close the US border, speak English as their first language, and share a great number of cultural similarities, American trends tend to be adopted by Canadians. Canadians differentiate themselves from Americans in legal issues, laws, and health care. For example, Canada has a health care system, funded by the provinces with financial help from the federal government that provides free services to its citizens. Canadians often point to this difference as a one of the defining elements of their culture that is different from the United States.

Figure 23. Quebec-built Bombardier CRJ900LR Jet Liner taking off from Heathrow Airport in London.

Canada is a great consumer of American popular culture. Canadians listen to, watch, and read tremendous quantities of American music; television and movies; and news, books, and other literature—so much so that some Canadians believed Canadian culture was in danger of being extinguished. In response to these concerns, a law was passed and a watchdog agency created so that a certain percentage of all radio and television broadcasts emanating from Canadian radio and television stations had to originate in Canada or have significant Canadian content. Others were less worried about the impact of American pop culture on Canadian culture. This segment of the Canadian population felt that Canadians have long identified themselves in contrast to Americans; therefore, consuming American books, newspapers, television shows, and movies would only give Canadian a greater basis of comparison and thus strengthen the Canadian identity and perception of Canadian culture.

Key Takeaways

1. Canada is a very large country with rich natural resources but a relatively small population that mostly lives in a narrow band in the southern part of the country.

2. Canada's English and French bilingualism is part of its British and French colonial past. The French culture is dominant in Quebec, where the population has considered seceding from Canada and becoming an independent country.

3. Canada's native population makes up less than 4 percent of the country's population but represents a great deal of cultural and linguistic diversity.

4. More than one in five Canadians is an immigrant, and most of the recent immigrants come from non-Western countries, especially those in Asia.

4.5 Regions of the United States and Canada

Learning Objectives

1. Name the key characteristics of the regions of the United States and Canada.
2. Understand the patterns of population growth or decline for the various regions.
3. Determine which regions have significant minority groups and why.
4. Examine the environmental and social costs to rapid growth in the West.
5. Explain how physical geography has contributed to economic activities.

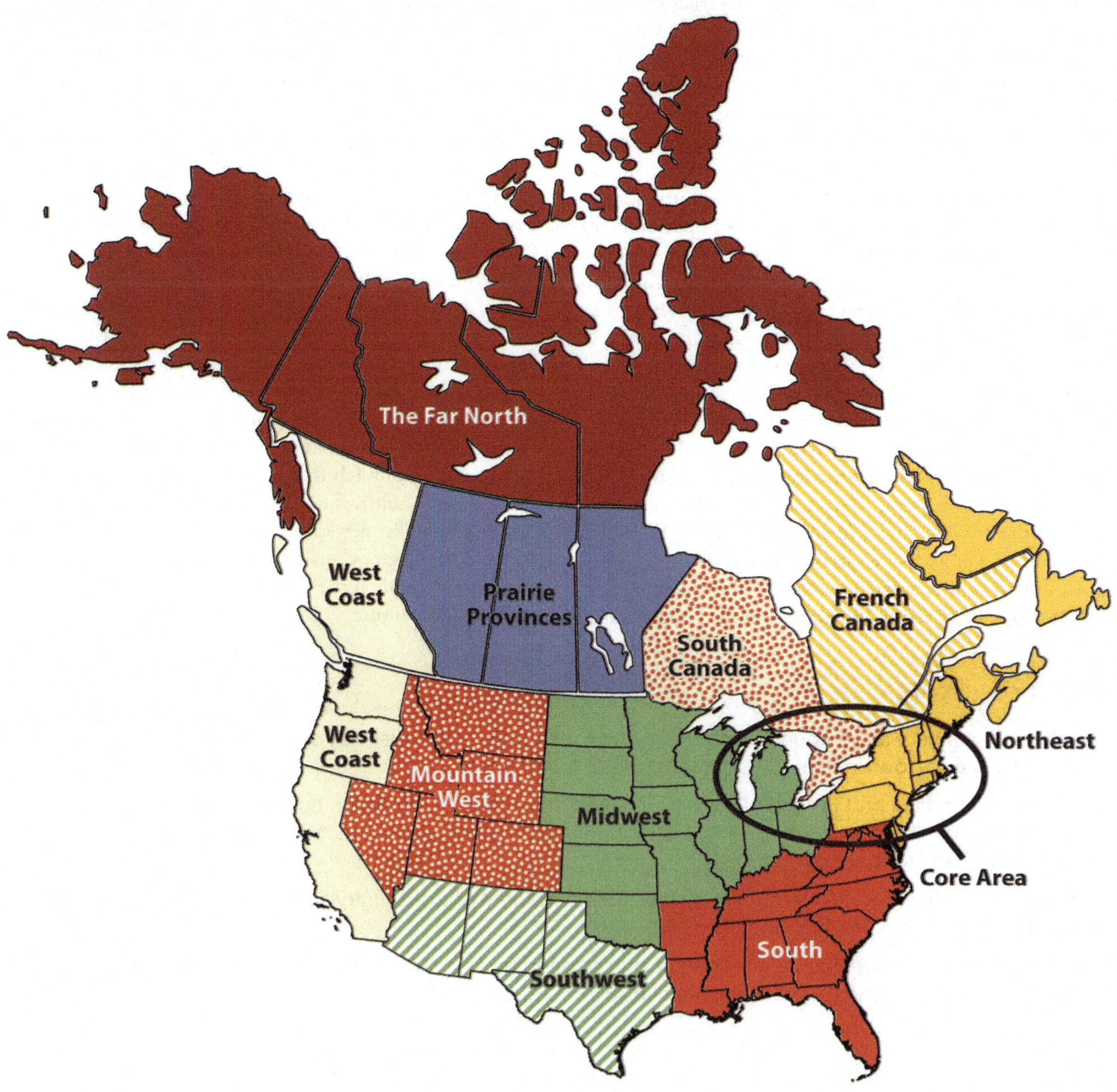

Figure 24. Main regions of the United States and Canada.

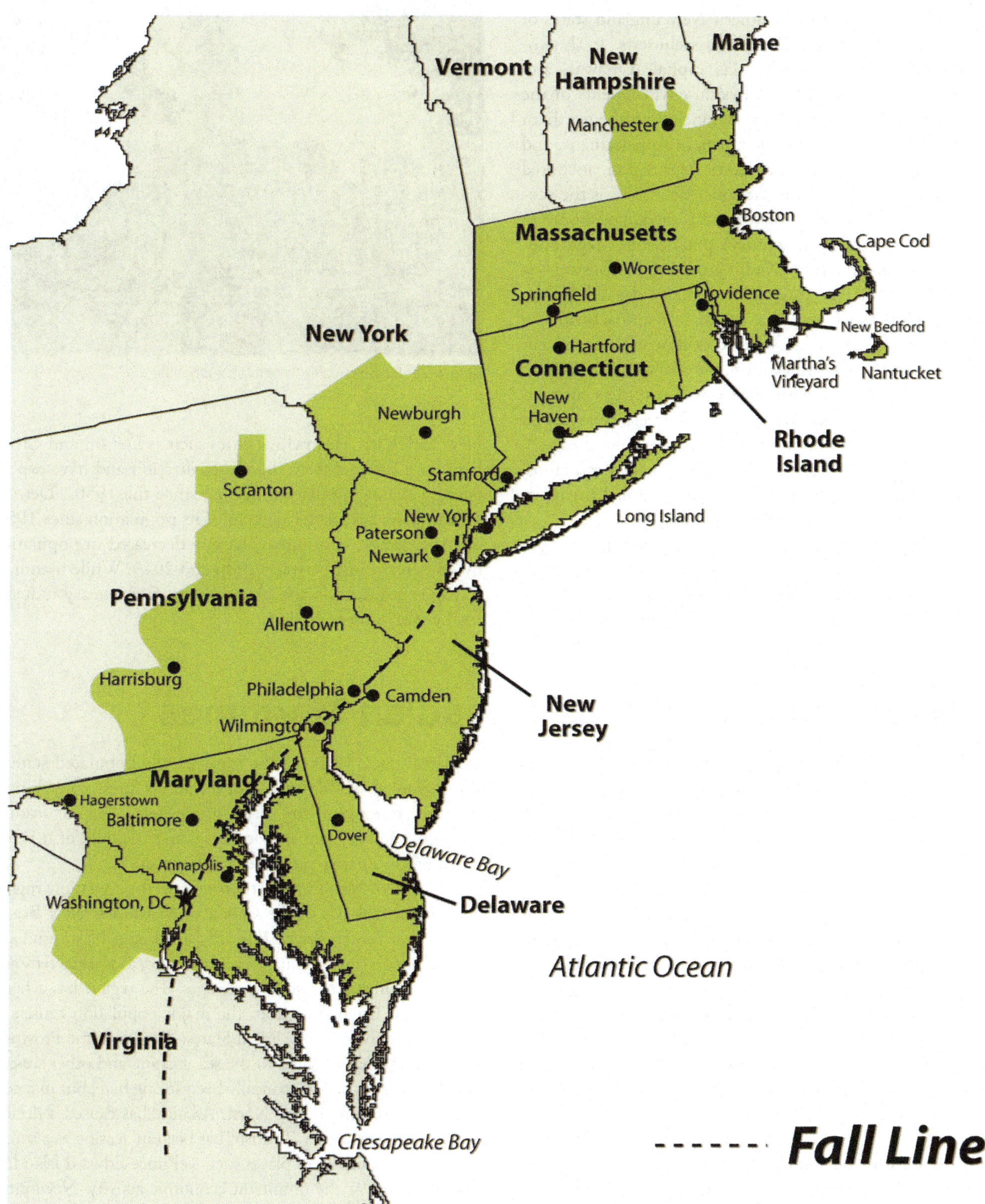

Figure 25. Megalopolis and the Fall Line.

A. The Northeastern Core

The Northeastern Core includes the upper Midwest (Illinois, Indiana, Ohio, and Michigan); the mid-Atlantic states of Pennsylvania, New York, Maryland, and New Jersey plus northern Virginia; and the southern New England states of Connecticut, Rhode Island, and Massachusetts. It also includes southern Ontario, Canada's capital (Ottawa), and its largest city (Toronto). The physical environments of the Northeastern Core are quite diverse, including the northern Gulf-Atlantic Coastal Plain, the northern Appalachians, and the area surrounding the Great Lakes. This region, anchored by North America's largest metropolis, New York, is the economic heart of the United States and Canada and home to more than a third of each country's population. **Megalopolis**—the built-up area from Washington, DC, to Boston—is part of this region. The core region contains the old **Manufacturing Belt**, which was once the main manufacturing region for North America but suffered decline with the advent of the information age (hence the name change to **Rust Belt**). The core region hosts the headquarters of countless corporations, banks, financial markets (e.g., Wall Street), colleges and universities (from community colleges to the Ivy League), cultural institutions (e.g., Broadway, world-class museums, dance and music organizations), and even global organizations such as the United Nations.

This large region includes geographic swaths of both wealth and economic suffering. Eight of the United States' ten wealthiest counties are in this region, most of them in the Washington, DC, area, and a number of billionaires live in

Figure 26. Downtown Philadelphia's city hall.

New York City. Meanwhile, cities such as Detroit and Cleveland have suffered from deindustrialization and have experienced a major population decline since the 1950s. Detroit, for instance, has lost 61 percent of its population since 1950, and the decline continues. The city decreased in population by 25 percent just between 2000 and 2010. While manufacturing is not dead in the Northeastern Core, heavy industry has been in long-term decline.

B. New England and the Canadian Maritimes

New England and the Canadian Maritimes overlap with the Northeastern Core because its major city—Boston—is considered the northern edge of Megalopolis. South of Boston, the low-lying states of New England were the center of colonial settlement in the region and were the birthplace of America's Industrial Revolution. Southern New England began as an agricultural and fishing colony, and as industry developed in the nineteenth century, the region attracted European immigrants from Ireland, Italy, and elsewhere to work in its factories. Today the region has a more diverse economic base, including recreation and tourism, finance, telecommunications, and health care. The mountains of western New England have been particularly attractive for the development of ski resorts, and the coasts of New England are popular for summer vacationing.

As you move north from Boston, the terrain becomes more rugged and the soil less fertile. There are fewer economic reasons for people to live in northern New England, and the states of Vermont, New Hampshire, and Maine have always been less densely populated than the southern New England states. Maine is the least densely populated state in this region; about 90 percent of its land is forested, making it the most forested of the fifty United States. Maine's leading economic activity is manufacturing, and the bulk of it is oriented around paper and other wood products.

Northern New England transitions to the even more rugged and remote uplands of the Canadian Maritimes: New Brunswick, Nova Scotia, Prince Edward Island, and Newfoundland and Labrador. The soil quality is quite poor, as glaciers removed most of it during the various ice ages. The region has a harsh climate and is removed from the major population centers of Canada and the eastern United States. The Maritime Provinces have always been oriented to the sea. Fishing and other sea-oriented businesses have historically been strong here, but in recent decades overfishing of the North Atlantic has caused a decline in the fishing economy. Tourism has been increasing as a source of revenue, especially in places such as Prince Edward Island, in which tourism is the dominant economic activity. Newfoundland and Labrador is Canada's poorest province.

C. French Canada

West of the Canadian Maritimes lies the province of Quebec, the heart of which is the St. Lawrence River valley, a lowland separating the Appalachian Mountains to the south from the inhospitable Canadian Shield to the north. As explained earlier, France was the first European country to colonize the coastal regions of what is now Canada, the St. Lawrence River Valley, most of the land surrounding the Great Lakes, and the Ohio and Mississippi River valleys, south to the Gulf of Mexico. Although Great Britain obtained all that land from France in 1763 following the French and Indian War, enough French inhabitants occupied part of that territory that the region did not automatically become English speaking. The core of French Canada today is the St. Lawrence Valley from Montreal to the Atlantic coast and west of Montreal to Ottawa and north to the Hudson Bay. These French speakers, the descendants of the early French settlers, created a vibrant

French-Canadian culture. About 21 percent of Canadians speak French as their mother tongue, including about 80 percent of Canadians living in the province of Quebec.

Throughout most of its history, the people of Quebec have been rural farmers, eking out a living on less-than-ideal land in a place with a short growing season. One unique characteristic of the farms in French Canada is their size and shape. Early on, the farms were laid out as long lots, maximizing the number of farms that would have access to the transportation artery—usually a river, but sometimes a road. Each farm was about ten times longer than it was wide and had a small access point to the river, some fertile riverfront land, and a woodlot at the rear of the farm. This land-use pattern was common throughout French Canada and can even be seen today in the United States in former French colonies such as Tennessee and Louisiana.

Since then, Quebec's economy has developed to include a manufacturing sector (fueled by abundant hydroelectricity), tourism, and a variety of tertiary and quaternary industries. Montreal, Canada's second-largest metropolitan area with 3.9 million residents, is the largest French-speaking city in the Western hemisphere. It developed as the region's most important city in the mid-nineteenth century, as it controlled access through the St. Lawrence River and the Great Lakes. It became a diverse industrial center, with oil refineries, steel mills, flour and sugar refineries, and shop yards for railroad companies. Montreal attracted **Anglophones** (English speakers) as well as the local **Francophones**, and at times in its history it has even had more English speakers than French speakers, despite being surrounded by a Francophone countryside. Most of northern Quebec is sparsely populated because of the lack of quality soil for agriculture, but a paper and pulp industry based on its forests has developed over the twentieth century, as well as hydroelectric power generation.

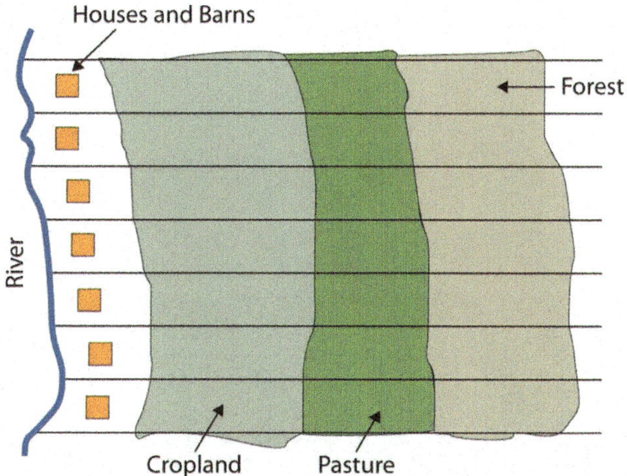

Figure 27. Long lot farms typical in French Canada.

D. The American South

The South includes the entire southeastern portion of the United States from Kentucky south to Louisiana, east to Florida, and north to Virginia. The South consists of most of the Gulf-Atlantic Coastal Plain and the southern portion of the Appalachian Highlands.

Before the Civil War, the coastal plain was dedicated to plantation agriculture using African slave labor. Land not used for plantation crops such as tobacco, cotton, and rice was typically farmed by poor whites and later by poor blacks. Some were sharecroppers, while others farmed their own small plots, especially on the lesser-quality land in Appalachia. The South had little urbanization or industrialization at

Figure 28. The Dixie Grill restaurant in Morehead, Kentucky. Names often reflect the cultural region of their location. Business names with "Dixie" in them can be found throughout the South.

the time of the Civil War. Well into the twentieth century, the region remained rural and economically deprived.

Coal mining was a major source of employment in places such as West Virginia and eastern Kentucky for the first two-thirds of the twentieth century, but increased mechanization of mining methods, as well as new mining techniques such as mountaintop removal mining, decreased the number of miners needed, even as coal production increased.

The Appalachian South is perennially plagued by high unemployment, poverty, and difficult social conditions. Other areas of the rural South are also among the poorest in the nation, including the Mississippi Delta and the lower Mississippi River valley. Despite the continued swaths of poverty in the South, parts of the region have prospered in the past generation as Sun Belt migrants have moved to southern places such as Atlanta, Charlotte, Tampa, Miami, and dozens of smaller cities. This has fueled a period of urbanization and economic growth, and the newfound prosperity has helped integrate the South into the nation's economy.

E. The Midwest and Great Plains

The center of the continent contains a relatively level agricultural region: the Midwest and the Great Plains, which includes the southern part of Canada's Prairie Provinces. This land includes some of the most fertile agricultural land in the entire world and has been dubbed **America's Breadbasket**. The climate gets progressively more arid as you move to the west within this region, and the type of agriculture changes with the decrease in precipitation. Closer to the Rocky Mountains, the land is typically used for raising cattle, but enormous grain farms are found where water is available (especially through irrigation). The water for irrigation comes from the continent's largest aquifer, the **Ogallala Aquifer**. Water is often pumped to the surface using a system called center pivot irrigation. The heart of the spring wheat belt is North Dakota, and the crop is also common in eastern Montana and in Canada's Prairie Provinces of Alber-

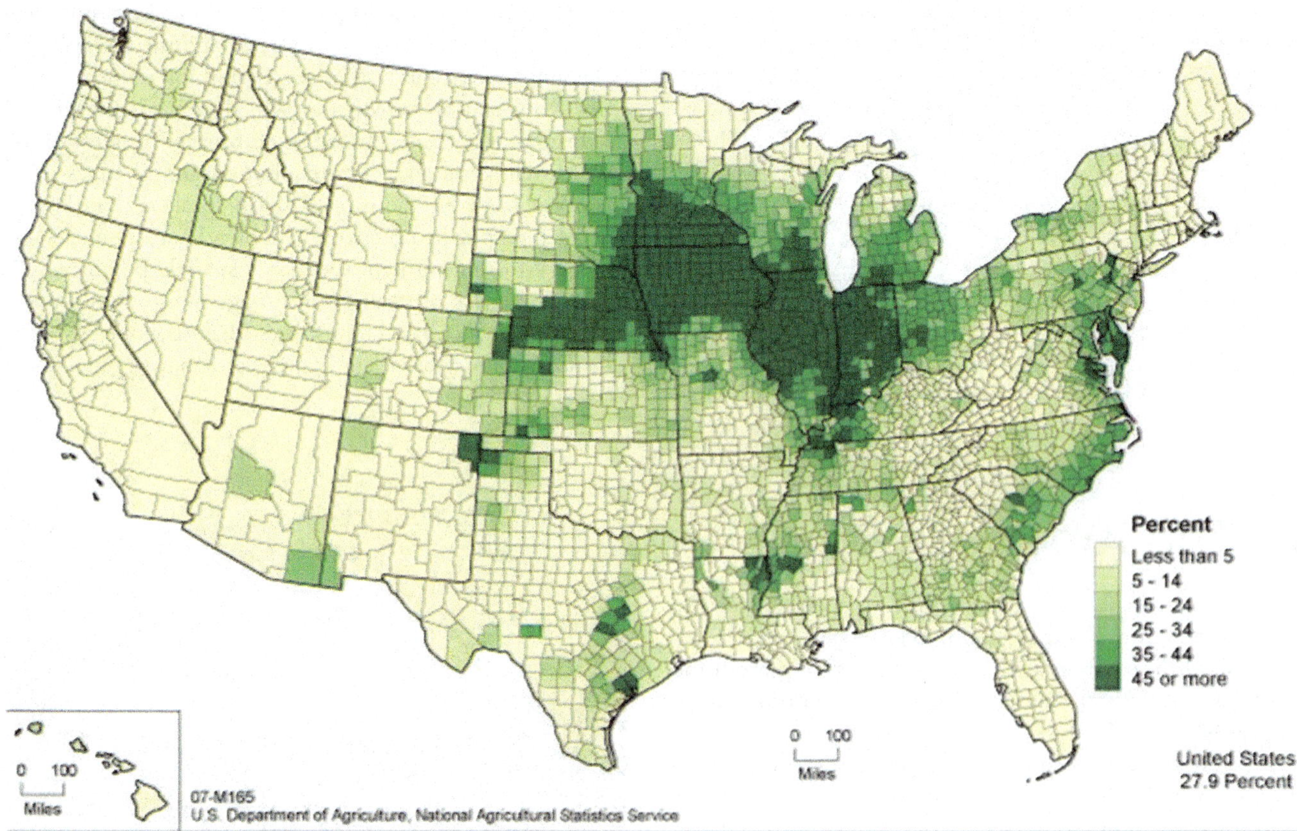

Figure 29. Acres of corn harvested for grain as a percentage of harvested cropland acreage, 2007. This map shows the extent of the Corn Belt.

ta, Saskatchewan, and Manitoba. Winter wheat is common in Kansas and surrounding states. Farther to the east, where precipitation is more abundant, is the **Corn Belt**, focused on Iowa and Illinois.

The dominant city in this region is Chicago, which developed as a market town for the livestock and grain produced in the surrounding states and was linked to its hinterland through a complex network of rail lines. In fact, nearly all the major cities of this region developed as places for the buying, selling, and processing agricultural products. Today the Midwest and the Great Plains remain the most important food-producing areas in North America, although as agriculture has become increasingly mechanized and farms have gotten larger, the number of farmers has decreased. This region, especially the Great Plains, is experiencing a period of long-term population decline and aging.

Figure 30. Satellite image of corn, sorghum, and wheat fields in Southwestern Kansas. The circular fields are between a half-mile and mile in diameter, and are characteristic of fields that use center-pivot irrigation.

F. The American Southwest

The states of Texas, New Mexico, and Arizona are considered the Southwest. The climate of the Southwest is more arid and receives a high amount of sunlight throughout the year. Desert conditions are integrated with higher elevations in the mountainous areas. Eastern Texas receives more rainfall from the Gulf of Mexico, and western Texas and the states of New Mexico and Arizona are quite arid and receive less rainfall. These conditions are more favorable to cattle ranching than to other agricultural activities. Large farming operations exist where water is available for irrigation. The warmer climate has been attractive for development and people emigrating from the colder regions of the north.

The Southwest has a strong Hispanic heritage and was part of Spain's Mexican colony. According to the 2010 census, about 46 percent of New Mexico's population was His-

panic or Latino, and in Texas and Arizona the figures were 37 percent and 30 percent, respectively.

The Southwest also has a strong Native American presence, especially in New Mexico and Arizona. Twenty-one federally recognized tribes with more than 250,000 people (five percent of the state's population) live in Arizona, and their reservations and traditional communities make up more than one-fourth of the state's land. The Navajo tribe is the largest in the United States, with more than 100,000 members in Arizona alone and others in surrounding states. The considerable Native American and Hispanic population in the Southwest means that non-Hispanic whites make up a minority of the population in New Mexico and Texas.

The three Southwestern states have been recipients of Sun Belt migrations over the past few decades, as people have moved to the Southwest for tertiary-sector jobs and for the re-

Figure 31. Typical home in a Phoenix suburb, where water is a valued commodity. There is no grass to mow, and cactus and palm trees are common.

Figure 32. Navajo dancer performing the Eagle Dance. One of the pollen trail dancers, a Navajo group near Joseph City, Arizona, performs the Eagle Dance on September 21, 2010, on the scenic South Rim of Grand Canyon National Park.

gion's warm climate. The region is quite urbanized, and most of the new migrants are moving to cities. Three-fourths of Arizona's population lives in the Phoenix or Tucson metropolitan areas. The most populous metropolitan area in the Southwest is Dallas–Fort Worth, Texas, with 6.4 million residents in 2010, making it the fourth-largest metropolitan area in the United States. The Houston metropolitan area is not far behind with 6.1 million residents. The economy of Texas used to be based on oil and natural gas, but it has since become more diversified. Residents of these cities work in high-tech manufacturing, health care, business, and information. One of the most famous high-tech industries in Texas is space: Houston is home to the National Aeronautics and Space Administration's (NASA) Lyndon B. Johnson Space Center, where astronauts and thousands of others work in the space industry.

G. The Mountain West

From the Rocky Mountains to the Sierra Nevada and Cascades and the Intermontane (literally, "between the mountains") Basins and Plateaus in between, this part of North America has gone from the old Wild West to an area of rapid economic and population growth. The region encompasses western Colorado; western Wyoming; western Montana; Idaho; Nevada; Utah; the eastern portions of Washington, Oregon, and California; and the southern portion of the Canadian Rockies.

The population of the Mountain West is growing much faster than the population of the United States as a whole. For example, Nevada's population grew 32.3 percent from 2000 to 2009, which is more than three times as much as the United States as a whole (9.1 percent). All US states in this region, except for Montana, grew at faster rates than the US average. What is fueling this growth? It is part of the larger pattern of Americans flocking to the Sun Belt, searching for an attractive climate and lifestyle. Jobs have been created in recreation (gambling, skiing), in high-tech firms, and in other tertiary sector industries. Many of the migrants come from southern California because the housing in the Mountain West is more affordable and the region is much less crowded. Nearly all the growth is occurring in urban and suburban areas.

However, the rapid growth of the West since 1990 has come at a cost. In some areas the large population is putting a strain on physical resources, such as water. Water is a hot-button political issue, particularly in the more arid states such as Nevada. Las Vegas, for example, is a desert city that gets 90 percent of its water from a Colorado River reservoir: Lake Mead. This water comes from snowmelt in the Rocky Mountains far to the east, and due to drought and high water demand, Lake Mead's water level has been dropping. If current patterns persist, Las Vegas will have a water crisis soon. The water shortage is happening even though Las Vegas has managed to reduce per capita water usage by raising prices for water and creating incentives to remove grass lawns. Las Vegas recycles 94 percent of all sewage water, which is the highest rate in the United States.

Figure 33. Rocky Mountains of Western Montana on US Highway 2.

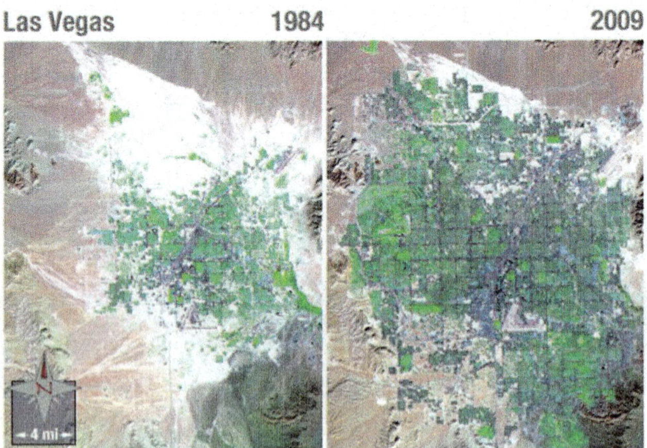

Figure 34. Urban Growth in Las Vegas, Nevada, from 1984 to 2009. Notice the dark purple of city streets and the bright green of irrigated vegetation.

H. The West Coast

The West Coast includes the coastal portions of California, Oregon, and Washington, plus the southwestern portion of British Columbia in Canada. This region is typically thought of as two subregions: California and the Pacific Northwest. The two areas are quite different from each other in terms of climate and economy. However, both areas are part of the so-called Ring of Fire that encircles the Pacific Ocean. The Ring of Fire is a zone of earthquakes and volcanoes that occur near where the Pacific tectonic plate meets the surrounding plates. In the United States, two areas of concern are the San Andreas Fault in California and Mount St. Helens in Washington. The 1906 earthquake that destroyed San Francisco was a result of activity on the **San Andreas Fault**, and scientists predict that strong earthquakes will reappear along the fault in the future. Thousands of small earthquakes occur along the fault every year. Mount St. Helens is a volcano in the Cascades that erupted in 1980, killing fifty-seven people and destroying hundreds of square miles of forest.

The West Coast represents a large population center a continent away from what we consider the North American core. Most of the region's population is urban, and Los Angeles and its metropolitan area is by far the largest area of settlement. Twelve percent of the US population lives in California (thirty-nine million people), and the greater Los Angeles metropolitan area has over eighteen million people. Los Angeles is the second-largest US city after New York. Los Angeles is the quintessential automobile city. It developed into a major

city in the mid-twentieth century at the time that automobile ownership had become common, and people who lived in the area tended to move to suburbs that were connected to each other by an extensive highway system. Los Angeles is a highly decentralized city, with more urban sprawl than cities in other parts of North America that formed during other transportation regimes.

The West Coast region is also famous for its agriculture. California's Central Valley lies between the Coast Ranges to the west and the Sierra Nevada to the east and is among the most productive agricultural areas in the world. The irrigated farmland in the valley produces all types of nontropical crops and is the largest US producer of tomatoes, grapes, almonds, and other foods. When other parts of the country are still frozen in the winter months, the fields of the Central Valley are already producing bountiful harvests. California is also famous for its wine production, especially in Napa Valley near San Francisco.

Besides agriculture, the economic base of the West Coast is quite diverse and rich. If it were an independent country, California would be the world's sixth-largest economy. Los Angeles is considered the capital of the US entertainment industry, and other major industries include aerospace, manufacturing, and foreign trade. The port of Los Angeles is the busiest in North America, receiving shipments of goods from China and other Asian countries. Silicon Valley, near San Francisco, is a key area for high-tech research and Internet commerce. The Pacific Northwest is home to major corporations such as Boeing (whose headquarters recently moved to Chicago), Microsoft, Nike, and other famous companies such as Starbucks, Amazon, REI, T-Mobile, Costco, and Eddie Bauer. One of the richest Americans, Bill Gates, lives near Seattle.

Across the border to the north, Vancouver is Canada's third-largest metropolitan area with over two million residents. Vancouver is unlike any other city in North America. Nearly one-third of its residents are of Chinese origin, and more than half its population speaks a language other than

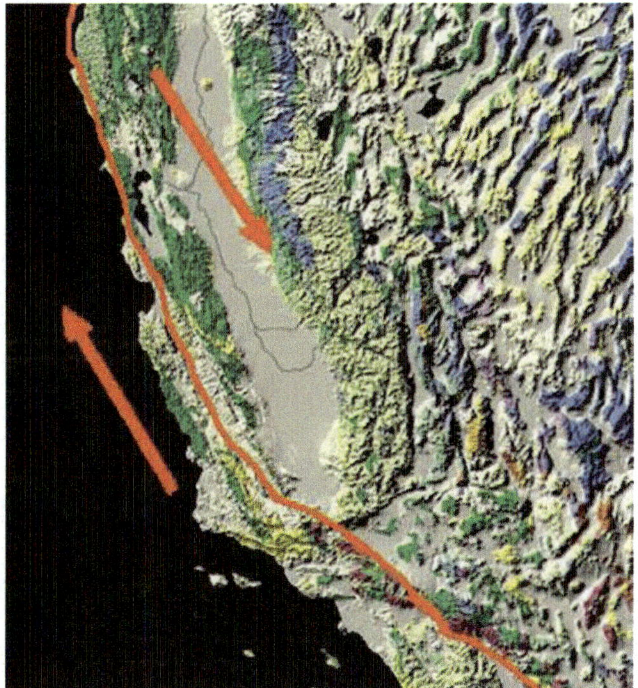

Figure 35. The San Andreas Fault. The Pacific tectonic plate is moving northward relative to the North American plate.

Figure 36. Vincent Thomas Bridge, San Pedro Harbor, Los Angeles, California.

English at home. Vancouver began as a logging town but developed into its position as the Asian gateway to Canada because of its port, the busiest in Canada. Vancouver is a popular location for the film industry and is sometimes dubbed as "Hollywood North." It is also growing in the biotechnology and software industries.

I. The Far North

The Far North is the least densely populated of any region in North America due to its brutally cold winters, short growing season, and poor soils. It includes the boreal forests of the upper Great Lakes region and the Canadian Shield and the territory to the north of the tree line that extends beyond the Arctic Circle. Physically, this region is immense, including the state of Alaska plus most of Canada. The climate is similar to that of Russia: cold continental and arctic climates, arctic air masses swooping down from the north, and long winters. Most inhabitants of the northern portions of North America live in the forested areas rather than in the frozen Arctic.

Two groups of people live in this region. First are the native peoples who have always lived there. They are small in number and traditionally make a living by hunting and fishing. More recently, the native populations such as the Inuit and the First Nations in Canada subsist by combining wage employment with their traditional means of living off the land. American Indians or Alaskan natives make up about 15 percent of Alaska's population, for a total of roughly 106,000 people. In Canada's Northwest Territories, First Nations people make up just over half the population, but the total population is quite small—only about 41,000 in the entire territory. In Nunavut, the native population is about 85 percent of the total 30,000 residents, living in a territory the size of Western Europe.

Figure 37. Radar Station at Point Lay, Alaska.

The other residents are more recent immigrants who are there to exploit the land's natural resources. The economy is dominated by the primary economic sector: forestry, oil and natural gas extraction, and mining. In the Canadian Shield, metallic ores such as copper, gold, nickel, silver, and uranium are found in the rocks and diamond mines are in operation, as are mines producing rare earth elements used in computer screens, electric car batteries, and computer hard drives. These elements include metals such as cerium, terbium, dysprosium, and neodymium. Alaska is an oil-producing state, and the decision of whether to open additional areas of Alaska's Arctic to oil drilling remains controversial and uncertain.

Key Takeaways

1. The economic core of North America has traditionally been in the US Northeast and its surrounding territory.

2. As you move north from Boston into the rest of New England, the Canadian Maritimes, and Quebec, the economy is increasingly based on primary industries such as forestry and fishing.

3. Agriculture and mining have been in decline in the South, while tertiary and quaternary industries have attracted new migrants to the region's urban and suburban areas.

4. The Midwest and the Great Plains make up North America's breadbasket. The climate gets more arid as you move west, but thr-ough irrigation, agricultural productivity remains high.

5. The Southwest is unique in its high proportion of Hispanic and Native American residents.

6. The Mountain West is growing rapidly, especially in its urban and suburban areas. This is putting stress on the physical environment (particularly its water resources) and made the region susceptible to the real estate bubble collapse.

7. The West Coast region is prone to earthquakes and volcanic activity, yet it is home to the second-largest metropolitan area in the United States and is known for its rich, diverse economic base.

8. The far northern stretches of North America are sparsely populated, with an economy based on primary industries such as forestry and extraction. The Far North, both Canada and Alaska, is also known for its large native populations.

End-of-Chapter Summary

- The United States and Canada are two countries with a great deal in common: their large territories, their histories of European colonization, their immigrant populations, and their high standards of living.

- Both the United States and Canada are becoming less European as immigrants arrive from outside Europe. In the case of the United States, the largest group of immigrants is from Latin America. For Canada, the largest group of immigrants is from Asia.

- The United States and Canada are both countries with a small native population, although in Canada native people have achieved more self-representation than in the United States, especially since the creation of Nunavut.

- Quebec, the French-speaking heart of Canada, has struggled for years to maintain its cultural uniqueness without risking its economic well-being.

- Both countries are postindustrial, with service- and information-oriented economies. The United States is the world's largest economy, and it has a history of spreading its culture, ideas, and military prowess around the globe.

- North America is made up of various regions with distinct cultural or physical features. Each region has majority and minority populations based on immigration or native heritage. Economic conditions vary from region to region. The Sun Belt is attracting an ever-growing number of information-based high-tech firms.

Chapter 5

Australia and New Zealand

5.1 Introducing the Realm

A. Isolation Geography

Australia and New Zealand have flora and fauna that are found nowhere else on Earth. Australia is distinctive because it is an island, a country, and a continent—the smallest of the world's continents. No other land mass can concomitantly make those three claims. Australia consists of a large mainland and the island of **Tasmania** to the south. The main physical area of New Zealand, on the other hand, consists of two main islands separated from Australia's southeastern region by the **Tasman Sea**. Australia is surrounded by various seas. The Indian Ocean surrounds its western and southern coasts. Indonesia and Papua New Guinea lie to the north, separated by the **Timor Sea** and the **Arafura Sea**. The **Gulf of Carpentaria** distinguishes **Cape York**, which extends north along Australia's eastern coast almost to Papua New Guinea. The **Great Barrier Reef** runs for

more than 1,600 miles off the continent's northeastern shores. The Coral Sea separates the Great Barrier Reef from the South Pacific. The southern side of Australia is the **Great Australian Bight** and the island of Tasmania. A bight is a large, wide bay. To the south of Australia and New Zealand is Antarctica. The two countries have distinct physical geographies. Australia is relatively flat with low elevation highlands and an extensive dry interior, while New Zealand has high mountains and receives adequate rainfall.

The historic isolation of New Zealand and Australia from the rest of the world has caused animals and organisms that are not found anywhere else to develop in these two countries. The unique biodiversity includes **marsupials**, or animals whose young are raised in the mother's pouch, such

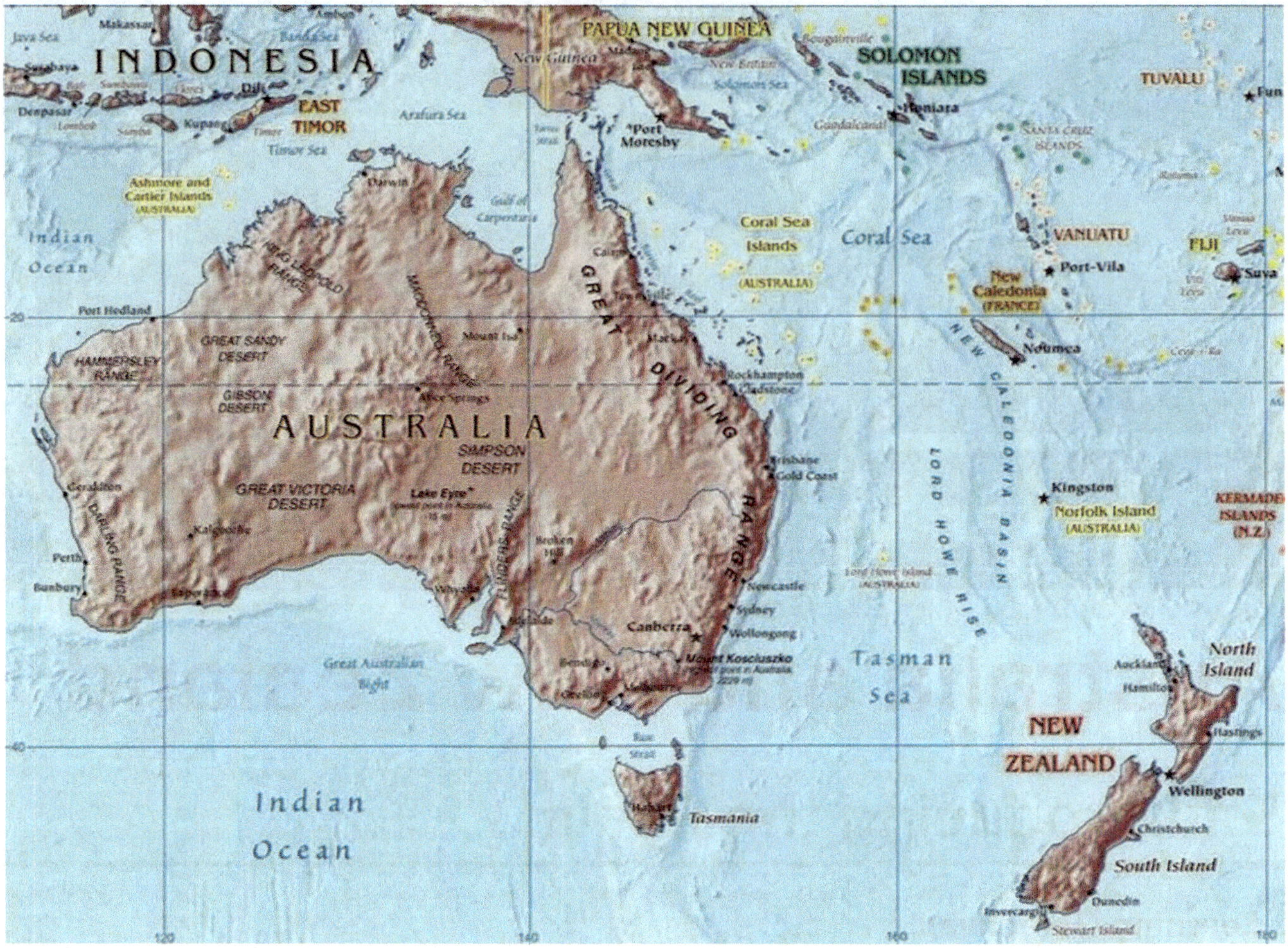

Figure 1. Australia and New Zealand. The tropic of Capricorn runs through the middle of Australia. The Tasman Sea seperates Australia from New Zealand.

as kangaroos, wallabies, koalas, and bandicoots. It is believed that these creatures developed separately after the continents broke away from each other more than two hundred million years ago. Many plant species are also unique to this realm. The biodiversity found here is separate from that of Asia. This has been explained by various biogeographers by drawing imaginary lines just north of Australia to indicate the line of division between the Asian realm and the Austral realm. **Wallace's Line** and **Weber's Line** are two such examples. Both examples attempt to establish the correct line of demarcation for the differences in species development between the two sides. During the ice ages, sea level was lower, and the many islands of Southeast Asia were connected by land to the mainland. Papua New Guinea was connected to Australia. Wallace and Weber believed that no land bridge connected the Asian side with the Austral side for animals to cross over. This separation caused the organisms to the south to develop independently of those in the north. For example, marsupials are not found on the Asian side of these lines but are found on the Australian side.

Figure 2. Wallace's and Weber's Lines were developed independently to account for the differences in biodiversity between the Austral realm and the Asian realm. Scientists continue to analyze the true boundary between the realms. These lines demarcate a clear environmental difference in species development between the two sides.

B. Colonialism

New Zealand and Australia were both inhabited before the era of European colonialism. For thousands of years before the Europeans arrived, the **Aborigines** carved out an existence in Australia and developed their cultural ways. Only about four hundred fifty thousand Aborigines remain in Australia today (about two percent of Australia's total population). New Zealand was inhabited by the Polynesian group called the Maori who established themselves on the islands in the tenth century. For hundreds of years they, too, established their culture and traditions in the region before the Europeans arrived. The Aborigines in Australia and the **Maori** in New Zealand were both confronted with the European invaders. From their standpoint, there was much to lose by the arrival of the Europeans. Lands were lost, new diseases killed many, and control of their methods of livelihood were taken over by Europeans. The Maori initiated a number of wars against British colonizers, but in the end the greater military power gained the advantage. At the present time, the Maori make up less than 10 percent of the population of New Zealand.

James Cook, a naval officer working for the British navy, mapped Australia's eastern coast in 1770. He made port at Botany Bay, just south of the current city of Sydney and claimed the region for Britain. He named the land New South Wales. The charting of the coast resulted in continued attention being paid to the region.

Meanwhile, England had a severe problem with overcrowding of its prisons. Its problem was exacerbated by the loss of Britain's American colonies. Upon Cook's return to England, interest was generated in the concept of relieving prison overcrowding by sending prisoners to Australia. In 1787, eleven ships with seven hundred fifty convicts sailed

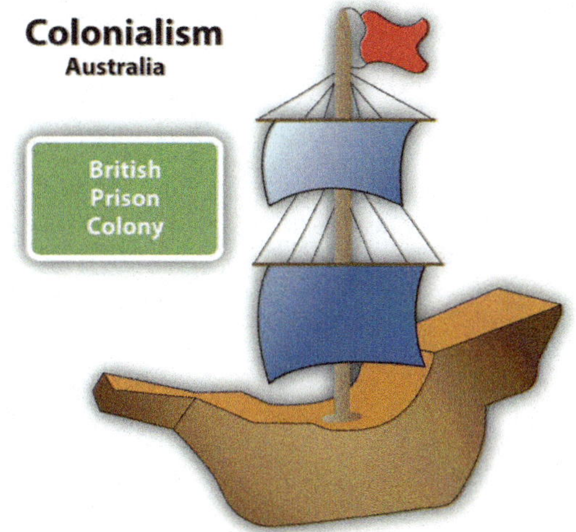

Figure 3. Australian Colonialism
Great Britain colonized Australia by establishing prison colonies. The prison colony of Botany Bay was located near the current city of Sydney, Australia.

from Great Britain to Botany Bay. Prison colonies were established in Australia. By the end of the eighteenth century, the entire Australian continent was under the British Crown. At the same time that the movement of prisoners from England to Australia was diminishing, the next wave of immigration was being fueled by the discovery of gold in the 1850s. The practice of transferring prisoners to Australia ended in 1868. The arrival of the Europeans had caused a serious demise in the Aboriginal population. Aborigines were completely decimated in Tasmania.

In 1901, the various territories and states of Australia came together under one federation called the Commonwealth of Australia. A new federal capital city of Canberra was proposed. By 1927, Canberra was ready for government activity. This commonwealth government still allowed for individual state differences. The British monarch is considered the head of state, though it is mainly a ceremonial position. There have been movements within Australia in recent years to separate from the British Crown, but they have not been approved. Australia has a democratically elected government.

British naval officer James Cook mapped the coastline of New Zealand in 1769. As the colonial era emerged, Great Britain took possession of New Zealand and included it with its colony of New South Wales. In the 1840s, New Zealand became a separate crown colony. The colony developed a local parliament and a representative government. By 1893, New Zealand made headlines as the first country in the world granting all women the right to vote. As a part of the British Empire, the country was made a commonwealth nation in 1947 and has been functioning independently ever since.

Key Takeaways

1. Australia is an island continent that was home to aboriginal people who have lived there for thousands of years. The British colonized Australia by first creating prison colonies for convicts from Great Britain.

2. New Zealand has two main islands and is home to the Maori, who were originally from Polynesia. The British colonized New Zealand and often were in conflict with the Maori.

3. Australia is relatively flat with low elevation highlands and an extensive dry interior, while New Zealand has high mountains and receives adequate rainfall.

4. The Austral realm was isolated by physical geography. Weber's Line and the Wallace Line were attempts to distinguish the location of the separation between biological environments.

5.2 Australia

Learning Objectives

1. Summarize the colonial exploitation and development of Australia.
2. Understand the basic characteristics of Australia's physical geography.
3. Outline how the core-periphery spatial relationship applies to Australia.
4. Describe the country's general cultural attributes.
5. Summarize the methods used for the country to gain national wealth.

A. Physical Geography

There is an international attraction to the island continent of Australia, and the attraction has grown in intensity in the past few decades. Tourism is now the number one economic activity in Australia. Just slightly smaller in physical area than the contiguous United States, Australia is a large country with many resources but few people relative to its size. The Tropic of Capricorn runs right through the middle of this country. Australia hosts many unique species of plants and animals, including marsupials and a host of poisonous snakes and insects. With the advent of European colonialism, new species were introduced to the country, which regrettably caused the extinction of some of the native species but also gave Australia a wide diversity of organisms and natural conditions.

Australia is a relatively low-lying island with low relief. It is the flattest of all the continents. The various highland ranges are pronounced, but are not high in elevation. The Great Dividing Range is a mountain chain extending from Melbourne in the south to Cape York in the north. This low-lying range of highlands averages about four thousand feet and reaches an elevation of just over seven thousand feet at its highest peaks in the south. The largest river in Australia is the **Darling-Murray River** system that starts in the highland of the Great Dividing Range and flows inward through New South Wales, Queensland, Victoria, and South Australia.

The great interior of the country is home to the massive **outback**. Extending west from the Great Dividing Range, the outback encompasses most of the interior. This region receives less rainfall than along the coast and its terrain consists of deserts and semiarid plateaus with rough grasses and scrublands. The outback is sparsely populated, but is home to a number of aboriginal groups. Many of the school-age children in the outback have traditionally received their school lessons through television or radio broadcasts because of their isolation. Mining and some agricultural activities can be found in the outback. **Alice Springs** is located in the center of the continent and has been given the designation of the middle of nowhere, or the center of everything.

The deserts of Australia's interior make up a large portion of the continent. Western Australia has three large deserts: the **Gibson Desert**, **Great Victoria Desert**, and **Great Sandy Desert**. The **Simpson Desert** is located in the border region between the Northern Territory, Queensland, and South Australia. These deserts are not all sand; course grasses and various species of spinifex, a short plant that grows in sandy soil, also grow in the deserts. The **Great Artesian Basin** on the western edge of the Great Dividing Range receives very little rainfall. It would be classified as a desert but for its underground water resources, which support extensive farming operations. Large livestock businesses exist in Australia's interior with massive herds of cattle and sheep. The grassy plateaus and scrublands provide grazing for domesticated livestock and even wild camels.

The **Great Barrier Reef**, the largest barrier reef in the world, extends for 1,600 miles off the northeastern coast of Australia. It is home to a host of sea creatures and fish that draw millions of tourists each year. The reef attracts scuba divers and water enthusiasts from around the world. The reef is a main tourism attraction and brings income to the Australian economy. The Great Barrier Reef has been designated as

Figure 4. Alice Springs. The remote town of Alice Springs is located at the center of the Australian outback.

a United Nations World Heritage Site. **Brisbane** is located on the **Gold Coast**, which gets its name from the beautiful sandy beaches. The beaches attract an important tourism market for the country.

A couple of large physical features of interest and significance to Australia are the two largest **monoliths** in the world. In western Australia, more than five hundred miles to the northeast of Perth, is Mt. Augustus National Park, which features the rock known as Mt. Augustus. It is considered to be the largest single rock in the world. Mt. Augustus rises 2,352 feet above the desert landscape. The single structure is about five miles long. Mt. Augustus is more than twice the size of the most famous Australian monolith of **Uluru** (**Ayers Rock**). Uluru is located about two hundred miles southwest of Alice Springs in the Northern Territory and is a well-known tourist attraction. Uluru rises 1,142 feet above the outback and is about 2.2 miles long. Both rocks hold significant cultural value to the aboriginal populations in Australia. They both have ancient petroglyphs, and both are considered sacred sites. Uluru has been more popularized through tourism promotions.

Figure 5. Aerial view of Uluru (Ayers Rock), located in the interior of Australia near Alice Springs. The rock rises 1,142 feet above the outback and is 2.2 miles long. The monolith is sacred to the Aborigines and is a major tourist attraction. It is listed as a World Heritage Site.

Figure 6. Australia's Provinces and Territories and their respective major cities. The two core areas, where most of Australia's population resides, are also noted in blue. Not unexpectedly, the core areas have a dominant type C climate, following the general principle that humans gravitate toward type C climates.

B. Climate Regions

Central and western Australia are sparsely populated. Large areas of the Northern Territory and the desert regions are uninhabited. Approximately 40 percent of Australia's interior is desert, where Type B climates dominate. The large land mass can heat up during the summer months, triggering high temperatures. Low humidity allows heat to escape into the atmosphere after the sun goes down, so there is wide temperature variation between day and night.

Along the northern coastal region there are more tropical Type A climates. Closer to the equator and with the sea to moderate temperatures, the northern areas around **Darwin** and Cape York have little temperature variation. Temperatures in Darwin average about 90 °F in the summer and 86 °F in the winter. Spring monsoons bring additional rainfall from February to March.

Tasmania, Victoria, and the core region of the southeast have a more moderate and temperate Type C climate. The main cities, such as Sydney, Melbourne, and Adelaide, are within this area. It is not surprising that there is a direct correlation between Type C climates and the major population areas. The Tropic of Capricorn cuts across the continent, indicating that the cities are not that far south of the tropics. Average winter temperatures in June and July do not usually fall below 50 °F and average summer temperatures in January and February remain around 70 °F. Since the seasons are reversed from that of the Northern Hemisphere, many Australians go to the beach for Christmas.

C. Population, Urbanization, and the Core-Periphery Spatial Relationship

Australia is divided politically into six states and two territories. They are the Northern Territory, Australian Capital Territory, Western Australia, Tasmania, Victoria, South Australia, Queensland, and New South Wales. Australian protectorates are composed of a number of small islands around Australia. Australian core areas are conducive to large human populations. To locate the core population areas in Australia, simply find the moderate Type C climates. Australia has two core regions. There is a small core region in the west, anchored by the city of Perth. Most of Australia's people live in the large core region in the east along the coast. This region extends from Brisbane to Adelaide and holds most of the country's population.

The total population of Australia in 2010 was only about twenty-two million. There are more people living in Mexico City than in all of Australia. More than 90 percent of this population has European heritage; most of this percentage is from the British Isles. English is the dominant language. Christianity is the dominant religion. The makeup of the people is a product of European colonialism and immigration.

About 24 percent of the current population was born outside Australia; most come from the United Kingdom, and another large percentage comes from New Zealand. Asian countries have also contributed to the Australian population, with measurable numbers of immigrants from China, Vietnam, and the Philippines. And lastly, people from Italy and India also make up a notable proportion of Australia's immigrant population.

Australia's population is not spread evenly across the landscape, since a large portion of the country is desert. The population is concentrated mostly in the urban areas. About 90 percent of the population inhabits the cities, which are mostly in coastal areas. The largest city, Sydney, is often referred to as the New York of Australia. Sydney is positioned at the heart of the main core area, the state of New South Wales. To the south of Sydney is the Australian Capital Territory, home to the capital city of Canberra. Other major Australian cities include Melbourne, Perth, Adelaide, and Brisbane. Hobart is the largest city on the island of Tasmania and Darwin is the largest city in the Northern Territory.

All the large cities of Australia—except the planned capital city of Canberra—are located on the coast. This pattern of urban distribution was a product of European colonial development. Most of Australia's population lives in the two economic core regions, so Australia exhibits a distinct core-periphery spatial pattern. The core areas hold the pow-

Figure 7. The Sydney Opera House viewed from the water with the city skyline.

er, wealth, and influence while the periphery region supplies all the food, raw materials, and goods needed in the core. Australia has never had a majority rural population since its Aboriginal times. There has been little rural-to-urban shift in Australia's population. This is similar to Japan's urban development pattern.

English is the first language of the vast majority of the population. Recently enacted policies and changing attitudes toward multiculturalism have spurred growth in the number of immigrants and their descendants who speak two languages fluently—English and the language of their birthplace or national heritage. Indigenous languages have not fared so well. As many as three hundred indigenous languages were spoken by Aborigines before the Europeans arrived, and just a few hundred years later, that number now stands at about seventy. Most aboriginal languages are in danger of dying out.

D. Culture and Immigration

Until 1973, Australia had a collection of laws and policies known as the White Australia policy, which served to limit the immigration of nonwhite persons to Australia. While the White Australia policies limited immigration from some areas, other policies sought to expand immigration from the United Kingdom. Subsidies were offered to British citizens to relocate to Australia. Between 1830 and 1940, more than a million British citizens took advantage of the offer.

Recent census data indicate that about a quarter of the population identifies itself as Roman Catholic and another 20 percent self-identifies as Anglican (the national religion of the United Kingdom). An additional 20 percent self-identify as Protestant, other than Anglican, and about 15 percent as having no religion. Regular church attendance is claimed by at about 7.5 percent of the population. Despite modern Australia having been settled by the British, Australian law decrees that Australia will have no national religion and guarantees freedom of religion.

Sports are an important part of Australian culture, perhaps owing to a climate that allows for year-round outdoor activity. About a fourth of the population is involved in some kind of organized sports team. Football (soccer) is popular, as is true in most European countries, and rugby and cricket are popular as well. The most popular spectator sport in Australia is Australian Rules Football, also known as Aussie Rules Football, or simply "footy." This uniquely Australian game has codified rules that date back to 1858 and is a variant of football and rugby. Other forms of entertainment include television, film, and live performances of every kind. Although Australia has a number of its own television stations, there are concerns that popular culture is beginning to be dominated by American influences. Australia's large cities have extensive programs in the arts. Sydney is becoming a center for world-class performances in dance, opera, music, and theatre.

E. Economic Geography

Most of Australia—especially the wide expanse of the arid interior known as the **Outback**—has immense open spaces, agricultural potential or excellent resource extraction possibilities. The extensive grasslands support tens of millions of domesticated animals—mainly cattle and sheep—which accounts for up to one-fifth of the world's wool production. Large agricultural businesses include thousands of acres under one operation. The western sector of the Great Dividing Range in New South Wales is an excellent region for commercial grain operations. The coastal region in Queensland, since it is warmer and receives more rainfall, is good for sugarcane and similar crops. Sheep and cattle ranches are common in central Queensland and Western Australia. Various regions of southern Australia are excellent for grape and fruit production. Australian wine production has risen to compete with the US and European markets. Only the dry central desert regions in the center of the continent are not favorable for agriculture. In the early portion of the twentieth century, Australia gained enormous wealth by exporting food products to the rest of the world. This is still true, but the profit margin on food goods is no longer what it used to be. The country has had to look elsewhere to gain wealth.

Australia has excellent food production capabilities. It also has an excellent mineral resource base. Different types of minerals can be found in different regions throughout Aus-

Figure 8. The agricultural region of the Barossa Valley in South Australia. The agricultural region of the Barossa Valley in South Australia grows grapes and produces wine. Agricultural production is a major source of economic wealth for Australia even though only 11 percent of the population lives in rural areas. Australia's wine production is expanding to compete in the global marketplace, with France and California as major competitors.

tralia. Western Australia has iron ore mines. The eastern region of Queensland and New South Wales has abundant coal reserves. Minerals such as zinc, copper, gold, silver, tungsten, and nickel can be found in various parts of the country, including Tasmania. Oil and gas fields can be found in the northwestern coastal waters and in the Tasman Sea east of Melbourne. The country is self-sufficient in natural gas but does have to import some petroleum products.

Are any Australian-manufactured products available where you live? What products can you think of? Australia does not export many manufactured goods. Its main exports are food and raw materials. If you remember how countries gain wealth, the method with the highest valued-added profit is manufacturing. Think about

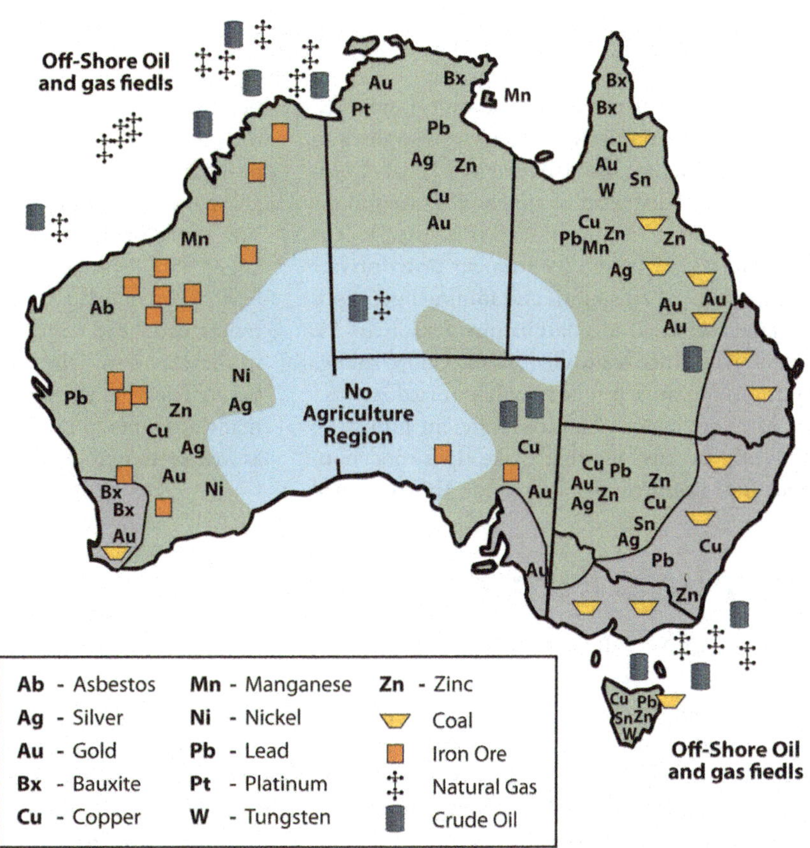

Figure 9. Australia's distribution of raw materials.

Japan and the four Asian economic tigers, and how they have gained their wealth. The economic tigers have few raw materials. Where do you suppose the economic tigers and Japan get their raw materials? With Japan's enormous manufacturing capacity, it has a high demand for imported iron ore, minerals, and raw materials. Though Australia is a former British colony, Great Britain is not considered Australia's largest trading partner. Australia is closer geographically to the Asian economic community than to the European Union. Japan has become Australia's biggest trading partner. When Australia is viewed in the news, in television programs, and in Hollywood movies, it is portrayed as a country with a similar standard of living to the United States or Europe. How do Australians have such a high standard of living if they don't manufacture anything for export? To evaluate this, think about the size of the population of Australia and consider the distribution of wealth. They export an immense amount of raw materials and have a relatively small population to share the wealth.

Australia is an attractive place to visit. The environment, the animals, and the culture make it inviting for tourism. As of the year 2002, tourism has become Australia's number one means of economic income. From the Great Barrier Reef and the Gold Coast to the vast expanse of the outback, Australia has been marketing itself as an attractive place to visit with

Figure 10. Bondi Beach.
The Gold Coast of eastern Australia draws tourists from the Northern Hemisphere throughout the winter season. It is called the Gold Coast because of the long stretches of golden sand beaches, the golden tanned bodies of beach goers, and the high level of income (gold) derived from the tourism industry.

great success. Tourism from Japan provides a large percentage of the tourist activity. Australia has moved through the initial stages of the index of economic development to become a society that is about 90 percent urban with small families and high incomes.

F. Mining and Aboriginal Lands

Territorial control of Australian lands has become a major issue in recent years. Large portions of western Australia and the outback have traditionally been Aboriginal lands. European colonialism on the Australian continent displaced many of the native people. Large sections of land once used by the Aborigines were taken over by the government or by private interests. Large agricultural operations and mining operations have used the lands without adequate compensation to the Aboriginal people who once controlled them. Court rulings aimed at reparation for native people have had mixed results.

There are as many as four hundred different groups of Aborigines currently in Australia that make up a total population of about four hundred fifty thousand. This is a small percentage of Australia's population but involves a large part of the physical area of the country. Their land claims include all of the Northern Territory, a large portion of western Australia, and parts of South Australia and Queensland. This is in addition to claims located within many urban areas, such as the largest city, Sydney. Mining operations on Aboriginal lands have become highly regulated. Concerns have arisen that Australia's extractive industries will diminish, causing a decline in the economy. The concern for the Aboriginal population has increased in the past few decades and the government has made attempts to mediate their political and economic issues as well as strengthen programs that address their social welfare.

G. Australia's Future

The economic future of Australia is complex. Though tourism has become a viable means of providing income, Australia must import manufactured products that it does not produce locally, including electronic goods, computers, and automobiles. Import dependence has increased its trade deficit. Trade agreements and protectionism have become a part of the economic puzzle of how to sustain a competitive standard of living. Australia is located next to the Asian realm. Its economy, culture, and future are becoming more Asian.

Immigration has been an issue in that the government has always restricted immigration to ensure a European majority. Millions of Asian people would like to migrate to Australia to seek greater opportunities and advantages, but they are legally restricted. It is becoming more difficult for Australians to hold to their European connections with such an Asian presence. How the country will handle this situation in the future will prove interesting.

Key Takeaways

1. Australia is an island continent that was home to aboriginal people before the British colonized it by first creating prison colonies for convicts from Great Britain.

2. Australia is a relatively flat continent with low elevation highlands, including the Great Dividing Range along its eastern coast. The interior outback lacks precipitation and has numerous deserts.

3. Two main core areas exist where Type C climates prevail and where most of the population lives: a large core area on the southeastern coast and a small core area around Perth on the western coast. The sparsely populated outback makes up the vast periphery, which has large amounts of mineral and agricultural resources.

4. Aboriginal people were in Australia for forty thousand years. The British colonial activity didn't heat up until the late 1700s. Today, most of the twenty-two million people are from the British Isles and Europe. Only about four hundred fifty thousand Aborigines remain.

5. Australia has few manufacturing enterprises for export profits. Tourism has become the number one method of gaining wealth, with the export of raw materials the second-largest method.

5.3 New Zealand

Learning Objectives

1. Outline New Zealand's main physical features. Understand how the North Island is different from the South Island.
2. Understand how tectonic plate activity has helped to isolate New Zealand from the rest of the world and still affects the island today.
3. Summarize the situation of the Maori in New Zealand. Learn about the Maoris' relationship with the dominant culture in New Zealand.
4. Describe the economic geography of New Zealand and how the country gains wealth.

A. Physical Geography

To the east of Australia across the Tasman Sea is the country of New Zealand. New Zealand is one of a number of sets of islands that make up Oceania, also referred to as the Pacific Islands, a region occupying the western and central Pacific Ocean. The Pacific Islands region is generally divided into three subregions: Micronesia, Melanesia, and Polynesia, with New Zealand being part of **Polynesia**. The Pacific Island region includes more than twenty-five thousand individual small islands representing twenty-five nations and territories. Most of these islands are very small. The South and North islands of New Zealand are the second- and third-largest islands, respectively. The North and South Islands of New Zealand are separated by a body of water known as the **Cook Straight**, which is only about thirteen miles wide at its narrowest point. The North and South Islands together are about the same size as the US state of Colorado. New Zealand also includes a number of smaller nearby islands.

New Zealand, like Australia, is in the Southern Hemisphere, which means that its seasons are the opposite of those in North America. In other words, the warmest summer months are January and February and the cooler winter months are June and July. New Zealand lies within the Temperate Zone. There are only very moderate seasonal differences, which are slightly more pronounced in the inland areas because the inland areas lack the moderating influence of the ocean. In general, the North Island has somewhat warmer average temperatures than the South Island. In summer, average low temperatures are about 50 °F, with daytime highs around 75 °F. In the winter months, low temperatures average about 35 °F and high temperatures are about 50 °F. The occurrence of more extreme temperatures is limited to the mountainous peaks of the Southern Alps. Snow is common in these mountainous regions but rarely occurs in coastal regions.

Rainfall is heaviest on the western coasts of both islands, but especially on the South Island. The prevailing westerly winds, carrying moisture from the ocean, come in contact with the mountains of the **Southern Alps** and high precipitation results. The mountains also have the opposite effect. On the eastern side of the mountains is a rain shadow where the westerly winds descend, warming and drying out. The eastern coasts are therefore substantially drier than the western coasts. The average annual rainfall in Christchurch, which is on the eastern coast of the South Island, is about twenty-five inches per year. Auckland, in the mid portion of the North Island, receives twice that amount and areas on the wetter western coast receive as much as one hundred fifteen inches per year.

As an island country, New Zealand's coastlines and oceans are some of its most important geographic features. New Zealand has one of the world's largest **exclusive economic zones**, an oceanic zone over which a nation has exclusive rights of exploration and exploitation of marine resources. New Zealand's exclusive economic zone covers more than one million square miles. The dramatic nature of New Zealand's landscape is well known to many moviegoers as the landscape of Middle Earth, as depicted in New Zealand film director Peter Jackson's version of J.R.R. Tolkien's *Lord of the Rings*.

Figure 11. Southern Alps, New Zealand.

Figure 12. New Zealand. Wellington is the capital, Auckland is the largest city in the country, and Christchurch is the largest city on the South Island. The Southern Alps extend along the western length of the south and reach elevations of twelve thousand feet.

B. Tectonic Plates and Mountains

The North Island of New Zealand features a rather rugged coastline with numerous harbors, bays, and inlets. The port cities of Auckland and Wellington are located on two of the largest bays. The coastline of the South Island is somewhat more regular, except along the southern portion of the eastern coastline, which has deep fjords. Though the North Island has lower elevations than its southern counterpart, its few mountains are volcanic in origin. The two main islands are accompanied by smaller islands around their shores. The North Island's highest peak is Mt. Ruapehu, which reaches almost 9,175 feet and is an active cone volcano. The volcanism associated with **Mt. Ruapehu** results from New Zealand's location atop two tectonic plates: the Pacific Plate and the Indo-Australian Plate. The boundary of these two plates forms a subduction zone under the North Island; consequently, New Zealand experiences tens of thousands of earthquakes per year. Though most of the earthquakes do not greatly disrupt human activity, some have registered higher than 7 on the Richter scale. New Zealanders have made use of the geothermal power generated by this and the tectonic features of this area and hence, New Zealand is home to several geothermal power plants.

A range of mountains, the Southern Alps, divides the South Island lengthwise. Many of the peaks reach over ten thousand feet. The highest peak is **Mt. Cook**, which reaches higher than twelve thousand feet. These mountains are also formed by the area's tectonic situation. However, the two plates meet in a different way under the South Island. Rather than creating a subduction zone, the plates move laterally. The lateral movement created the South Alps. New Zealand's location places it along one of the edges of the Ring of Fire, which encircles the Pacific Ocean.

C. Biodiversity in New Zealand

New Zealand's geological history has laid the groundwork for more than 2,000 indigenous plant species, about 1,500 of which are found nowhere else in the world. The biomes of the North Island include a subtropical area, including mangrove swamps, an evergreen forest with dense undergrowth of mosses and ferns, and a small grasslands area in the central volcanic plain. The South Island biomes include extensive grasslands in the east, which are excellent for agricultural pursuits; forest areas, dominated by native beech trees in the west; and an alpine vegetation zone in the Southern Alps.

In terms of fauna, the most influential factor may be the relative absence of predatory mammals, again related to New Zealand's geological history. With few ground predators and a favorable climate, bats, small reptiles, and birds were able to thrive and flourish. Without predators, many species of birds became flightless, such as the noted Kiwi. New Zealand's most famous bird, the moa, was similar to the ostrich but is now extinct. Moa could grow to more than twelve feet high and weigh more than five hundred pounds. New Zealand is known for its large number of species of wild birds. The Kiwi is the most noted and is often used to refer to people from New Zealand, as it is the national symbol of the country.

New Zealand has a variety of landscapes that have been attractive for economic activity and tourism. The South Island is larger than the North Island and is more mountainous. Large livestock-raising operations and agricultural activities can be found on the vast grasslands of the South Island. Millions of sheep and cattle are raised on the grassy highlands and the valley pasturelands. The North Island has more low-lying terrain, which is also good for agriculture and is home to a large dairy industry. The central highlands of the north offer some rugged relief and provide for a diverse physical landscape.

Figure 13. Kiwi Bird.

D. Cultural Dynamics and the Maori

New Zealand is home to many Polynesian groups. Its original inhabitants were the Maori, who came to the islands around the tenth century. They grew crops of gourds and sweet potatoes. Fur seals were hunted regularly, as were moa, which were hunted to extinction before the Europeans arrived. The Maori had created extensive trading networks with other island groups and developed a heritage of traditional rituals and cultural ways. The Maori culture thrived for hundreds of years and was well established in New Zealand before the arrival of the colonial ships from Europe.

Britain was the main colonizer of the islands. The British settled in to establish their presence and gain control. In 1840, the British colonizers and the Maori signed the Treaty of Waitangi, which granted British sovereignty over the islands but allowed the Maori certain rights over tribal lands. The actual language in this treaty has been debated between the English version and the Maori version. Over time the tribal lands were codified into legal arrangements by the European colonizers. Since the Treaty of Waitangi, the situation has evolved, with subsequent land exchanges, some legal and others questionable. The Maori have complained about unfair treatment and the loss of land and rights in the process. These issues have finally reached a point of negotiation in the past couple of decades. Starting in the 1990s, treaty settlements have been made to help correct the actions of the colonial activity and compensate the Maori for the conditions they were subjugated to.

Figure 14. A Maori man with traditional topknot and tattoos many Maori participate in performances in New Zealand both for tourism and to maintain their heritage and traditions.

Claimed by Great Britain in their colonial empire, the country of New Zealand became independent of Britain in 1901. In 2010, the estimate of the population of the country was about 4.3 million, with Europeans making up 60 percent of the population and the Maori making up about 8 percent. There are also many people who are of mixed ethnic background, including Maori and other groups. Asians, Polynesians, and other ethnic minorities make up the rest. New Zealand's main religion is Christianity and English is the official language.

E. Economic Conditions

Land and climate could be said to be New Zealand's most important natural resources. Fertile soils and a mild climate, complete with thousands of hours of sunshine annually, create ideal conditions for agriculture. Grass continues to grow throughout the year, which means that sheep and other livestock can be well grazed. Wool and other agricultural products, notably meat and butter, are important exports for New Zealand's economy. Healthy forests produce timber products, which are important to the economy as well. Some of New Zealand's natural resources are found underground, including coal, natural gas, gold, and other minerals.

Wellington is the capital of the country and is located on the southern end of the North Island. Wellington is one-fourth the size of the primary city, Auckland, which has 1.2 million people and is located to the north. The major cities are located along the coastal regions and provide a connection to sea transportation. Christchurch is the largest city of the South Island and is located along the eastern seaboard on the productive **Canterbury Plain**. The soils and conditions on the Canterbury Plain are excellent for productive agriculture of all types. Coastal plains also provide access to building transportation systems of highways and railroads that are more costly to construct in the mountainous regions of the Southern Alps or the northern highlands.

New Zealand has a market economy. New Zealand does not gain a large part of its national income from mining or manufacturing, though these industries do exist. The mainstay of the economy is, and has been for many years, a productive agricultural sector. The modern cities are home to a multitude of processing centers preparing the abundant agricultural products for domestic consumption and for export. The ever-growing populations of Asia and the rest of the world continue to place a high demand on food products and welcome New Zealand's agricultural exports. A "wool boom" in the 1950s furthered the emphasis on agricultural products as tremendous profits accrued in the wool production and export industry.

Today, New Zealand's economy is still heavily focused on the export of agricultural products, though the economy has diversified into other areas such as tourism and exploitation of natural resources, especially natural gas. The development of hydroelectricity generation in recent years has also been important to the economy. The high standard of living that exists in New Zealand is similar to that of Australia in that the population is not very large, so that the national wealth can be distributed via the private sector economy to accommodate a relatively good lifestyle and provide for a comfortable standard of living.

Key Takeaways

1. New Zealand has been isolated by the separation of continents through tectonic plate action. Shifting plates continue to create earthquakes and volcanic activity in the region.

2. New Zealand has high mountains, with the Southern Alps along the western coast of the South Island and highlands along the eastern side of the North Island. Adequate rainfall and good soils provide for excellent agricultural production, which has been the traditional economic activity.

3. The Maori were established in New Zealand before the British colonized it. Various agreements were made to work out common arrangements, with varying degrees of success. The Maori continue to be a minority population and have not acculturated into the mainstream society of the country.

4. New Zealand has historically relied on agricultural exports for national income. Shifting global markets and changes in government policies and structures have highly affected the economy of New Zealand. The country continues to work its way through the transition to a more global economy.

End-of-Chapter Summary

- Australia is an independent country and a relatively low-elevation island continent that is distinguished by its large semiarid interior region called the outback. Two core areas with moderate Type C climates along the eastern and western coasts hold most of the modest population of about twenty-two million.

- Great Britain colonized Australia and first used it as a prison colony. European traditions and heritage prevail, with English as the main language and Christianity as the main religion. The Aborigines had been there for tens of thousands of years before the Europeans arrived but only about four hundred fifty thousand remain.

- The traditional method of gaining wealth in Australia has been through agricultural production. In the latter half of the twentieth century, the export of minerals and raw materials became a major method of gaining wealth. In the past decade, tourism has risen to Australia's number one means of gaining wealth. Physical features such as the Great Barrier Reef, the outback, and the many unique animals make Australia a destination for world travelers.

- New Zealand is about 1,500 miles to the east of Australia across the Tasman Sea. The two main islands of New Zealand have high relief, with the Southern Alps located on the South Island. Moderate climate conditions and adequate rainfall make New Zealand an appealing place for agriculture and tourism. The remote locations create an element of isolation that is a cost consideration for travel and trade.

- The Maori people from Polynesia inhabited New Zealand before the British colonized it. The Maori had conflicts with the British and, later, the government over the loss of lands and legal arrangements. The Maori make up less than 10 percent of the 4.3 million people in New Zealand. The Maori situation has common ground with the situation with the Aborigines in Australia, who also have been working to regain land rights and legal settlements.

- New Zealand's economic situation has evolved to accommodate international market prices and demands as well as the internal business and political climate of the country. The main economic activity has been agricultural exports, but the country has been diversifying into industrial processes and tourism. Natural gas development has also been expanding.

Chapter 6

The Pacific and Antarctica

Introduction

The immense tropical Pacific realm and the ice-covered continent of Antarctica have almost opposing physical characteristics, but they are similar in that they are remote and isolated from the rest of the world. Understanding the geographic qualities of these two realms will help in comprehending the unique traits that humans have developed to survive in diverse environments. Both places include large physical areas with vast open spaces between human settlements. In the Pacific, human settlements are on islands. The only human settlements in Antarctica are isolated research stations. Historically, the South Pacific required a water-based transportation network, and in Antarctica, humans traveled across the snow and ice exploring the earth's southern extremes. Air travel is now available to connect both places with the rest of the world.

Almost all of Antarctica rests south of the Antarctic Circle. Antarctica is a continent surrounded by the Southern Ocean. The next nearest continent is South America. Many countries have laid claim to sections of Antarctica, but the continent remains off limits to industrial development and many other activities. The hundreds of islands of the South Pacific are surrounded by the Pacific Ocean and make up the largest geographic area in the world. The primary realm includes the island groups in the tropics between the Tropic of Cancer and the Tropic of Capricorn. There has been little industrial development in the South Pacific. Just as Antarctica has been divided up and claimed by other countries, though it was not colonized by them, most of the islands in the South Pacific were claimed or colonized by the imperial powers of Europe, Japan, or the United States. Both areas are considered to be peripheral realms in the overall scheme of the global economy. Tourism is the major activity in the South Pacific, and research and tourism are the major activities in Antarctica. Both areas have opportunities for greater economic development in the future. However, the difference is that Antarctica is not a country, and any benefit will go to the countries with claims on the continent or to the businesses taking tourists there.

Both the Pacific realm and Antarctica would be heavily impacted by increased climate change. Rising temperatures would continue to melt the ice in the polar caps, which in turn could raise sea levels. Changes in precipitation patterns could seriously alter the biodiversity of tropical islands in the Pacific, and changes in temperatures or precipitation could affect agricultural activity and tourism on many islands. Climate change in Antarctica could cause a further decline in the populations of penguins or other organisms.

6.1 The Pacific Islands

Learning Objectives

1. Outline the three main areas of the South Pacific: Melanesia, Micronesia, and Polynesia.
2. Distinguish between low islands and high islands.
3. Determine which islands remain under the auspices of France, the United Kingdom, New Zealand, or the United States.
4. Describe the primary economic activities of the islands in the realm.
5. Summarize the main environmental concerns of the islands in each region.

A. Introducing the Realm

The Pacific realm is home to many islands and island groups. The largest island is New Guinea, which is home to most of the realm's population. Many of the Pacific islands have become independent countries, while others remain under the auspices of their colonial controllers. The Pacific Theater of World War II was a battleground between the Japanese and American forces and had a large impact on the current conditions of many of the islands. The United States has been a major player in the post–World War II domination and control of various island groups. The Hawaiian Islands became the fiftieth US state in 1959.

The many islands can be divided into three main groups based on physical geography, local inhabitants, and location: Melanesia, Micronesia, and Polynesia. Indigenous cultural heritage remains strong in the South Pacific, but Western culture has made deep inroads into people's lives. The globalization process bears heavily on the economic conditions that influence the cultural dynamics of the Pacific. Islands or island groups that remain under outside political jurisdiction are the most influenced by European or American cultural forces. Western trends in fast food, pop music, clothing styles, and social customs often dominate television, radio, and the cinema. Invasive

Western cultural forces take the focus away from the traditional indigenous culture and heritage of the people who inhabited these isolated islands for centuries.

Traditionally, the islands were economically self-sufficient. Fishing and growing crops were the main economic activities, and nearby islands often established trade and exchanged natural resources. Fishing has been one of the most common ways of supporting the economy. There have been changes in the national boundaries to protect offshore fishing rights around each sovereign entity. Many waters have been

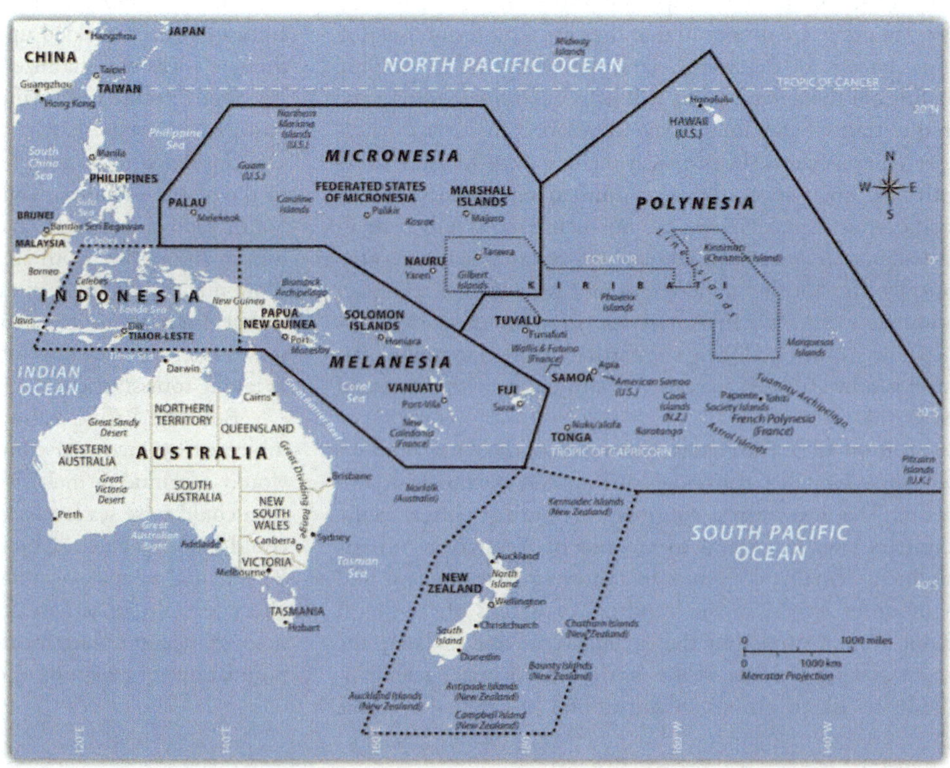

Figure 1. The tropical realm of the South Pacific with the three main regions of islands.

overfished, reducing the islands' ability to provide food for their people or to gain wealth. An increase in population and the introduction of modern technologies has brought about a dependency on the world's core areas for economic support.

The Pacific is an extreme peripheral realm with little to offer to the core areas for economic exploitation. In recent decades, some national wealth has been gained from the mining of substances such as phosphates on a few of the islands. The main resources available are a pleasant climate, beautiful beaches, and tropical island terrain, all of which can be attractive to tourists and people from other places. Tourism is a growing sector of the service industry and a major means of gaining wealth for various island groups. To attract tourism, the islands must invest in the necessary infrastructure, such as airports, hotels, and supporting services. Long distances between islands and remote locations make tourism transportation expensive. Not every island has the funding to support these expenditures to draw tourists to their location.

B. Melanesia

The region of the Pacific north of Australia that borders Indonesia to the east is called Melanesia. The name originally referred to people with darker skin but does not adequately describe the region's current ethnic diversity. The main island groups include Fiji, New Caledonia, Vanuatu, the Solomon Islands, and Papua New Guinea. All are independent countries except New Caledonia, which is under the French government. The island of New Guinea is shared between Papua New Guinea and Indonesia. Many islands on the eastern side of Indonesia share similar characteristics but are not generally included in the region of Melanesia.

Papua New Guinea

Papua New Guinea is the largest country in the Pacific realm and therefore the largest in Melanesia. It is diverse in both physical terrain and human geography. The high mountains of the interior reach 14,793 feet. Snow has been known to fall in the higher elevations even though they are located near the equator. Many local groups inhabit the island, and more than seven hundred separate languages are spoken, more than in any other country in the world. Indigenous traditions create strong centripetal forces. Many islands of Melanesia are recently independent of their European controllers; Papua New Guinea received independence in 1975 and is working toward fitting into the global community.

Papua New Guinea is a diverse country that still has many mysteries to be revealed in its little-explored interior. The country's large physical area provides greater opportunities for the exploitation of natural resources for economic gain. The interior of the island has large areas that have not been exploited by large-scale development projects. In the past few decades, oil was discovered and makes up its largest export item. Gold, copper, silver, and other minerals are being extracted in extensive mining operations, often by outside multinational corporations. Subsistence agriculture is the main economic activity of most of the people. Coffee and cocoa are examples of agricultural exports.

A number of islands off Papua New Guinea's eastern coast—including Bougainville—have valuable mineral deposits. Bougainville and the islands under its jurisdiction are physically a part of the Solomon Island archipelago but are politically an autonomous region of Papua New Guinea. Volcanic vents deep under the sea continue to bring hot magma

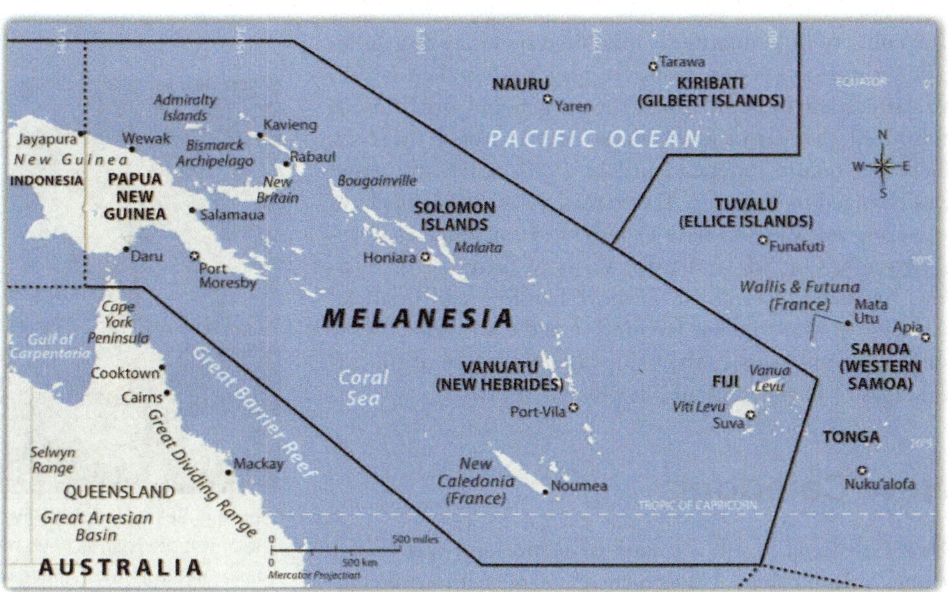

Figure 2. The region of Melanesia.

and minerals to the surface of the ocean floor, creating valuable exploitable resources. Papua New Guinea has laid claim to these islands and the underwater resources within their maritime boundaries. Rebel movements have pushed for the independence of the Autonomous Region of Bougainville but have been unsuccessful. The islands remain under the government of Papua New Guinea.

Solomon Islands

To the east of the island of Guinea are the Solomon Islands, a group of more than one thousand islands. About eighty of them hold most of the population of more than a half a million. The island of Guadalcanal was the site of some of the fiercest fighting in World War II between Japan and the United States. Honiara, the capital city, is on Guadalcanal. The Solomon Islands were a colony of Great Britain but gained independence in 1978. Colonialism, World War II, and ethnic conflict on the islands created serious centrifugal cultural forces, divisions, and political tensions over the past few decades. In 2003, military and police troops from other islands and Australia intervened to restore order after ethnic tension erupted into civil unrest.

Shifting tectonic plates are the source of environmental problems. Active seismic activity has created earthquakes and tsunami conditions that have brought devastation to the region. An earthquake of 8.1 magnitude hit the Solomon Islands in 2007, bringing high waves and many aftershocks. The tsunami killed at least fifty-two people, and as many as one thousand homes were destroyed. The islands contain several active and dormant volcanoes. Tropical rain forests cover a number of the islands and are home to rare orchids and other organisms. There is concern these resources might be harmed by deforestation and the exploitation of resources for economic gain.

Figure 3. Malaitan Chief on the Solomon Islands. The heritage and history of these islands includes local cultures complete with isolated traditions and ethnic organization.

Vanuatu

The country of Vanuatu was inhabited by many South Pacific groups; as a result, many languages are spoken within a relatively small population. The French and British both colonized the island archipelago. It was called the New Hebrides before independence in 1980, when the name was changed to Vanuatu. These small volcanic islands have an active volcano and have experienced earthquakes in recent years. One of Vanuatu's means of bringing in business has been to establish offshore banking and financial services, similar to what is found in the Caribbean. Many shipping firms register their ships there because of the advantages of lower taxes and flexible labor laws.

New Caledonia

New Caledonia is still a colony of France and was once a French prison colony. Under a current agreement, sovereignty is slowly being turned over to the local island government.

Figure 4. Saint Joseph's Bay on the Isles of Pines, New Caledonia These remote islands have moderate climates and beautiful coastal settings. Many have been unduly romanticized by works of fiction. These islands are isolated and can lack resources; life can be more difficult than is often portrayed.

Periodic reevaluations of the local government will be conducted to see if independence can be granted.

New Caledonia has historically relied on subsistence agriculture and fishing for its livelihood. About 25 percent of the world's known nickel resources are located here. Nickel resources will substantially affect the economy, bring in foreign investments, and raise the standard of living.

Fiji

Fiji is located in the eastern sector of Melanesia and has almost one million people. The country includes more than one hundred inhabitable islands, but two are home to most of the population. Colonialism heavily impacted the population's ethnic makeup. During British colonial rule, thousands of workers from South Asia were brought in by the British to work on the sugar plantations. After a century of British rule, Fiji became independent in 1970. The people of South Asian descent remained in Fiji and now make up more than one-third of the population. Ethnic conflicts erupted on the political scene between the Melanesian majority and the South Asian minority. Political coups and coalition governments have attempted to work out political solutions with limited success. Fiji is quite well developed and has a substantial tourism industry that augments the other agricultural and mining activities.

C. Micronesia

North of the Solomon Islands and Papua New Guinea is the large region of Micronesia. The "micro" portion of the name refers to the fact that the islands are small in size—often only one square mile or so in physical area. The region has more than two thousand islands. Most of the islands are composed of coral and do not extend above sea level to any large ex-

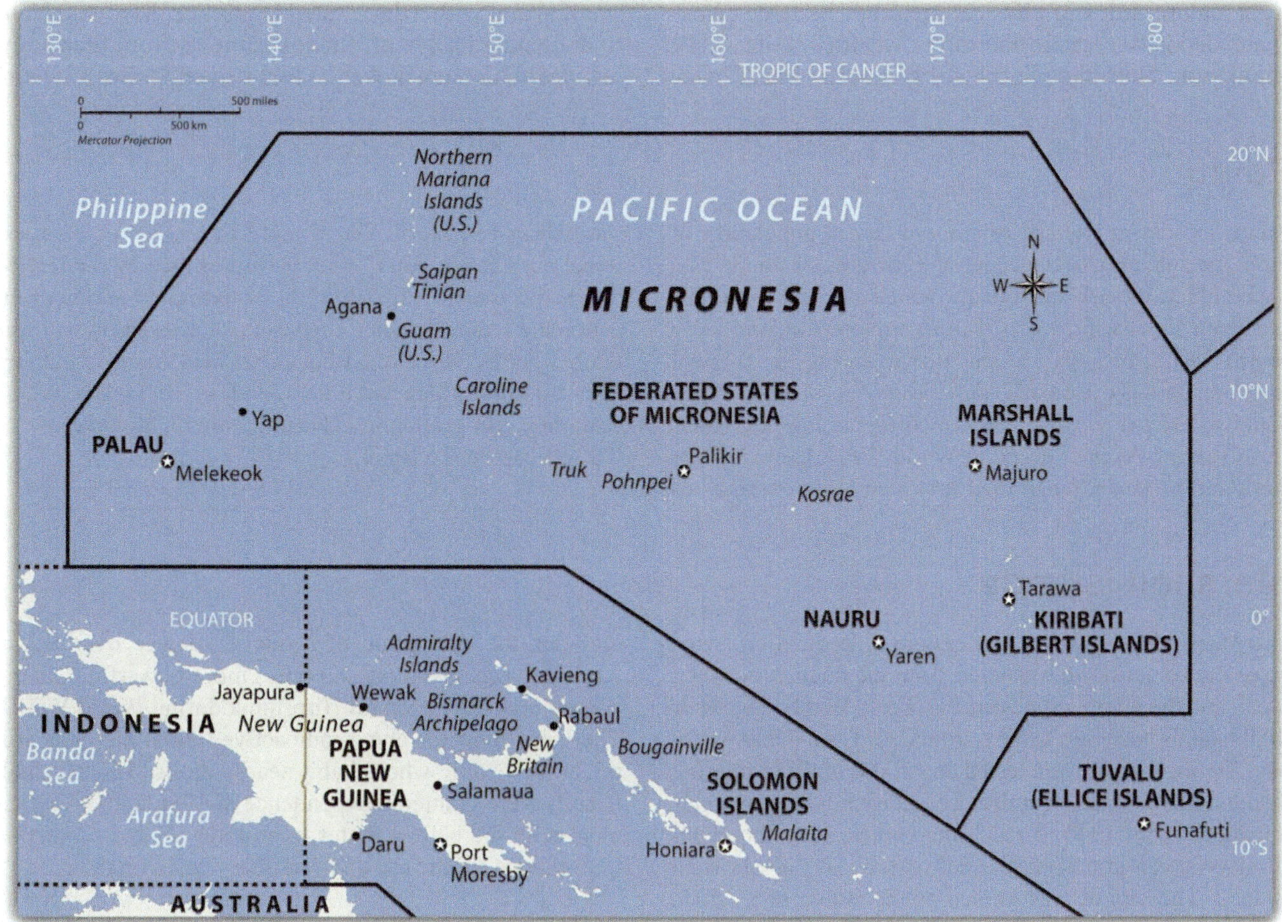

Figure 5. Micronesia.

tent. These low islands dominate the high islands. The high islands are usually of volcanic origin and reach elevations in the thousands of feet. The independent countries in Micronesia include the Federated States of Micronesia, Kiribati (western portion), the Marshall Islands, Nauru, and Palau. Other island groups include Guam, Wake Island, and the Northern Mariana Islands (all US possessions).

Guam and the Northern Mariana Islands

The largest island in Micronesia is Guam. It is only 210 square miles in area and reaches an elevation of 1,335 feet at its highest point. Coral reefs surround Guam's volcanic center. Guam is not an independent country but a US possession. The island was a strategic location during World War II, and the United States has major military installations located on the island. The Northern Mariana Islands, together with Guam to their south, comprise the Mariana Islands archipelago, and are also current US possessions, along with Wake Island in the northeast.

Figure 6. American Flags in Guam. Guam is a US possession. This photo shows the front of the University of Guam's School of Nursing. The university has more than three thousand students and is accredited by the Western Association of Schools and Colleges in the United States.

Nauru

The independent island country of **Nauru** is only about eight square miles in physical area, but its large phosphate deposits created enormous wealth for its small population. Once the phosphates had been mined, however, there was little means to gain wealth on such a small island with a devastated landscape. Many on Nauru are trying to live off the investments from their mining wealth or have moved to find a livelihood elsewhere.

Palau

Palau, located in western Micronesia, has a population of about twenty thousand people and an area of about 177 square miles. Its early inhabitants included people from Asia and from the Pacific realm. British explorers arrived early on the island, but Spain dominated it during the colonial era. After losing the Spanish-American War, Spain sold the island to Germany, which implemented mining operations on the island. After its defeat in World War I, Germany lost the island to Japan. Japan used it as a strategic outpost but was defeated in World War II and had to give up all its external possessions. After 1945, Palau was held by the United States and the UN. In 1994, the island opted for independence and retained an agreement of free association with the United States. The United States has held strategic military installations on Palau and other islands in Micronesia. Palau's economic and geopolitical dynamics are highly reflective of US activities in the region.

Marshall Islands

The Marshall Islands, on the eastern side of Micronesia, experienced serious devastation from the conflict between Japan and the United States during World War II. The Marshall Islands became a testing ground for US nuclear weapons. Atomic bombs were tested on various atolls, rendering them uninhabitable. An atoll is a coral island that surrounds a lagoon. From 1946 to 1958, the United States conducted sixty-seven atmospheric nuclear tests in the Marshall Islands. The largest was known as the Bravo test, which included the detonation of a nuclear device over Bikini Atoll that was one thousand times more powerful than the atomic bomb dropped on Hiroshima during World War II. There are concerns about radioactive fallout that may still affect the people who inhabit nearby atolls. The Marshall Islands were granted independence in 1986 with an agreement with the United States to provide aid and protection in exchange for the use of US military bases on the islands.

D. Polynesia

The largest region of the Pacific is Polynesia, a land of many island groups with large distances between them. The root word *poly* means "many." Numerous groups of islands have come together under separate political arrangements. The region includes the **Hawaiian Islands** in the north and the **Pitcairn Islands** and **Easter Island** to the east. New Zealand is now studied as a part of the Austral realm, but the Maori living there are originally from Polynesia. Polynesia has a mixture of island types ranging from the high mountains of Hawaii, which are more than 13,800 feet, to low-lying coral atolls that are only a few feet above sea level. Islands that have enough elevation to condense moisture from the clouds receive adequate precipitation, but many islands with low elevations have a shortage of fresh water, making habitation or human development difficult.

Polynesian culture stems from island resources. Fishing, farming, and an understanding of the seas created a way of life that gave Polynesia its identity. Polynesians created innovative maps that provided a means of sailing across large expanses of open seas to connect with distant is-

lands. Their lifestyle revolved around natural resources and the creative use of natural materials. Polynesian art, music, and language reflect a diversity of cultural trends derived from a common heritage. The warm climate and beautiful islands contrast with violent destructive storms and a lack of fresh water or resources, which can make life difficult.

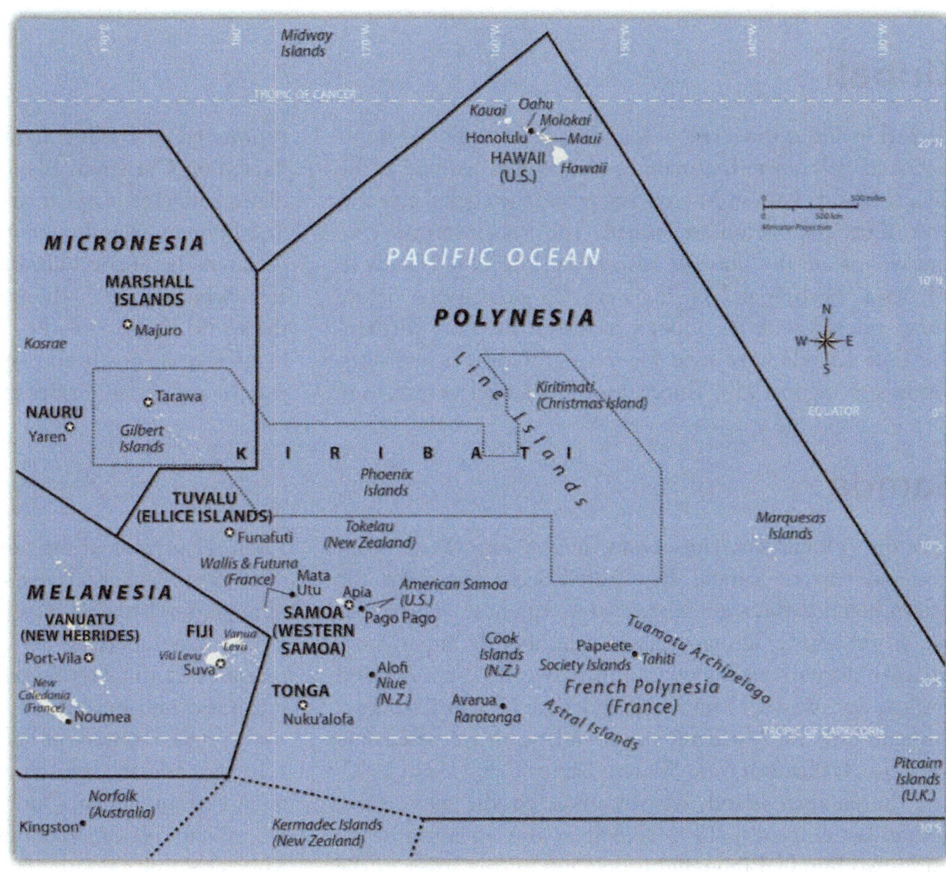

Figure 7. Polynesia. The region of Polynesia has island groups that are high islands with mountainous interiors.

Hawaii

At the beginning of the twenty-first century, Polynesia only had four independent island groups: Kiribati (eastern portion), Samoa, Tonga, and Tuvalu. The rest of the many islands and island groups in Polynesia are claimed by or are under the control or jurisdiction of other countries: mainly the United States, France, Great Britain, or New Zealand. Hawaii was a sovereign and independent kingdom from 1810 to 1893, when the monarchy was overthrown and the islands became a republic that was annexed as a US territory. Hawaii became the fiftieth US state in 1959. Hawaii's development pattern is modern, based on tourism from the continental United States and the US military base on Pearl

Harbor. According to the US Census, Hawaii had a population of 1.3 million in 2010. More than one-third of the people are of Asian descent, and at least 10 percent are native Hawaiians or Pacific Islanders. The United States has a number of additional possessions in Polynesia that include various small islands, atolls, or uninhabited reefs.

The Hawaiian Islands include more islands than the few usually listed in tourist brochures. Approximately 137 islands and atolls are in the Hawaiian chain, which extends about 1,500 miles. Hawaii is one of the most remote island groups in the Pacific. The islands of the Hawaiian archipelago are a product of volcanic activity from an undersea magma source called a hotspot, which remains stationary as the tectonic plate over it continues to shift creating new volcanoes. Mt. Kilauea, an active volcano on Hawaii, the largest island in the Hawaiian chain, is considered by geologists to be one of the most active volcanoes in the world. The active volcano of Mauna Loa and two dormant volcanoes, Mauna Kea and Hualālai, are on the same island. Mauna Kea is Hawaii's highest mountain at 13,796 feet above sea level, and is over 33,000 feet tall (taller than Mt. Everest) if measured from its base on the ocean floor.

Hawaii, like most islands of the Pacific realm, has a tropical type A climate, but snow can be found on the tops of its highest mountains during the winter months. The island of Kauai receives more than 460 inches of rain per year and is one of the wettest places on Earth. The orographic uplift created by Mt. Wai'ale'ale is the reason for the high level of precipitation. Abundant rain falls on the windward (eastern) side of the mountain, creating a rain shadow on the leeward (western) side of the mountain, which has a semi-desert steppe climate.

Kiribati

Kiribati includes three sets of islands located in both Micronesia and Polynesia. The main component of Kiribati is the Gilbert Island chain in Micronesia, where the capital city and most of the population are located. The other two minor island chains are the Phoenix Islands and the Line Islands in Polynesia. Both island chains were US possessions before being annexed with the Gilbert Islands to become Kiribati. The Line Islands were used for testing of British hydrogen bombs starting in 1957. Three atmospheric nuclear tests were

conducted by the British on Malden Island, and six were conducted on Christmas Island. There is concern about how radiation affected people present during the tests and thereafter. The Phoenix Islands have few inhabitants. In 2008, Kiribati declared the entire island group a protected environmental area, which made it the largest protected marine habitat in the world. Kiribati is the only country with land in all four hemispheres: north and south of the equator and on both eastern and western sides of the 180° meridian.

Samoa

After the colonial era, Samoa was divided into Western Samoa and Eastern Samoa. The United States controlled the eastern islands, which are referred to as American Samoa. Before World War I, Germany gained control of the larger, more extensive western islands only to lose them to New Zealand after the war. Western Samoa was under the New Zealand government until 1962, when it gained independence. The name was officially changed from Western Samoa to Samoa in 1997.

The Samoan Islands are volcanic, and the most active volcano last erupted in 1906. In Samoa, three-quarters of the nearly two hundred thousand people live on the larger of the two main islands. Colonialism has had a major impact on the culture, especially in the case of religion. Christianity became widespread once it was introduced and is now the religion of

about 99 percent of the population. American music and societal trends are also a major influence on the islands because of migration between Hawaii and the US mainland. Many Samoans have moved to the United States and established communities. Cultural traditions have been preserved and are often integrated into modern society. Samoa has some of the oldest history and traditions of Polynesia. For many years, the United States has held an extensive naval station in the bay of Pago Pago on American Samoa. During World War II, there were more US military personnel on the islands than Samoans. American Samoa became a key military post for the United States. American Samoa remains a US possession; however, Samoans are not US citizens unless one of their parents is a US citizen.

Tonga

South of Samoa is an archipelago that is home to the Kingdom of Tonga. Only about 36 of the 169 islands are inhabited by a total population of about one hundred twenty thousand people. Tonga is ruled by a monarchy that never lost its

governance powers throughout the colonial era. Tonga is the only monarchy in the Pacific. The two main methods of gaining wealth are by remittances from citizens working abroad and tourism.

Tuvalu

The island nation of Tuvalu comprises four reef islands and five atolls for a total land area of about ten square miles. In 2008, it had a population of about twelve thousand people. These statistics indicate that Tuvalu is one of the four smallest countries in the world. Nauru is only about eight square miles in area. Only the Vatican and Monaco are smaller. The low elevation of the islands of Tuvalu make them susceptible to damage from rising sea levels. The highest point is only fifteen feet in elevation. Any increase in ocean levels as a result of climate change could threaten the existence of this country.

French Polynesia

The South Pacific is home to many islands and island groups that are not independent countries. The biggest and most significant group in the southern region is French Polynesia. France colonized a large number of islands in the South Pacific and has continued to hold them in its control or possession as external departments or colonies. In western Polynesia, the French maintain control over the islands of Wallis and Futuna. French Polynesia consists of four main island groups: the Society Islands, the Austral Islands, the Tuamotu Islands, and the Marquesas. There are around 130 islands in French Polynesia, and many are too small or lack resources to be inhabited.

Tahiti, located in the Society Islands, is the central hub of French Polynesia. Papeete is the capital and main city with a population of almost thirty thousand. Tahiti is a major tourist destination with a mild climate that stays at 75 °F to 85 °F year-round and receives adequate rainfall to sustain tropical forests. Most of the people live along the coastal areas; the interior is almost uninhabited. The Society Islands include the island of Bora Bora, which is considered by many to be a tropical paradise and one of the most exotic tourist destinations in the world.

The volcanic Marquesas Islands to the northeast are the second-most remote islands in the world after the Hawaiian Islands. The weather pattern in the Pacific does not bring enormous amounts of precipitation to the Marquesas, a reality that restricts human expansion in the archipelago. The higher elevations in the mountains—the highest is 4,035 feet—draw some precipitation from orographic uplift, giving rise to lush rain forests on portions of the islands. With fewer than ten thousand people, the Marquesas do not have a large population to support and rely on financial support from outside to sustain them.

French Polynesia also includes the Tuamotu Archipelago, between the Society Islands and the Marquesas, which consist of about 75 atolls and an uncounted number of coral reefs that extend for about nine hundred miles. The French government used islands in this archipelago as test sites for nuclear weapons. From 1966 to 1974, the French tested 41 atomic devices above ground in the atmosphere, and from 1974 to 1996, they tested 137 atomic devices below ground.

Radiation concerns are the same here as on the Marshall Islands, where the US tested atomic weapons. Scientific testing monitored by the World Health Organization has determined the humans living closest to the atolls are not presently in danger of radioactive materials either in the environment or in their food supply. The long-term effects of the underground tests continue to be monitored.

Figure 8. The Moorea Ferry in Papeete Harbor, Tahiti
The only ways to get to the islands are by aircraft or by ship. Transportation costs can be high for imported goods or for tourism development.

Figure 9. Bora Bora in the Society Islands in French Polynesia with Mount Otemanu in the background. Bora Bora is a world-class tourist destination catering to the international traveler.

The Pitcairn Islands, Easter Island, and the Cook Islands

To the east of French Polynesia are the four Pitcairn Islands, controlled by Great Britain. The main island, Pitcairn, is the only inhabited island in this chain and is one of the least inhabited islands in the world; the total population is fewer than fifty people. Mutineers from the HMS Bounty escaped to Pitcairn in 1790 after taking various Tahitians with them.

Even farther east than Pitcairn, on the edge of Polynesia, is Easter Island. Now under the government of Chile, Easter Island was historically inhabited by Polynesians who built large stone heads that remain somewhat of a mystery. At the center of Polynesia are the fifteen small Cook Islands, which are controlled by New Zealand and are home to about twenty thousand people, many of whom claim Maori ethnicity.

Key Takeaways

1. Melanesia includes the islands from Papua New Guinea to Fiji. Micronesia includes small islands located north of Melanesia. Polynesia includes island groups from the Hawaiian Islands to the Pitcairn Islands. Papua New Guinea is the largest country in the Pacific, approximately seven hundred languages are spoken by the many local groups that live there.

2. Low islands in this region are usually composed of coral and low in elevation. High islands are usually volcanic in origin and mountainous with high elevations. Micronesia consists mainly of low islands, while Polynesia consists of many high islands, such as Hawaii.

3. Tourism is the main economic activity in the Pacific, but minerals and fossil fuels provide some islands with additional wealth. Fishing and subsistence agriculture have been the traditional livelihoods. Offshore banking has also been established in the region.

4. The United States, the United Kingdom, and France used various islands for nuclear testing. Radiation fallout continues to be an environmental concern. Typhoons, tsunamis, volcanic activity, earthquakes, and flooding create devastation on the islands. Fresh water can be a valuable resource, as it is in short supply on many islands.

6.2 Antarctica

Learning Objectives

1. Summarize the layout of the continent's main physical features, including the ice shelves and volcanic activity.

2. Understand the political nature of the various claims held on sections of Antarctica and how the continent is managed by the international community.

3. Outline the dynamics of the principle of global warming and describe what changes would occur in Antarctica and the rest of the world if the ice sheet covering Antarctica were to melt.

4. Describe how good ozone is depleted and understand the role Antarctica plays in the seasonal cycle of changes in the amount of ozone in the atmosphere above the South Pole.

Figure 10. The Southern Ocean and Antarctica.

A. The Southern Ocean

The Southern Ocean, which surrounds the continent of Antarctica, is often misunderstood or not included on many maps of the Southern Hemisphere. The cold waters off the coast of Antarctica move from west to east in a clockwise rotation around the continent in a movement called the West Wind Drift, or the Antarctic Circumpolar Current. The Southern Ocean's northern boundary does not border a land mass but meets up with the Atlantic, Pacific, and Indian Oceans. Most geographers accept the northern boundary of the Southern Ocean to be located south of 60° latitude even though the actual limit has not yet been firmly agreed upon.

The Southern Ocean's northern boundary has more to do with the marine conditions of the realm. There is a transition called the Subtropical Convergence in which the cold, dense waters of the Southern Ocean meet up with the warmer waters of the Pacific, Atlantic, and Indian Oceans. The cold, dense water from the south sinks below the warm waters from the north to create a zone of upwelling and mixing that is conducive to high levels of productivity for organisms such as phytoplankton and krill. The zone of Subtropical Convergence can be visually observed by the grayish, cold southern waters meeting up with the bluish-green, warm northern waters. The krill, which thrive on phytoplankton, are an important link in the food chain for marine organism such as fish, penguins, seals, albatrosses, and whales in the Southern Ocean.

B. Physical Geography

The world has seven focal continents. Rated by physical area from the largest to the smallest, they are Asia, Africa, North America, South America, Antarctica, Australia, and Europe. Antarctica, which is larger than Australia and 1.3 times larger than Europe, is located entirely south of 60° latitude and is surrounded by the Southern Ocean. Antarctica has the highest average elevation of any continent; there are many mountain ranges. The two-thousand-mile-long Transantarctic Mountain range divides Antarctica into a small western region and a larger eastern region. At both ends of the Transantarctic Mountains are the two main ice shelves: the Ross Ice Shelf and the Ronne Ice Shelf. The Ellsworth Mountains are located in the western region and are home to Mt. Vinson (or the Vinson Massif), which is the highest peak on the continent, reaching an elevation of 16,050 feet. This is higher than any mountain in the contiguous United States, Europe, or Australia.

The Antarctic Peninsula is actually an extension of the Andes Mountains of South America and is home to active volcanoes. The peninsula is the location of a volcano on Deception Island that erupted in the late 1960s and destroyed research stations in the area. There was an additional large eruption in 1970. The volcano continues to show activity, and sightings of lava flow continue to be reported. There may be more volcanic activity than what has been recorded. An underwater volcano in the Antarctic Peninsula was discovered in 2004. Mt. Erebus (12,448 feet), located on Ross Island on the other side of Antarctica from the Antarctic Peninsula, is the world's southernmost active volcano. Mt. Erebus has been active since 1972 and has a large lava lake in its inner crater.

About 98 percent of Antarctica is covered by an ice sheet that is, on average, up to a mile deep. In some areas, it is nearly three miles deep. In the winter season, the ice sheet's area might double as it extends out from the coastline. The

Rank of 7 Continents in Square Miles	
1. Asia	17,139,445
2. Africa	11,677,239
3. North America	9,361,791
4. South America	6,880,706
5. Antarctica (*with ice sheet*)	5,400,000
6. Europe (*with Russian region*)	3,997,929
7. Australia	2,967,124
The United States (*land area*)	**3,537,438**

Figure 11. Size Comparison of Antarctica and the United States. Antarctica without the ice sheet would be considerably smaller; some estimate it would be only about one hundred thousand square miles in land area, or the equivalent of the physical area of the US state of Oregon.

Antarctic ice sheet holds about 70 percent of the earth's fresh water. If the ice sheet were to melt, the sea level could rise considerably and cover many of the earth's low-lying islands, peninsulas, and coastal regions with low elevations. Antarctica is considered a desert because it usually averages fewer than ten inches of precipitation per year. Coastal regions annually receive as much as four feet of snow, while the interior near the South Pole might only receive a few inches.

There are areas in Antarctica that are not covered with ice but have a landscape of bare ground. This non-ice portion of the continent protrudes above the ice sheet and only covers a combined physical area equivalent to about half the US state of Kentucky. The only plant life that exists here are the many different mosses and lichens that grow during short periods of the year. Below the giant ice sheet are dozens of subglacial lakes. Lake Vostok, the largest lake discovered in the Antarctic so far, was found two miles below the ice sheet and is the size of Lake Ontario. It is unknown what aquatic life might exist in these lakes. If all the ice and snow were removed from the continent, the total land area would be considerably smaller and would consist mainly of mountain ranges and islands. Some estimate that this land area altogether would only equate to about one hundred thousand square miles, roughly equivalent to the physical area of the US state of Oregon. This does not account for the fact that if all the ice were to melt, the sea level would rise and cover more land area. The land portion of the continent would also expand

Figure 12. Mt. Erebus (12,448 feet), located on Ross Island, is Antarctica's most active volcano and has a lava lake in its inner crater.

upward because of the loss of the weight of the ice, which has been compressing the continent.

Not only is Antarctica the driest continent with the least average annual precipitation and the highest continent in average elevation; it is also the coldest of the continents. The lowest temperature ever recorded on Earth was –128 °F in 1983 at a Russian research station in Antarctica. Temperatures reach a minimum of less than –110 °F in winter in the interior and greater than 55 °F near the coast in summer. No permanent human settlements exist in Antarctica other than research stations from a number of countries.

C. The Antarctic Treaty

The continent is not politically controlled by any one government. Early seafaring explorers sailed in these waters, and various countries laid claim to sections of the continent. The continent was first sighted by explorers in 1820, and the South Pole was first reached in 1911. Land claims to the continent were established by the home countries of early explorers. Forty-six countries are now included in the Antarctic Treaty, which was originally signed by twelve countries in 1959. The treaty, designed to protect the environment and encourage scientific research, prohibits military activities, mineral mining, and the disposal of waste products. All land claims were suspended when the Antarctic Treaty was initiated, but the claims are not without political ramifications. Antarctica is divided into pie-shaped sections, and each of the original claimant countries is allocated a portion, according to their claim. The countries with original claims are Norway, New Zealand, France, Chile, Australia, and Argentina. Other countries, including Brazil, Peru, Russia, South Africa, Spain, and the United States, have reserved their right to submit claims on the continent in the future if the issue of territorial claims becomes significant. A large sector of West Antarctica called Marie Byrd Land remains unclaimed.

Research stations account for the entire human population in Antarctica. Approximately one thousand people live

Figure 13. Emperor Penguins, Ross Sea, Antarctica.

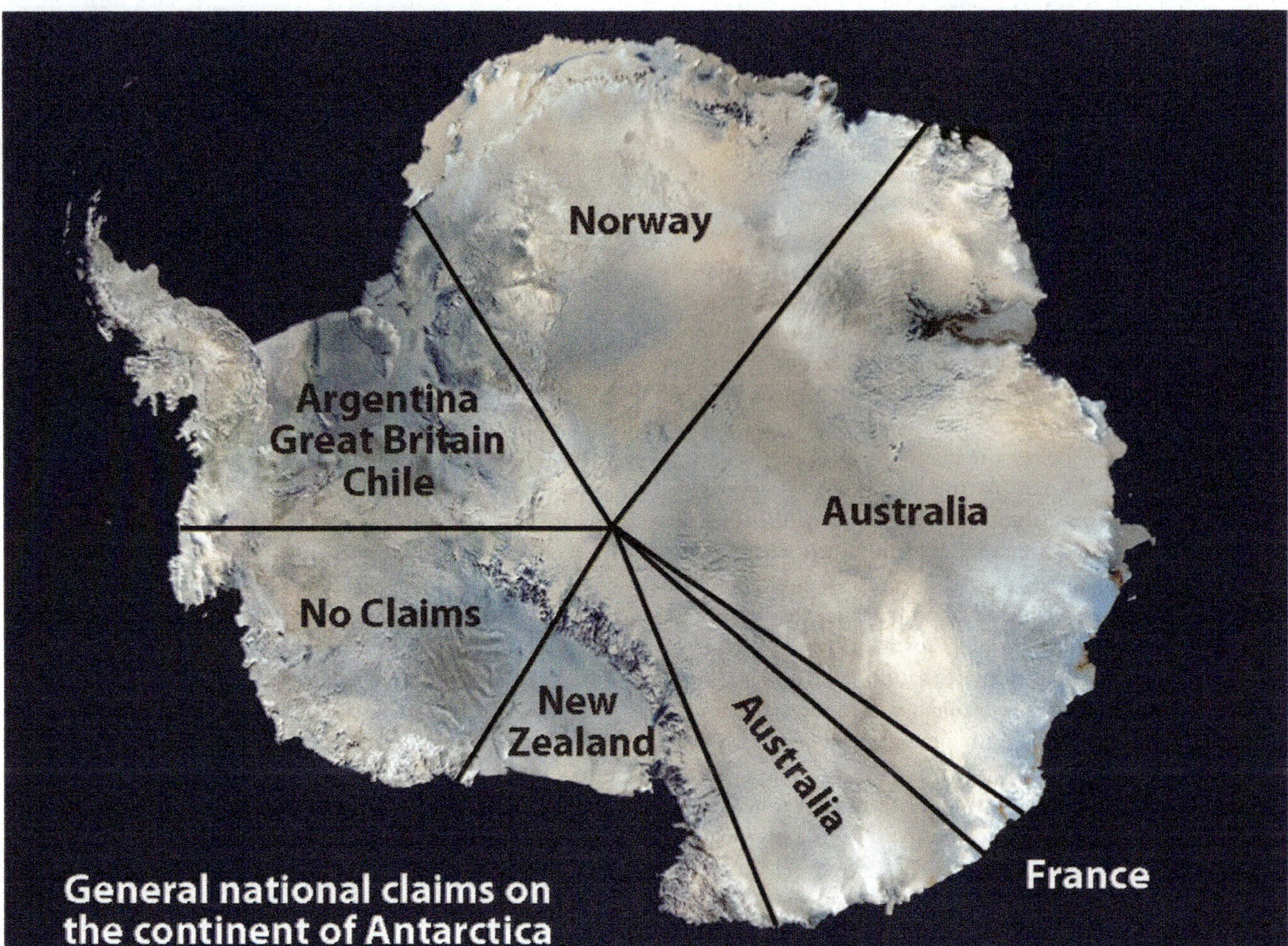

Figure 14. Antarctic territorial claims.

in Antarctica year-round, and up to five thousand or more live there during the summer months. Many of the research stations rotate their personnel, and tours of duty last anywhere from a few months to a year or more. Various family groups have worked there as well as other service workers, including Russian Orthodox priests, who have rotated every year at one of the Russian research stations.

Tourism brings the largest number of additional people to the continent. Tourists come for short-term visits to experience the conditions or see the many species of penguins or fauna that exist here. More than forty-five thousand tourists visit the Antarctic Treaty area yearly. Most arrive on commercial ships that specialize in tours of the region. Tours only last one or two weeks.

Research has revealed that mineral resources are to be found under the ice in Antarctica, and oil and natural gas are found in offshore deposits. Antarctica is a frontier for economic development that is not under the jurisdiction of

any one government. The Antarctic Treaty has been the determinant of the level of human activity. The current treaty restricts any extractive activity. Fishing is also regulated within the treaty, but without enforcement procedures, there have been questions about its effectiveness. Whaling was once a major industry in this realm. Whaling stations were established on the Antarctic Peninsula and nearby South Georgia Island. However, the increased use and extraction of petroleum reduced the need for whale oil and the industry collapsed. Some whaling continues in the waters of the Southern Ocean, which has led to questions about how to manage these natural resources. In 1998, negotiations between interested countries met in Madrid, Spain, and created the Protocol on Environmental Protection to the Antarctic Treaty (known as the Madrid Protocol). The protocol designates Antarctica as a natural reserve that can only be used for peaceful purposes and for science. All mining or economic activity is banned.

D. Climate Change

Climatic conditions on Earth have varied widely during the planet's history. There have been long periods of heating or cooling. The last ice age, which ended about ten thousand years ago, created large ice sheets that covered much of the Northern Hemisphere. The earth then entered into the current interglacial period with warmer temperatures that melted the ice sheets; the polar regions have the last remaining ice on the planet. The earth has experienced large fluctuations in its temperature at various times in its past. Natural changes in the conditions that affect climate can include but are not limited to the dynamics of the sun, changes in the earth's orbit, and volcanic eruptions.

Human activity has impacted conditions both locally and globally. Since the Industrial Revolution, humans have been pumping enormous amounts of carbon dioxide into the atmosphere, which affects the planet's climate and temperature. The Industrial Revolution introduced the burning of coal as a fuel to boil water to operate steam engines. This allowed power to become more versatile and mobile. The introduction of the automobile increased the burning of petroleum, which released carbon dioxide into the atmosphere in the form of engine exhaust.

Large-scale deforestation and the burning of fossil fuels have increased the quantity of heat-trapping "greenhouse gases" in the atmosphere. Nitrous oxide, methane, carbon dioxide, and similar gases act like the glass panels of a greenhouse that allow short-wave radiation from the sun to enter but do not allow the long-wave radiation of heat to escape into space. Deforestation reduces the number of trees that use carbon dioxide and store carbon in plant fibers. The burning of wood or carbon-based energy sources such as oil or coal releases the carbon back into the atmosphere. Fossil fuels are created when dead plant and animal life have been under pressure and decay for long periods and have retained their carbon component. This all leads to a rise in the activity of the carbon cycle. Carbon is a key component to the regulation of the earth's temperature. Life on Earth is dependent on

temperature conditions that are regulated by the atmosphere. This natural cycle has been augmented by human activity.

Changing global temperatures are one aspect of climate change that has received attention in recent years. Technically speaking, global warming is an average increase in the temperature of the atmosphere near the earth's surface. Few environmental effects could impact Antarctica as much as the phenomenon of changing temperatures. There could be significant impacts around the world if temperatures increase and melt some of the Antarctic ice sheet. Sea levels would rise, and many areas of the planet with large urban populations could be flooded. Many of the largest urban centers in the world are port cities that rely on the import and export of goods and materials. Portions of these cities would be in danger of being under water if the sea level were to rise even a few feet. Climate change might also affect agricultural production. Global changes in temperature would alter ecosystems and the habitats of organisms, changing the balance of nature in many biomes.

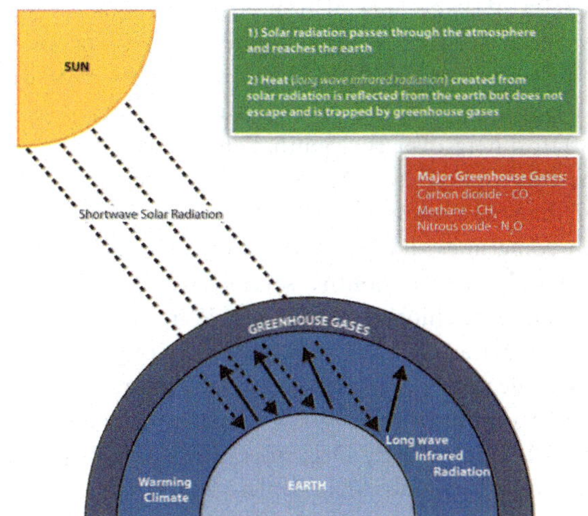

Figure 15. The Greenhouse Effect and climate change.

E. Ozone Depletion

Climate change can have a major impact on atmospheric conditions. It should be noted that ozone depletion in the stratosphere (the second layer of the earth's atmosphere) has different causes and conditions than temperature change in the troposphere (the layer just above the earth's surface). The two concepts have separate and distinct dynamics that are not directly related. Ozone (O3) is a simple molecule consisting of three oxygen atoms. Common oxygen gas molecules have two oxygen atoms (O2). Depending on where ozone is located in the atmosphere, it can be either a protective safeguard from ultraviolet (UV) radiation from the sun or an element

in smog that causes health problems. Good ozone in the stratosphere provides a protective shield preventing harmful UV radiation from reaching the earth. UV rays from the sun are known to cause skin cancers, eye damage, and harm to organisms such as plankton. Bad ozone molecules in the troposphere mix with various chemicals to create smog, which reduces visibility and can cause respiratory health problems.

Chemicals such as chlorine and bromine interact with protective ozone molecules in the stratosphere and break them down in a chain reaction that depletes the stratosphere of ozone molecules and stops the cycle that absorbs the UV radiation. One chlorine molecule can destroy one hundred thousand ozone molecules, and bromine atoms can destroy ozone molecules at a rate of many times that of chlorine. Chlorine and bromine enter the stratosphere through the discharge of chlorofluorocarbons (CFCs), hydrofluorocarbons (HCFCs), and other chemicals that deplete ozone molecules. These chemicals have been mainly used in industrial processes such as refrigeration and air conditioning and in solvents and insulation foam. In the stratosphere, radiation from the sun breaks HCFCs and CFCs apart, releasing chlorine atoms that destroy ozone molecules.

This process also occurs naturally when volcanoes erupt and emit sulfur aerosols into the atmosphere. The sulfur aerosols break down CFCs and halon; this results in the release of chlorine and bromine, which deplete ozone molecules. In recent years, nitrous oxide has become a major chemical that can reach the stratosphere to act as an agent in the ozone depletion process. Nitrous oxide can be released into the atmosphere from human activities such as the use of nitrogen fertilizers in agriculture or from vehicle exhaust from burning fossil fuels. Nitrous oxide is also released naturally from the soil or from ocean water.

Ozone depletion has been especially noticeable over Antarctica. A large area of ozone depletion in the upper atmosphere is often called an ozone hole. The ozone hole over Antarctica is not exactly a total depletion of ozone but a major reduction in the amount of ozone found in the stratosphere over the South Pole. Recent ozone levels in the stratosphere over Antarctica have decreased and are lower than they were in 1975. Polar stratospheric clouds, which sometimes develop over Antarctica during the extremely cold winter months, severely reduce ozone levels. The clouds trap chlorine and nitric acid in their ice crystals. As the circulation of westerly winds starts in the spring, an atmospheric vortex is created over the polar area. The ozone hole increases during the spring when sunlight increases—from September to early December. The sunlight speeds up the chemical reactions that deplete the ozone molecules. During this time, as much as half of the lower stratospheric ozone can be destroyed, creating an ozone hole.

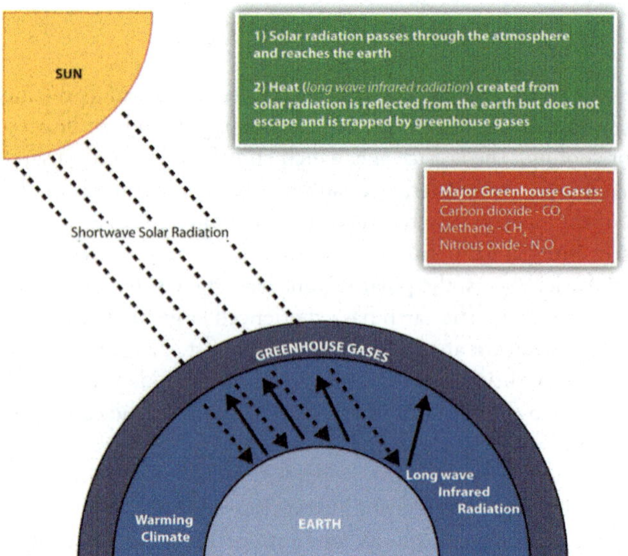

Figure 16. Ozone in the stratosphere protects the earth from harmful ultraviolet radiation from the sun.

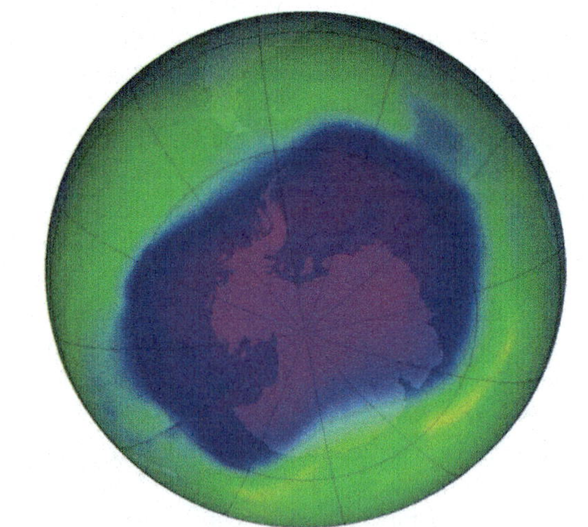

Figure 17. One of the largest ozone holes was observed in 2006 over Antarctica.

F. Global Impacts and Organizations

Many governments around the world have established agencies to address environmental issues and have invested resources in continued research and scientific discovery. Many nongovernmental agencies and organizations such as Greenpeace International have also been established to address the development or management of Antarctica and to address environmental concerns. International organizations such as the United Nations Environment Programme (UNEP) have been formed to enhance cooperation between countries and concerned entities. In 1998, the Intergovernmental Panel on Climate Change (IPCC) was established by the UNEP and the World Meteorological Organization to address the work of the United Nations Framework Convention on Climate Change (UNFCCC). Through an international treaty, the agency focuses on the harmful effects of climate change. One outcome of the UNFCCC's work was the Kyoto Protocol developed in 1997 in Kyoto, Japan, which created a legal commitment by participants to reduce greenhouse gases and address climate change issues. As of 2010, 191 countries have signed on to the initiative. The United States has not yet ratified the protocol. Enforcing the Kyoto Protocol is a matter of debate. The 2009 United Nations Climate Change

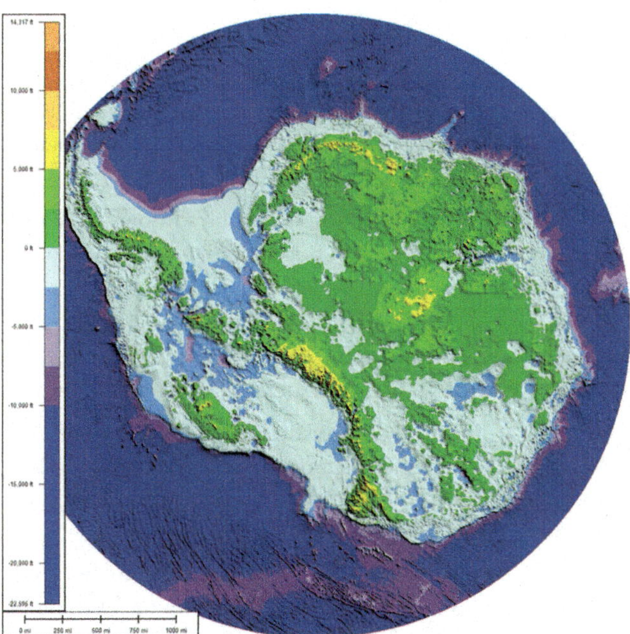

Figure 18. One projection of the subglacial topography of Antarctica below the ice sheet. This is what Antarctica might look like if all the ice were removed. The physical area remaining is estimated by some to be about one hundred thousand square miles, which is about the same physical area as the US state of Oregon. It is unclear what the actual physical area would be if the ice melted.

Conference (also referred to as the Copenhagen Summit) was held in Denmark to address some of the same issues that were discussed in the Kyoto Protocol.

Depletion of the stratospheric ozone over the polar regions or the thinning of the ozone layer over the midlatitudes would have worldwide implications for human activity and environmental conditions. Climate change is not restricted to one country or one government; these and other environmental issues affect the whole planet and impact everyone, whether they are contributing to the problem or not. If the sea level were to rise to the level predicted because of global warming, it would affect every country with a coastline in the world. It would not be restricted to any one category of country, developed or undeveloped. Issues of increased UV radiation or climate change are not restricted by political boundaries or economic conditions; they require global thinking and awareness and cannot be solved by one country alone—if they can be solved at all.

Key Takeaways

1. Antarctica has the highest average elevation of any continent and has many mountains. The continent includes active volcanoes, high mountains, and the Transantarctic Mountain range. The continent is surrounded by the Southern Ocean.

2. The ice sheet that covers Antarctica is more than a mile thick and holds about 70 percent of the earth's fresh water. Antarctica is still considered a desert because it receives so little precipitation on an average basis. Large bodies of water are also located below the ice sheet.

3. The concept of global warming is one aspect of climate change that indicates an increase in greenhouse gasses such as carbon dioxide, which help regulate the temperature of the earth's atmosphere. The end result is warmer temperatures on the earth's surface. The burning of fossil fuels is a significant source of carbon dioxide that enters the atmosphere. Climate change may result in the melting of the polar ice sheet over Antarctica, which could raise sea levels.

4. Ozone in the stratosphere protects the earth from harmful UV radiation from the sun. Various chemicals reduce the amount of protective ozone in the stratosphere, which allows more UV radiation to reach the earth. A seasonal cycle varies the amount of ozone in the stratosphere over the South Pole, causing an ozone hole when ozone is not abundant.

End-of-Chapter Summary

- The low coral islands and the high volcanic islands of the South Pacific can be divided into three regional groups: Polynesia, Micronesia, and Melanesia. The Pacific islands cover the world's largest geographic realm, and there is a great distance between many of the island groups.

- The Pacific has a high level of ethnic diversity; there are hundreds of different languages and cultural traditions. Papua New Guinea is the largest island and is host to more than seven hundred languages. Various Pacific islands remain possessions of France, the United Kingdom, New Zealand, or the United States.

- The Pacific islands are subject to natural forces such as typhoons, volcanic eruptions, earthquakes, tsunamis, and tropical storms. Some of the islands have radiation issues because they were used as nuclear test sites by the United States, France, and Great Britain. Other environmental issues relate to deforestation and the lack of fresh water.

- Antarctica is an ice-covered continent that is larger than the continents of Europe or Australia. The Southern Ocean surrounds Antarctica, which is located entirely south of 60° latitude. Antarctica's ice sheet is more than a mile thick but does not cover the entire land surface. Extensive mountain ranges exist with volcanic peaks that have erupted in recent years. The only permanent human settlements in Antarctica are research stations.

- The concept of global warming may impact the continent of Antarctica if temperatures rise and the ice sheet melts rapidly. Temperature increases in the atmosphere might be a result of an increase in greenhouse gases caused by the burning of carbon-based fossil fuels.

- Ozone in the stratosphere protects the earth from ultraviolet (UV) radiation. Ozone depletion in the stratosphere can be caused by the release of industrial chemicals. A large ozone hole sometime exists above Antarctica during a seasonal increase in ozone depleting elements

Chapter 7

Middle America

Introduction

Middle America, the geographic realm between the United States and the continent of South America, consists of three main regions: the Caribbean, Mexico, and the Central American republics. The **Caribbean** region, the most culturally diverse of the three, consists of more than seven thousand islands that stretch from the Bahamas to Barbados. The four largest islands of the Caribbean make up the Greater Antilles, which include Cuba, Jamaica, Hispaniola, and Puerto Rico. Hispaniola is split between Haiti in the west and the Dominican Republic in the east. The smaller islands, extending all the way to South America, make up the Lesser Antilles. The island that is farthest south is Trinidad, just off the coast of Venezuela. The Bahamas, the closest islands to the US mainland, are located in the Atlantic Ocean but are associated with the Caribbean region. The Caribbean region is surrounded by bodies of salt water: the Caribbean Sea in the center, the Gulf of Mexico to the west, and the North Atlantic to the east.

Central America refers to the seven states south of Mexico: Belize, Guatemala, Honduras, El Salvador, Nicaragua, Costa Rica, and Panama. Panama borders the South American country of Colombia. During the colonial era, Panama was included in the part of South America controlled by the Spanish. The Pacific Ocean borders Central America to the west, and the Caribbean Sea borders these countries to the east. While most of the republics have both a Caribbean and a Pacific coastline, Belize has only a Caribbean coast, and El Salvador has only a Pacific coast.

Mexico, the largest country in Middle America, is often studied separately from the Caribbean or Central America. Mexico has an extensive land border with the United States, its neighbor to the north. The Baja Peninsula, the first of Mexico's two noted peninsulas, borders California and the Pacific Ocean and extends southward from California for 775 miles. The **Baja region** is mainly a sparsely populated desert area. The **Yucatán Peninsula** borders Guatemala and Belize and extends north into the Gulf of Mexico. The Yucatán was a part of the ancient Mayan civilization and is still home to many Maya people.

Middle America is not a unified realm but is characterized by a high level of political and cultural diversity. A diverse mix of people—with **Amerindian** (people native to the Americas), African, European, and Asian ethnic backgrounds—make up the cultural framework. This realm is often associated with the term "Latin America" because of the dominance of colonialism from European countries such as Spain, France, and Portugal, where the people speak a Latin-based language. The truth is that Latin is not an active language, and Middle America has created its own cultural identity in spite of the impact of colonialism, and the realm can be defined by its people and their activities as much as by its physical environment.

Figure 1. Middle America: Caribbean, Mexico, and Central America. Central America includes the countries south of Mexico through Panama.

7.1 Introducing the Realm

Learning Objectives

1. Define the differences between the rimland and the mainland.
2. Summarize the impact of European colonialism on Middle America.
3. Distinguish between the Mayan and Aztec Empires and identify which the Spanish defeated.
4. Describe how the Spanish influenced urban development.

A. Physical Geography

Middle America has various types of physical landscapes, including volcanic islands and mountain ranges. Tectonic action at the edge of the **Caribbean Plate** has brought about volcanic activity, creating many of the islands of the region as volcanoes rose above the ocean surface. The island of Montserrat is one such example. The volcano on this island has continued to erupt in recent years, showering the island with dust and ash and making habitation difficult. Many of the other low-lying islands, such as the Bahamas, were formed by coral reefs rising above the ocean surface. Tectonic plate activity not only has created volcanic islands but also is a constant source of earthquakes that continue to be a problem for the Caribbean community.

The republics of Central America extend from Mexico to Colombia and form the final connection between North America and South America. The **Isthmus of Panama**, the narrowest point between the Caribbean Sea and the Pacific Ocean, serves as a **land bridge** between the continents. The backbone of Central America is mountainous, with many volcanoes located within its ranges. Much of the Caribbean and all of Central America are located south of the Tropic of Cancer and are dominated by tropical type A climates. The mountainous areas have varied climates, with cooler climates located at higher elevations. Mexico has extensive mountainous areas with two main ranges in the north and highlands in the south. There are no landlocked countries in this realm, and coastal areas have been exploited for fishing and tourism development.

B. Rimland and Mainland

Using a regional approach to the geography of a realm helps us compare and contrast a place's features and characteristics. Location and the physical differences explain the division of Middle America into two geographic areas according to occupational activities and colonial dynamics: the **rimland**, which includes the Caribbean islands and the Caribbean coastal areas of Central America, and the **mainland**, which includes the interior of Mexico and Central America.

Colonialism thrived in the rimland because it consists mainly of islands and coastal areas that were accessible to European ships. Ships could easily sail into a cove or bay to make port and claim the island for their home country. After an island or coastal area was claimed, there was unimpeded transformation of the area through **plantation agriculture**.

On a plantation, local individuals were subjugated as servants or slaves. The land was planted with a single crop—usually sugarcane, tobacco, cotton, or fruit—grown for export profits. Most of these crops were not native to the Americas but were brought in during colonial times. European diseases killed vast numbers of local Amerindian laborers, so slaves were brought from Africa to do the work. Plantation agriculture in the rimland was successful because of the import of technology, slave labor, and raw materials, as well as the export of the harvest to Europe for profit.

Plantation agriculture changed the rimland. The local groups were diminished because of disease and colonial subjugation, and by the 1800s most of the population was of African descent. Native food crops for consumption gave way to

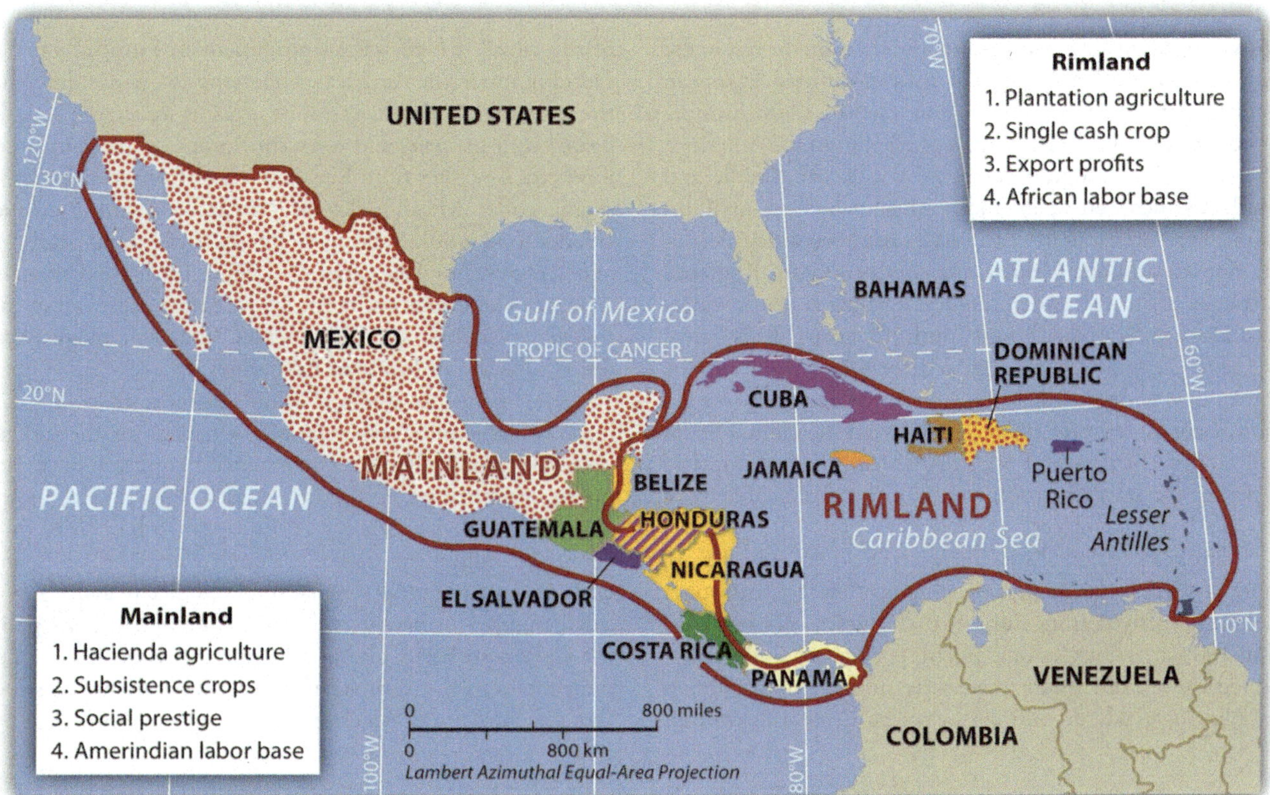

Figure 2. Mainland and Rimland Characteristics of Middle America Based on Colonial-Era Economic Activities. The rimland was more accessible to European ships, and the mainland was more isolated from European activity.

cash crops for export. **Marginal lands** were plowed up and placed into the plantation system. The labor was usually seasonal: there was a high demand for labor at peak planting and harvest times. Plantations were generally owned by wealthy Europeans who may or may not have actually lived there.

The mainland, consisting of Mexico and the interior of Central America, diverged from the rimland in terms of both colonial dynamics and agricultural production. The interior lacked the easy access to the sea that the rimland enjoyed. As a result, the **hacienda** style of land use developed. This Spanish innovation was aimed at land acquisition for social prestige and a comfortable lifestyle. Export profits were not the driving force behind the operation, though they may have existed. The indigenous workers, who were poorly paid if at all, were allowed to live on the haciendas, working their own plots for subsistence. African slaves were not prominent in the mainland.

In the mainland, European colonialists would enter an area and stake claims to large portions of the land, often as much as thousands or even in the millions of acres. Haciendas would eventually become the main landholding structure in the mainland of Mexico and many other regions of Middle America. In the hacienda system, the Amerindian people lost ownership of the land to the European colonial masters. Land ownership or the control of land has been a common point of conflict throughout the Americas where land transferred from a local indigenous ownership to a colonial European ownership.

The plantation and hacienda eras are in the past. The abolition of slavery in the later 1800s and the cultural revolutions that occurred on the mainland challenged the plantation and hacienda systems and brought about land reform. Plantations were transformed into either multiple private plots or large corporate farms. The hacienda system was broken up, and most of the hacienda land was given back to the people, often in the form of an **ejidos system**, in which the community owns the land but individuals can profit from it by sharing its resources. The ejidos system has created its own set of problems, and many of the communally owned lands are being transferred to private owners.

The agricultural systems changed Middle America by altering both the systems of land use and the ethnicity of the population. The Caribbean Basin changed in ethnicity from being entirely Amerindian, to being dominated by European colonizers, to having an African majority population. The mainland experienced the mixing of European culture with the Amerindian culture to form various types of **mestizo** groups with Hispanic, Latino, or Chicano identities.

C. The European Invasion

European colonialism had an immense effect on the rest of the world. Among other things, colonialism diffused European languages and the Christian religion. Consider how European colonialism altered language and religion in the Americas. The two main European countries to colonize Middle and South America were Spain and Portugal, which are predominantly Roman Catholic. Latin Mass has been a tradition in the Roman Catholic Church, and the Latin-based Romance languages of Spanish and Portuguese are now the most widely used languages in Middle and South America. This is where the term *Latin America* originated for this realm, and the name is still widely used. Today, however, *Middle America* is a more accurate term for the region between the United States and South America, and South America is the appropriate name for the southern continent, in spite of the connection to Latin-based languages.

European colonialism impacted Middle America in more ways than language and religion. Before Christopher Columbus arrived from Europe, the Americas did not have animals such as horses, donkeys, sheep, chickens, and domesticated cattle. This meant there were no large draft animals for plowing fields or carrying heavy burdens. The concept of the wheel, which was so prominent in Europe, was not found in use in the Americas. Food crops were also different: the potato was an American food crop, as were corn, squash, beans, chili peppers, and tobacco. Europeans brought other food crops—either from Europe itself or from its colonies— such as coffee, wheat, barley, rice, citrus fruits, and sugarcane. Besides food crops, building methods, agricultural practices, and even diseases were exchanged. Since it occurred after the arrival of Columbus in the late 15th century, this widespread transfer of plants, animals, diseases, and technology is referred to as the **Columbian Exchange**.

The Spanish invasion of Middle America following Columbus had some devastating consequences for the indigenous populations. It has been estimated that fifteen to twenty million people lived in Middle America when the Europeans arrived, but after a century of European colonialism, only about 2.5 million remained. Through warfare, disease, and enslavement, the local populations were decimated. Only a small number of people still claim Amerindian heritage in the Caribbean Basin, and some argue that these few are not indigenous to the Caribbean but are descendants of slaves brought from South America by European colonialists.

D. The Maya and the Aztec

Though the region of Mexico has been inhabited for thousands of years, one of the earliest cultures to develop into a civilization with large cities was the **Olmec**, which was believed to be the precursor to the later **Mayan Empire**. The Olmec flourished in the south-central regions of Mexico from 1200 BCE to about 400 BCE. Anthropologists call this region of Mexico and northern Central America **Mesoamerica**. It is considered to be the region's cultural hearth because it was home to early human civilizations. The Maya established a vast civilization after the Olmec, and Mayan stone structures remain as major tourist attractions. The classical era of the Mayan civilization lasted from 300 to 900 CE and was centered in the Yucatán Peninsula region of Mexico, Belize, and Central America. Guatemala was once a large part of this vast empire, and Mayan ruins are found as far south as Honduras. During the classical era, the Maya built some of the most magnificent cities and stone pyramids in the Western Hemisphere. The city-states of the empire functioned through a sophisticated religious hierarchy. The Mayan civilization made advancements in mathematics, astronomy, engineering, and architecture. They developed an accurate calendar based on the seasons and the solar system. The extent of their immense knowledge is still being discovered. The descendants of the Maya people still exist today, but their empire does not.

The Toltec, who controlled central Mexico briefly, came to power after the classical Mayan era. They also took control of portions of the old Mayan Empire from the north. The **Aztec** federation replaced the Toltec and Maya as the dominant civilization in southern Mexico. The Aztec, who expanded

Figure 3. Mayan Site of Uxmal in the Yucatán region of Mexico. The classical Mayan era lasted from 300 to 900 CE. Many magnificent cities were built with stone and remain today as major tourist attractions.

outward from their base in central Mexico, built the largest and greatest city in the Americas of the time, **Tenochtitlán**, with an estimated population of one hundred thousand. Tenochtitlán was located at the present site of Mexico City, and it was from there the Aztec expanded into the south and east to create an expansive empire. The Aztec federation was a regional power that subjugated other groups and extracted taxes and tributes from them. Though they borrowed ideas and innovations from earlier groups such as the Maya, they made great strides in agriculture and urban development. The Aztec rose to dominance in the fourteenth century and were still in power when the Europeans arrived.

E. Spanish Conquest of 1519–21

After the voyages of Columbus, the Spanish conquistadors came to the New World in search of gold, riches, and profits, bringing their Roman Catholic religion with them. Zealous church members sought to convert the "heathens" to their religion. One such conquistador was **Hernán Cortés**, who, with his 508 soldiers, landed on the shores of the Yucatán in 1519. They made their way west toward the Aztec Empire. The wealth and power of the Aztecs attracted conquistadors such as Cortés, whose goal was to conquer. Even with metal

armor, steel swords, sixteen horses, and a few cannons, Cortés and his men did not challenge the Aztecs directly. The Aztec leader Montezuma II originally thought Cortés and his men were legendary "White Gods" returning to recover the empire. Cortés defeated the Aztecs by uniting the people that the Aztecs had subjugated and joining with them to fight the Aztecs. The Spanish conquest of the Aztec federation was complete by 1521.

F. The Spanish Colonial City

As the Spanish established urban centers in the New World, they structured each town after the Spanish pattern, with a **plaza** in the center. Around the plaza on one side was the church (Roman Catholic). On the other sides were government offices and stores. Residential homes filled in around them. This pattern can still be seen in almost all the cities built by the Spanish in Middle and South America. The Catholic Church not only was located in the center of town but also was a supreme cultural force shaping and molding the Amerindian societies conquered by the Spanish.

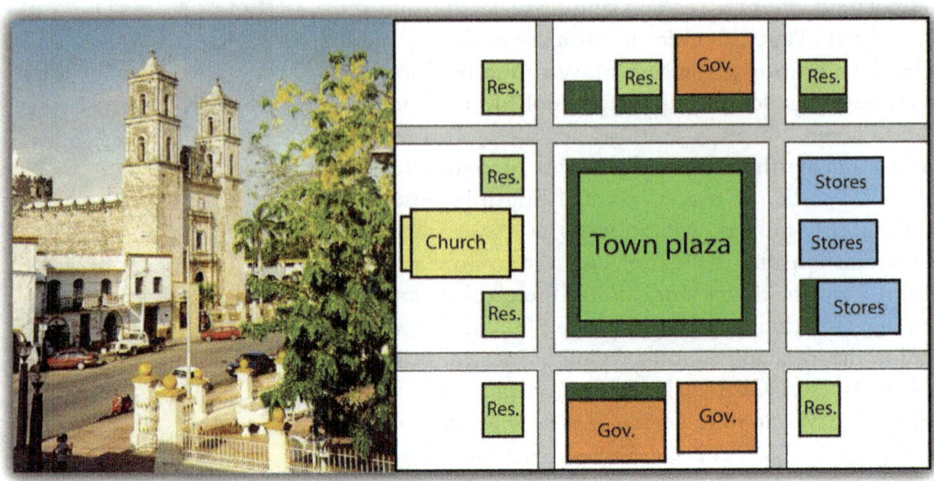

Figure 4. Catholic cathedral across from a plaza in the Yucatán City of Valladolid (left); Model of a Spanish colonial urban pattern (right). The Spanish colonial urban pattern had a plaza in the center of the city with government buildings around the square and a Catholic church on one side.

In Spain, the cultural norm was to develop urban centers wherever administration or military support was needed. Spanish colonizers followed a similar pattern in laying out the new urban centers in their colonies. Extending out from the city center (where the town plaza, government buildings, and church were located) was a commercial district that was the backbone of this model. Expanding out on each side of the spine was a wealthy residential district for the upper social classes, complete with office complexes, shopping districts, and upper-scale markets.

Surrounding the **central business district (CBD)** and the spine of most cities in Middle and South America are concentric zones of residential districts for the lower, working, and middle classes and the poor. The first zone, the **zone of maturity**, has well-established middle-class residential neighborhoods with city services. The second concentric zone, the **zone of transition (in situ accretion)**, has poorer working-class districts mixed with areas with makeshift housing and without city services. The outer zone, the **peripheral zone** (Spanish, **periférico**), is where the expansion of the city occurs, with makeshift housing and squatter settlements. This zone has little or no city services and functions on an informal economy. This outer zone often branches into the city, with slums known as **favelas** or **barrios** that provide the working poor access to the city without its benefits. Impoverished immigrants that arrive in the city from the rural areas often end up in the city's outer periphery to eke out a living in some of the worst living conditions in the world.

Cities in this Spanish model grow by having the outer ring progress to the point where eventually solid construction

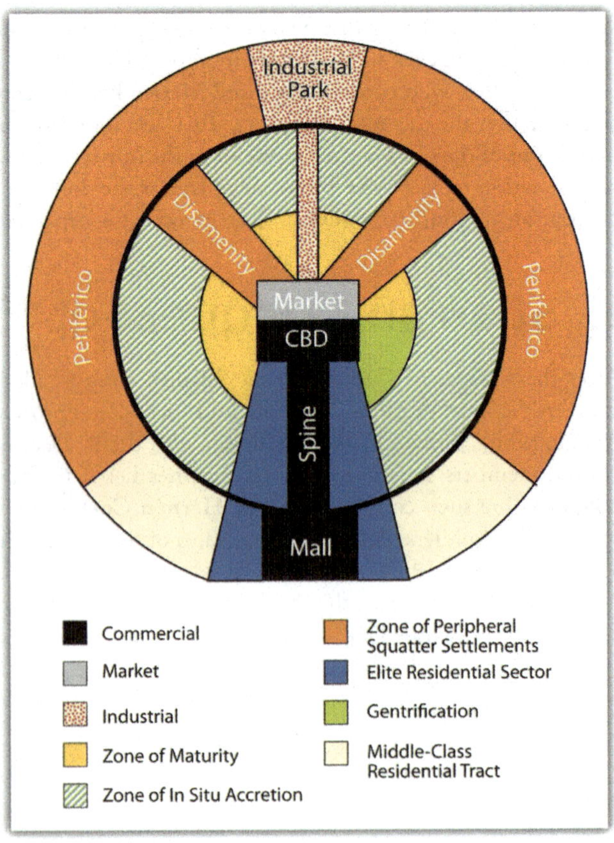

Figure 5. Spanish-American city structure according to the Ford-Griffin model.

takes hold and city services are extended to accommodate the residents. When this ring reaches maturity, a new ring of squatter settlements emerges to form a new outer ring of the city. The development dynamic is repeated, and the city continues to expand outward. The urban centers of Middle and South America are expanding at rapid rates. It is difficult to provide public services to the outer limits of many of the cities. The barrios or favelas become isolated communities, often complete with crime bosses and gang activities that replace municipal security.

Key Takeaways

1. Haciendas were located chiefly in the mainland and plantations were located mainly in the rimland.

2. Both the hacienda and the plantation structures of agriculture altered the ethnic makeup of their respective regions. The rimland had an African labor base, and the mainland had an Amerindian labor base.

3. In their quest for wealth, Spanish conquistadors destroyed the Aztec Empire and colonized the Middle American mainland. Much historical knowledge was lost with the demise of the learned class of the Aztec Empire.

4. Europeans introduced many new food crops and domesticated animals to the Americas and in turn brought newly discovered agricultural products from America back to Europe. This is known as the Columbian Exchange.

5. The Spanish introduced the same style of urban planning to the Americas that was common in Spain. Many cities in Middle and South America were patterned after Spanish models.

7.2 Mexico

Learning Objectives

1. Describe the physical geography of Mexico, identifying the core and peripheral areas.
2. Outline the socioeconomic classes in Mexico and explain the ethnic differences of each.
3. Explain how the North American Free Trade Agreement (NAFTA) and maquiladoras have influenced the economic and employment situations in Mexico.
4. Understand how the drug cartels have become an integrated part of the Mexican economy and political situation.

A. Physical Characteristics

Mexico is the eighth-largest country in the world and is about one-fifth the size of the United States. One of Mexico's prominent geographical features is the world's longest peninsula, the 775-mile-long **Baja California Peninsula**, which lies between the Pacific Ocean and the Gulf of California (also known as the Sea of Cortez). The Baja California Peninsula includes a series of mountain ranges called the Peninsular Ranges.

The Tropic of Cancer cuts across Mexico, dividing it into two different climatic zones: a temperate zone to the north and a tropical zone to the south. In the northern temperate zone, temperatures can be hot in the summer, often rising well above 80 °F, but considerably cooler in the winter. By contrast, temperatures vary very little from season to season in the tropical zone, with average temperatures hovering very close to 80 °F year-round. Temperatures in the south tend to vary as a function of elevation.

Mexico is characterized by a great variety of climates, including areas with hot humid, temperate humid, and arid climates. There are mountainous regions, foothills, plateaus, deserts, and coastal plains, all with their own climatic conditions. For example, in the northern desert portions of the

Figure 6. Major Volcanoes of Mexico.

country, summer and winter temperatures are extreme. Temperatures in the Sonoran and Chihuahuan Deserts exceed 110 °F, while in the mountainous areas snow can be seen at higher elevations throughout the year.

Two major mountain ranges extend north and south along Mexico's coastlines and are actually extensions of southwestern US ranges. The **Sierra Madre Occidental**, an extension of the Sierra Nevada range, runs about 3,107 miles along the west coast, with peaks higher than 9,843 feet. The **Sierra Madre Oriental** is an extension of the Rocky Mountains and runs 808 miles along the east coast. Between these two mountain ranges lies a group of broad plateaus, including the Mexican Plateau, or **Mexican Altiplano** (a wide valley between mountain ranges). The central portions, with their rolling hills and broad valleys, include fertile farms and productive ranch land. The Mexican Altiplano is divided into northern and southern sections, with the northern section dominated by Mexico's most expansive desert, the Chihuahuan Desert.

Another prominent mountain range is the **Cordillera Neovolcánica** range, which as its name suggests, is a range of volcanoes that runs nearly 620 miles east to west across the central and southern portion of the country. Geologically speaking, this range represents the dividing line between North and Central America. The peaks of the Cordillera Neovolcánica can reach higher than 16,404 feet in height and are snow covered year-round.

Copper Canyon, in the northern Mexican state of Chihuahua, is about seven times larger than the Grand Canyon. Copper Canyon was formed by six rivers flowing through a series of twenty different canyons. Besides covering a larger area than the Grand Canyon, at its deepest point, Copper Canyon is 1,462 feet deeper than the Grand Canyon.

Though sandy beaches often come to mind when thinking about Mexico, the mountainous regions are home to pine-oak forests. More than a quarter of Mexico's landmass is covered in forests; as a result, timber is an important natural resource. Mexico ranks fourth in the world for biodiversity; it has the world's largest number of reptile species, ranks second for mammals, and ranks fourth for the number of amphibian and plant species. It is estimated that more than 10 percent of the world's species live here. Forest depletion is

Figure 7. Mexican economic core area centered on urban areas around Mexico City. The periphery is the northern region, including the border area, and the southern region, including the Yucatán Peninsula.

a key environmental concern, but timber remains an important natural resource.

Three tectonic plates underlie Mexico, making it one of the most seismically active regions on earth. In 1985, an earthquake centered off Mexico's Pacific coast killed more than ten thousand people in Mexico City and did significant damage to the city's infrastructure.

Many of Mexico's natural resources lie beneath the surface. Mexico is rich in natural resources and has robust mining industries that tap large deposits of silver, copper, gold, lead, and zinc. Mexico also has a sizable supply of salt, fluorite, iron, manganese, sulfur, phosphate, tungsten, molybdenum, and gypsum. Natural gas and petroleum also make the list of Mexico's natural resources and are important export products to the United States. There has been some concern about declining petroleum resources; however, new reserves are being found offshore in the Gulf of Mexico.

Though only about 13 percent of Mexico's land area is cultivated, favorable climatic conditions mean that food products are also an important natural resource both for export and for the feeding the country's sizable population. Tomatoes, maize (corn), vanilla, avocado, beans, cotton, coffee, sugarcane, and fruit are harvested in sizable quantities. Of these, coffee, cotton, sugarcane, tomatoes, and fruit are primarily grown for export, with most products bound for the United States.

Mexico has very pronounced wet and dry seasons. Most of the country receives rain between June and mid-October, with July being the wettest month. Much less rain occurs during the other months: February is usually the driest month. More importantly, Mexico lies in the middle of the hurricane belt, and all regions of both coasts are at risk for these storms between June and November. Hurricanes along the Pacific coast are much less frequent and less violent than those along Mexico's Gulf and Caribbean coasts. Hurricanes can cause extensive damage to infrastructure along the coasts where major tourist resorts are located. Mexico's extensive and beautiful coastlines provide an important boon to the nation's tourism industry.

B. The Core versus the Periphery

The Mexican economy is a mix of modern industry, agriculture, and tourism. Current estimates indicate that the service sector makes up about 60 percent of the economy, followed by the industrial sector at 33 percent. Agriculture represents just above 4 percent. Per capita income in Mexico is about one-third of what it is in the United States. The Mexican labor force is estimated at forty-six million individuals; 14 percent of the labor force work in agriculture, 23 percent in the industrial sector, and 62 percent in the service sector.

Mexico is an example of a country with a clear **core-periphery spatial relationship**. Mexico City and its surrounding metropolitan centers represent the country's core: the center of activity, industry, wealth, and power. Industries and manufacturing have been traditionally located in this region. The core region has most of the country's 110 million people (as of 2010). Mexico's population is about 77 percent urban, with the largest urban areas found in the core region.

Mexico City is one of the largest cities in the world and anchors the core region of Mexico. In 2010, the official population of Mexico City was about eighteen million, but unofficial population estimates are as high as thirty million. The actual population of Mexico City is unknown because of the hundreds of slums that surround the city on the slopes of the central valley. Mexico City is growing at a rate of more than one thousand people per day through a combination of the number of births and the number of migrants. The

Figure 8. Mexico City on a clear day, with the ridge of the mountains visible in the background. A day with extensive air pollution will restrict the view of the horizon.

Figure 9. Mayan home in the rural village of Yachachen in the Yucatán Peninsula. This is located in a peripheral region of Mexico.

lure of opportunities and advantages still pulls migrants to the city in search of a better life. Higher populations tax the resources in rural areas, where jobs and opportunities are hard to find. This push-pull relationship creates a strong rural-to-urban shift in Mexico. This same trend is found throughout the developing world.

Mexico City is a historic and vibrant city, but is not without problems. At higher than seven thousand feet in elevation, it is located between two mountain ranges. Air pollution is severe and is augmented by frequent temperature inversions that trap pollution over the city. Fresh water is in short supply, and wastewater from sewage is discharged into lakes down the valley. Amerindians who live by these lakes or on the islands have to deal with the pollution. Because about four to five million inhabitants of Mexico City have no utilities, human waste buildup has become a challenge. Fresh water is pumped into the city through pipelines from across the mountains. Leakage and inadequate maintenance cause a large percentage of the water to be lost before it can be used in the city. Water is also drawn from underground aquifers beneath the city, which has caused parts of the city to sink as much as two feet, causing serious structural damage to historic buildings.

C. Mexican Social Order

The early European control of the land, the economy, and the political system created conflict for the people of Mexico. The country has experienced domination followed by revolution at various times, starting with colonial domination, then economic domination, and lastly political domination. The result was a mixing of Europeans and Amerindians, creating the current **mestizo** mainstream society. Mestizos make up about 60 percent of the current population, Europeans make up about 9 percent, and Amerindians make up about 30 percent. More than sixty indigenous languages spoken by Amerindian groups are recognized in Mexico.

Mexican society is regionally and ethnically diverse, with sharp socioeconomic divisions. Many rural communities have strong ties with their regions and are often referred to as *patrias chicas* ("small homelands"), which helps to perpetuate the cultural diversity. The large number of indigenous languages and customs, especially in the southern parts of Mexico, further emphasize cultural diversity. *Idigenismo* ("pride in the indigenous heritage") has been a unifying theme of Mexico since the 1930s. However, daily life in Mexico can be dramatically different according to socioeconomic class, gender, ethnicity, rural or urban settlements, and other cultural differences. A peasant farmer in the rainforests of the Yucatán will lead a very different life than a museum curator in Mexico City or a lower-middle-class auto factory worker in Monterrey.

Those of European descent are at the top of the pyramid and control a higher percentage of the wealth and power even though they are a minority of the population. The small middle class is largely mestizo, including managers, business people, and professionals. The working poor make up most of the population at the bottom of the pyramid. The lower class contains the highest percentage of people of Amerindian descent or, in the case of the Caribbean, African descent.

The most desirable type of social structure is illustrated by a diamond shape: in the middle is a large, employed middle class that can pay most of the taxes and purchase consumer goods that help bolster the economy. The narrow top is made up of the richest, and the narrow bottom is made up of the poorest.

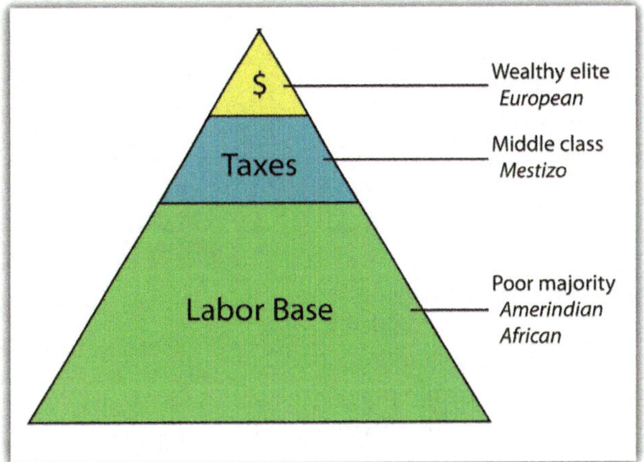

Figure 10. Socioeconomic classes in Mexico and most of Latin America. The current social status of Mexican society can be illustrated by a pyramid shape.

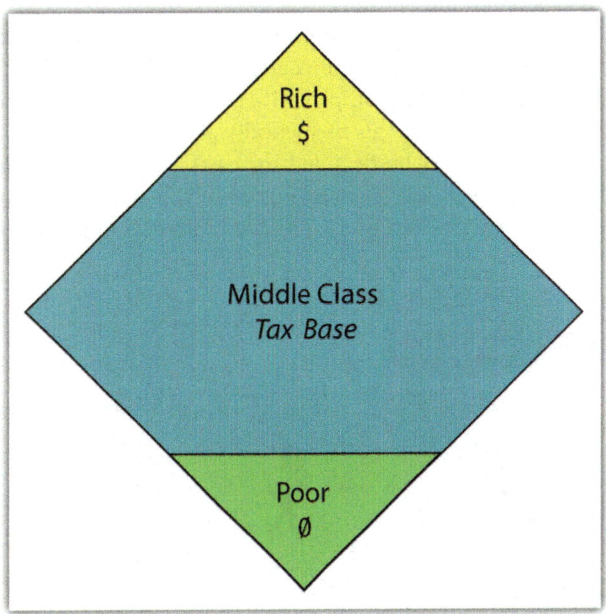

Figure 11. A more ideal socioeconomic class structure with a large middle-class tax base.

Unfortunately, this optimal type of social structure does not always materialize in the manner hoped for. As an example, a goal of the economic planners of the United States has been to create a wide social profile. Unfortunately, in recent decades the US middle class has been declining, and the wealthy class and the working-class poor have increased. In Mexico, about 40 percent of the population lives in poverty.

Over the course of the past century, the people of Mexico have been working through a demographic transition. As the rural regions of Mexico continued to have a high fertility rate, death rates declined, and the country's population grew exponentially. In 1970, the population of Mexico was about fifty million. By the year 2000, it had doubled to more than one hundred million. However, the population estimate for 2010 was just greater than 110 million. As Mexico urbanizes and industrializes, family size and fertility rates have been in decline, and population growth has slowed.

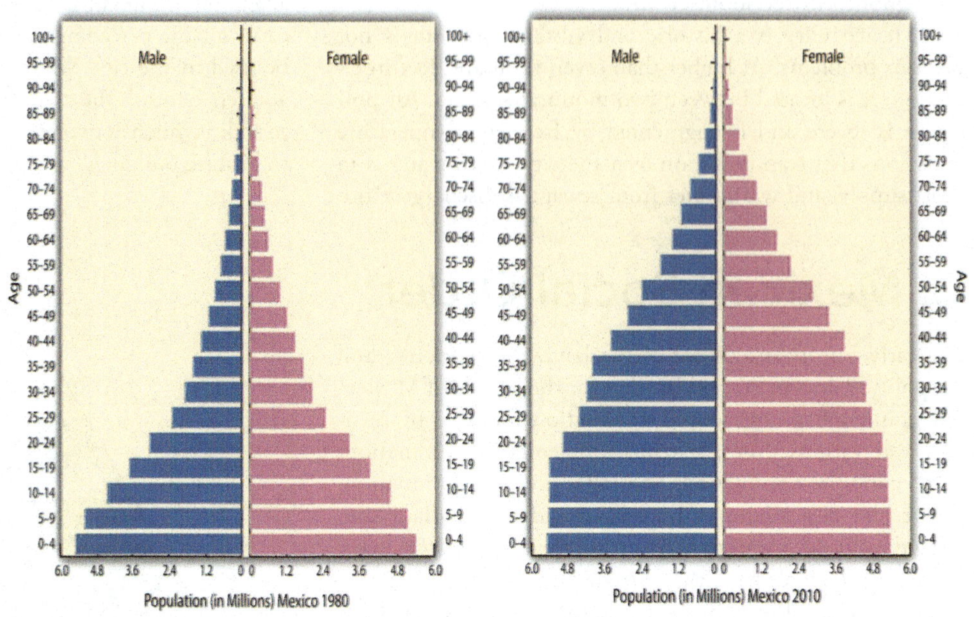

Figure 12. Population pyramids for Mexico in 1980 and 2010. The 1980 pyramid indicates rapid population growth. The 2010 pyramid illustrates a slight decline in the past few years.

D. NAFTA and Maquiladoras

The **North American Free Trade Agreement** (**NAFTA**), is a 1994 economic agreement between Canada, the United States, and Mexico that eliminated or reduced the tariffs, taxes, and quotas between the countries to create the world's largest trading bloc to compete with the European Union and the global economy. This theoretically allows more corporate investments across borders and increases foreign ownership of business facilities. It stimulated a shift in the location of industrial activity and in the migration patterns of people in Mexico. Capitalizing on the old industrial locations of northern Mexico, such as Monterrey, corporations started to relocate manufacturing plants from the United States to the Mexican side of the border to take advantage of Mexico's low-cost labor. The aspect of cheaper labor was a benefit understood to bolster corporate profits and reduce product costs. The United States is one of the world's largest consumer markets, so these manufacturing plants, called maquiladoras (also known as *maquilas*), could benefit both countries.

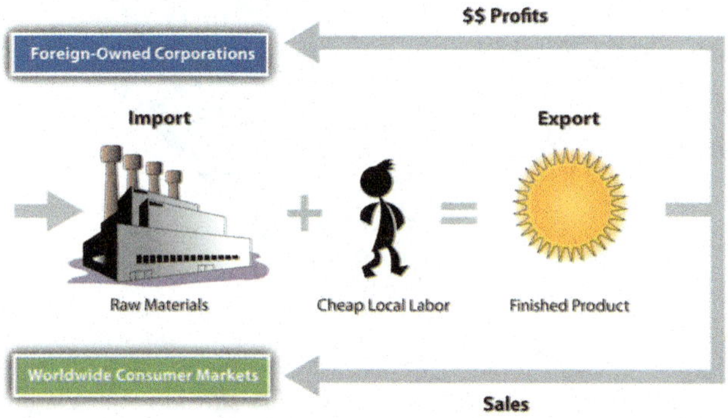

Maquiladoras are foreign-owned factories that import most of the raw materials or components needed for the products they manufacture, assemble, or process with local cheap labor, and then they export the finished product for profit. US corporations own more than half the maquiladoras in Mexico, and about 80 percent of the finished goods are exported back to the United States. Although most maquiladoras are located near the US-Mexican border, additional factories are located around Monterrey and other cities with easy access to the United States. A major trade corridor is developing between Monterrey and Dallas/Ft. Worth, which acts as a doorway to the US markets.

Figure 13. Import/export system of a Maquiladora operation

Figure 14. US-Mexican border. Opportunities and advantages drive the push-pull of migrants searching for improved economic conditions.

southern regions of Mexico to the border region to look for work. When they do not find work, they are tempted to cross the US border illegally. The United States is considered a land of opportunity and attracts immigrants—both legal and illegal—from Mexico.

Critics of NAFTA claim that the term *free trade* really means corporate trade. NAFTA is also viewed as a component of globalization in the form of corporate colonialism, which only benefits those wealthy enough to hold investments at the corporate level. The exploitation of cheap labor has caused undue immigration across the US-Mexican border, bringing millions of illegal workers into the United States. The Mexican government has not adequately addressed Mexico's economic conditions to provide jobs and opportunities for the people or to use the wealth held or controlled by the elite minority to enhance economic opportunities for the middle- and lower-class majority.

Thousands of maquiladoras flourish along the US-Mexican border, although the Mexican government has also promoted maquiladoras in other parts of Mexico. Maquiladoras provide jobs for workers in Mexico and provide cheaper goods for US consumers. However, this system has inherent problems. Labor unions in the United States complain that the high-paying industrial jobs that support the US middle class are being lost to cheap Mexican labor. Labor laws in Mexico are less rigorous than US laws, allowing for longer work hours and fewer benefits for maquiladora employees. In addition, pollution standards in Mexico are not as restrictive as those in the US, giving rise to environmental concerns. With the rapid increase in employment along the border, many of the people who work in the factories do not have adequate housing or utilities. Extensive slum areas have grown around maquiladoras, which have little law enforcement, high crime, and few services.

The US-Mexican border region has become a strong pull factor, enticing poor people who seek greater opportunities and advantages to move from Mexico City and other

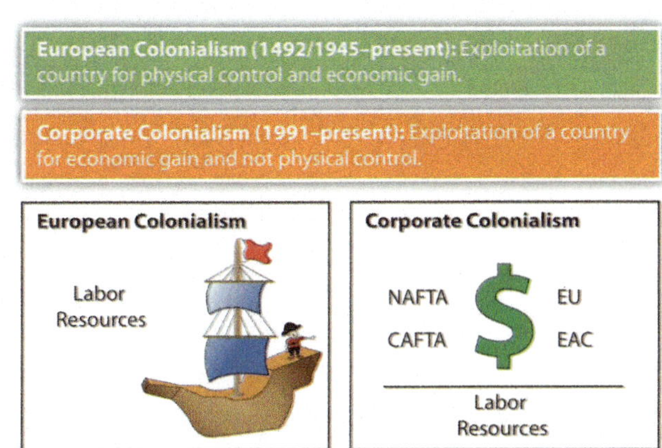

Figure 15. Labor and resources in globalization. Slave labor was prominent during European colonial era, whereas cheap labor is the target in neocolonial activity—that is, corporate colonialism.

E. Illegal Drug Trafficking in Mexico

The illegal drug trade is a multibillion-dollar industry, and Mexico has traditionally been the transitional area or stop-off point between the South American drug producing areas and entrance into US markets. Cocaine, marijuana, and more recently heroin were produced in the Andes Mountains of South America and shipped north to the United States. Colombian cartels were once the main controllers of illegal drugs in the Western Hemisphere, but in recent decades, organized crime units in Mexico have muscled in on the control of drugs coming through Mexico, making deals with their South American counterparts to become the main traffickers of drugs into the United States, and the influence and power of Mexican drug cartels has increased immensely since the demise of the Colombian cartels in the 1990s. Enormous profits fuel the competition for control.

Just as the United States has declared a war on drugs and has used its Drug Enforcement Administration (DEA) as a main arm in combating the industry, the Mexican government has been engaged in its own internal war against the illegal drug trade. The battles between the drug cartels and the Mexican government have created a serious internal conflict in the country, killing thousands of innocent bystanders in the cross fire. Bribes, payoffs, and corruption have been difficult to battle in a country with a high percentage of the population living in poor conditions.

Key Takeaways

1. Mexico possesses extensive natural resources that provide for a wide range of biodiversity and economic activities.

2. Mexico portrays a clear core-periphery spatial relationship. Mexico City and its urban neighbors anchor the core, while the northern border region and states such as Chiapas represent the rural periphery.

3. People of European heritage continue to hold positions of power and privilege in Mexico's socioeconomic class structure. Amerindian populations exist at the lowest level with the fewest economic opportunities.

4. Economic reforms that coincide with NAFTA have greatly enhanced the industrial capacity of Mexico and helped integrate the country into the global economy.

5. Mexican drug traffickers have become the major controllers of illegal drugs entering the United States from the south. Drug cartels in Mexico reap enormous profits and have become a major problem for the Mexican government and the country.

7.3 Central America

A. Physical Environment

Central America is a land bridge connecting the North and South American continents, with the Pacific Ocean to its west and the Caribbean Sea to its east. A central mountain chain dominates the interior from Mexico to Panama. The coastal plains of Central America have tropical and humid type A climates. In the highland interior, the climate changes with elevation. As one travels up the mountainsides, the temperature cools. Only Belize is located away from this interior mountain chain. Its rich soils and cooler climate have attracted more people to live in the mountainous regions than along the coast.

The volcanic activity along the central mountain chain over time has provided rich volcanic soils in the mountain region, which has attracted people to work the land for agriculture. Central America has traditionally been a rural peripheral economic area in which most of the people have worked the land. Family size has been larger than average, and rural-to-urban shift dominates the migration patterns as the region urbanizes and industrializes. Natural disasters, poverty, large families, and a lack of economic opportunities have made life difficult in much of Central America.

Altitudinal Zonation

High mountains ranges run the length of Central and South America. The Andes Mountains of South America are the longest mountain chain in the world, and a large section of this mountain range is in the tropics. Tropical regions usually have humid type A climates. What is significant in Latin America is that while the climate at the base of the Andes may be type A, the different zones of climate and corresponding human activity vary as one moves up the mountain in elevation. Mountains have different climates at the base than at the summit. Type H highland climates describe mountainous areas that exhibit different climate types at varying degrees of elevation.

Human activity varies with elevation, and the activities can be categorized into zones according to **altitudinal zonation**. Each zone has its own type of vegetation and agricultural activity suited to the climate found at that elevation. For every thousand-foot in-

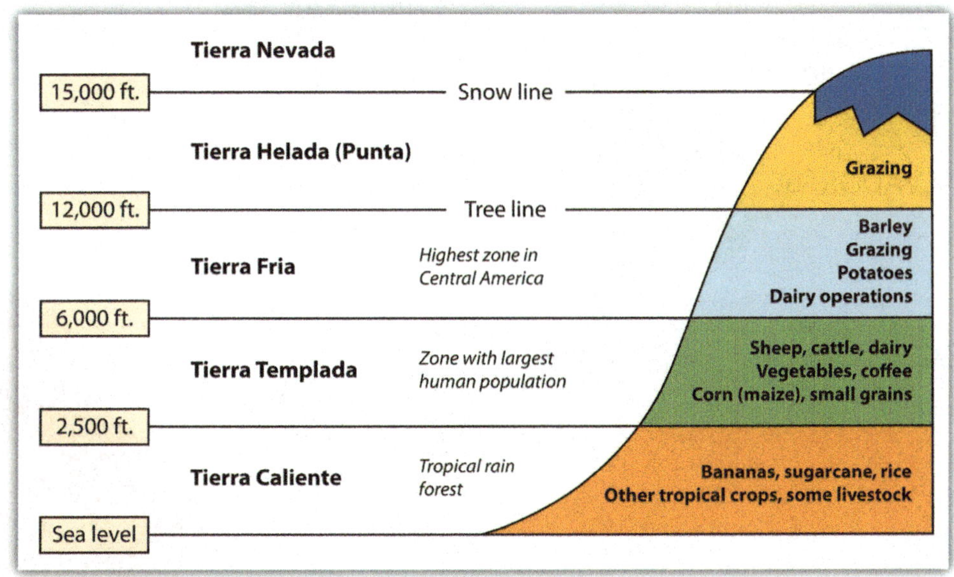

Figure 16. Altitudinal zonation system in latin America for locations near the equator.

crease in elevation, temperature drops 3.6 °F. In the tropical areas of Latin America, there are five established temperature-altitude zones. These five zones and their corresponding elevation ranges are listed below. Note that these elevation ranges are for locations near the equator. The upper elevation limit for each zone generally decreases as you move further from the equator.

From sea level to 2,500 feet, **Tierra Caliente** (*Hot Land*) are the humid tropical lowlands found on the coastal plains. The coastal plains on the west coast of Middle America are quite narrow, but they are wider along the Caribbean coast. Vegetation includes tropical rain forests and tropical commercial plantations. Food crops include bananas, manioc, sweet potatoes, yams, corn, beans, and rice. Livestock are raised at this level, and sugarcane is an important cash crop. Tropical diseases are most common, and large human populations are not commonly attracted to this zone.

From 2,500 to 6,000 feet, **Tierra Templada** (*Temperate Land*) is a zone with cooler temperatures than at sea level. This is the most populated zone of Latin America. Four of the seven capitals of the Central American republics are found in this zone. Just as temperate climates attract human activity, this zone provides a pleasant environment for habitation. The best coffee is grown at these elevations, and most other food crops can be grown here, including wheat and small grains.

6,000 to 12,000 feet **Tierra Fria** (*Cold Land*) is the highest zone found in Middle America. This zone is usually the limit of the tree line; few trees grow north of this zone. The shorter growing season and cooler temperatures found at these elevations are still adequate for growing agricultural crops of wheat, barley, potatoes, or corn. Livestock can graze and be raised on the grasslands. The Inca Empire of the Andes Mountains in South America flourished in this zone.

Some classify **Tierra Helada** (*Frozen Land*) as the "Puna" zone. At this elevation, there are no trees. The only human activity is the raising of livestock such as sheep or llama on any short grasses available in the highland meadows. Snow and cold dominate the zone. Central America does not have a *tierra helada* zone, but it is found in the higher Andes Mountain Ranges of South America.

There is little human activity in the Tierra Nevada (*Snowy Land*) above 15,000 feet. Permanent snow and ice is found here, and little vegetation is available. Many classification systems combine this zone with the *tierra helada* zone.

Figure 17. Central American republics.

B. European Colonialism

Amerindian groups dominated Central America before the European colonial powers arrived. The Maya are still prominent in the north and make up about half the population of Guatemala. Other Amerindian groups are encountered farther south, and many still speak their indigenous languages and hold to traditional cultural customs. People of European stock or upper-class mestizos now control political and economic power in Central America. Indigenous Amerindian groups find themselves on the lower rung of the socioeconomic ladder.

During colonial times, the Spanish conquistadors dominated Central America with the exception of the area of Belize, which was a British colony called British Honduras until 1981. Guatemala, El Salvador, Honduras, Nicaragua, and Costa Rica were Spanish colonies and became independent of Spain in the 1820s. Panama was a part of Colombia and was not independent until the United States prompted an independence movement in 1903 to develop the Panama Canal. As is usually the case with colonialism, the main religion and the lingua franca of the Central American states are those of the European colonizers, in this case Roman Catholicism and Spanish. In some locations, the language and religion take on variant forms that mix the traditional with the European to create a unique local cultural environment.

C. People and Population

About 50 percent of the people of Central America live in rural areas, and because the economy is agriculturally based, family size has traditionally been large. Until the 1990s, family size averaged as high as six children. As the pressures of the postindustrial age have influenced Central America, average family size has been decreasing and is now about half that of the pre-1990s and is declining. For example, the World Bank reports that in Nicaragua the average woman has 2.68 children during her lifetime. Rural-to-urban shift is common, and as the region experiences more urbanization and industrialization, family size will decrease even more.

During the twentieth century, much of Central America experienced development similar to stage 2 of the index of economic development. An influx of light industry and manufacturing firms seeking cheap labor has pushed many areas into stage 3 development. The primate cities and main urban centers are feeling the impact of this shift.

Over the years, larger family sizes have created populations with a higher percentage of young people and a lower percentage of older people. Cities are often overwhelmed with young migrants from the countryside with few or no places to live. Rapid urbanization places a strain on urban areas because services, infrastructure, and housing cannot keep pace with population growth. Slums with self-constructed housing districts emerge around the existing urban infrastructure. The United States has also become a destination for people looking for opportunities or advantages not found in these cities.

CAFTA and Neocolonialism

Just as Canada, the United States, and Mexico signed the North American Free Trade Agreement (NAFTA) into law in 1994, the United States and five Central American states signed the **Central American Free Trade Agreement** (**CAFTA**) in 2006. The agreement was signed by trade representatives from El Salvador, Honduras, Nicaragua, Guatemala, and the United States. The CAFTA-DR agreement, which includes the Dominican Republic, was ratified in 2007. In 2010, Costa Rica's legislature approved a measure to join the agreement. CAFTA is supported by the same forces that advocated neocolonialism in other regions of the world.

CAFTA's purpose is to reduce trade barriers between the United States and Central America, thus affecting labor, human rights, and the flow of wealth. During negotiations for CAFTA, US political forces cited CAFTA as a top priority and argued that it would help move forward the possibility of the larger **Free Trade Area of the Americas** (**FTAA**), which would create a single market for the Americas.

Countries gain national wealth in the three main ways: by growing it, extracting it, or manufacturing it. These methods, however, contribute to a nation's wealth only if the

Figure 18. Protests against CAFTA in Central America. The banner reads, "For the sovereignty of the people...we demand the repeal of CAFTA (Central American Free Trade Agreement)."

wealth stays within the country. With free-trade agreements such as NAFTA and CAFTA, the wealth gained from manufacturing, which has the highest value-added profits, does not stay in the country of production. Instead, the profits are carted off to the foreign corporation that controls the industrial factory. Multinational corporations see Central American countries as profitable sites for industrial; they can exploit cheap labor sources and at the same time provide jobs for local people. These advantages should result in lower product costs for consumers.

There have been protest marches and anti-CAFTA activities in many Central American countries. One of the primary arguments opponents to CAFTA make is that the wealth generated by the exploitation of the available cheap labor will not stay in Central America; instead, it will be removed by the wealthy core nations, just as European colonialism removed the wealth generated by the conquistadors and shipped it back to Europe.

Supporters of CAFTA claim that it provides jobs, infrastructure, and opportunities to the developing countries of Central America. In return, cheap consumer goods are available to the people. The globalized economy is a mixed game: on the one hand, consumer goods are inexpensive to purchase; on the other hand, the world's wealth flows into the hands of a few people at the top and is not always shared with most of the people who contribute to it.

Key Takeaways

1. Central America shares a similar climate type and physical features. It has enormous potential for tourism development. The political history of the region is quite diverse, with each republic experiencing different political and economic conditions.

2. High population growth and rapid rural-to-urban shift has created higher unemployment rates and fewer economic opportunities. CAFTA was implemented to help multinational corporations tap into the cheap labor pool.

3. The United States has had a major impact on this region both politically and economically. The United States has intervened in civil wars and invaded Panama. US companies have dominated much of the region's fruit and coffee production. Most recently, the United States has supported industrial activities and the implementation of CAFTA.

7.4 The Caribbean

Learning Objectives

1. Describe how the physical environment has affected human activity in the region.

2. Outline the various ways in which colonialism has impacted the islands.

3. Explain why the United States has an economic embargo against the socialist country of Cuba.

4. Explain how tourism has become the main means of economic development for most of the Caribbean.

5. Identify the main music genres that have emerged from the Caribbean.

Figure 19. Caribbean regions of the Greater Antilles, the Lesser Antilles, and the Bahamas

The regions of Middle America and South America, including the Caribbean, follow similar colonial patterns of invasion, dominance, and development by outside European powers. The Caribbean Basin is often divided into the Greater Antilles and the Lesser Antilles (the bigger islands and the smaller islands, respectively). The **Greater Antilles** includes the four large islands of Cuba, Jamaica, Hispaniola, and Puerto Rico. The **Lesser Antilles** are in the eastern and southern region. The Bahamas are technically in the Atlantic Ocean, not in the Caribbean Sea, but they are usually associated with the Caribbean region and are often affiliated with the Lesser Antilles. Middle America can be divided into two geographic areas according to occupational activities and colonial dynamics. The rimland includes the Caribbean islands and the Caribbean coastal areas of Central America. The mainland includes the interior of Mexico and Central America.

Many of the Caribbean islands experience the rain shadow effect. Jamaica has as much as a twenty-inch difference in rainfall between the north side and south side of the island because most of the rain falls on the north side, where the prevailing winds hit the island. The Blue Mountains in the eastern part of the island provide a rain shadow effect. Puerto Rico has a tropical rain forest on the northeastern part of the island, which receives a large amount of rainfall. The rain shadow effect creates semi-desert steppe conditions on the southwestern side of Puerto Rico because the southwestern side receives little rainfall. Low elevation islands such as the Bahamas do not receive as much rain because they are not high enough to affect the precipitation patterns of rain clouds.

A. European Colonialism in the Caribbean

The Spanish were not the only Europeans to take advantage of colonial expansion in the Caribbean: the English, French, Dutch, and other Europeans followed. Most of the European colonial countries were located on the west coast of Europe, which had a seafaring heritage. This included smaller countries such as Denmark, Sweden, and Belgium. The Caribbean Basin became an active region for European ships to enter and vie for possession of each island.

Many of the Caribbean islands changed hands several times before finally being secured as established colonies (Table 7.1). The cultural traits of each of the European colonizers were injected into the fabric of the islands they colonized; thus, the languages, religions, and economic activities of the colonized islands reflected those of the European colonizers rather than those of the native people who had inhabited the islands originally. The four main colonial powers in the Caribbean were the Spanish, English, Dutch, and French. Other countries that held possession of various islands at different times were Portugal, Sweden, and Denmark. The United States became a colonial power when they gained Cuba and Puerto Rico as a result of the Spanish-American War. The US Virgin Islands were purchased from Denmark in 1918. Sweden controlled the island of St. Barthelemy from 1784 to 1878 before trading it back to the French, who had been the original colonizer. Portugal originally colonized Barbados before abandoning it to the British.

Colonialism drastically altered the ethnic makeup of the Caribbean; Amerindians were virtually eliminated after the arrival of Africans, Europeans, and Asians. The current social hierarchy of the Caribbean can be illustrated by the pyramid-shaped graphic that was used to illustrate social hierarchy in Mexico. Those of European descent are at the top of the pyramid and control a higher percentage of the wealth and power even though they are a minority of the population. In the Caribbean, the middle class includes mulattos, or people with both African and European heritage, many of which include managers, businesspeople, and professionals. In some countries, such as Haiti, the minority mulatto segment of the population makes up the power base and holds political and economic advantage over the rest of the country while the working poor at the bottom of the pyramid make up most of the population. In the Caribbean, the lower economic class contains the highest percentage of people of African heritage.

Not only was colonialism the vehicle that brought many Africans to the Caribbean through the slave trade, but it brought many people from Asia to the Caribbean as well. Once slavery became illegal, the colonial powers brought indentured laborers to the Caribbean from their Asian colonies. Cuba was the destination for over one hundred thousand Chinese workers, so Havana can claim the first Chinatown in the Western Hemisphere. Laborers from the British colonies of India and other parts of South Asia arrived by ship in various British colonies in the Caribbean. At the present time, about 40 percent of the population of Trinidad can claim South Asian heritage and a large number follow the Hindu faith.

Colonizer	European Colonies
Spain	Cuba, Dominican Republic, Puerto Rico
British	Bahamas, Jamaica, Cayman Islands, Turks and Caicos Islands, Antigua, Dominica, St. Lucia, St. Vincent, Grenada, Barbados, Virgin Islands, Trinidad and Tobago, Montserrat, Anguilla, St. Kitts and Nevis
Dutch	Curacao, Bonaire, Aruba, St. Eustatius, Saba and Sint Maarten (south half)
French	Haiti, Guadeloupe, Martinique, St. Martin (north half), St. Barthelemy
United States	Puerto Rico, Virgin Islands, Cuba

Table 1. Historical Caribbean Colonizers.

B. The Greater Antilles

Cuba: A Rimland Experience

The largest island in the Greater Antilles is Cuba, which was transformed by the power of colonialism, the transition to plantation agriculture, and a socialist revolution. Cuba is less than half the size of the US state of Oregon, but it has more than eleven million people, while Oregon has just over four million. The elongated island has the Sierra Maestra mountains on its eastern end, the Escambray Mountains in the center, and the Western Karst region in the west, near Viñales. Low hills and fertile valleys cover more than half the island. The pristine waters of the Caribbean that surround the island make for some of the most attractive tourism locations in the Caribbean region.

With the defeat of Spain in the Spanish-American War, the United States gained possession of the Spanish possessions of Cuba, Puerto Rico, Guam, the Philippines, and various other islands and thus became a colonial power. Cuba technically became independent in 1902 but remained under US influence for decades. Sugar plantations and the sugar industry came to be owned and operated by US interests, and wealthy Americans bought up large haciendas (large estates), farmland, and family estates, as well as industrial and business operations. Organized crime syndicates operated many of the nightclubs and casinos in Havana. As long as government leaders supported US interests, things went well with business as usual.

Figure 20. US "Colonial" Influence in Cuba. The old capitol building in Havana, a replica of the US Capitol, was built by the United States during its control of Cuba. The building is a tourist area and no longer used for the government. The old US cars in the photo were made before the Cuban Revolution (1958) but are still used and make up about half the motor vehicles in Havana.

The Cuban Revolution

In January of 1934, with the encouragement of the US government, Fulgencio Batista led a coup that took control of the Cuban government. **Fidel Castro**, once a prisoner under Batista and having fled to Mexico in exile for a number of years, returned to Cuba to start a revolution. Joining him were his brother **Raúl Castro** and revolutionaries such as **Che Guevara**, an Argentinean doctor turned comrade-in-arms. Starting in the remote and rugged Sierra Maestras in the east, Castro rallied the support of the Cuban people. By the end of 1958, the Cuban Revolution brought down the US-backed Batista government. Castro gained power and had the support of most of the Cuban population.

Castro worked to recover Cuba for Cubans. The government cleared rampant gambling from the island, forcing organized crime operations to shut down or move back to the United States. Castro nationalized all foreign landholdings and the sugar plantations, as well as all the utilities, port facilities, and other industries. Foreign ownership of land and businesses in Cuba was forbidden. Large estates, once owned by rich US families, were taken over and recovered for Cuban purposes.

Figure 21. Fidel Castro (left); billboard in Havana promoting the virtues of revolutionaries Antonio Maceo and Che Guevara (right).

The US Embargo Era

Castro's policy of seizing (nationalizing) businesses and property raised concerns in the United States. As a result, US president Dwight D. Eisenhower severed diplomatic relations with Cuba in 1960 and issued an executive order implementing a partial trade embargo to prohibit the importation of Cuban goods. Later presidents implemented a full-scale embargo, restricting travel and trade with Cuba.

To deter any further US plans of invading or destabilizing Cuba, Castro sought economic and military assistance from the Soviet Union. The collapse of the Soviet Union in 1991 caused a downturn in Cuba's economy. With the loss of Soviet aid, the 1990s were a harsh time for Cubans, a period of transition. Castro turned to tourism and foreign investment to shore up his failing economy. Tensions between the United States and Cuba did not improve. In 1996, the United States strengthened the trade embargo.

At the turn of the twenty-first century, Cuba emerged as the lone Communist state in the Americas. Castro was the longest-governing leader of any country in the world. He never kept his promises of holding free elections; instead, he cracked down on dissent and suppressed free speech. He turned over power to his brother Raúl in 2006. Fidel Castro died in late 2016.

A Post-Castro Cuba

With Fidel Castro no longer in power, Cuba's future looks more positive but difficult. The island has natural resources, a great climate, and an excellent location but is also struggling economically. Cuba has a high literacy rate and has standardized health care, though medical supplies are often in short supply. The Cubans who live in dire poverty look to the future for relief. Personal freedoms have been marginal, and reforms are slowly taking place in the post-Fidel era. As the largest island in the Caribbean, Cuba has the potential to become an economic power for the region. There is vast US interest in regaining US dominance of the Cuban economy, and corporate colonialists would like to exploit Cuba's economic potential. Keeping corporate colonialism out is what Fidel's socialist experiment worked so hard to achieve, even at the expense of depriving the Cuban people of civil rights and economic reforms.

Cuba today is in transition from a socialist to a more capitalist economy and is counting on tourism for an added economic boost. With some of the finest beaches and the clearest waters in the Caribbean, Cuba is a magnet for tourists and water sports enthusiasts. Its countryside is full of wonders and scenic areas. The beautiful **Viñales Valley** in western Cuba has been listed as a UNESCO World Heritage Site for its outstanding karst landscape and traditional agriculture as well as for its architecture, crafts, and music. **Karst topography** is made up of soluble rock, such as limestone, which in the Viñales Valley results in unusual bread loaf–shaped hills that create a scenic landscape attractive for tourism. This region is also one of Cuba's best tobacco-growing areas and has great potential for economic development. The Cuban economy is banking on tourism to forge a path to a more prosperous future.

Figure 22. Dump truck taxi. Cubans use all available resources and opportunities to get by. These people are catching a ride on a dump truck to get where they want to go.

Figure 23. Viñales Valley in Western Cuba. In 2008, hurricanes Gustav and Ike devastated the tobacco crops, but the region is recovering and is a major tourist area.

The Commonwealth of Puerto Rico

Populated for centuries by Amerindian peoples, the island of Puerto Rico was claimed by the Spanish Crown in 1493, following Columbus's second voyage to the Americas. In 1898, after four hundred years of colonial rule, during which the indigenous population was nearly exterminated and African slave labor was introduced, Puerto Rico was ceded to the United States as a result of the Spanish-American War. Puerto Ricans were granted US citizenship in 1917. Popularly elected governors have served since 1948. In 1952, a constitution was enacted providing for internal self-government. In elections held in 1967, 1993, and 1998, Puerto Rican voters chose to retain the commonwealth status, although they were almost evenly split between total independence and becoming a US state. In a nonbinding referendum in 2012, for the first time a majority of voters in Puerto Rico favored becoming a US state, but that has not transpired as of 2016.

Figure 24. US government building in San Juan, Puerto Rico, with both US and Puerto Rican flags.

Puerto Rico is the smallest of the four islands of the Greater Antilles and is only slightly larger than the US state of Delaware. Puerto Rico's population is about four million, similar to the population of Oregon. As US citizens, Puerto Ricans have no travel or employment restrictions anywhere in the United States, and about one million Puerto Ricans live in New York City alone. The commonwealth arrangement allows Puerto Ricans to be US citizens without paying federal income taxes, but they cannot vote in US presidential elections. The Puerto Rican Federal Relations Act governs the island and awards it considerable autonomy.

Puerto Rico has one of the most dynamic economies in the Caribbean Basin; still, about 60 percent of its population lives below the poverty line. A diverse industrial sector has far surpassed agriculture as the primary area of economic activity. Encouraged by duty-free access to the United States and by tax incentives, US firms have invested heavily in Puerto Rico since the 1950s, even though US minimum wage laws apply. Sugar production has lost out to dairy production and other livestock products as the main source of income in the agricultural sector. Tourism has traditionally been an important source of income, with estimated arrivals of more than five million tourists a year. San Juan is the number one port for cruise ships in the Caribbean outside Miami. The US government also subsidizes Puerto Rico's economy with financial aid.

The future of Puerto Rico as a political unit remains unclear. Some in Puerto Rico want total independence, and others would like to become the fifty-first US state; the commonwealth status is a compromise. Puerto Rico is not an independent country as a result of colonialism. Many of the islands and colonies in the Caribbean Basin have experienced dynamics similar to Puerto Rico in that they are still under the political jurisdiction of a country that colonized it.

Hispaniola: The Dominican Republic and Haiti

Sharing the island of **Hispaniola** are the two countries of Haiti and the Dominican Republic. The island became a possession of Spain under European colonialism after it was visited by Columbus in 1492 and 1493. French buccaneers settled on the western portion of Hispaniola and started growing tobacco and agricultural crops. France and Spain finally agreed to divide the island into two colonies: the western side would be French, and the eastern side would be Spanish.

The Dominican Republic holds the largest share of Hispaniola. A former Spanish colony, the Dominican Republic has weathered the storms of history to become a relatively stable democratic country. It is not, of course, without its problems. The Dominican Republic has long been viewed primarily as an exporter of sugar, coffee, and tobacco, but in recent years the service sector has overtaken agriculture as the economy's largest employer. The mountainous interior and

Figure 25. UN peacekeeping troops in Haiti.

the coastal beaches are attractive to the tourism market, and tourism remains the main source of economic income. The economy is highly dependent on the United States, which is the destination for nearly 60 percent of its exports. Remittances from workers in the United States sent back to their families on the island contribute much to the economy. The country suffers from marked income inequality; the poorest half of the population receives less than one-fifth of the gross domestic product (GDP), while the richest 10 percent enjoys nearly 40 percent of GDP. High unemployment and underemployment remains an important long-term challenge. The **Central American-Dominican Republic Free Trade Agreement** (**CAFTA-DR**) came into play in March 2007, boosting investment and exports and reducing losses to the Asian garment industry. In addition, the global economic downturn has not helped the Dominican Republic.

Plantation agriculture thrived in Haiti during the colonial era, producing sugar, coffee, and other cash crops. The local labor pool was insufficient to expand plantation operations, so French colonists brought in thousands of African slaves to work the plantations, and people of African descent soon outnumbered Europeans. Haiti became one of the most profitable French colonies in the world with some of the highest sugar production of the time. A slave revolt that began in 1792 finally defeated the French forces, and Haiti became an independent country in 1804. It was the first country ever to be ruled by former slaves. However, the transition to a fully functional free state was difficult. Racked by corruption and political conflicts, few presidents in the first hundred years ever served a full term in office.

Haiti has had a difficult time finding political and economic stability. Haiti is the poorest nation in the Western Hemisphere, and many Haitians live in dire poverty with few employment opportunities. An elite upper-class minority controls the bulk of the nation's wealth.

C. Tourism and Economic Activity in the Rimland

The physical geography of the Caribbean region makes it a prime location for tourism. Its beautiful coastal waters and warm tropical climate draw in tourists from all over North America and the world. Tourism is the number one means of economic income for many places in the Caribbean Basin, and the tourist industry has experienced enormous growth in the last few decades. There has been strong growth in the number of cruise ships operating in the Caribbean. Cruise ships from the southern coasts of the United States ply their trade around the islands and coastal regions. Even the poorest country in the Caribbean, Haiti, has tried to attract cruise ships to its ports. The main restriction on cruise ship travel is the hurricane season, from June to November.

Even though tourism has become a vital economic component of the Caribbean Basin, in the long term, tourism creates many problems. Large cruise ships can overtax the environment; there have been occasions where there were actually more tourists than citizens on an island. An increase in tourist activity brings with it an increase in environmental pollution.

Most people in the Caribbean Basin live below the poverty line, and the investment in tourism infrastructure, such as exclusive hotels and five-star resorts, takes away resources that could be allocated to schools, roads, medical clinics, and housing. However, without the income from tourism, there would be no money for infrastructure. Tourism attracts people who can afford to travel. Most of the jobs in the hotels, ports, and restaurants where wealthy tourists visit employ people from poorer communities at low wages. The disparity between the rich tourist and the poor worker creates strong centrifugal cultural dynamics. The gap between the level of affluence and the level of poverty is wide in the Caribbean. In the model of how countries gain wealth, tourism is a mixed-profit situation. Local businesses in the Caribbean do gain income from tourists who spend their money there; however, the big money is in the cruise ship lines and the resort hotels, which are mainly owned by international corporations or the local wealthy elite.

Figure 26. Carnival victory cruise ship in San Juan Harbor. Large cruise ships in the Caribbean can hold up to four thousand passengers and crew members. The major cruise lines do not operate in the Caribbean during hurricane season.

Key Takeaways

1. Colonialism created a high level of ethnic, linguistic, and economic diversity in the Caribbean. The main shifts were the demise of indigenous groups and the introduction of African slaves.

2. The Caribbean Basin faces many challenges, including natural elements such as hurricanes, earthquakes, and volcanic activity. Economic conditions are often hampered by environmental degradation, corruption, organized crime, or the lack of employment opportunities.

3. The Cuban Revolution led by Fidel Castro created a socialist state that nationalized foreign-owned assets and brought about a trade embargo by the United States. Cuba lost its aid from the Soviet Union after the USSR's collapse in 1991 and has been increasing its focus on tourism and capitalistic reforms.

4. Tourism can bring added economic income for an island country, but it also shifts to the service sector resources that are needed for infrastructure and services. A high percentage of tourism income goes to external corporations.

End-of-Chapter Summary

- The Caribbean, Mexico, and Central America make up the realm of Middle America. Two types of development patterns emerged with European colonialism. The rimland, with its plantation agriculture, dominated the Caribbean and coastal regions. The mainland, with its haciendas, dominated Mexico and interior regions of Central America.

- European colonialism decimated the Amerindian population of the Caribbean and conquered the Aztec Empire of the mainland. Colonialism altered the food production, building methods, urbanization, language, and religion of the realm.

- African slave labor became prominent in the Caribbean and altered the ethnic makeup of most islands. Amerindians make up most of the lower working class on the mainland. A minority of wealthy Europeans continue to be at the top of the socioeconomic class structure. Most of Mexico's population is of mestizo heritage.

- Mexico has transitioned from a Spanish colony to a partner in the North American Free Trade Agreement (NAFTA). Trade relations have helped industrialize Mexico's economy and provide employment, especially in maquiladoras. Mexico has many natural resources but struggles to provide economic opportunities for its entire population. Wealth and power is controlled by an elite minority with a European heritage.

- Various geographic concepts and principles can be applied to this realm: rural-to-urban shift, core-periphery spatial relationship, altitudinal zonation, and the impact of climate types on human habitation.

- Population growth and the lack of employment opportunities have contributed to the high poverty levels in many areas. There is a wide disparity between the income levels of the wealthy and the poor. Haiti, for example, is the poorest country in the Western Hemisphere.

- The United States has had a major impact on this region, both politically and economically. The US military has intervened in many places to control its interests. US companies have dominated the region's economies. Most recently, the United States has supported industrial activities and the implementation of free-trade agreements to take advantage of cheap labor.

- Central America is a diverse and fragmented realm with every country, island, or republic possessing a different geography.

- Tourism is an important economic sector that has mixed impacts on the local situation. Every part of the Middle American realm has sought to improve their tourism draw to help bolster their economy.

- The global economy has prompted the political entities of the region to work more closely together to advance their economic interests. Trade associations such as NAFTA and the Central American-Dominican Republic Free Trade Agreement (CAFTA-DR) are attempts to develop a greater level of economic integration. Some argue that multinational corporations stand to benefit the most from free-trade agreements.

Chapter 8

South America

8.1 Introducing the Realm

Learning Objectives

1. Summarize the main physical features and characteristics of South America.

2. Explain how European colonialism dominated the realm and divided up the continent.

3. Describe the ethnic, economic, and political patterns in the Guianas.

4. Outline the main cultural realms of South America. Describe each realm's main majority and explain how colonialism impacted each region.

5. Summarize how the South American countries are attempting to integrate their economies.

Europeans called the Western Hemisphere the New World. South America is the realm consisting of the southern portion of the New World. This realm includes the entire continent of South America, which is smaller in physical area than North America. As a continent, South America is larger in physical area than Europe, Antarctica, or Australia but is smaller than Africa or Asia. Almost the entire landmass of South America lies to the east of the same meridian that runs through Miami, Florida. The Atlantic Ocean borders the continent to the east and the Pacific Ocean borders the continent on the west. The narrow Isthmus of Panama creates a natural break between the South American continent and its neighbors to the north. The Caribbean Sea creates the northern boundary.

South America covers an extensive range of latitude. The equator cuts through the northern part of the continent directly through the mouth of the mighty **Amazon River.** The country of Ecuador is located on the equator—hence its name. The equatorial region is dominated by the tropical climates of the immense Amazon Basin. The Tropic of Capricorn runs directly through the latitude of São Paulo, Brazil, and Chile's **Atacama Desert**, which reveals that most of the continent is in the zone of the tropics to the north. South of the Tropic of Capricorn is the Southern Cone of South America, home to the physical regions of the Pampas and Patagonia. Tierra del Fuego is the southern tip of the realm with territory in both Argentina and Chile. On the south side of the Tierra del Fuego archipelago is **Cape Horn**, which is the southernmost land point of the continent.

The continent of South America has a wide diversity of physical landscapes, from the high Andes Mountains to

Figure 1. South America: Political map of the countries and various physical regions. The main two physical features of South America are the Andes Mountains and the Amazon Basin

the tropical forests of the **Amazon Basin**. This assortment of physical features offers many resources, allowing people to engage in economic activity, gain wealth, and provide for their needs. **The Andes** hold mineral riches that have been extracted since ancient times. Precious metals have been mined from the mountains to grant great opportunities for those fortunate enough to be recipients of its wealth. Fossil fuels have been found in abundance in the far northern regions of Venezuela and Colombia. The **Amazon Basin** has been a source of hardwood lumber and, more recently, extensive mineral wealth. Some of the largest iron-ore mines in the world are located here. The massive plains of Brazil and the rich soils of the Pampas allow for enormous agricultural operations that provide food products for the continent and for the world. Even the inhospitable Atacama region in northern Chile holds some of the world's largest copper reserves. In addition, the wide variety of climate zones allowed a diverse range of species to develop.

Before the era of European colonialism, many local groups organized themselves into states or empires. The Inca Empire was the largest in existence at the time the Europeans arrived. European colonialism altered the continent in several ways. Not only did the Europeans defeat and conquer indigenous Amerindian groups such as the Inca, but cultural exchanges also took place that altered the way of life for countless South Americans. Colonialism created many of the current country borders and influenced trade relationships with the newly created colonies. The plantation system and the introduction of slaves from Africa drastically changed the ethnic makeup of the people living along the eastern coast. After slavery was abolished, indentured servitude brought workers from Asia to support the labor base.

Indentured servants were usually poor individuals who agreed to work for an agreed upon period of time, usually less than seven years, in return for the necessities of life such as lodging, food, and transportation or clothing. These individuals did not usually receive a salary but may have received a lump sum payment upon completion of the agreed upon service. Under favorable conditions indentured servants were treated like relatives and gained important experience and job skills to provide for their future. Many situations were much less favorable and resembled a form of slavery where individuals suffered from disease, harsh conditions, or even succumbed to death.

The continent can be divided into regions by ethnic majorities influenced by early colonial development. The mixing of ethnic groups from Europe, Africa, and Asia with each other or with the indigenous population has created a diverse cultural mosaic. For example, most people in Guyana and Suriname are from Asia, most people in Argentina and Uruguay are from Europe, most people in Peru and Bolivia are Amerindian, and most people in many areas along the eastern coast of Brazil are of African descent. A large percentage of the population of South America is of a mixed ethnic background.

South America's modern economic development has helped integrate it with the global economy. The levels of economic development vary widely within the realm. There are clear indications of core-peripheral spatial patterns within various regions of the continent, and rural-to-urban shift has been strong in many areas. The rural regions in the Andes or the interior suffer from a lack of economic support needed to modernize their infrastructure. At the same time, metropolitan areas are expanding rapidly and are integrated with global markets and the latest technologies. Most of the large cities are located along the coastal regions. This pattern of urbanization is mainly a result of colonial activity and influence. The countries of South America are working among themselves to network trade and commerce activities. Trade agreements and economic unions have become standard methods of securing business partnerships to enhance the realm's economic opportunities.

A. Physical Geography

The far-reaching Andes Mountains and the massive Amazon River system dominate South America's physical geography. The five-thousand-mile-long Andes Mountain chain extends along the entire western region of the continent from Venezuela to southern Chile. The Andes are the longest mountain chain on Earth and the highest in the Americas. The Andes Mountain range has more than thirty peaks that reach at least twenty thousand feet in elevation, many of which are active volcanoes. The Andes has provided isolation to the Inca Empire, mineral wealth to those with the means for extraction, and a barrier to travelers crossing the continent. The Andes' minerals include gold, silver, tin, and other precious metals. Mining became a major industry in the colonial era and continues to the present.

At the core of the continent is the mighty Amazon River, which is more than 4,300 miles long and has an enormous drainage basin in the largest tropical rain forest in the world. The Amazon's many tributaries are larger than many other world rivers. The Amazon and its many tributaries drain the entire interior region of the continent, covering 40 percent of South America. The Amazon has the greatest discharge of any river in the world, carrying about one fifth of all river water on the planet, and is probably the longest river in the world (only the Nile River may be longer, depending on how the exact length is measured). During the rainy season, the Amazon River can be more than one hundred miles wide. No bridges span the Amazon River. Its source is a glacial stream located high in the Peruvian Andes, only about 150 miles from the Pacific Ocean.

The Amazon's extended tributaries—such as the **Rio Negro,** the **Madeira,** and the **Xingu**—move massive amounts of water through the Amazon Basin and are major rivers in their own right. The Amazon has more than 1,100 tributaries; a dozen are more than one thousand miles long. Hydroelectric dams are located on the tributaries to produce electricity for the region's fast-growing development. South America has additional large rivers that drain the continent, including the Orinoco, which flows through Venezuela; the Sao Francisco, which flows through southeast Brazil; and the **Paraguay** and the **Paraná** Rivers, which flow south from Brazil into the **Rio de la Plata** between Argentina and Uruguay.

The **Altiplano Region** is a wide basin between two main Andean mountain ranges. The word altiplano means "valley" in Spanish. There are a substantial number of altiplanos in South America. They provide for agricultural production and human habitation. Lake Titicaca rests in the middle of the Altiplano Region of the Central Andes on the border between Peru and Bolivia. Lake Titicaca is a large freshwater lake about 120 miles long and 50 miles wide. The surface is at an elevation of about twelve thousand feet above sea level, and the lake is more than nine hundred feet deep in some areas. Usually at such high elevations, the temperature would dip below freezing and restrict agriculture. However, the large lake acts as a solar energy collector by absorbing energy from the sun during the day and giving off that energy in the form of heat during the night. The energy redistribution allows for a moderate temperature around the lake that is conducive to growing crops. With abundant fresh water and the ability to grow food and catch fish, the Altiplano Region has supported human habitation for thousands of years.

Across the Andes Mountains from the Altiplano Region is the Atacama Desert. The Atacama is one of the driest places on Earth: in some parts, no rain has fallen in recorded history. In normal circumstances, the Atacama would be a desolate region without human activity, but that is not the case. Some of the world's largest copper reserves are found here. Nitrates, which are used in fertilizers, are also found in large quantities. Mining the Atacama has brought enormous wealth to people fortunate enough to be on the receiving end of the profits. The rain shadow effect is responsible for the extraordinary dryness of the Atacama. The Andes are quite high at this latitude, and the winds blow in rain clouds from the east. When the clouds reach the mountains, they ascend in elevation, releasing their precipitation without ever reaching the western side of the Andes.

South America has large agricultural plateaus east of the Andes, such as the Mato Grosso Plateau, which includes a portion of the great **Cerrado** agricultural region of central

Figure 2. Lake Titicaca with traditional reed boat made by Amerindian locals. Lake Titicaca is the highest-elevation navigable lake in the world.

Brazil. The Cerrado is a vast plain that has been developed for agriculture and produces enormous harvests of soybeans and grain crops. Bordering the **Cerrado** to the southeast are the **Brazilian Highlands,** an extensive coffee-growing region. **The Pampas** in eastern Argentina, Uruguay, and southernmost Brazil is another excellent agricultural region with good soils and adequate rainfall. Farming, cattle ranching, and even vineyards can be found here, making the Pampas the breadbasket of the Southern Cone. To the south of the Pampas is the lengthy expanse of **Patagonia,** which covers the southern portion of Argentina east of the Andes. Patagonia is a prairie grassland region that does not receive much rainfall because of the rain shadow effect of the Andes to the west. The main activities in Patagonia are the raising of cattle and other livestock. The region is starting to attract attention for the extraction of natural resources such as oil, natural gas, and valuable minerals.

To the northern part of the continent in Venezuela and Colombia, sandwiched between the Andes Mountains and the **Guiana Highlands,** is a grassland region with scrub forests called the Llanos. The human population is small because of the remoteness of the region along the Orinoco River basin. The Guiana Highlands of southeast Venezuela and the Guianas are an isolated set of mountainous plateaus mixed with rugged landscapes and tropical climates. Angel Falls, the highest waterfall in the world, with a free fall of more than 2,647 feet and a total drop of about 3,212 feet (more than half a mile), is located here. To the northwest of the **Llanos** and the Guiana Highlands in Venezuela is Lake Maracaibo, a large inland lake open to the Caribbean Sea. A coastal lake, Maracaibo rests atop vast oil reserves that provide economic wealth for Venezuela.

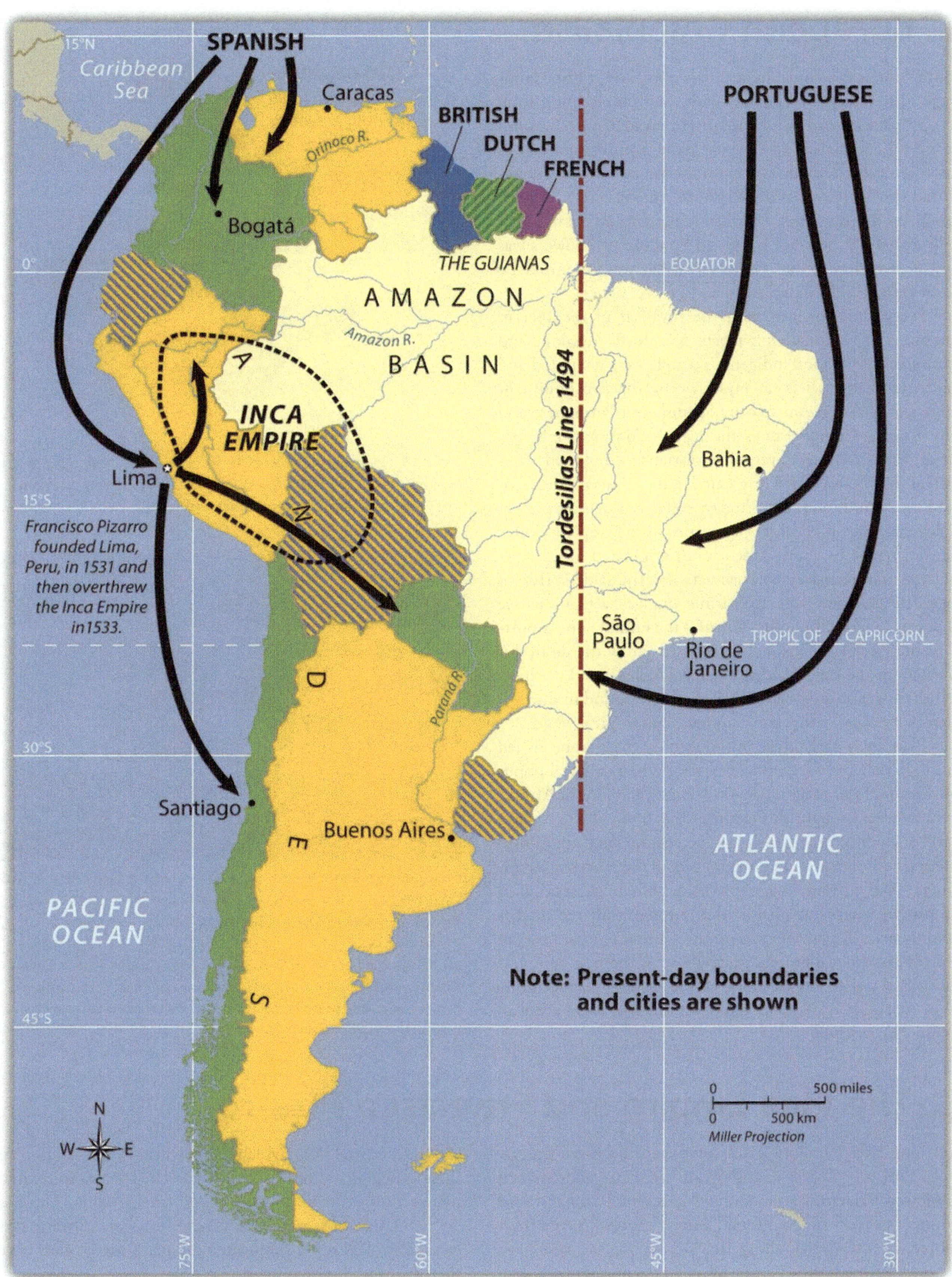

Figure 3. Colonial activity in South America.

B. European Colonialism

South America's colonial legacy shaped its early cultural landscape. The indigenous people, with their empires and local groups, were no match for the Iberian invaders who brought European colonialism to the continent. South America was colonized exclusively by two main Iberian powers: Spain colonized the western part of the South America, and Portugal colonized the east coast of what is present-day Brazil. The only region that was not colonized by those two powers was the small region of the **Guianas**, which was colonized by Great Britain, the Netherlands, and France.

Everything changed with the invasion of the Iberian colonizers. The underlying tenets of culture, religion, and economics of the local indigenous people were disrupted and forced to change. It is no mystery why the two dominant languages of South America are Spanish and Portuguese and why Roman Catholicism is the realm's dominant religion. Colonialism also was responsible for transporting food crops such as the potato, which originated in the Peruvian Andes, to the European dinner table. Today, coffee is a main export of Colombia, Brazil, and other countries in the tropics. Coffee was not native to South America but originated in Ethiopia and was transferred by colonial activity. The same is true of sugarcane, bananas, and citrus fruits; oranges were not native to South America, but today Brazil is the number one exporter of orange juice. Colonialism was driven by the desire for profit from the quick sale of products such as gold or silver, and there was a ready market for goods not found in Europe, such as tobacco, corn, exotic animals, and tropical woods.

Plantation agriculture introduced by the Europeans led to a high demand for manual laborers. When the local populations could not meet the labor demand, millions of African workers were brought through the slave trade. These African slaves introduced their own unique customs and traditions, altering the culture and demographics of the Western Hemisphere. The current indigenous Amerindian population, a fraction of what it was before the Europeans arrived, makes up only a small percentage of South America's total population. Europeans colonizers generally took the best land and controlled the economic trade of the region. The acculturation in South America is directly related to the European colonial experience.

Figure 4. Colonialism in South America. The two main colonizers in South America were Spain and Portugal. The Spanish conquistador Francisco Pizarro defeated the Inca Empire.

Figure 5. The Jesuit Church of La Compañia de Jesús on the Plaza de Armas in Cuzco, Peru. The original church was constructed in 1571 on the site of the ancient Incan palace. The earthquake of 1650 caused severe damage to the building, so it had to be recon-structed in about 1688. Many Catholic cathedrals in Latin America were built with stones from ancient sites. This cathedral is claimed to be the Western Hemisphere's most ideal example of colonial baroque architecture.

C. The Inca Empire and Francisco Pizarro

Not long after Hernán Cortés conquered the Aztec Empire of Mexico in 1521, a young Spanish conquistador named **Francisco Pizarro**, stationed in what is now Panama, heard rumors of silver and gold found among the South American people. He led several sailing excursions along the west coast of South America. In 1531, he founded the port city of Lima, Peru. Since 1200, the Inca had ruled a large empire extending out from central Peru, which included the high-elevation Altiplano Region around Lake Titicaca. **The Inca Empire** dominated an area from Ecuador to Northern Chile. The Inca were not the most populous people but were a ruling class who controlled other subjugated groups. Pizarro, with fewer than two hundred men and two dozen horses, met up with the Inca armies and managed to defeat them in a series of military maneuvers. The Inca leader was captured by the Spanish in 1533. Two years later, in 1535, the Inca Empire collapsed.

The Inca Empire was significant thanks to the high volume of gold and silver found in that region of the Andes. The mineral wealth of the Andes made the conquistadors rich. Lima was once one of the wealthiest cities in the world. Europeans continued to dominate and exploit the mining of minerals in Peru and Bolivia throughout the colonial era. European elites or a Mestizo ruling class has dominated or controlled the local Amerindian groups in the Andes since colonial times.

D. The Iberian Division of the Continent

The Spanish conquistadors were not the only European invaders to colonize South America. Colonial influence—which forced a change in languages, religion, and economics—also came from the small European kingdom of Portugal. Portuguese ships sailed along the eastern coast of South America and laid claim to the region for the king. The Portuguese did not find large gold or silver reserves, but they coveted the land for the expansion of their empire. Soon the Spanish and the Portuguese were fighting over the same parts of South America. In 1494, the issue was brought before the Roman Catholic Church. **The Tordesillas Line** was drawn on a map to divide South America into the Spanish west and the Portuguese east. The region that is now Brazil became the largest Portuguese colonial possession in South America, a center for plantation agriculture similar to that in the Caribbean. For this reason, a large African population lives in Brazil, and most of the people in Brazil speak Portuguese and are Roman Catholics.

Independence did not come for the Spanish colonies until 1816 and 1818, when Chile and Argentina broke away in an independence movement in the south. **Simón Bolívar** led liberation movements in the north. By 1824, the Spanish were defeated in South America. Brazil did not gain independence from Portugal until 1822, when the prince of Portugal declared an independent Brazil and made himself Brazil's first emperor. It was not until 1889 that a true republic was declared and empire was abolished.

E. Cultural Regions of South America

It is impossible to understand the current conditions in South America without first understanding what occurred to create those conditions. This is why studying European colonialism is so important. Colonialism changed the ethnicity, religion, language, and economic activities of the people in South America. The past five hundred years have tempered, stretched, and molded the current states and regions of the South American continent. To identify standards of living, ethnic majorities, and economic conditions, it is helpful to map out South America's various **cultural regions.**

In South America, five main cultural regions indicate the majority ethnic groups and the main economic activities: **Tropical Plantation Region, Rural Amerindian Region, Amazon Basin, Mixed Mestizo Region, and European Commercial Region (Southern Cone).**

These are generalized regions that provide a basic understanding of the whole continent. Technological advancements and globalization have increased the integration of the continent to the point that these regions are not as delineated as they once were, but they still provide a context in which to comprehend the ethnic and cultural differences that exist within the realm.

Figure 6. Cultural regions of South America.

Tropical Plantation Region

Located along the north and east coast of South America, the **Tropical Plantation Region** resembles the Caribbean rimland in its culture and economic activity. The region, which extends as far south as the Tropic of Capricorn, has a tropical climate and an agricultural economy. Europeans opened up this area for plantation agriculture because of coastal access for ships and trade. The local people were forced into slavery, but when the local people died off or escaped, millions of African slaves were brought in to replace them. After slavery was abolished, indentured servants from Asia were brought to the Guianas to work the plantations. The Tropical Plantation Region has a high percentage of people of African or Asian descent.

Rural Amerindian Region

The **Rural Amerindian Region** includes the countries of Ecuador, Peru, and Bolivia. The ruling Mestizo class that inherited control from the European conquistadors mainly lives in urban areas. Most of the rural Amerindian population lives in mountainous areas with type H climates and ekes out a hard living in subsistence agriculture. This is one of the poorest regions of South America, and land and politics are controlled by powerful elites. The extraction of gold and silver has not benefited the local Amerindian majority, which holds to local customs and speaks local languages.

Amazon Basin

The **Amazon Basin**, which is characterized by a type A climate, is the least-densely populated region of South America and is home to isolated Amerindian groups. Development has encroached upon the region in the forms of deforestation, mining, and cattle ranching. Large deposits of iron ore, along with gold and other minerals, have been found in the Amazon Basin. Preservation of the tropical rain forest of this remote region has been hampered by the destructive pattern of development that has pushed into the region. The future of the basin is unclear because of development patterns that are expected to continue as Brazil seeks to exploit its interior peripheral region. Conflicts over land claims and the autonomy of Amerindian groups are on the rise.

Mixed Mestizo Region

The **Mixed Mestizo Region** includes the coastal area of the west and the interior highlands of the north and east. This region between the Tropical Plantation Region and the Rural Amerindian Region includes a majority of people who share a mixed European and Amerindian ethnicity. It is not as poor as the Rural Amerindian Region and yet not as wealthy as

Figure 7. Young Women in Salvador, Brazil. Salvador, Brazil, is located along the coastal region of South America where the Tropical. Plantation Region was prominent. Most people in this region are of African descent.

Figure 8. Amerindians. The Amerindian woman and child in this photo live in the Sacred Valley of the Andes in Peru.

Figure 9. Amazon River Drainage Basin. The Amazon has more than 1,100 tributaries.

the European-dominated region to the south. Paraguay falls into the Mixed Mestizo Region, as do other portions of other South American countries such as parts of Brazil, Colombia, and Venezuela. Paraguay is mainly Mestizo, but its economic qualities resemble that of the Rural Amerindian Region to the north, even though Paraguay is not located in the mountains.

European Commercial Region (Southern Cone)

The southern part of South America, called the European Commercial Region or the Southern Cone, includes Chile, Argentina, Uruguay, and parts of Brazil. European ethnic groups dominate this region and include not only Spanish and Portuguese but also German, Austrian, Italian, and other European ethnic heritages. Fertile soils and European trade provided early economic growth, and the region attracted industry and manufacturing in the later decades of the twentieth century. There are not many Amerindians or people of African descent here. More than 90 percent of all the people in Argentina, Chile, and Uruguay are of European descent and live in urban areas. With a highly urbanized population and with trade connections to a globalized economy, it is no surprise that the Southern Cone is home to South America's most developed economies.

F. Globalization and Trade

South America has been fragmented by European colonialism, which established colonies and economic dependence on its European masters. The colonial economic patterns did not encourage the South American countries to work together to create an integrated continental trade network. Countries outside the continent have promoted trade partnerships to benefit from South America's natural resources and agricultural exports. The establishment of the European Union and the North American Free Trade Agreement (NAFTA) created globalized trading blocs that challenged the South American countries to consider how to take advantage of trading opportunities within their realm to protect and support their own economic interests.

Since the 1990s, cooperation and business ventures have started to form within the realm to create a more integrated network of trade and commerce to benefit the countries of South America. Transportation and communication systems are being developed through joint ventures by internal investment groups. River and road systems continue to be managed and developed for improved transport of people and goods throughout the continent. Free-trade agreements have been implemented to support the integration of internal economic networks and competition in the global marketplace.

In 2008, the South American countries formed the **Union of South American Nations (UNASUR)** to oversee the customs unions and trade agreements within the realm. One of the more established trade associations is **Mercosur (the Southern Cone Common Market)**, created in 1995 by the southern countries. It has evolved to include most countries in South America and is the most dominant trade agreement in the realm. Full members of Mercosur include Argentina, Uruguay, Paraguay, Brazil, and Venezuela (although Venezuela's membership was suspended in late 2016). Chile, Bolivia, Peru, Ecuador, and Colombia are associate members. **The Andean Community** (Colombia, Ecuador, Peru, and Bolivia) was established in 1969 but did not gain ground until 1995, when it established stronger trade measures. Multinational corporations have supported the creation of a **Free Trade Area of the Americas (FTAA)** to include all of the Western Hemisphere in one unified trade association. It has not been approved and has received strong opposition from Mercosur and economic forces that support a more localized economy controlled by local people.

South America faced division and competition during the colonial era between the Spanish and the Portuguese. Today's new era of corporate colonialism has created similar fragmentation and divisions. The level of trade between the countries of South America and the United States and Europe varies widely. Countries such as Colombia and Chile have well-established trade relationships with the United States and are unwilling to jeopardize those trade connections to strengthen ties with their neighbors that have less-supportive political relationships with the United States. External global trade arrangements often provide financial benefits to individual countries that might not be shared by the bordering countries in the same region. South America's historical fragmentation has not made it easy to unify the continent under a singular trade agreement to compete against the European Union or NAFTA.

Key Takeaways

1. The extensive Andes Mountain chain and the massive Amazon River dominate the realm's physical geography.

2. The Spanish and the Portuguese were the two main colonial powers that dominated South America. The Guianas were the only part of the continent not dominated by these two European powers.

3. Identifying the majority ethnic groups in South America can be helpful in classifying the various cultural regions of the realm. Colonial activities and ethnic backgrounds are consistent enough to formulate regions with similar characteristics.

4. Globalization and the creation of economic or political units such as the European Union and NAFTA have prompted the South American countries to work together to implement cooperative trade agreements and create the Union of South American Nations.

Figure 10. Northern South America and Venezuela. Notice that most of the main cities are located along the northern coast.

8.2 Urban North and Andean West

Learning Objectives

1. Understand the dynamics of Venezuela's urban society and why Venezuela has not experienced a robust rural-to-urban shift to the extent that other countries have.

2. Summarize the production of the three main export products of Colombia and explain the US role in their export.

3. Compare the three main countries in the Andean West region of South America and understand how they gained their wealth and who has benefited the most over the years from that wealth.

4. Outline how Paraguay's geographical setting has allowed it to gain wealth and provide opportunities for its people.

A. Venezuela: Oil, Politics, and Globalization

Bordering the Caribbean is the large urban country of Venezuela. The Andes Mountains reach into the northern part of the country and make up the terrain of the northern coastal region all the way to the capital city of Caracas. The large grassland plains of the Llanos extend farther south from the Colombian border to the Orinoco River delta. The Llanos is a large, sparsely populated region that makes up about one-third of the country. It is remote, susceptible to flooding, and used mainly for raising cattle. In the southeast of Venezuela are the Guiana Highlands, which make for a spectacular physical landscape of tropical forests and rugged mountainous terrain. The highlands include **Angel Falls**, the tallest waterfall in the world. Angel Falls drops 2,647 feet and is Venezuela's most popular tourist attraction. Lake Maracaibo, a large inland sea located in the western region of the country, is not a true lake in that it is open to the Caribbean Sea, but it is considered the largest inland body of water in South America. Lake Titicaca, located in the Andean region of the Altiplano on the border between Peru and Bolivia, is considered the continent's largest freshwater lake.

Venezuela has an assortment of physical regions, but most of the population lives along the northern coast. About 90 percent live in urban areas, and the capital Caracas has the highest population. Less than 5 percent of the population lives south of the Orinoco River, and Amerindian groups live in the interior and along the river.

Included in the Mixed Mestizo Cultural Region, Venezuela has a heavy Spanish influence laid over an Amerindian base in a plantation region known for its African infusion. There is also a strong Caribbean cultural flavor, which is evident in the region's music and lifestyle. The official language is Spanish, but more than thirty indigenous languages are still spoken in the country.

Figure 11. Angel Falls, Venezuela. The total falls is estimated at 3,212 feet; the largest free fall is estimated to be about 2,647 feet—about half a mile. Angel Falls is the country's number one tourist attraction.

As much as 90 percent of Venezuela's export earnings are from the export of oil. Venezuela's national oil company, **CITGO**, has made extensive inroads into the US gas station market. The country was one of the founding members of **OPEC (Organization of Petroleum Exporting Countries)**, which is usually associated with the oil-rich states of the Persian Gulf. In the past decade, Venezuela has been one of the top five countries exporting oil to the United States. The other four are Canada, Mexico, Saudi Arabia, and Nigeria.

The country's enormous oil revenues and its current political climate have increased Venezuela's visibility in the global arena, but how this will play out over the long term is unclear. In the past, Latin America has had a greater number of elected political leaders with more progressive or socialist views. These trends continue to shape the economic trade agreements between countries. Venezuela has been working to increase sales of oil to countries in **Mercosur (the Southern Cone Common Market)**, which is the most significant trade association in South America.

Figure 12. The Two Sides of Caracas, the Capital City of Venezuela. The photo on the left is of the main urban core, with upscale neighborhoods. The photo on the right is a barrio located on a steep mountainside. Barrios are usually self-constructed slum areas and are similar to favelas in Brazil.

B. Colombia: Drugs, Coffee, and Oil

Three ranges of the Andes Mountains run from north to south through Colombia, which is larger than the nine most southeastern US states. With a land area covering about 440,839 square miles, Colombia is more than four times the size of the US state of Oregon and close to twice the size of France. Colombia borders five countries, with the Caribbean to the north, the Pacific Ocean to the west, the Orinoco River to the east, and a short segment of the Amazon River to the far south. Even though agriculture has been a mainstay of the country's economic activities, because of the influence of the mountainous terrain, about 75 percent of the population lives in urban areas.

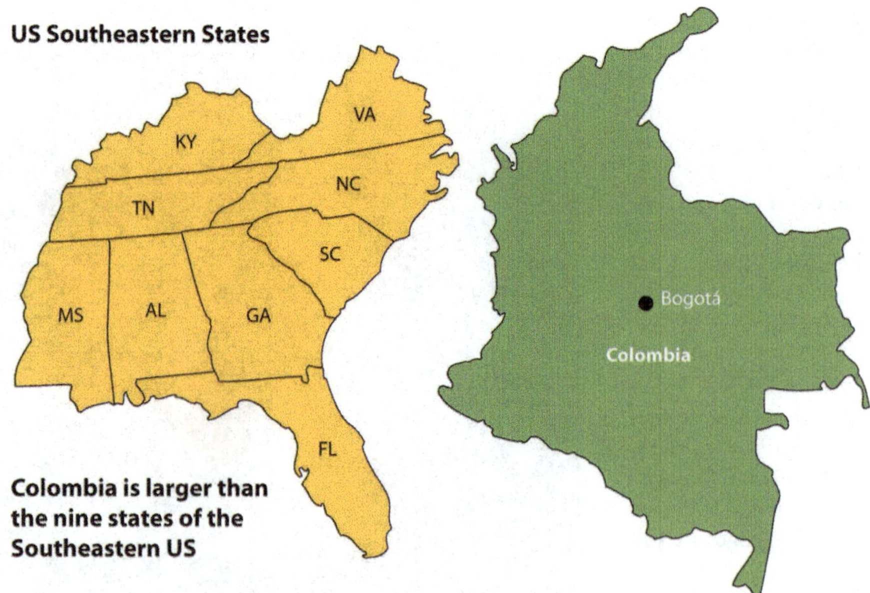

Figure 13. Physical size of Colombia: 440,839 Square Miles.

Colombia and the Drug Trade

Colombia's tropical climate and its many remote areas contributed to its development as a major coca-growing region. By the 1970s, extensive drug smuggling had developed, and powerful drug cartels became major political brokers within the country, competing against the government for control of Colombia. The largest and most organized cartels operated out of **Medellin and Cali**, the second- and third-largest cities in the country after the capital city of **Bogotá**.

The coca plant grows throughout the slopes of the Andes, from Colombia to Bolivia. Historically, locals have chewed it or brewed it into tea. Coca can alleviate elevation sickness and act as a mild stimulant. Using modern methods and strong chemicals, the coca leaves can be converted into coca paste and then into cocaine, a powerful narcotic. (It must be noted that the short, leafy coca plant that cocaine comes from is not the same as the cacao tree that produces the beans that chocolate or cocoa comes from. They are two completely different plants with separate processes.)

The United States is the largest cocaine market. Secret airfields and private boats transport the cocaine from Colombia to distribution centers in Mexico, Central America, or the Caribbean. From there, the drugs are smuggled into the United States. Colombian drugs are a multibillion-dollar industry that makes up a large portion of the Colombian economy. The effect of the drug industry on the people of Colombia is extensive—from the gunfire on the streets to the corruption of government officials. In recent years, the same drug cartels that have operated the cocaine industry have imported opium poppies, which grow well on the higher and more arid slopes of the Andes. Opium poppies are native to Asia but have been transported to South America. Opium is extracted from the seedpod and can be further refined into heroin. Colombian drug cartels, with a Mexican distribution network, have muscled into as much as 20 percent of the US heroin market. The US government has supported the Colombian government in the fight against the drug cartels and the trafficking of illegal drugs out of Colombia.

Colombian Coffee and Oil

Colombia's two main legal exports to the United States are coffee and oil. Coffee is only grown in the tropics, since coffee trees must be grown in a frost-free environment. Coffee trees can grow in elevations from sea level to six thousand feet, but most of the best specialty coffee is grown at elevations of three thousand to six thousand feet. Colombia has ideal conditions for growing coffee and was once the world's largest coffee producer; now Brazil and Vietnam each produce more.

Oil has now become Colombia's number one legal export. Oil is found in fields in the northern and central regions of Colombia. Immense quantities of coal are also found in the same regions, but oil is more valued on the export market. Pipelines connect the interior oil fields of Colombia with the northern ports. The US market size and population make it the world's largest oil consumer. US oil companies have been investing in the development of Colombian oil for many years. Colombia has been a developing oil source even though its total extractable resources are not as vast as in other countries. For example, in 2006 the United States imported more oil from Colombia than from Kuwait, Oman, the United Arab Emirates (UAE), Bahrain, Qatar, and Yemen combined.

The relationship between Colombia and the United States is often conflicting. The US consumer supports the Colombian drug cartels by being the largest consumer of illegal drugs. The US government, under the Drug Enforcement Administration (DEA), has declared a war on drugs and has supported the Colombian government with billions of dollars in foreign aid to fight that war. On another front, US oil corporations have paid insurgent groups to protect their oil assets. Oil is exported to the United States, bringing billions of dollars into the Colombian economy. The chaos in Colombia is directly related to the exploitation and marketing of their resources. It is the people of Colombia that suffer in the cross fire from this civil war of corruption, crime, death, and destruction. The United States is a counterforce partner in this situation but operates from the consumer end of the resource pipeline. The largest consumer market for Colombia's export of oil, drugs, and coffee is the United States, which is also the largest contributor of foreign aid to Colombia.

Figure 14. Colombian Exports. The three main export products of Colombia are illegal drugs, oil, and coffee. The United States is the largest consumer of all three.

C. Rural Amerindian States of Ecuador, Peru, and Bolivia

The Central Andes, which includes Ecuador, Peru, and Bolivia, were home to the Inca Empire. The empire had gone through some internal divisions and was working on unifying the region when Francisco Pizarro's small army defeated the Incan warriors and brought about colonial rule beginning in the 1530s. Many cultures lived in the Central Andes before the Inca, and their legacy continues in the customs and the ways of the Amerindian people who still live there today. Spanish is the official language, or the lingua franca, but indigenous languages are widely spoken and dominate in the rural areas and remote villages. Ecuador, Peru, and Bolivia make up the core of the Rural Amerindian Region of South America. There have been border disputes among the three countries, and also with their neighbors. Nevertheless, they all share the Andes and have many things in common.

Figure 15. Machu Picchu. High in the Peruvian Andes, the Lost City of the Incas, Machu Picchu, was rediscovered in 1911 by Yale archaeologist Hiram Bingham and is one of the most beautiful and enigmatic ancient sites in the world. The ruins are located at about 7,970 feet in elevation and are surrounded by higher peaks of the Andes.

Physical Geography

The physical geography of the Central Andes includes more than just the high Andes Mountains, although they dominate the landscape. The coastal region to the west of the Andes is generally warmer than the cooler climate of the mountains. The equatorial region is rather humid. The coastal region in southern Peru is dry and arid because of the ocean currents and the rain shadow effect of the Andes, which creates the Atacama Desert that extends up from northern Chile. Southwest Bolivia has some of the world's largest salt flats in this dry and barren region. In the interior, on the eastern side of the mountain ranges, is the huge expanse of the Amazon Basin. Tropical and humid with heavy precipitation is generally the climate rule. Rain forests and jungle fauna can be found on the eastern slopes. The Altiplano region has the high-elevation Lake Titicaca. The variations in physical terrain provide extensive biodiversity in animal and plant species. It also supports a variety of economic activities to exploit the bountiful natural resources.

Even though the Altiplano region borders the Pacific Ocean, it also links directly to the Atlantic Ocean. The headwaters that create the Amazon River start in Peru, and by the time the water reaches the Peruvian city of Iquitos, the river is large enough to accommodate large shipping vessels. Iquitos is a port city for the Atlantic Ocean with access to Europe, Africa, or eastern North America. The port also links the region with Brazil's free-trade zone in Manaus, which has access to large oceangoing shipping and an international market.

Economic Geography

The region's main income comes from exports of minerals, fossil fuels, and agricultural products. Oil is the number one means of gaining national wealth in Ecuador and Peru; natural gas is the number one export of Bolivia. Gold, silver, tin, and other minerals are also abundant.

Poverty and the exploitation of natural resources usually result in environmental degradation unless proper measures are taken to prevent it. The area's heavy reliance on oil and gas extraction to gain national wealth has come at a great cost to their environment. Many oil spills have caused oil to enter the freshwater supplies of local residents and pollute the rivers and streams of the Amazon Basin. Mining has traditionally devastated the land because large portions of earth are removed to extract the ore or mineral. Pollution is causing a loss of habitats and destroying ecosystems, and few measures are being taken to prevent it. Deforestation is being caused by the timber industry and by clearing for agriculture. Overgrazing and the removal of the trees leave the soil open to erosion.

Figure 16. Quito, Ecuador, is an urban center high in the Andes, with a population of more than 1.5 million. This photo shows the central business district with the mountains in the background surrounding the city. Quito is at about 9,200 feet in elevation and is considered the second-highest-elevation capital in the world after La Paz in Bolivia.

Tourism

Tourism is expanding to connect travelers with opportunities to explore Incan and pre-Incan sites, which are the main attractions. One of the main tourist attractions in Peru is the ancient city of **Machu Picchu** in the Andes not far from Cuzco. In 2010, Peru gained over two billion dollars from the tourist activities of about two million foreign tourists. Ecuador's major tourist attraction is the **Galapagos Islands,** which aided Charles Darwin in understanding natural selection and the evolutionary process. Bolivia has a number of ancient sites that predate the Inca and have become major tourism destinations. The ancient city of Tiahuanaco and the enigmatic Lake Titicaca are good examples.

Figure 17. Indigenous women on their way home from the MAS congress in Bolivia, January 2009. MAS is the Movement for Socialism, which has been active in Bolivian politics.

Tourism can be a great source of economic income but it can come at a cost to the environment. There is always concern that high-traffic tourism sites like Machu Picchu can be degraded by the sheer mass of people visiting the site. The environmental imprint may be extensive. The term ecotourism has been used to indicate the activity of people traveling to experience and enjoy the natural world with an aim not to damage the environment in the process. The main objective was to make the tourism activity sustainable, which promoted stewardship of the land and respect for its attractions. Jonathan Tourtellot, director of the National Geographic Society's Center for Sustainable Destinations, coined and prefers the term **geotourism**, which can be translated as the stewardship of place and the preservation of its essential character. These concepts are becoming more integrated with the tourism industry to promote a sustainable model for high traffic sites like the Galapagos Islands with fragile ecosystems.

Population and Culture

Population growth is a major factor in the future of Peru, Ecuador, and Bolivia. In 2010, Bolivia had more than ten million people, Ecuador had more than fourteen million, and Peru had about twenty-nine million. More than 30 percent of the population of Ecuador and Bolivia resides in rural areas and make a living from subsistence agriculture. All three countries have large populations in relation to the production of adequate food. Peru and Bolivia are large countries in physical area but do not have a high percentage of arable land. Rural-to-urban shift is increasing and the major cities are continuing to expand, overtaxing public works and social services.

The culture of the Central Andes is heavily influenced by its rural Amerindian heritage. The foundation of the traditional agrarian society has been subsistence agriculture. One-third of the population in Ecuador and Bolivia and up to one-fourth of the population in Peru continue to live a traditional way of life. Local cuisine reflects the connection to the land. Potatoes, maize, guinea pigs, and fish are common fare in rural areas. The cities are encountering international influences that are changing the demands in local cuisine and culture. Traditional food, arts, and local crafts still thrive in the local districts and for the tourism market.

Figure 18. Rural Amerindian States of Ecuador, Peru, and Bolivia This map also displays adjacent states and part of the Amazon Basin, the Atacama Desert, Altiplano, and the Mato Grosso Plateau.

Key Takeaways

1. Most of the people in Venezuela live in cities along the Caribbean coast. The country has always been urban and has not developed its agricultural production. Venezuela depends on oil exports to gain most of its national wealth.

2. The United States is the largest consumer of products exported from Colombia. Illegal drugs, oil, and coffee and are the three main export products of Colombia. The United States directly or indirectly supports the three main factions vying for power within Colombia.

3. The Andean countries of Ecuador, Peru, and Bolivia share similar developmental dynamics, including a large percentage of Amerindian people and reliance on the export of natural resources to gain national wealth. The wealth gained from exports is concentrated in the hands of elites rather than filtering down to most of the people.

8.3 Brazil

Learning Objectives

1. Summarize the ethnic composition of Brazil and learn why the population is so diverse.

2. Explain how the core-periphery spatial relationship applies to the country. Be able to clarify how the dynamics of the country shape the core and the periphery.

3. Describe the main activities that are involved in the development and exploitation of the Amazon Basin. Learn how deforestation affects the tropical rain forest and environmental conditions.

4. Outline the main characteristics and economic activities of the main regions of Brazil.

A. Portuguese Colony

Brazil, the largest country in South America, is similar in physical area to the contiguous United States (the United States without Alaska or Hawaii). Catholicism is the dominant religion and Portuguese is the primary language. Once a Portuguese colony, the country's culture was built on European immigration and African slave labor, making for a rich mixture of ethnic backgrounds.

In colonial times, Brazil was a part of the **Atlantic Trade Triangle**, which functioned as a transportation conveyor, moving goods and people around the regions bordering the Atlantic Ocean. Colonial merchant ships financed by Europe's wealthy elite brought goods and trinkets to the African coast to trade for slaves, who were shipped to the Americas and the Caribbean to diminish the labor shortage for the colonies. The last leg of the Atlantic Trade Triangle moved food crops, sugar, tobacco, and rum from the colonies back to the European ports. The merchant ships never sailed with an empty hold, and their successful voyages provided enormous profits to the European financiers.

The total number of individuals taken as slaves from Africa is unclear and often debated. It is esti-

mated that more than ten million African slaves survived the Middle Passage from Africa to the Western Hemisphere, which is more than the current population of Bolivia. Slavery supplied cheap labor for the plantations and agricultural operations in the New World. Brazil took in more African slaves than any other single country—at least three million. Colonial Brazil thrived on early plantation agriculture. When

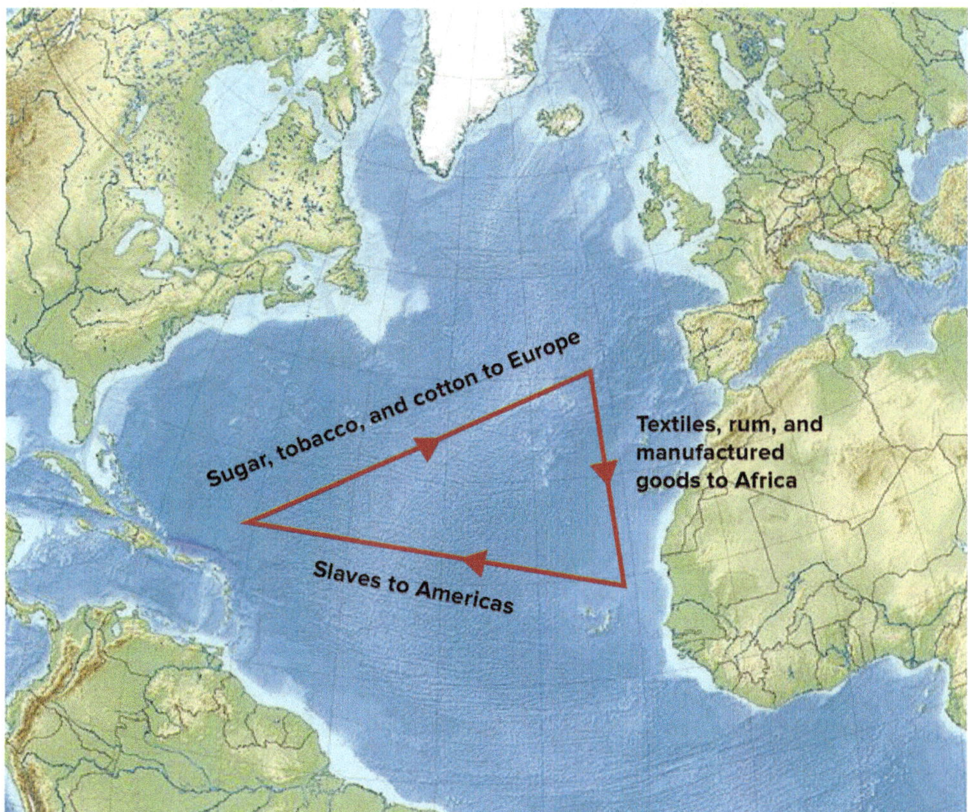

Figure 19. Atlantic Trade Triangle. Manufactured trinkets were sent to Africa from Europe, slaves were sent to the Americas, and plantation products and rum were sent to Europe. The Atlantic slave trade was responsible for bringing more than ten million African slaves to the Americas. Brazil received the largest number of slaves.

slavery was abolished in Brazil in 1888, the freed slaves found themselves on the lower end of the socioeconomic hierarchy. People of mixed African descent now make up more than one-third of Brazil's population. The Afro-Brazilian heritage remains strong and dominates the country's east coast. The African influence is evident in everything from the samba schools of the Brazilian carnival to the music and traditions of the people. In spite of Brazil being a culturally diverse country, Africans still have not found themselves on an even playing field in terms of economic or political opportunities in positions of power in the country.

B. Southeast Core: Urban and Industrial

Brazil's human development patterns are an example of the core-periphery spatial relationship. The main economic core area is located in the southeast region of Brazil, an area that is home to the largest cities of the realm and acts as the hub for industrial and economic activities. Political and economic power is held by elites residing in the urban core areas. The rural northern Amazon Basin is the heart of the periphery, providing raw materials and resources needed in the core. The periphery has a small population density, and most are Amerindian groups that make a living from subsistence agriculture, mining, and forestry. Rural-to-urban shift has prompted many of the rural poor to migrate to the large cities.

Founded in 1554 as a Catholic mission, the city of **São Paulo** rests at the heart of the core region. Its pattern of development is similar to that of Mexico City. Coffee production was the early basis of the local economy. São Paulo is located about thirty miles inland from the coast. It has grown to be the center of trade and industry for Brazil, but **Rio de Janeiro** receives greater attention for tourism. With more than twenty million people, São Paulo is the largest industrial metropolis south of the Equator, the largest city in South America, and among the largest cities on the planet.

Rio de Janeiro, Brazil's best-known city, is a travel and international business center with a population of more than ten million. The city is renowned for its carnival festivities and famous coastline. Tourists are attracted by its cultural attractions and coastal setting, with beautiful sandy beaches and the landmark Sugarloaf Mountain located in an open bay. **Salvador**, located on the coast, was Brazil's first capital. Rio de Janeiro became the capital in 1763, but to further develop Brazil's interior, in 1960 the capital was moved from Rio to the **forward capital** of Brasilia. Forward capitals are created to either shift development or to safeguard a geographical region. Brazil has an enormous interior region that it wishes to continue to develop for economic gain and the creation of the forward capital of Brasilia is in line with this objective.

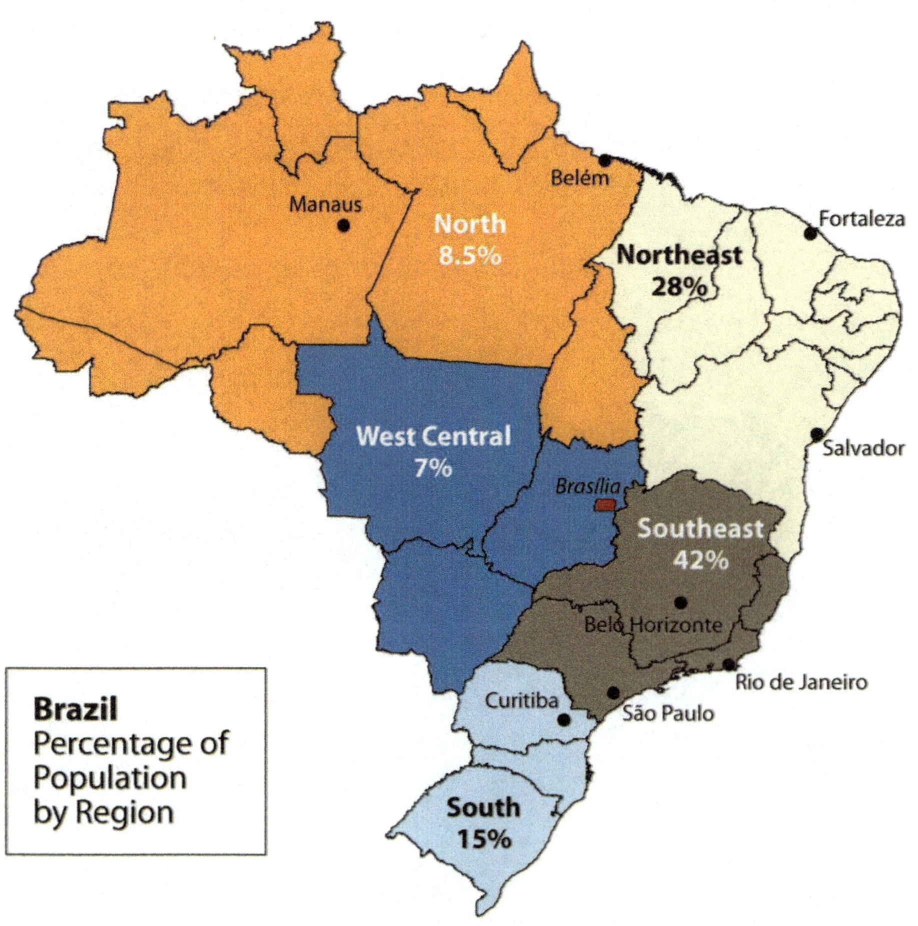

Figure 20. Population of the various regions of Brazil indicating the core region around the cities of the southeast. The peripheral region includes the large Amazon Basin of the north. Notice the many small states in the Northeast from Fortaleza to Salvador; these states are densely populated.

The three cities of São Paulo, Rio de Janeiro, and Brasilia, along with their urban neighbors, anchor the core region in the south. Brazil's internal migration to these cities follows a rural-to-urban or periphery-to-core pattern, and these three cities continue to grow at an unprecedented rate. São Paulo is more than sixty miles across. As migrant workers from the countryside and from the rural northeast migrate to the cities looking for work, they expand the city through self-construction. Slums, called favelas, extend out from the central city for miles.

The cultural fabric of Brazil has been built by immigrants from many countries. Brazil's diverse population has linked the country to the homelands of its immigrants and established trade and cultural connections that benefit the country in the global marketplace. The more than two hundred million people that make up Brazil's population are diverse and yet integrated, even if many still hold to the traditions or cultural heritage of their country of origin. After World War II, a large number of Japanese migrated to São Paulo. Today, Brazil boasts the largest Japanese population outside Japan. Many people from both Western and Eastern Europe have made Brazil their home, and large numbers of people from Lebanon, Syria, and the Middle East have immigrated here.

In spite its cultural diversity, Brazil has two overarching cultural forces that have helped hold the country together: the Portuguese language and Catholicism. These two centripetal forces help establish a sense of nationalism and identity. The Portuguese language has been adapted to the Brazilian society to reflect a slightly different dialect than the Portuguese spoken in Portugal. The Portuguese language has more of a unifying effect than religion. Though about 70 percent of the population claims to be Roman Catholic, additional religious affiliations in Brazil range from the African influence of Umbanda to the Muslim minority. Protestant denominations are the second-largest religious affiliations in Brazil. Secularism is on the rise, and many do not actively practice their specified religion.

In both population and physical area, Brazil ranks fifth in the world. Brazil is as urban as the United States or countries in the European Union. Average family size has dropped significantly, from 4.4 children in 1980 to 2.4 children in 2000, and continues to decline. Religious traditions give way to the urban culture and secular attitudes of modernity when it comes to family size.

Approximately 87 percent of Brazil's population lives in urban areas. Urbanization seldom eliminates poverty. Though the index of economic development indicates that urbanization will increase incomes for the population as a whole, poverty is a standard component of any large urban area in most places of the world, and the favelas of Brazil are similar to slums elsewhere.

The favelas of Brazil's urban centers are dynamic places, where land ownership, law and order, and public services are questionable. Millions of people live in the favelas of Brazil's large cities. Poverty and the search for opportunities and ad-

Figure 21. Christ the Redeemer. This giant statue overlooks the immense urban landscape of Rio de Janeiro, with Sugarloaf Mountain in the background.

Figure 22. Favela La Rocinha, Rio de Janeiro, Brazil. Favelas are usually self-constructed and start out as slums. Many lack ownership rights, police protection, or public services. With time, some become established neighborhoods.

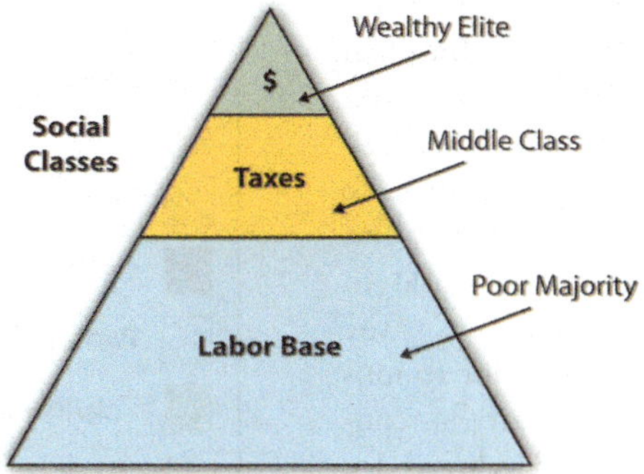

Figure 23. Socioeconomic Classes in Brazil, also common throughout Middle and South America

vantages are common elements of urban life. The core cities of Brazil suffer from the same problems as other developing megacities: overcrowding, pollution, congestion, traffic jams, crime, and increased social tensions. The energy generated by the sheer magnitude of people, industry, and commerce continues to fuel Brazil's vibrant growth and its many large, dynamic cities.

The pattern of wealth developing in Brazil is quite common: wealth and the ability to obtain wealth are held in the hands of those at the higher end of the socioeconomic ladder. The pyramid-shaped graphic used to illustrate Mexico's social layers in the chapter on Middle America can also be applied to Brazil. The minority wealthy elite own most of the land and businesses and control more than half the wealth. Corporate colonialism is quite active in Brazil. Multinational corporations take advantage of the country's development activities and muscle in on the profits, which seldom reach the hands of the majority at the lower end of society.

Brazil economically dominates and anchors South America and is an emerging power in the world marketplace. Brazil has the largest economy in South America and is a parallel force with the United States in the Northern Hemisphere. The country has urbanized and industrialized to compete with the global economic core areas in many ways. Brazil is among the ten largest economies in the world.

Brazil has favorable resources and labor to complete in the global marketplace. Its agricultural output has grown immensely over the past few decades. Brazil is a major exporter of soybeans, coffee, orange juice, beef, and other agricultural products. Brazil is the largest coffee producer in South America, but coffee only constitutes about 5 percent of its current annual exports. Coffee production is extensive in the Brazilian Highlands just inland from the coast. In addition, Brazil exports more orange juice than any other country and is second in the world in soybean production. The vast central interior regions such as the cerrado continue to be developed for industrial farming of massive food crops. However, exports of industrial manufactured products surpassed agricultural exports in 2010.

The availability of abundant minerals and iron ore has supported an expanding steel industry and automotive manufacturing, and its industrial activity continues to develop. Competitive high-tech companies continue to emerge, and production has increased in semiconductors, computers, petrochemicals, aircraft, and a host of other consumer-based products that provide economic growth.

Brazil is not only emerging on the world's stage on the economic front, but it also has a strong social and cultural presence in the world. The large metropolitan areas of the country are a Mecca for fashion and the arts. Carnival is the most well-known of Brazil's cultural festivals, but the country has been host to a number of other international events. Brazil is active in world sports competition and hosted the 2007 Pan American Games. Rio de Janeiro hosted the 2016 Summer Olympics and the 2014 FIFA (soccer) World Cup.

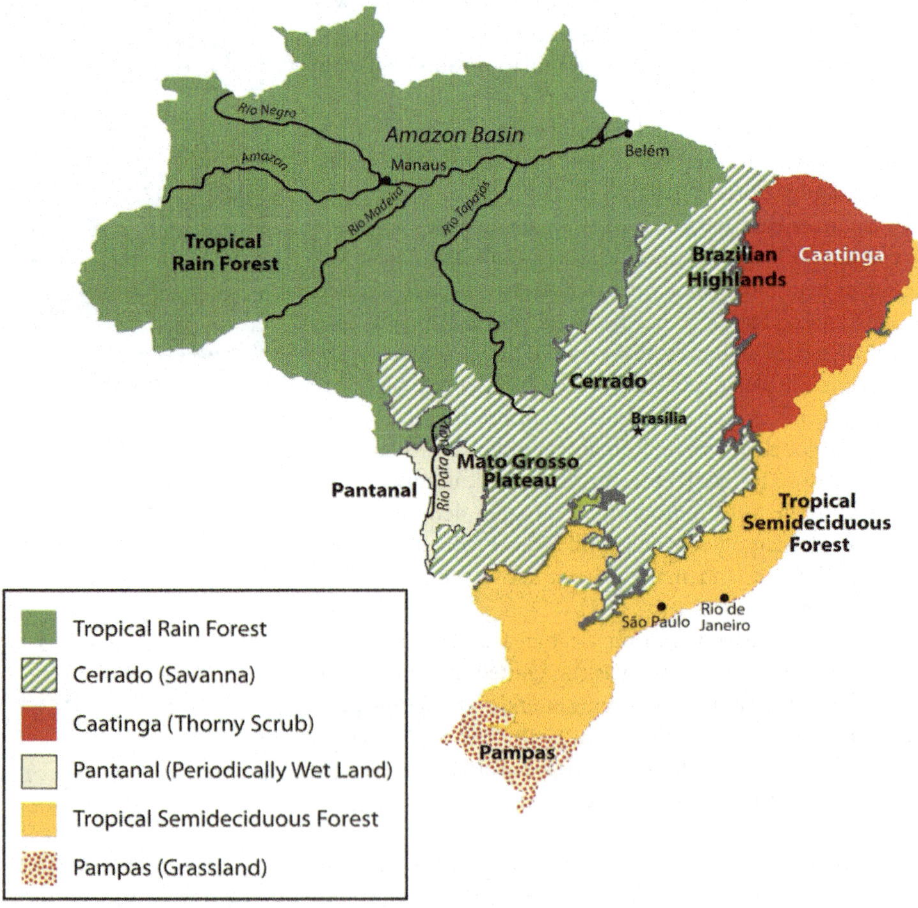

Figure 24. Natural Vegetation of Brazil. The immense area of the tropical rain forest can be seen on this map. The extensive cerrado region is noted on this map between the tropical rain forests of the Amazon Basin and the deciduous forests of the southeast. The soils of the thorny scrub region along the northeast are not as conducive to large-scale agriculture as is the cerrado.

C. Northern Periphery: The Amazon Basin

Just as the megacities of southern Brazil anchor the core of the country, it is the rural expanse of the Amazon North that makes up the periphery. A region the size of the US Midwest, the Amazon River basin is a frontier development area that has been exploited for its natural resources. Rubber barons of early years traveled up the Amazon River and established the port city of Manaus to organize rubber plantations for automobile tires. The Amazon River is large enough for oceangoing vessels to travel to Manaus. Today, **Manaus** has a free-trade zone with an entire industrial complex for the production of electronic goods and an ultramodern airport facility. Smaller ships can continue up the Amazon River all the way to Iquitos in Peru, which makes Manaus an ideal core city for economic trade; smuggling; and transshipment of illegal goods, including exotic animals from the region, such as monkeys, beautifully colored parrots, and other birds.

The other regions of the Amazon Basin have not been as fortunate as Manaus. Deforestation from cattle ranching, logging, and mining have devastated parts of the tropical rain forests of the Amazon Basin. The Amerindian populations have also suffered from encroachment into their lands. Only about two hundred thousand Amerindians are estimated to remain in Brazil, and most reside in the Amazon interior. This region boasts one of the world's leading reserves of iron ore; as much as one-third of Europe's iron ore demands are met through extensive mining southwest of Belém. The rapidly expanding development activities in the Amazon basin have boosted the region's economic situation, but at the same time there is growing concern about the preservation of the natural environment.

Deforestation has reduced the habitat critical to the survival of native species. An estimated 50 percent of the earth's species live in tropical rain forests, which only cover about 5 percent of the earth's surface. Tropical rain forests in the Amazon Basin are being cut down at an unsustainable rate, creating serious environmental problems. Loggers cut down the large trees, and the rest are usually burned to allow the ash to provide nutrients for other plants. The cleared areas are most often used by cattle ranchers until the soil is no longer viable. Then more forest is cut down and the process continues.

The forest has many layers of habitat. Soils in the tropics are extremely low in nutrients, which have been leached out by the abundant rainfall. Leaching is a process in which the nutrients dissolve in rain water, and then drop below the root zone of crops as the water percolates into the ground. Most of the nutrients are on the surface layer of the ground built up from falling leaves, branches, and debris decomposing on the forest floor. The removal of the forests removes these nutrients and also results in serious soil erosion.

Forest studies have indicated that tropical rain forests are actually quite resilient and can recover with proper forest management. However, clear-cutting large, wide areas for timber leaves an area devastated for an extended period. Clear-cutting could be replaced by strip cutting, which would harvest trees in narrower strips, leaving rows of trees standing. Strip cutting allows for more edges to be available for young plants to get their start to replenish the forest.

Figure 25. Deforestation in the Amazon. Deforestation in the Amazon is caused mainly by logging, agriculture, and mining. Agricultural practices can include slash-and-burn farming and cattle ranching.

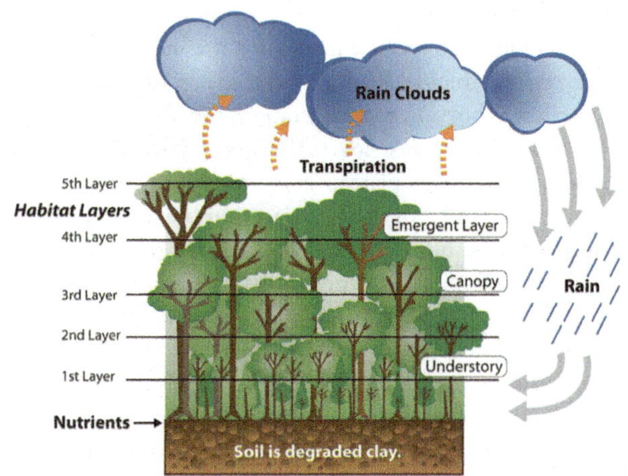

Figure 26. Dynamics of a Tropical Rain Forest. This image illustrates that the nutrients of the tropical rain forest ecosystem rest on the forest floor because tropical soils are degraded. The forest canopy has many habitat layers providing homes to a diversity of organisms. Dense tropical rain forests also contribute to the process of transpiration, which cycles precipitation from the ground back into the atmosphere where it can collect and return in the form of rain.

D. Regional Brazil

The core-peripheral spatial relationship can help us understand the power of Brazil's southeast core and the natural resource base of the country's peripheral north. This can explain the basic layout of Brazil's economic geography, but the northeast, the west central, and the south are three additional regions to consider that contribute to the geographic understanding of the country. Each has specific qualities that make it unique in Brazil's diversity of landscapes.

The **northeast** region is anchored along the coast, where plantation agriculture thrived during the colonial era. African slaves became the main labor base, and once freed, they made up most of the population. This agricultural region still grows sugarcane along the coast and other crops inland where the rainfall is reduced. The inland region includes parts of the Brazilian Highland, which runs parallel to the coast. Agriculture has traditionally required a large labor base, and family sizes in rural areas have been larger than their urban counterparts. This trend has given rise to a regional population of more than fifty million. However, the agriculture-based economy has not been able to supply the number of jobs and opportunities required for this large population. Poverty and unemployment in the northeast are high and have a devastating effect on the people. Some farms are not large enough for families to make a living on. Individuals in the northeast have developed a pattern of migration to the larger cities of the core area, looking for opportunities and employment—a pattern that fuels the self-constructed housing in the favelas of Rio de Janeiro and São Paulo.

The west central region has been opening up for development since the last part of the twentieth century and has experienced enormous advancements in industrial agriculture. In this region, vast cerrado grasslands are being plowed and converted into colossal fields of soybeans, grains, and cotton. The northwest portion of the cerrado is the huge Motto Grosso Plateau, which almost covers the largest state of the region. The landscape of this plateau is mainly scrub forest and savanna. Traditionally, the savanna portions have been used for grazing livestock, but in recent years more of the land is being plowed for growing agricultural crops. The entire west central region of Brazil is a giant breadbasket for the country. Its full agricultural potential has not been reached. Industrial agriculture requires infrastructure and transportation networks to transport the harvests to market, but the government has not kept up with the development of infrastructure at the same rate as the private sector has been developing industrial agricultural production. This region has enormous potential for agricultural expansion in the future.

Located on the eastern edge of the west central region is the forward capital of Brasilia. Its creation was prompted by the need to further develop the country's interior. Built in 1960 and now a metropolitan center with more than 3.8 million people, the city was planned and designed to be a capital city. Because of its rapid growth and development, the city faces issues similar to those found in Rio de Janeiro or São Paulo. Rural-to-urban shift has brought many rural people to Brasilia in search of opportunities and employment. Just as Rio de Janeiro and São Paulo have favelas, Brasilia has slums and self-constructed districts, too. Moving the capital to Brasilia moved the focus from the coastal region of the south to the interior. However, a large percentage of the interior lies beyond the city of Brasilia. Highways that reach the Amazon

Figure 27. Metropolitan Cathedral of Brasilia. The forward capital of Brasilia, built in 1960, showcases many architectural styles.

Figure 28. Miolo Vineyard. The Miolo Vineyard is located in the interior of the Brazil's affluent south region. Many Italian immigrants settled here and developed a wine industry that continues today.

can be accessed from the capital, but large areas of Brazil's interior remain a frontier unexploited by development. Developing these areas may bring great opportunities and benefits to the country; however, recent development activities have also brought devastation to the tropical rain forests in many parts of the frontier.

The three states of the south region are located well south of the Tropic of Capricorn and extend south to the border with Uruguay. This extrusion is often called South Brazil. Brazil was colonized by the Portuguese, but many of the immigrants to the south were from other parts of Europe, including Germany and Italy. The Italian immigrants developed a growing wine industry, and the German immigrants raised cattle and grew other crops. The region's good soils and moderate climate support many types of agriculture, which have dominated the early development of the region. Success in growing rice and tobacco and raising livestock has brought prosperity to the people who live here. Brazil has been one of the world's main producers of tobacco for many years, and the tobacco from the south is highly regarded for its nicotine content.

The south is one of Brazil's most affluent regions, and it has gained much wealth from agricultural activities. Farming is only one of the region's highly developed economic activities. The area is also blessed with natural resources such as coal that is shipped north to the main cities where steel is produced. The wealth of the region has provided support for high-tech industries, which are attracted to the region because of the supply of skilled labor, access to quality transportation, and communication links. Computer companies have established software firms that have in turn attracted other new companies. A technology center similar to California's Silicon Valley has combined with the manufacturing complex that has sprouted up along the coast of the south. The government and the business community have both provided economic incentives for these and other emerging enterprises.

Key Takeaways

1. Brazil was a Portuguese colony during the colonial era. This is why the Portuguese language and the Catholic religion are dominant components of the Brazilian culture. The strong African heritage comes from the many African slaves brought in during the colonial era to work the plantations. Immigrants from many other countries have settled in Brazil as well.

2. The urban southeast, with its large cities of São Paulo and Rio de Janeiro, represents the core economic area of the country. The large cities of Brazil continue to expand to accommodate the influx of new immigrants. The outer edges of the cities expand through self-construction in slums called favelas.

3. The northern regions of the Amazon Basin represent the periphery of the country. The north is being exploited for its natural resources. Development patterns have caused extensive deforestation of the tropical rain forest of the region. The main causes are agriculture, logging, and mining.

4. Other regions of Brazil include the west central, with extensive agricultural activities; the northeast, heavily populated with serious poverty; and the south in the protruded area bordering Uruguay, which is an affluent region with high standards of living.

8.4 The Southern Cone

Learning Objectives

1. Summarize the main physical features and regions of the Southern Cone.

2. Identify and locate the urban centers and understand the pattern of population distribution in the three countries of the region.

3. Describe the ethnic consistency of each country. Explain the pattern of immigration that created the region's heritage.

4. Explain why Argentina has great potential for economic growth.

5. Outline how Chile has emerged as a strong and stable country and discuss its human and natural resources.

The Southern Cone region of South America includes the countries of Uruguay, Argentina, and Chile. The name is an indication of the physical shape of the southern portion of the continent. The region is located south of the tropics. The Tropic of Capricorn runs just north of Uruguay and cuts across the northern regions of Argentina and Chile. The Southern Cone has more moderate temperatures than the tropics. Type C climates dominate in Uruguay, the Pampas region of Argentina, and central Chile. The region has extremes in weather and climate. The Atacama Desert and Patagonia both have type B climates because of a lack of precipitation, which stems from the rain shadow effect of the Andes Mountains. Highland type H climates follow the Andes chain through the region and exhibit their typical pattern of warmer temperatures at lower elevations and colder temperatures at higher elevations.

The countries of the Southern Cone share similar economic and ethnic patterns. Agriculture has been a major focus of the region's early development pattern, but today a large percent-

age of the population is urbanized. The European heritage of most of the population ties this region to Europe as an early trading partner. The global economy has given cause for these

Figure 29. The Three Main Regions of Chile (left); Argentina and Uruguay with the Regions of Argentina Outlined and Labeled (right).

countries to form trade relationships with many countries. The physical geography has provided many opportunities for human activities. The mountains, plains, and coastal areas provide a diversity of natural resources that have been exploited for national wealth. All three countries have primate cities that hold a high percent of the country's population.

Primate cities are usually twice as large as the next largest city and usually are exceptionally expressive of the national feeling and culture. In this case, all three primate cities are also the capital cities of each country. The Southern Cone is an urban region with higher incomes and higher standards of living than many other parts of South America.

A. Argentina

South America's second-largest country is Argentina. In physical area, Argentina is ranked eighth in the world. The Andes Mountains span its entire western border with Chile. At the southern end of the continent is Tierra del Fuego. Argentina is a land of extremes. **Mt. Aconcagua** is the highest mountain in the Western Hemisphere at 22,841 feet in elevation, and **Laguna del Carbon** is the lowest point in the Western Hemisphere at −344 feet below sea level. Parts of the northern region have a tropical climate; the southern region extends into tundra-like conditions with treeless plains.

Physical Regions

Argentina can be categorized into a number of regions that correlate roughly with the varied physical and cultural landscapes of the country. The main regions include Chaco, Northern, Mesopotamia, Cuyo, Pampas, and Patagonia. The Northern region of Argentina has one of the highest average elevations because of the Andes Mountain Ranges. The Andes ranges widen as they proceed northward to the west of Chaco and are home to fertile river valleys. The northern ends of the ranges extend into Bolivia and enter the Altiplano of the Central Andes.

The Chaco region, which is formally called the Gran Chaco, extends from northern Argentina into western Paraguay. Scrublands and subtropical forests dominate the landscape. There is a wet season as well as a dry season suitable for raising livestock and some farming. Western Chaco, which is closer to the Andes, is drier with less vegetation and is known for its high temperatures during the summer months. To the east, the Chaco region receives more rainfall and has

better soils for agriculture. The agrarian lifestyle dominates the cultural heritage of this region. In the 1920s and 1930s, the Chaco region attracted a large number of Mennonite immigrants from Canada and Russia who established successful farming operations mainly on the Paraguay side of the border and also extending into Argentina.

To the east of the northern region—on the other side of the Paraná River and reaching to the banks of the Uruguay River—is the region called Mesopotamia, whose name means "between rivers." This unique region has a variety of features, from flatlands for grazing livestock to subtropical rain forests. The most noteworthy feature is the expansive Iguazú Falls on the Iguazú River, located on the border of Brazil and Argentina. It is a series of 275 parallel waterfalls that are just short of two miles across. It has the greatest average annual flow of any waterfall in the world. Most of the falls are more than 210 feet high; the tallest is 269 feet. The spectacular Iguazú Falls is a major tourist attraction, drawing people from all over the world.

The **Cuyo** region is located along the Andes Mountains in the west central part of the country. Mt. Aconcagua is located here, along with other high mountain peaks. This arid region gets most of its moisture from melting snow off the mountains, which irrigate the rich agricultural lands that produce fruits and vegetables. The Cuyo is a major wine-producing region; it accounts for up to three-quarters of the country's wine production. Picturesque vineyards and farms make the Cuyo a favorite tourist destination in Argentina. Mendoza is the country's fourth-largest city. Low mountain ranges form the eastern border between the Cuyo and the Pampas.

The Pampas is a large agricultural region that extends beyond Argentina and includes a large portion of Uruguay and the southern tip of Brazil. With adequate precipitation and a mild type C climate, the Pampas is well suited for both agriculture and human habitation. The rich agricultural lands of the Pampas include the largest city and the country's capital, Buenos Aires, which is home to up to a third of the nation's population. The

Figure 30. The Iguazú Falls on the Iguazú River on the Border between Brazil and Argentina. The headwaters of the Iguazú River are near Curitiba in Brazil. The river converges with the Paraná River about 14 miles downstream from the Iguazú Falls at the point where Brazil, Paraguay, and Argentina meet—called the Triple Frontier.

Pampas provides some of the most abundant agricultural production on the planet. The western grasslands host large haciendas (prestigious agricultural units) with cattle ranching and livestock production. This area has elevated Argentina to its status as a major exporter of beef around the world. Agricultural production has been a major part of the nation's economy. One hundred years ago, the export of food products made Argentina one of the wealthiest countries in the world. In today's global economy, the profit margins in agricultural products are not as lucrative, and industrialized countries have turned to manufacturing for national wealth. Argentina continues to have a strong agricultural sector but has been increasing its industrial production in order to secure a strong economy.

Patagonia is a large expanse of the south that is semiarid because of the rain shadow effect. This area possesses enormous natural resources, including large amounts of oil and natural gas. Deposits of gold, silver, copper, and coal are found here. Raising livestock has been the main livelihood in Patagonia, which is otherwise sparsely populated. Patagonia includes the southern region of Tierra del Fuego and the rugged Southern Andes, which have some of the largest ice sheets outside Antarctica and many large glaciers that provide fresh water that feeds the region's streams and rivers. Patagonia also has a number of scenic lakes. Abundant wildlife can be found along the Atlantic coast, including elephant seals, penguins, albatrosses, and a host of other species.

Figure 31. Rain Shadow Effect. The rain shadow effect in southern South America creates the Atacama Desert and an arid Patagonia.

Population and Culture

Argentina, with a population of about forty million, is a country of immigrants and a product of the colonial transfer of European culture to the Western Hemisphere. During the colonial era, millions of people immigrated to Argentina from Western European countries such as France, Germany, Switzerland, Portugal, Greece, the British Isles, and Scandinavia. Additional immigrants came from Eastern Europe and Russia. Eighty-five percent of the population is of European descent; the largest ethnic groups are Spanish and Italian. The Mestizo population is only at about 8 percent. A small number of people from the Middle East or East Asia have immigrated and make up about 4 percent of the population. Less than 2 percent of Argentines declare themselves to be Amerindians.

Old World European customs mix with New World Latin American traditions to form a cultural heritage unique to Argentina. This cultural heritage can be experienced in the metropolitan city of Buenos Aires, where all facets of society and culture can be found. With a population of about thirteen million—one-third of Argentina's total population—Buenos Aires is a world-class city. Argentina is an urban country: more than 90 percent of the population lives in cities. The rural side of the culture has often been characterized as the traditional gaucho (cowboy) image of the self-reliant rancher who herds cattle and lives off the land. Beef is a mainstay of the cuisine in much of the country. The urban culture includes the traditional Argentine tango with music and cama-raderie in upscale night clubs. These traditional images may be stereotypes, but the cultural scene in Argentina is heavily invested in the international trends of the modern world. The cultural landscape has become integrated with fashions and trends from across the globe.

Figure 32. Palermo District, Buenos Aires. This photo illustrates the enormous expanse of the most populous area of Buenos Aires, Argentina.

Figure 33. Mt. Fitz Roy's Rugged Landscape. Mt. Fitz Roy is in the Andes on the border between Chile and Argentina. The mountain range borders Patagonia on the Argentinean side in the remote region of the south. The village of El Chaltén, which is the main access to the mountain, can be seen in the lower right of the photo. The unique terrain of the mountain is often photographed but seldom climbed.

B. Chile

Chile is a long, narrow country on the western edge of southern South America. Chile is 2,500 miles long and only 90 miles wide on average. This country borders the Pacific Ocean on one side and the Andes Mountains on the other. Chile has a variety of environmental zones, administrative districts, and climate patterns. Temperatures are cooler as one moves south toward Tierra del Fuego, which is split between Chile and Argentina. Rain has never fallen in select areas of northern Chile, which includes the Atacama Desert, one of the driest places on Earth and home to one of the world's greatest copper and nitrate reserves. The sodium nitrates found in the Atacama Desert are used in plant fertilizers, pottery enamels, and solid rocket fuel.

The Rain Shadow Effect

The climate is due to the rain shadow effect. In northern Chile and the **Atacama** region, prevailing winds reach northern Chile from the east and hit the Andes Mountain chain, which are some of the highest mountains on the continent. The height of the Andes causes any moisture from rain clouds to precipitate on the eastern slopes. The western side of the Andes Mountains at that latitude receives little to no precipitation, causing extreme desert conditions in the Atacama region of Chile. Southern Chile receives a large amount of rainfall because the prevailing winds at that latitude come from the west. Here the winds, which have picked up moisture over the South Pacific Ocean, hit the western side of the Andes. The air then precipitates out its moisture as it rises up the mountainsides of the western slopes of the Andes. Less moisture reaches the eastern side of the mountains, creating a rain shadow with arid and dry conditions for the region called Patagonia in southern Argentina. The Andes are not as high in elevation in the south, which allows some precipitation to fall on the rain shadow side. Chile can be divided into three regions:

1. Northern Chile, with the dry Atacama Desert

2. Central Chile, with a mild type C climate, adequate rainfall, and good farmland

3. Southern Chile, with lots of rainfall, rural, isolated islands, and mountains

Central Chile is the core region because it has a valuable port in Valparaiso and the country's capital city, Santiago, which is also Chile's most populous city. Central Chile is home to more than 90 percent of the country's population.

Socioeconomic Conditions

The people of Chile are 95 percent European and Mestizo. They have worked to establish a good educational system and an increasing standard of living. The political system is faced with the unequal distribution of wealth that is common in Latin America and many other countries of the world. Half the country's wealth is concentrated in the hands of about 10 percent of the population. About 50 percent of the population is on the lower end of socioeconomic scale. Dire poverty exists in Chile, but it is not as prevalent here as it is in the Central Andes, Paraguay, or Northeast Brazil. Chile has a thriving middle class that has made good use of the opportunities and education that Chile has offered them.

Chile is blessed with natural resources that include the minerals of the Atacama Desert, extensive fishing along the coast, timber products from the south, and agricultural products from central Chile. All these factors have brought about an emerging development boom and have attracted international trading partners. The stable government and growing economy have successfully kept inflation low and employment high, reduced poverty, and brought in foreign investment. In the globalized economy, Chile has managed to work with various trading partners to increase its advantages and opportunities in the international marketplace. Chile has increased its trading activities with its counterparts in the Pacific Rim in Asia and North America. Chile has an abundance of fish in its coastal waters, copper and minerals in the Atacama Desert, and timber products in its southern region. The United States is one of Chile's main trading partners. Chile's main exports to the United States include paper, minerals, metals, and copper. Major agricultural products shipped to the United States include processed fruits, to-

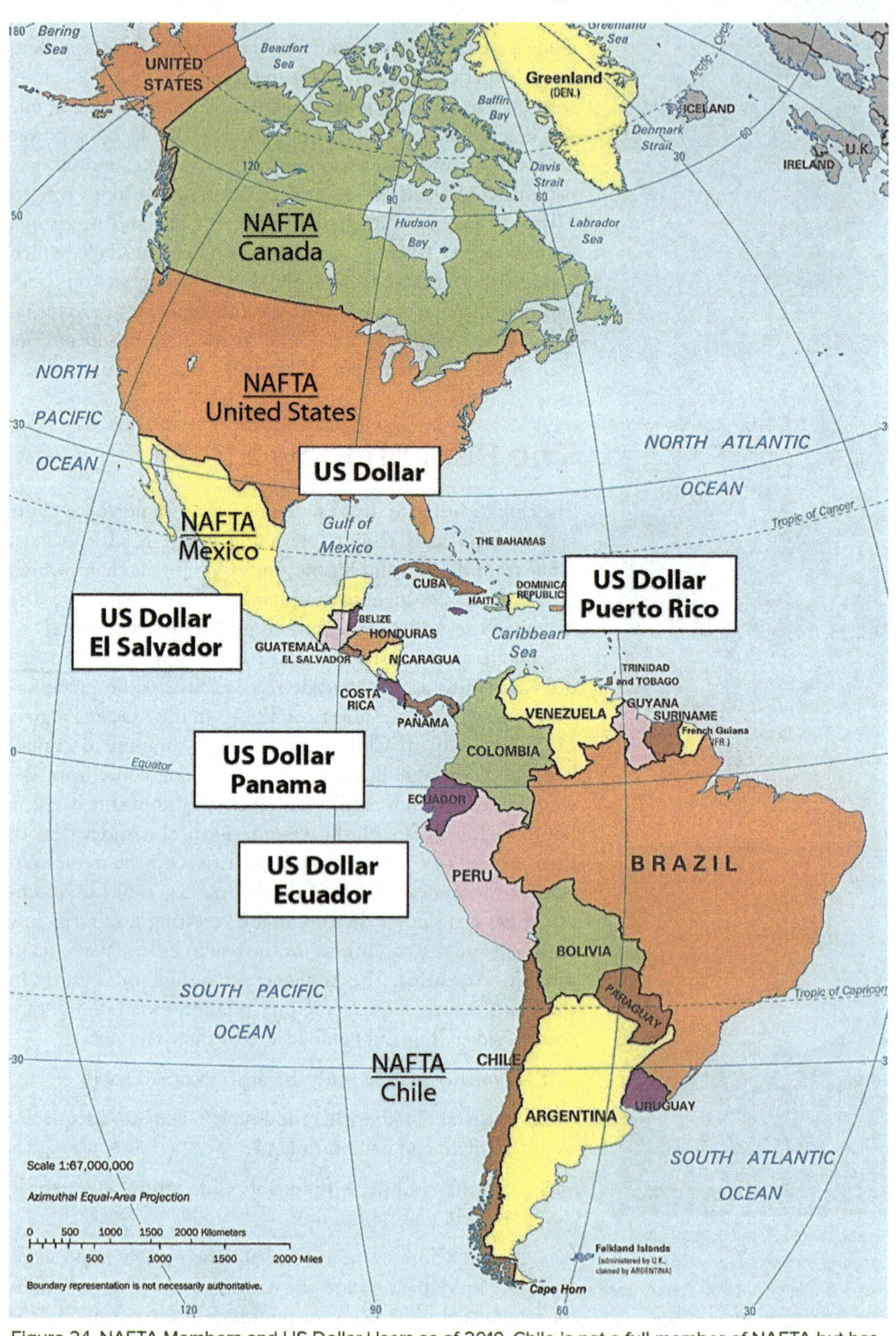

Figure 34. NAFTA Members and US Dollar Users as of 2010. Chile is not a full member of NAFTA but has separate free-trade agreements with all three NAFTA countries and is considered an unofficial member.

matoes, grapes, vegetables, and wines. One reason why the United States trades so much with Chile is due to the seasonal variations between the northern and southern hemisphere. When it is winter in the north, it is summer in the south. Each has an opposing growing season for fruits and vegetables that can complement the other.

Chile has strong ties to the economies of North America, but in spite of close ties with the north, Chile has retained its unique status in the Southern Cone. Chile still has its own currency even though countries with smaller economies, such as Ecuador and El Salvador, have adopted the US dollar as their medium of exchange. As of 2016, the NAFTA members of Canada, the United States, and Mexico, along with their trading partner of Chile, all used different currencies. The US dollar is the most widely used currency in the world and is also used in other Latin American countries. There has been talk of creating a similar currency within NAFTA called the Amero that would replicate the European Union's solution, which was to create a single currency, the euro. World currencies fluctuate in value, and a stable currency would increase the capacity for creating larger trading blocs that could do business on a more even economic playing field. Chile has individual free-trade agreements with all three members of NAFTA, so any change in currency with the NAFTA countries would also affect Chile.

Key Takeaways

1. The Southern Cone possesses large, diverse physical landscapes. Variations in terrain include tall Andean mountain peaks, desert conditions, prairie grasslands, and semitropical regions to the north.

2. The Southern Cone is an urban realm with high percentages of the population living in large cities. Primate cities dominate each country. Large sections of each country's interior make up the extensive rural periphery with activities based on natural resources.

3. The Southern Cone is a European commercial area, and more than 85 percent of the population is of European ancestry. There are few Amerindian minorities or immigrants from Africa or Asia in the Southern Cone.

4. Argentina is a large country in physical area and has a number of important regions that provide opportunities for economic prosperity. The country, however, has lacked a stable government and a consistent business climate to take advantage of Argentina's favorable geographic situation.

5. Chile has emerged on the global stage as a model for economic growth and stability. Its people have a higher standard of living, and the country has abundant natural resources. Chile has been a trading partner with NAFTA countries.

End-of-Chapter Summary

- The equator runs through the center of South America at its widest point and parallels the Amazon River. The two main physical features are the mighty Amazon River and the extensive Andes Mountains. The Andes, the longest mountain chain in the world, run from Venezuela to southern Chile. The Amazon River has the largest flow of water of any river on Earth, and the Amazon Basin is home to the world's largest tropical rain forest. Rain forests cover less than 5 percent of the earth's surface but have the richest biodiversity of any biome (environmental region), holding up to 50 percent of the world's organisms.

- All of South America except for the Guianas was colonized by Spain and Portugal. The Guianas were located along the coast, where plantation agriculture was prominent. African slaves were the main labor base on the plantations. When slavery was abolished, slaves were replaced by indentured servants from the Asian colonies. Most of the population in Guyana and Suriname is from Asia. The cultural geography of South American can be understood by identifying the cultural regions formed by the majority ethnic groups the human activity in which they are engaged.

- Venezuela and Colombia dominate the countries of northern South America. Venezuela is an urban country, and much of its wealth is generated from the export of oil. Colombia is mountainous with vast tropical forests bordering the Amazon. Illegal drugs, oil, and coffee are Colombia's three main export products. The United States is the main buyer of these products. The control of the physical territory of Colombia is divided between the government, drug cartels, and insurgent groups.

- South America's Andean West region was the home to the ancient Inca Empire, which was conquered by Spanish conquistador Francisco Pizarro in 1533. Ecuador, Peru, and Bolivia have regions with high elevations in the Andes, including the Altiplano region with Lake Titicaca. Most of the population is Amerindian and holds to the customs and traditions of their heritage. Oil and mineral resources have been a main source of wealth for the region, but the profits have not always reached the majority population; instead, the wealth is often held by a minority at the top of the socioeconomic layer.

- Brazil was colonized by Portugal, which gave the country the Portuguese language and the Catholic faith. During the colonial era, a large number of African slaves were introduced to the country, which added to the country's multicultural society. Many other Europeans and Japanese also immigrated to Brazil. The country's large size allows for diverse human activities. The core urban area is located along the southeast coast, where the large cities of São Paulo and Rio de Janeiro are located. The forward capital of Brasilia is positioned inland on the edge of the core area.

- The rural periphery of Brazil includes the large Amazon Basin, with tropical rain forests and large undeveloped regions. Originally inhabited by Amerindian groups, the Amazon Basin is being developed for agriculture, mining, and the timber industry. This type of development is devastating large tracts of rain forest and placing increasing pressure on the few remaining indigenous groups. Large mining operations have fueled development, causing serious environmental concerns over habitat loss and the destruction of the tropical rain forest.

- The Southern Cone region of South America has a wide range of physical landscapes, from the semitropical areas of the north to the arid grasslands of the south in Patagonia. The Atacama Desert on the west coast is a product of the rain shadow effect from the high Andes. The rich agricultural lands of the Pampas are the breadbasket of the region, producing large amounts of grain and livestock. Coastal areas are good for fishing and allow access to foreign markets.

- The southern regions of South America also have heavily European populations. There are few Amerindians or people of African heritage. Most people live in urban areas, and one-third of Argentineans live in Buenos Aires. Chile has been a major trading partner with the North American Free Trade Agreement (NAFTA) nations and has a stable and growing economy. Argentina is a large country in physical area and has great potential for economic development.

Chapter 9

Sub-Saharan Africa

Introduction

Sub-Saharan Africa includes the African countries south of the Sahara Desert. The **African Transition Zone** cuts across the southern edge of the Sahara Desert at the widest portion of the continent. Many of the countries in the African Transition Zone are included in the realm of Sub-Saharan Africa. The realm can be further broken down into regional components: **Central Africa, East Africa, West Africa,** and **Southern Africa**. At the eastern end of the African Transition Zone is the **Horn of Africa,** which is often included in the region of East Africa. Maps vary in terms of which countries are included in each region, but this general geographic breakdown is helpful in identifying country locations and characteristics. **Madagascar** is a large island off the southeastern coast of Africa and is usually not included with the other regions because its geographic qualities and biodiversity are quite different from the mainland.

The continent of Africa is surrounded by salt water. The Indian Ocean borders it on the east, and the Atlantic is on the west. The southern tip of the continent—off South Africa—is often referred to as the **Cape of Good Hope**, where the Atlantic Ocean meets the Indian Ocean. The continent of Africa has a number of small island groups that are associated with the realm and are independent countries. Approximately 350 miles off the coast of West Africa in the North Atlantic are ten islands that make up the independent country of Cape Verde. Just south of Nigeria on the eastern side of the Gulf of Guinea near the equator are the two islands that make up the independent country of São Tomé and Príncipe, a former Portuguese colony. The small country of Equatorial Guinea also includes an island off the coast of Cameroon where its capital is located. Three island groups in the Indian Ocean around Madagascar include the independent countries of the Seychelles, Comoros, and Mauritius.

Figure 1. The Continent of Africa.

9.1 Introducing the Realm

There is considerable variation with respect to how the regions of Sub-Saharan Africa are delineated or identified on maps. The debate is not about what regions are in Sub-Saharan Africa but rather about which countries are to be included within each region. The regions have both similarities and differences. The cultural geography varies widely from country to country and from one ethnic group to another, but at the same time, there are shared cultural patterns across all Sub-Saharan African regions. For example, colonialism has been a major historical factor in the shaping of the countries. Families are large, and rapid rural-to-urban shift is occurring in all regions. Every region has large urban centers—often port cities that act as central core locations supported by a large peripheral rural hinterland.

Globalization has entered into the dynamics connecting these once-remote regions with the rest of the world. Advancements in communication and transportation technology have created networks connecting Africa with global markets. Sub-Saharan Africa has a young population that is on the move, seeking to gain from any opportunities or advantages it can find. The political arena is dynamic: changes in political leadership through coups or military takeovers are common, as is authoritarian rule. Sub-Saharan Africa is home to some of the poorest countries in the world. Poverty is evident in the countryside and in the urban slums of the largest cities. Bitter civil wars are a part of every region's history. Violence and conflicts continue in some areas, while other areas exhibit political stability and thriving economies. The diversity in human geography is the most noteworthy dynamic in Sub-Saharan Africa. The variety of ethnic groups along with the multiplicity of languages and religious affiliations create strong centripetal and centrifugal forces that interact in a thriving sea of cultural diversity.

Most of the population lives an agrarian lifestyle, but there are people who are developing the skills necessary to adapt to the rapid globalization wave that is importing new technology and new ideas to the continent. The urban core areas of the continent are the main focus of the global trends in technology and communication. These urban core areas exhibit the typical dynamics of the core-periphery spatial relationship. Sub-Saharan Africa has many core areas and many peripheral areas. The core urban centers have political power thanks to the social elites who have connections to the global economy and often dominate political activities.

These core urban areas are often magnets for people from large families in rural peripheral areas seeking employment. Millions of people in Africa who seek employment are willing to migrate to the cities or even other countries to find work. The rural immigrants are often not of the same ethnic group as those in power, which sets up the basis for discriminatory policies that disadvantage the many minority groups that are not affiliated with the government. These dynamics can fuel protest activities with a goal of overthrowing the powerful elite. Various ideas have been proposed to help level the socioeconomic playing field. One of the more prominent options is the implementation of a democratic government, where most of the people have a stake in electing those that hold positions of leadership and power.

Broad patterns and dynamics of people and places are replicated throughout the Sub-Saharan realm. The regions share common demographic trends of large family sizes, agrarian lifestyles, and low income levels. The patterns of an economy based on agricultural production and mineral extractive activities as well as disruptive changes in political leadership are common throughout the continent. Each region has diverse ethnic groups and an array of different languages. South of the African Transition Zone, the most prevalent belief systems are Christian based and animist, while north of the zone, Islam is widespread. Division and civil unrest can occur where the different religions meet and compete for political control. These concepts will bear repeating throughout the chapter. The cultural mosaic of Sub-Saharan Africa is vast and complex, and this chapter will outline the basic trends and patterns with specific examples that will help place it all in perspective.

A. Physical Geography of Sub-Saharan Africa

The African Transition Zone divides North Africa from the rest of Africa because of climatic and cultural dynamics. Cultural conflicts and desertification are common in the zone. Dry, arid type B climates, common in the Sahara Desert, are dominant north of the zone. Tropical type A climates prevail south of the zone. Global climate changes continue to shape the continent. The shifting sands of the Sahara are slowly moving southward toward the tropics. Desertification in the zone continues as natural conditions and human activity place pressure on the region through overgrazing and the lack of precipitation. Type B climates resurface again south of the tropics in the southern latitudes. The **Kalahari** and **Namib Deserts** are located in Southern Africa, mainly in the countries of Botswana and Namibia.

Africa does not have extended mountain ranges comparable to the ranges in North or South America, Europe, Asia, or Antarctica. There are, however, smaller mountain ranges and individual mountains of great height. The **Ethiopian Highlands** reach as high as 15,000 feet in elevation, and the region is sometimes referred to as the "Roof of Africa" due to its height and relatively large area. East Africa also has a number of well-known volcanic peaks that are high in elevation. The tallest point in Africa—**Mt. Kilimanjaro** in Tanzania near the border with Kenya—is 19,340 feet high. Nearby in Kenya, Mt. Kenya is 17,058 feet high. The **Rwenzori Mountains** on The Congo/Uganda border reach more than 16,000 feet in elevation and create a rain shadow effect for the region. Permanent glaciers exist on these ranges even though they are near the equator. On the western side of the continent, **Mt. Cameroon** in Central Africa is more than 13,000 feet in elevation. South Africa's **Cape Ranges** are low-lying mountains no higher than about 6,000 feet. The continent of Africa generally consists of basins and plateaus without long mountain chains. The plateaus can range more than 1,000–2,500 feet in elevation. The only extensive continuous feature is the **Great Rift Valley** that runs along the tectonic plate boundaries from the Red Sea southward to South Africa.

The main rivers of Africa include the **Nile, Niger, Congo,** and **Zambezi**. The Nile River competes with the Amazon for the status as the longest river in the world; the **White Nile** branch begins in **Lake Victoria** in East Africa, and the Blue Nile branch starts in Lake Tana in Ethiopia. The Niger flows through West Africa; its mouth is in Nigeria. The Congo River crosses the equator with a large tropical drainage basin that creates a flow of water second only to the Amazon in volume. The Zambezi River in the south is famous for the extensive Victoria Falls on the Zambia and Zimbabwe border. Victoria Falls is considered the largest waterfall in the world. Other significant rivers exist such as the **Orange River**, which makes up part of the border between South Africa and Namibia.

There are a number of large lakes in Sub-Saharan Africa. The largest is **Lake Victoria**, which borders

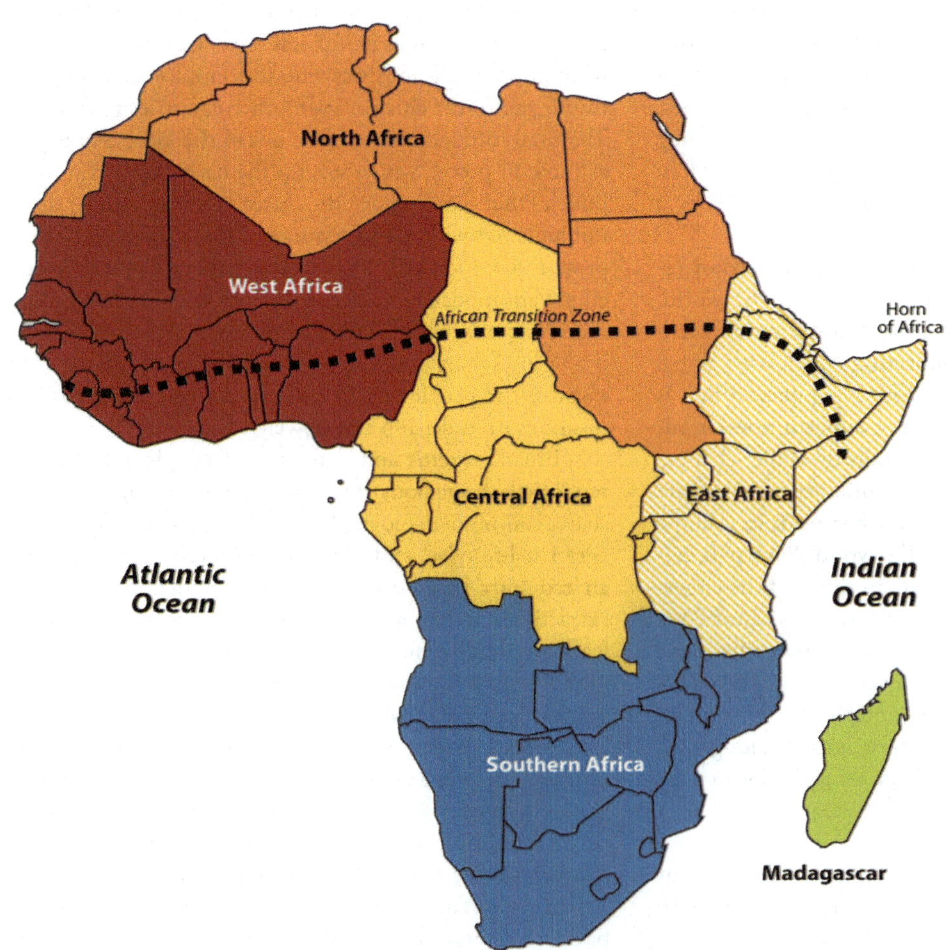

Figure 2. Regions of Africa. To help identify places discussed in this book Africa has been divided into regions along traditional boundaries.

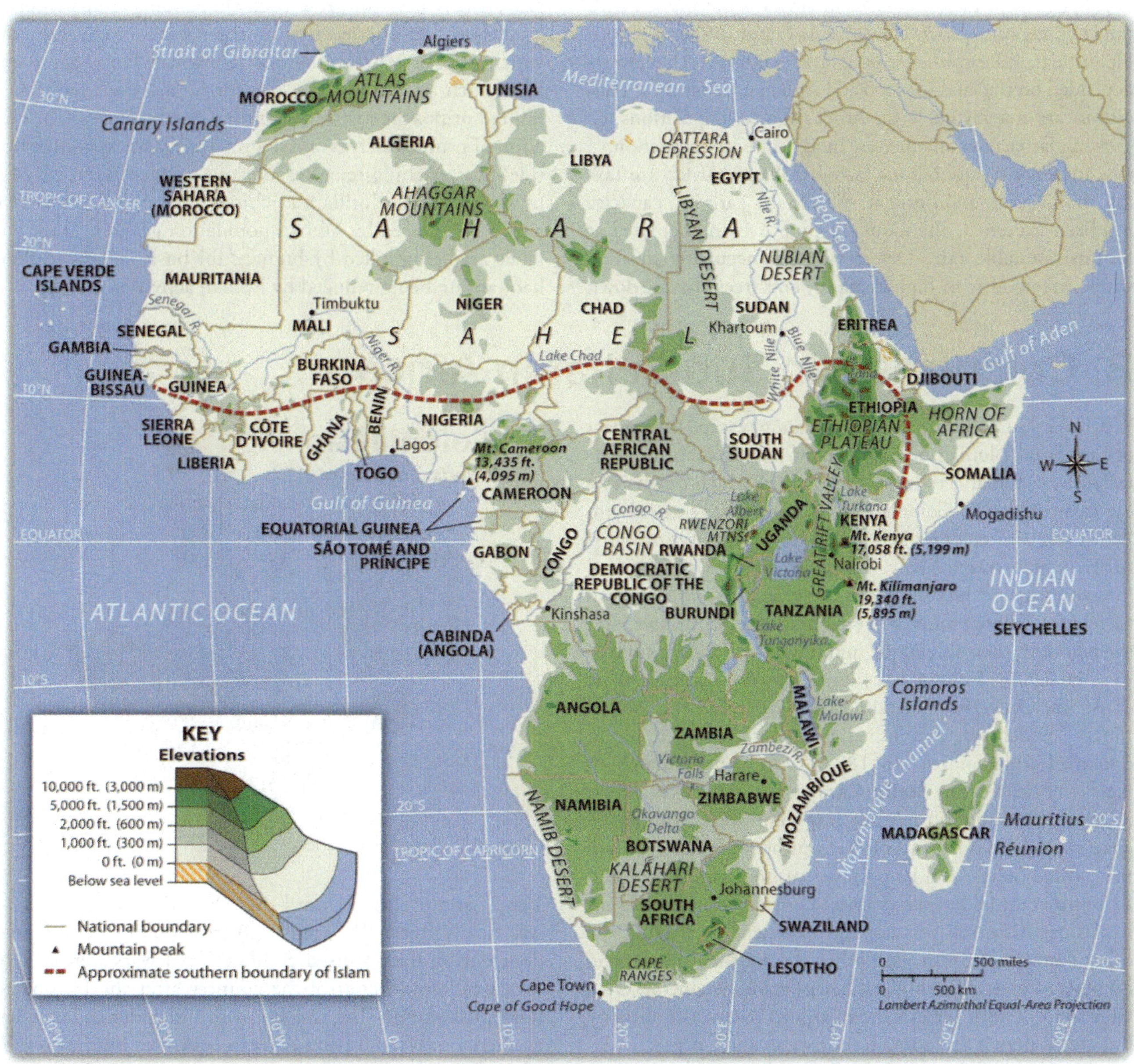

Figure 3. Physical features of Sub-Saharan Africa. Sub-Saharan Africa does not have long mountain chains such as those found in the other continents. The Nile, Congo, and Niger Rivers are the main waterways. The Namib and Kalahari are the main deserts south of the African Transition Zone. The Great Rift Valleys are the longest physical features in East Africa.

several East African countries and is considered to be the second-largest lake in the world in surface area. Only Lake Superior in North America has a greater surface area. A number of large lakes are located in the rift valleys of the east. Three of the largest lakes along the western rift are **Lake Malawi**, **Lake Tanganyika**, and **Lake Albert**. To the northeast of these in Kenya is **Lake Turkana**, which reaches to the Ethiopian border. **Lake Chad** is located in the African Transition Zone on the border between Chad, Mali, and Nigeria. Lake Chad has been severely reduced in size in recent years.

The equator runs through the center of Sub-Saharan Africa, providing tropical type A climates. These regions usually have more rainfall, resulting in lighter, leached-out soils that may not be as productive as regions with richer volcanic soils, such as those found in the rift valleys. Root crops are common in Africa, as are millet and corn (maize). The savanna regions of the east and south have seasonal rains that affect the growing season. Soils in savanna areas are usually not as productive and cannot be depended on to fulfill the agricultural needs of growing populations. Savannas are usually grasslands or scrub forests with a seasonal precipitation pattern. Cattle and livestock grazing are common in savannas, and migrations are frequent to follow seasonal grazing conditions. In specific areas of Southern Africa, larger farming operations exist in type

C climates. However, Sub-Saharan Africa is not blessed with the large regions of rich alluvial soils found in the Northern Hemisphere. The ever-growing agrarian population has always depended on the land for food and sustenance, but these conditions are not favorable for Africa's future. Populations are growing faster than any increase in agricultural production.

Increasing populations in Sub-Saharan Africa are taxing the natural environment. Where the carrying capacity has been exceeded, the natural capital is being depleted at an unsustainable rate. Deforestation is occurring in areas where firewood is in high demand, and trees are cut down faster than they can grow back. Expanding human populations are also encroaching on the natural biodiversity for which the African continent is renowned. Large game animals such as rhinoceroses, elephants, and lions have been hunted or poached with devastating consequences. The creation of game preserves and national parks has stemmed this tide, but poaching remains a serious problem even in these protected areas. Gorilla and chimpanzee populations have also been stressed by human population growth. These animals are being killed by humans for bush meat, and their habitats are being reduced by human activities.

B. Colonialism in Sub-Saharan Africa

Wherever you go in the world, you cannot escape the impact of European colonialism. Sub-Saharan Africa was broadly affected by colonial activities, the legacy of which can be recognized to this day. Colonial powers of Europe ventured into Africa to claim colonies and obtain slaves. Slavery impacted Africa in many ways. Many African groups were instrumental in capturing and holding slaves to trade with the European merchants. These groups still exist and have had to live with the fallout of their role in the supply side of the slave trade.

By 1900, European colonial powers claimed most of Africa. Only the Kingdom of Ethiopia and the area of Liberia, which was established as a home for freed slaves, remained independent. In 1884, Otto von Bismarck of Germany hosted the **Berlin Conference**, which to a great degree determined how Africa was colonized. This conference included fourteen colonial European countries and the United States, and its purpose was to divide Africa and agree on colonial boundary lines. Germany had few claims in Africa; Bismarck was hoping to benefit by playing the other countries against each other. At the time, more than 80 percent of Africa remained free of colonial control. On a large map of Africa, claims were argued over and boundary lines were drawn according to European agreements.

There was little regard for the concerns of local ethnic groups. Colonial boundaries divided close-knit communities into separate colonies. Ethnic boundaries were disregarded. Often-warring groups were placed together in the same colony. The Europeans, seeking profits from cheap labor and resources, did not consider the local people or culture and often employed brutal measures to subdue the local people. Most of the current borders in Africa are a result of the Berlin Conference, and many of the geopolitical issues that confront Africa today can be traced back to this event specifically and to colonialism in general.

European colonialism remained strong in Africa until the end of World War II, which left many European countries economically exhausted. In 1945, the United Nations (UN) was established. One of the primary UN objectives was to oversee the decolonization of European colonies. Still, colonialism in Sub-Saharan Africa lingered on. Most African countries did not gain their independence until the

Figure 4. Colonialism in Africa.

1960s, and it was not until the 1990s that the last colony was finally freed. The transition from colony to independent nation itself caused conflict. Civil wars were fought over who would control the country after the Europeans were pushed out. The transition to full independence has exacted a heavy toll from African countries but has resulted in stronger political structures and greater democratic liberties in many cases. The first country to gain independence in Sub-Saharan Africa was Ghana in 1957.

During the era of independence after World War II, the Cold War between the United States and the Soviet Union enticed many of the African countries to support one or the other of the superpowers. The political pressure that divided African countries during the Cold War has persisted and is the basis of some of the current political problems. The European countries extracted raw materials and minerals from their African colonies, as well as slave labor. Varying degrees of attention were given to education, medical care, or infrastructure development. The dependency that a colony had on a European country for economic income in some cases continued long after independence and may still continue today. On the other hand, major technology transfers from Europe to Africa infused greater efficiencies into Africa's economic activities.

Figure 5. European Colonies in Africa, 1884.

C. From Colonialism to Independence

The transition from European colony to independent state has not necessarily been a civil transition for African countries; likewise, independent African countries have struggled to create stable governments or peaceful conditions—though stable governments and peaceful conditions have been established in some cases. In nearly all cases, removing the colonial powers from Africa was only half the battle toward independence. The other half was establishing a functional, effective government. Though each country has taken a slightly different path, most former colonies have endured civil unrest, conflict, or warfare in their push for stable governments. Coups, fraudulent elections, military regimes, and corruption have plagued the leadership in a number of African countries. Civil unrest usually precedes a change in leadership: political power regularly changes hands by a military coup or an overthrow of the current leader.

The realm has had difficulty developing and maintaining effective political systems with democratic leadership. Political leaders have come and gone, many have been replaced before their terms were over, and a few have stayed long after their terms ended. When a leader is elected democratically, it is common to have widespread accusations of corruption, voter fraud, or ballot box scandals. Such government mismanagement and corruption have been common problems.

The vacuum left by retreating colonial powers at times has been filled by authoritarian dictators or by leaders who assumed control of the government and then proceeded to pillage and plunder the state for personal gain or for the enrichment of their clan or political cronies. After the colonial era, it was not uncommon for new leaders to be connected to the old colonial power and to adopt the language and business arrangements of the European colonizer. This gave them an advantage over their competitors and provided a means of income gained by the support of their former European colonial business partners, which often wanted to keep ties with their colonies to continue to exploit their resources for economic profits. Many of these leaders stayed in power because of military backing or authoritarian rule funded by the profits from selling minerals or resources to their former colonial masters.

A few countries are still struggling to bring about some type of order and unity, and in spite of all the negative issues, there have been democratic and relatively stable progressive governments in Africa that have emerged from the transition in recent years, which is a hopeful trend for the long term.

The objective of colonialism was to connect a colony with the mother country, not to connect African countries to each other. As a result, little cooperation occurs between African countries. Each individual country interacts more with its European colonial counterpart with regard to trade, economics, and cultural dynamics. European colonialism has isolated African countries from their neighboring countries and does not contribute to unity within or among African regions. Colonial powers often built new port cities to extract goods and resources from their colonies and transportation systems from the new ports to the interior to collect the resources and bring them to the port. However, the colonial powers did not build a network of transportation systems that connected the region as a whole. Colonial powers were intent on continuing dependence on the mother country so their colonies could be controlled and dominated. The current wave of globalization based on corporate colonialism continues to encourage the same pattern of discouraging interaction between counties and instead encouraging trade with more economically powerful non-African countries. Most of the interaction between countries is a result of crisis or warfare, in which case large numbers of refugees cross the border into neighboring countries for personal survival or security. To begin to work together and promote trade, common transportation, security, and industry, fifteen West African states came together in 1975 to create the **Economic Community of West African States (ECOWAS)**. Since then, additional political and economic agreements have been signed by various African countries. In 2001, the more expansive African Union was created to help African states compete in the international marketplace.

Figure 6. Freedom Square in Ghana. This monument honors Ghana's independence in 1957 from the British. Ghana was the first country in Sub-Saharan Africa to gain independence.

Key Takeaways

1. Sub-Saharan Africa includes the African Transition Zone and the regions south of the Sahara. The African Transition Zone is a transitional zone between type A and type B climates and between the religions of Islam and Christianity.

2. Sub-Saharan Africa is a realm of plateaus and basins with four main river systems. Mountain ranges, volcanic peaks, and large lakes are found in or along the rift valleys of eastern Africa. The rift valleys were created by tectonic activity.

3. Before the colonial era, many African kingdoms dominated regions of Sub-Saharan Africa. West Africa had a number of great kingdoms. Kingdoms along the coast contributed to the slave trade.

4. European countries colonized almost the entire realm of Sub-Saharan Africa. European powers made agreements during the 1884 Berlin Conference to divide the realm and create boundary lines between their African colonies. Many of the lines remain as the current borders today.

5. Colonies were designed to provide labor and resources to the mother country. Port cities, transportation systems, and other infrastructure were implemented to benefit the imperial power without regard for development of the colony.

6. African colonies received independence after World War II, with most becoming independent in the 1960s. The transition from colonies to independent countries has been plagued by civil war, political chaos, or economic devastation.

9.2 Human Geography of Sub-Saharan Africa

A. Economic Geography

The demographic data for each country in each region of Sub-Saharan Africa indicate the region's human geography. Urban percentages, family size, income levels, and the other data that can indicate the lifestyle or development level are acutely helpful in understanding the trends in Sub-Saharan Africa. The index of economic development illustrates the dynamics and conditions that exist in the realm. The data indicate the consistency of economic and development trends across Africa. The data does not, however, indicate differences in cultural dynamics and uniqueness in the ways that local people live. The interesting part of studying Sub-Saharan Africa is the many ethnic and cultural groups in each country that bring to the surface a wide array of global diversity in our human community. Within each and

Figure 7. Informal and Formal Sectors. The photo on the left is of side-street vendors in Senegal, who might contribute to the informal market system that is prevalent in many Sub-Saharan African countries. Cash and bartering are the main methods of payment The photo on the right is of a shopping mall in Zimbabwe, which is part of the formal market system that is regulated and taxed by the government.

every country are microcosms of human societies that hold particular customs that may be thousands of years old. Globalization and technological advancements challenge every cultural group to adapt and innovate to make a living yet provide continuity in their heritage. The remote cultural groups of Sub-Saharan Africa are most susceptible to the volatile nature of globalization, which threatens their current way of life.

Sub-Saharan Africa is a **peripheral** world region with neocolonial economic patterns. Peripheral regions usually supply raw materials, food, and cheap labor to the core industrial countries. Most of the population in Sub-Saharan Africa works in subsistence agriculture to make a living and feed their families. Families are large. In recent decades, there has been enormous rural-to-urban migration to the major cities, which are extremely overcrowded. Sub-Saharan Africa has more than 750 million people, and most earn the US equivalent of only $1–3 per day.

There is no single major core economic area in Sub-Saharan Africa. This realm has many core cities and the rest is periphery. Many of the core cities are improving their technology and infrastructure and entering into the globalized economy. Even so, as much as 70 percent of the people still work in agriculture, leaving little time to develop a large educated group of professionals to assist with social services and administrative responsibilities. The realm depends heavily on outside support for technical and financial assistance. Computers, medical equipment, and other high-tech goods are all imported. African states have formed trade agreements and have joined the African Union to assist each other in economic development and trade.

Sub-Saharan Africa has nearly forty urban areas of more than one million people. At the center of the central business districts (CBDs) are modern high-rise business offices well connected to the global economy. Outside the CBD are slums with no services and miserable, unsanitary conditions. The **informal sector of the economy**—that which is not regulated, controlled, or taxed—has become the primary system of doing business in most of the cities. The informal sector comprises trading, street markets, and any other business without financial records for cash transactions.

The lack of government regulation and control prevents taxes from being assessed or collected, which in turn diminishes support for public services or infrastructure. The formal sector of the economy—that which the government can regulate, control, and tax—is forced to foot the bill to operate the government and support public services such as education, security, and transportation. In spite of the misery and unhealthy conditions of the slums where millions of people already live, more migrants from the countryside continue to shift to the city in search of jobs and opportunities. African cities are growing rapidly, many without organized planning.

If the index of economic development were applied to Sub-Saharan Africa, a clear pattern would emerge. Rural areas would be in the lower stages of development, and only a few developing areas would be higher than stage 3. A large percentage of this realm would be in stage 1 or 2 of this model. Countries with higher standards of living, such as South Africa or Botswana, would be working through stage 3 or 4. This region has one of the fastest-growing populations of the world and economically lags behind countries in the Northern Hemisphere, which have transitioned into the higher stages of the index of economic development.

B. Incomes, Urbanization, and Family Size

The socioeconomic data illustrate well the conditions for people in Africa in comparison to the rest of the world. African countries are at the lowest end of the statistics for development prospects. The larger cities are showing promise for advancement into the higher stages of development. Family sizes in the rural countries are some of the largest in the world. The average fertility rate for much of Africa is about 5; in Mali and Niger, the rate is higher than 7. One-third to half of the populations of these countries are under the age of fifteen. Children make up most of the population in many areas—an indication that heavier burdens are placed on women, meaning that women are not easily able to get an education or work outside the home.

The populations of West African countries are increasing rapidly and will double in about thirty years at the current rate. This trend places an extra burden on the economy and on the environment. It is fueling one of the fastest rural-to-urban shifts in the world. West Africa is now only about 32 percent urban, and Burkina Faso and Niger are less than 20 percent urban—clear indications that agriculture is the largest sector of their economies and that most of the people live in rural areas or small villages.

Personal income levels in West African countries are among the lowest in the world; as far as standard of living is concerned, these are poor countries. Few economic opportunities exist for the millions of young people entering the employment market.

The United Nations (UN) **Human Development Index (HDI)** lists all but two of the West African countries in the lowest category of development. Sierra Leone is the lowest in the world. As a peripheral world region, the economic base is structured around agriculture with supportive extractive activities. These countries are in a subsistence mode with a rapidly expanding population and few industrial or postindustrial activities to gain income.

No Sub-Saharan African country is in stage 5 of the index of economic development. Nevertheless, the fact that they are not consumer societies does not negate the rich cultural values and heritage of the realm. The energy of the people does not revolve around consumerism but is instead focused on the people themselves. High levels of social interaction and community involvement bring about different cultural standards than those of a consumer society, which focuses more on the individual and less on community. Unless there is social unrest or open warfare, which does exist in various places, the people work hard to bring about a civil society based on family and tradition.

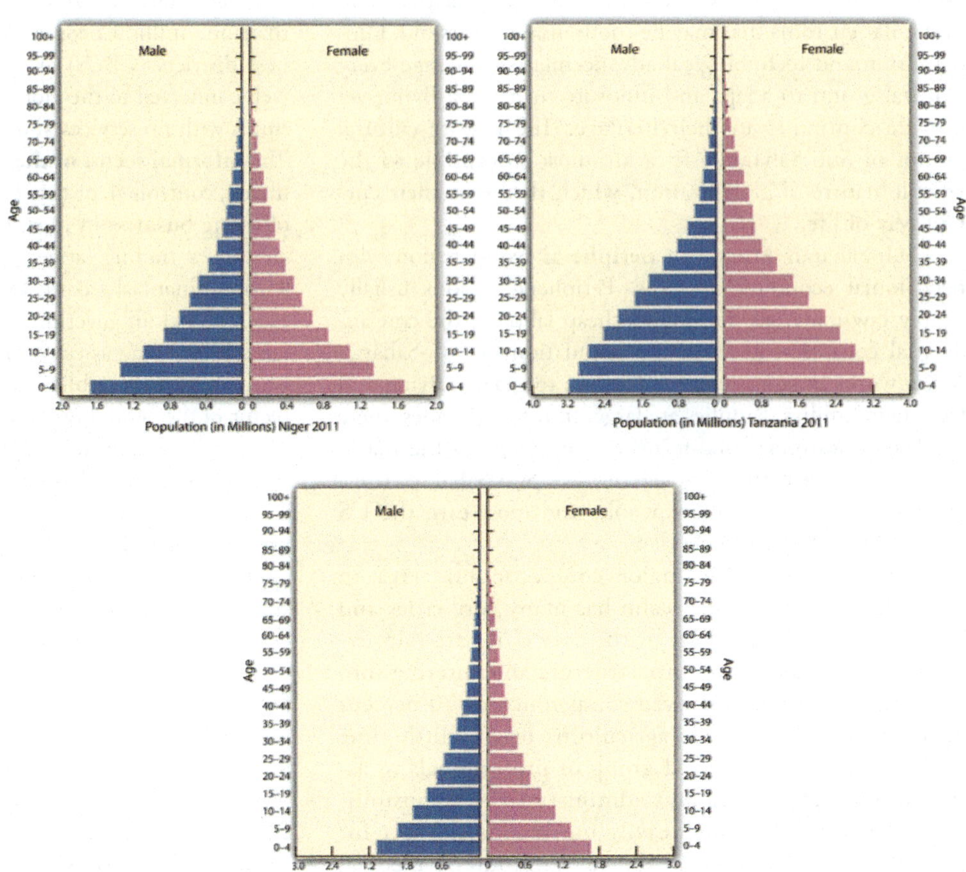

Figure 8. Population Pyramids of Some African Countries. The population pyramids of Niger, Tanzania, and South Africa illustrate the population growth dynamics of Sub-Saharan Africa. Niger's population pyramid illustrates large family size and rapid population growth. Tanzania's pyramid is similar but shows signs of slowing population growth. South Africa, which is more urbanized and industrialized, shows signs of declining family sizes and fewer children.

C. Languages in Sub-Saharan Africa

Sub-Saharan Africa covers a large land area more than 2.3 times the size of the United States. Thousands of ethnic groups are scattered throughout the realm. There is immense diversity within the 750 million people in Sub-Saharan Africa, and within each country are cultural and ethnic groups with their own history, language, and religion. More than two thousand distinct languages are spoken in all of Africa. Forty are spoken by more than a million people. Many local languages are not written down and have no historical record or dictionary. Local languages without a written history are usually the first to be lost as globalization affects the realm. Nigeria, with more than 130 million people, is the most populous country in Africa. The African Transition Zone cuts through the country's northern portion. More than five hundred separate languages are spoken in Nigeria alone. Three of the six dominant languages in Sub-Saharan Africa—spoken by at least ten million

people or more—are spoken in Nigeria: Hausa, Yoruba, and Ibo. The three remaining major languages of Sub-Saharan Africa are Swahili, Lingala, and Zulu.

Colonial activity changed much of how the African countries operated economically, socially, and politically. Language is one aspect of culture that indicates a colonial relationship. Many African countries today speak a European language as the official language. Mauritania is the only country that has Arabic as its official language. Nigeria has English plus other local languages. The official languages of most of West Africa are either French or English, and Guinea-Bissau's official language is Portuguese.

This vestige of colonial power would seem inconsistent with the desire to be free of foreign domination. However, because often dozens to hundreds of local languages are spoken within the country, choosing the colonial language as the official language produces less of an advantage for one group wishing to dominate the political arena with its own local language and heritage. For example, a language problem arises when a government needs to print material for the country. What language do they use? In Nigeria, there would be more than five hundred possible languages. What if the leadership used a language only spoken by a few people? The language of those in power would provide an advantage over those that could not understand it. This is why many African countries have chosen a colonial language as their country's official language.

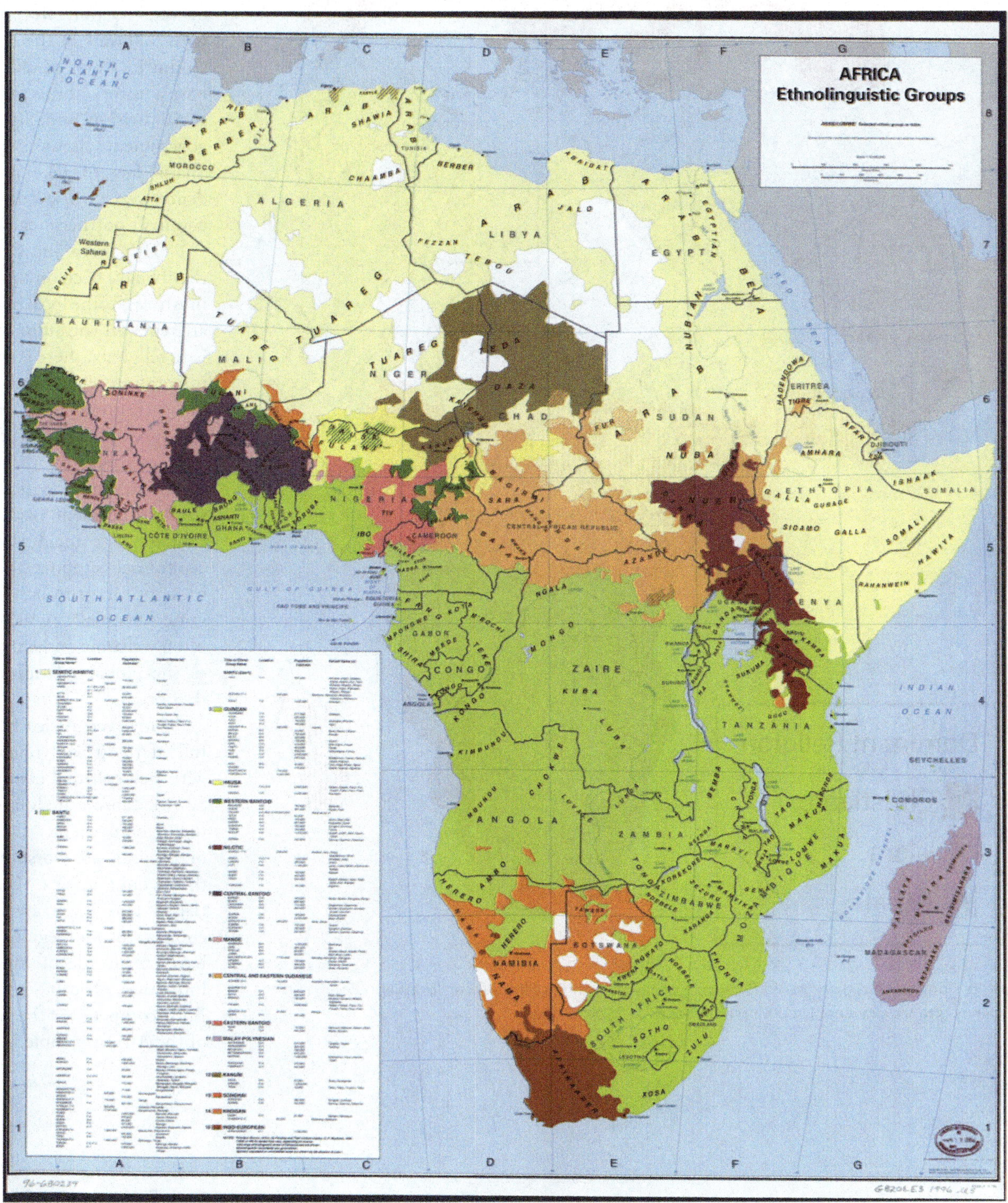

Figure 9. Language Families of Africa. There are six main language families with many variations of each. It is estimated that more than two thousand languages are spoken in all Africa.

D. Religion in Sub-Saharan Africa

The current religions in Africa follow the pattern of the African Transition Zone. Most of the population north of the zone follows Islam, and most of the population living south of the zone follows Christianity. Large percentages of people in some parts of the region still follow a wide array of traditional or animist beliefs.

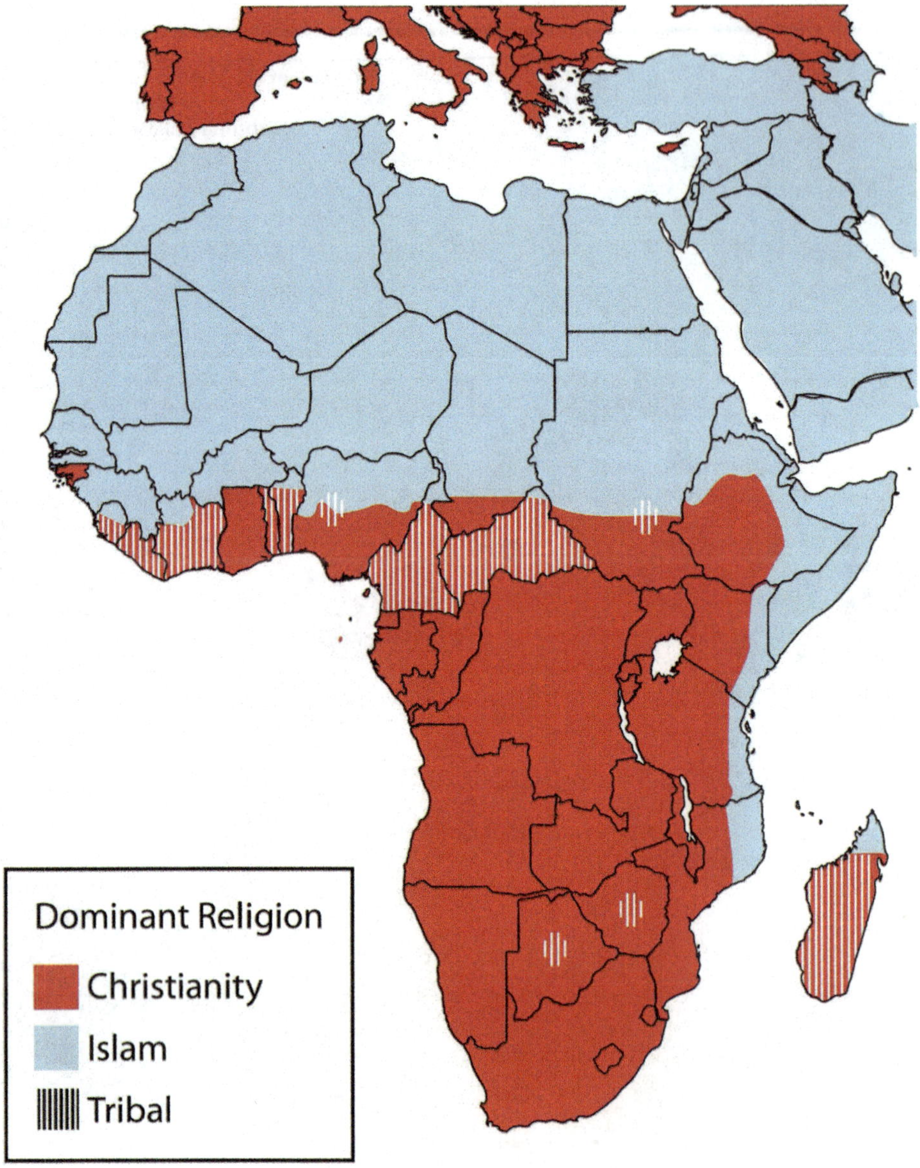

Along the African Transition Zone, followers of one religion will clash with followers of the other. Countries such as Nigeria have a history of this type of social division, and Nigeria's government allows Islamic Sharia law to take precedence over civil law in the country's northern regions.

Both Islam and Christianity have been on the rise in Africa. As the local beliefs are replaced with monotheistic religions, there is more integration with either the West (Europe and America) or the Middle East. Religious activity through Christian missionaries or the advancement of Islam sometimes coincides with economic support being brought in through the same channels, which is often welcome and enhances the global relationships that occur.

However, Africa is still full of traditional religions with rich spiritual histories. Spiritual forces are found in the environment. Deities of all kinds are worshipped throughout Africa. Christianity and Islam are latecomers to the region but have made deep inroads into the African culture. Both compete for the souls of the African people.

Figure 10. Dominant Religions in Africa. Islam is prominent north of the African Transition Zone, and Christianity is more prominent south of the African Transition Zone.

E. Ethnic Divisions and Civil Wars

Sub-Saharan Africa is home to thousands of ethnic or traditional groups. Each has a separate identity and history, and often one group is in conflict with another. The slave trade and the establishment of colonial political boundaries or policies exacerbated historical ethnic hostilities. Major civil wars have been fought throughout Sub-Saharan Africa and continue at the present time. Central Africa has recently endured brutal conflicts. More than five million people died as a result of the civil war in The Democratic Republic of Congo (formerly Zaire). Fighting continues between various factions over political control or control over natural resources, such as diamonds or gold. Civil wars have devastated some African countries. Many other countries, such as Zimbabwe, Chad, and the Central African Republic, have also suffered economic disintegration as a result of severe political unrest.

Rwanda's Tutsi-Hutu conflict was historic in its violence and in the senseless killing of innocent people. In 1994, the centuries-old conflict erupted into violence of unprecedented proportions. Hutu militias took revenge on the Tutsis for years of suppression and massacred anyone who did not support the Hutu cause. Tutsi rebels finally gained strength, fought back, and defeated the Hutu militias. More than a million people were killed, and more than a million defeated Hutus fled as refugees to neighboring countries, where many died of cholera and dysentery in refugee camps.

The civil war in Rwanda and the many refugees it created destabilized the entire Central African region. The shift in population and the increase in military arms along the Zairian border resulted in an extensive civil war in The Congo (Zaire) that has resulted in the deaths of more than five million people, many by disease or starvation. Over three million deaths are estimated to have been related directly to the war and another two million by the harsh conditions in the region. The civil wars in The Congo from 1996 to 2003 changed the cultural and political landscapes and destroyed valuable infrastructure. One of the driving forces in these wars was the control of valuable mineral resources found in the Great Rift Valley along the eastern boundary of The Congo. Diamonds, gold, copper, zinc, and other minerals are abundant in this region, and wealth that can be gained from the mining of these products attracts political forces to compete for their control.

Civil wars have also wreaked havoc on the countries of Angola, Sierra Leone, Liberia, and Somalia. Many of the civil wars were not reported by the news media worldwide, even though the number of people affected, injured, and killed was deplorable.

Creating stability in parts of Africa has been challenging, as civil unrest and political corruption continue in many African countries. The core industrialized countries have been hesitant to step in or invest in the peace and stability of Africa. Governments of more than a few African countries have been unable to bring stability or to provide for their people. For example, Somalia has no central government; rather, it is ruled by warlords and village chiefs. Corruption, dictatorial rule, and military force have been major components of government rule in these cases.

Figure 11. Tragedy in Rwanda. The bones of victims killed in the Rwandan genocide are kept in this school in Rwanda. The total number of people killed in the war is unknown.

F. HIV and AIDS in Sub-Saharan Africa

As if Sub-Saharan Africa did not have enough to work through to achieve stability, it must also cope with high numbers of people infected with HIV and AIDS. From South Africa to Kenya, there is a line of countries with some of the highest percentages of HIV-infected people in the world. The HIV virus has infected more than 24 percent of adults in Botswana, and some villages have lost an entire generation of adults to AIDS. The AIDS pandemic has become a major health crisis for Sub-Saharan Africa. According to the World Health Organization, as many as thirty million people in this realm are HIV-infected.

It's possible to live with HIV for years before dying of AIDS. HIV-infected individuals can pass the virus to others without knowing they have it, and millions of people die of AIDS in Sub-Saharan Africa without ever knowing they were infected. The lack of education, HIV testing, and medical services hinder progress toward stopping this deadly disease. Prevention is an ideal that has not materialized yet. Many people do not want to be tested out of fear of rejection by their families and friends if they are infected. AIDS will surely kill millions more in Africa before a solution is found.

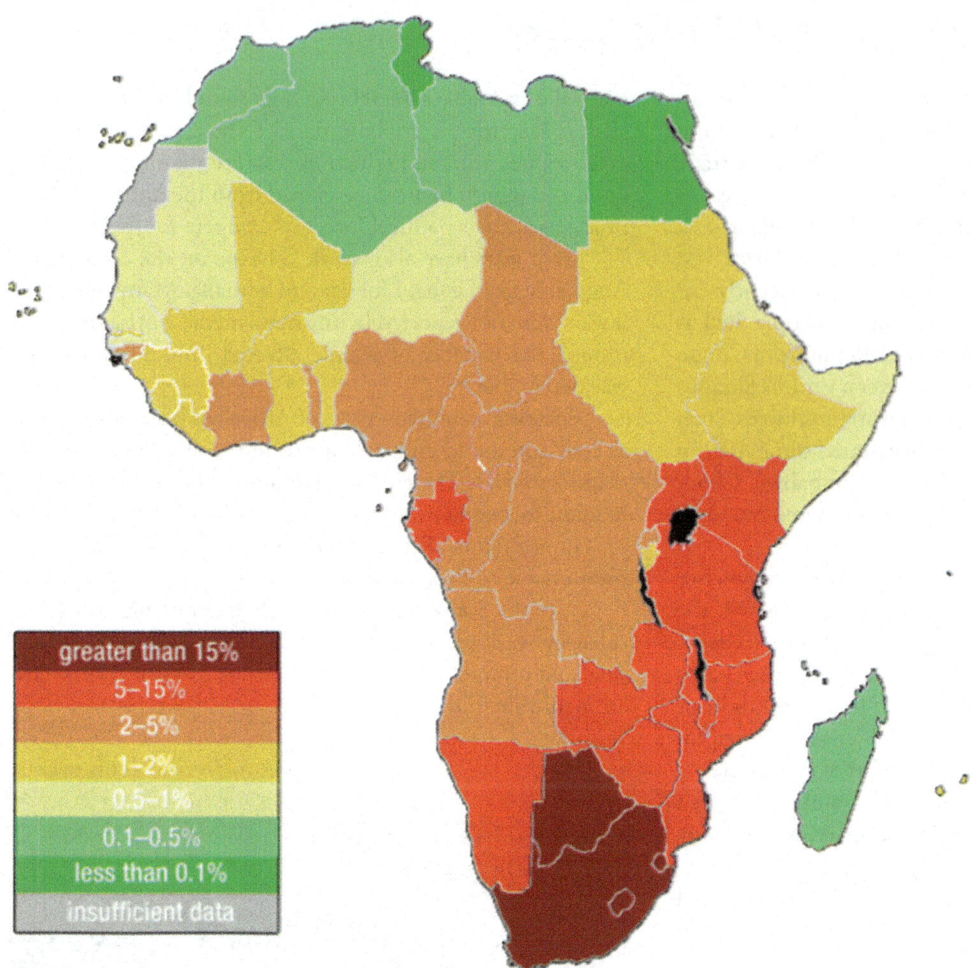

Figure 12. HIV and AIDS Prevalence in African Nations by Percentage.
This map shows the estimated percentage of adults with HIV in Sub-Saharan Africa for each state based on best estimates by the World Health Organization and the UN.

Many other diseases are common in Sub-Saharan Africa. Some are spread by vectors in the environment. Mosquitoes spread illnesses such as malaria and yellow fever, which are common throughout the realm. Sleeping sickness is spread by the tsetse fly, which can also infect cattle and livestock. Hepatitis is widespread. Schistosomiasis, tuberculosis, and typhoid are also common. Unsanitary conditions and polluted water are breeding grounds for microbes that cause diseases. People living in less-than-sanitary conditions are more likely to come in contact with and become infected with these diseases and are also less likely to obtain medical attention at an early stage of the disease.

Key Takeaways

1. Sub-Saharan Africa's many core cities and peripheral regions create a high rural-to-urban shift. Much of the realm's economic activity is conducted in the informal sector.

2. Most countries in this realm are in the earlier stages of the index of economic development. The integration into global markets has brought urbanization and development to many of the major core cities.

3. The informal sector dominates many of the economic activities in Sub-Saharan Africa. The emergence of stable governments and economic structures usually results in a stronger formal sector of the economy, which in turn provides improved national infrastructure.

4. Sub-Saharan Africa is realm with a high diversity of languages and religions. In an attempt to provide an equal footing for all languages, many countries have chosen their colonial language as their official language to avoid giving speakers of one language any political advantage over another.

5. Parts of Sub-Saharan Africa continue to experience social unrest and civil war. Armed conflicts in Rwanda and The Congo (Zaire) have either killed or affected millions of people.

6. A high percentage of people in Sub-Saharan Africa are infected with HIV—up to one-fourth of the population of some countries. Health care, education, and cultural forces are important factors in combating AIDS and the tropical diseases common in the realm.

7. Sub-Saharan Africa has enormous potential to exploit their physical features and cultural heritage to expand their tourism market. It is often difficult to balance the investments for adequate infrastructure and services required for tourism with other needs, such as education and health care.

9.3 West Africa

The region of West Africa includes the southern portion of the bulge of the continent, which extends westward to the Atlantic Ocean. This region is bisected by the African Transition Zone, which borders the southern edge of the Sahara Desert. The main physical features include the Sahara Desert and the Niger River. The Cameroon Highlands are located on the eastern border between Nigeria and Cameroon. At 4,100 miles long, the Nile River is the longest, while the Congo River is Africa's second longest at 2,922 miles in length. The Niger River is Africa's third-longest river and extends more than 2,600 miles from the Guinea Highlands through Mali, Niger, and Nigeria before reaching the Atlantic Ocean in the Gulf of Guinea.

Some geographers include the country of Chad or portions of it within the region of West Africa. In this textbook, Chad is listed with Central Africa. The portions of Chad located north of the African Transition Zone share similar characteristics with North Africa. Off the coast of Mauritania are the Cape Verde Islands, which are united as an independent country associated with Africa. Cape Verde was once a Portuguese colony but received its independence in 1975. Western Sahara has been in conflict with Morocco over independence and is most often associated with the region of North Africa because of the influence of Islam and because of its connection to Morocco.

The African Transition Zone cuts across the region of West Africa, indicating a division between Islam and Christianity and between the Sahara Desert and the tropics. This diversity in religion and climate is usually exhibited with a north/south division. Islam is the dominant religion on the north side of the African Transition Zone; Christianity is more dominant to the south. The two religions often clash in the areas where they meet. Traditional beliefs and animist religions are also practiced in the African Transition Zone.

The terms "state" and "country" are often used interchangeably by the world community outside of the United States. Both are meant to refer to a physical unit with a sovereign and independent government. In Sub-Saharan Africa the term "state" is commonly used to refer to a country in any one of the various regions.

Figure 13. West Africa. This map shows the region of West Africa as defined in this chapter. The African Transition Zone crosses the middle of this region.

A. Niger

The former French colony of Niger is landlocked, with the Sahara Desert making up its largest portion. Niger is a land of subsistence farmers, and most of the population lives in the southern regions. The country is less than 20 percent urban. Other economic activities include uranium mining, which is the country's main export. The world demand for uranium has not been strong in recent years. Oil exploration has begun and international oil corporations have garnered contracts for drilling.

The Sahara Desert is moving southward, and the agrarian culture at the base of Niger's society is often plagued by drought and famine. The Niger River flows through the southwestern region and provides fresh water, but the northern region is mainly the Sahara Desert, and large portions are covered with sand dunes.

The country has extreme demographics. Niger has the highest fertility rate (7.6) in the world, and half the population is under the age of fifteen, causing a population explosion that taxes the sparse natural resources and brings even more poverty to a country at the bottom end of the index of economic development. Infant mortality rates in Niger have been the highest in the world.

The mainly Sunni Muslim country has a rich cultural base but suffers from economic problems that appear to increase with the increase in population and desertification. Heavy national debt has hindered social services and has required a considerable amount of foreign aid from a number of sources.

The political conditions in Niger are typical of the region. For the first thirty years after independence from France, the country was ruled by a single political party and military rule. There have been several coups and various political leaders have been in power. A dispute remains with Libya over its northern border. Ethnic infighting with a minority Taureg group has emerged in recent years, bringing conflict and discord.

B. Sierra Leone

Sierra Leone has been devastated in the past decades due to brutal civil war and political turmoil; as a result, the country was the lowest in the world on the human development index in 2010. There has been little political stability. Military factions have wreaked havoc on the country since it gained independence from Britain in the early 1960s. After independence, the country's contested elections and multiple coups resulted in a ban on all but one political party and established military rule. At the core of the conflicts was competition for the control of the diamond industry, which was also a primary factor in the civil war.

The abuses of power escalated to a breaking point in the late 1980s, which set the stage for a decade-long civil war in the 1990s. A group calling themselves the Revolutionary United Front (RUF) pushed to take control of the eastern diamond mining sector through terrorist tactics and brute force. Thousands of people died and thousands more were mutilated by having their arms, legs, or other body parts cut off by machetes. Whole villages were destroyed and the residents killed, tortured, or maimed.

The government fell under various military coups and eventually resorted to hiring mercenaries to help push back the RUF forces. By 1998 the whole country resembled a military camp with arms and ammunition being trafficked to all sides. There was a total breakdown in state structures and institutions. In 2000, British troops were employed to help evacuate foreign nationals and establish order.

UN forces with US support eventually established a sense of control of the country. By 2003, major fighting was over and attempts were made to establish a civilian government. Approximately fifty thousand people were killed in this civil war and as many as two million people were displaced, many of them refugees. Sierra Leone set up a special court to address crimes against humanity, and the country has been working to establish a stable government and maintain a sense of order.

The diamond trade still dominates and political factions still vie for control. Much of the diamond mining is uncontrolled, allowing considerable smuggling operations to operate. After the civil war, revenues from diamond mining increased from less than $10 million in 2000 to an estimated

Figure 14. Diamond Fever. A miner in Kono District, Sierra Leone, searches his pan for diamonds. Diamonds have been used as currency to fund armed conflicts in this region, hence the term *conflict diamonds or blood diamonds. The term blood diamonds was first used in the wars in Angola.*

$130 million in 2004. The civil war destroyed the infrastructure of the country, and most of the resources were looted by forces on one side or another. The economy has had a difficult time recovering. Most of the people make their living in subsistence agriculture. Medicine, food, and goods have been in a short supply. Many people have died due to the lack of these items during and after the civil war. Today the country struggles to recover and work toward stabilization.

C. Nigeria

Africa's most populous country is Nigeria. The exact population has been difficult to determine, but 2010 estimates reported the population to be just short of 150 million. This is a country of more than 250 different ethnic groups with twice that many separate languages or dialects. English is the official language, along with **Hausa, Yoruba,** and **Igbo (Ibo),** all of which are spoken by ten million people or more. The distribution of the major ethnic groups is illustrated by the different regions of the country. Hausa groups are found mainly in the northern region, **Kanuri** groups in the northeast, Yoruba in the southwest, and Igbo (Ibo) in the southeast. Ethnic division has also caused serious confrontations and violence. Political instability in Nigeria has resulted in an almost endless number of military coups and government leaders being removed from office.

The size and diversity of this country create a host of **centrifugal forces** that can bring about divisions along any number of cultural lines. Religious issues add to the political instability. The north is mainly Muslim, as it is in the African Transition Zone. The south is mainly Christian. A large percentage of the population follows animist religions with many different traditional beliefs. Clashes have erupted in the streets that pit Muslims and Christians against each other. Several northern provinces have pushed to have the Sharia criminal code made into the area's civil law.

Family size has been large in Nigeria, which has caused an exploding population. Statistically, Nigeria has more people of African heritage than any other country in the world. The population density is equivalent to having half the population of the United States pushed into an area the size of Texas and Oklahoma. Most of the population makes their living on subsistence agriculture, but millions are employed in the growing urban service sector. Nigeria's main economic engine is the oil industry, which accounts for up to 80 percent

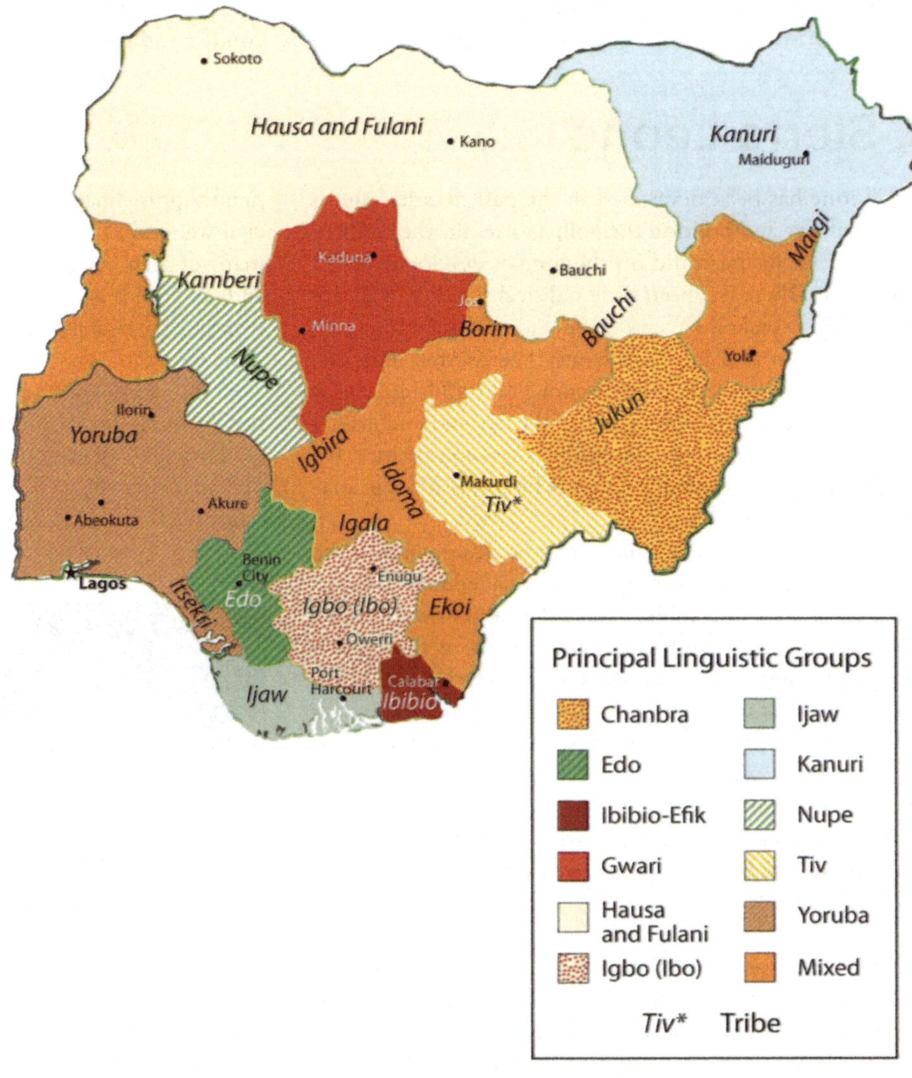

Figure 15. Linguistic Groups in Nigeria. There are many linguistic groups in Nigeria but only four main ethnic groups: Yoruba, Hausa, Kanuri, and Igbo (Ibo), each in the four corners of the country.

of government revenues and is the number one export product. Nigeria is a member of the Organization of Petroleum Exporting Countries (**OPEC**) and is one of the world's top ten oil-exporting countries.

Poverty, a low standard of living, the lack of opportunities and advantages, a poor educational system, or the lack of social services does not diminish the human spirit. Confronting all these issues and more, the people of West Africa and Nigeria are vibrant, energetic, and hardworking and value the institutions of family and religion. Just because they have not transitioned to a consumer society does not mean they cannot find fulfillment and happiness in their lives. A survey conducted by social scientists in 2001 and reported by BBC News in 2003, indicated that Nigeria had the highest percentage of happy people of any country in the world at that time. The status of each country may change from year to year, but the interesting part is that the survey confirmed that money or income level does not always equate to happiness. Countries such as Nigeria, with a low level of per-capita income, can still highly value their heritage and the traditions that revolve around their family and community and transcend the global push for econom-

ic gain and the possession of consumer items. Much can be said for the vibrant cultural attributes of the people of West Africa and its developing countries.

Figure 16. Making Music in Nigeria. Yoruba drummers perform at a local event (left). A Hausa harpist (right). The Hausa are located in the northwest part of the country, which is mainly Muslim. The Yoruba are located in the south and are mainly Christian.

Key Takeaways

1. Most of West Africa lies in the African Transition Zone, with portions north and south of the transitional region. Dry, desert conditions exist to the north and tropical conditions exist to the south.

2. The main economic activity in the region is subsistence agriculture. Minerals, diamonds, or oil are also extracted in varying amounts in West Africa.

3. West Africa has a large number of independent states that share similar economic qualities of poverty with rapidly expanding populations. Foreign aid and international assistance have been extensive.

4. The transition from colony to independent nation has not been without serious political conflicts. Bloody civil wars, military coups, and political unrest have plagued the region.

5. In spite of the political difficulties, poverty and the lack of resources, the people of West Africa hold vibrant cultural values that revolve around family and religion. The diversity of languages, religious beliefs, and rich cultural traditions provide an African heritage that is celebrated and valued by its people.

9.4 Central Africa

A. Central African Republics

Central Africa covers a large area ranging from desert conditions in the north to tropical rain forests and mountains in the equatorial region of The Congo. The entire region is roughly the same size as the United States west of the Mississippi River. In this textbook, the countries included in the region include Cameroon, Congo, Chad, the Central African Republic, Gabon, Equatorial Guinea, The Congo (former Zaire), Rwanda, and Burundi. Located off the west coast of Central Africa is the small twin island country of São Tomé and Príncipe. Burundi and Rwanda are often included in the region of East Africa, but their connection to The Congo makes them more relevant to the Central African region for the purposes of this textbook.

The equator runs through the middle of Central Africa. Type A climates are dominant in the region, complete with tropical rain forests and jungle environments. In the north is the African Transition Zone, which runs through Chad. In Chad, the arid region of the Sahara Desert transitions into the humid tropics. Central Africa usually only includes portions of Sub-Saharan Africa south of the African Transition Zone. Southern Chad exhibits qualities similar to Central Africa, and the northern areas exhibit qualities similar to North Africa or the northern regions of Niger, Mali, or Mauritania.

Figure 17. Central Africa. Chad is often included in other regions, but in this text it is included with Central Africa. The Congo River is the main river of this region.

The physical geography varies with each country in Central Africa. The most prominent physical landscape is the tropical rain forests of the equatorial region. Highlands can be found in both the western and eastern regions of Central Africa. The **Cameroon Highlands** are a product of a geologic rift in tectonic plates that created São Tomé and Príncipe, the island portion of Equatorial Guinea, and the mountainous portions of the mainland on the border between Nigeria and Cameroon. There are a number of volcanoes in Cameroon. The only currently active volcano and the highest in elevation is Mt. Cameroon, at more than thirteen thousand feet. Mt. Cameroon emitted a cloud of carbon dioxide in 1986, killing more than 1,700 people. The volcano last erupted in 2000.

Lake Chad in the north is a large, shallow body of water that lies on the border of Chad, Niger, Nigeria, and Cameroon. Lake Chad provides water for local livestock and fishing for millions of people. Its location on the border of four countries has caused political infighting over local water rights, which are valuable commodities in such an arid climate. Climatic conditions and diversion of the lake for human purposes have caused the water to recede. If these conditions continue, the lake might virtually disappear by the end of the twenty-first century, which would have disastrous effects on both the human population around the lake and the biodiversity. Waterfowl, crocodiles, fish, and a host of other creatures depend on Lake Chad for their survival, and its loss would create an environmental catastrophe.

On the eastern border of the Congo is a portion of the Great Rift Valley, which extends from Southern Africa in Mozambique to Lebanon in the Middle East. Lake Tanganyika and **Lake Albert** are two of the larger lakes along the western section of the Great Rift Valley. Lake Tanganyika is more than 418 miles long and runs the entire length of the boundary between The Congo and Tanzania. These are deepwater lakes. Lake Tanganyika is the world's second-deepest lake, with a depth of 4,800 feet. Because of its depth, it is also the world's second-largest lake by volume after Lake Baikal in Russia, which has the record for both volume and depth.

At the heart of Central Africa are the massive Congo River and all its tributaries. It is the deepest river system in the world and has some stretches that run more than seven hundred feet deep, providing habitats for a wide range of organisms and fish species. The **Congo River basin** is second in size only to the Amazon basin in South America. Home to Africa's largest tropical rain forest, this region is host to a massive variety of plant and animal species, which create an extensive environmental resource base. Human activity has been encroaching on this valuable environmental region filled with extensive biodiversity. Logging, slash-and-burn agriculture, and civil war have devastated large areas of the Congo basin, resulting in the loss of habitat for many tropical species.

Figure 18. Living on the Edge. This mountainside home is in the Rwenzori Mountain range in the eastern region of the Congo. Elevations reach as high as 16,761 feet. Some peaks have permanent glaciers.

B. Conflicts and Unrest in Central Africa

The two landlocked countries of **Chad** and the **Central African Republic** have endured unstable conditions in their transitions to independent, stable democratic states. Chad had been in dispute with Libya over the **Aozou Strip** bordering their two countries, an area deemed rich in minerals and uranium, but in 1994, Chad was awarded sovereignty over the Aozou Strip by the United Nations (UN) International Court of Justice. Chad was a temporary home for more than 250,000 refugees from the ethnic cleansing campaign in the Darfur region of the Sudan. Thousands more from the Central African Republic sought refuge in Chad. Meanwhile, Chad's government has been plagued with corruption and mismanagement, a state of affairs that has hampered its efforts to offer humanitarian aid to its refugees. Similarly, the Central African Republic is troubled by a history of unstable and short-lived democratic governments. Military coups and

Figure 19. A Tailor in Chad. Chad is a poor country with existing political conflicts and many refugees from Sudan. The man in this photo is using an old manual sewing machine to make clothes.

Figure 20. Refugee Camp. An estimated 1.2 million Rwandan refugees fled to the Congo (Zaire) after a civil war erupted in Rwanda. Many are still in the Kibumba refugee camp, shown here covered in a smoky haze.

Figure 21. Soldiers in The Congo. In this photo, General Kisempia, former chief of staff for Joseph Kabila, inspects his troops. About five million people died in warfare in the Congo from 1996–2003.

transitional governments are frequent. Civil unrest erupts into chaos. Rebel groups control large parts of the countryside. It is difficult for the people to access reliable public services such as health care, education, and transportation systems when the government is not functioning adequately.

Rwanda has been severely affected by the difficulties typically associated with the transition between colony and independent nation. Ethnic divisions manipulated by the colonial masters erupted to challenge the country's stability and future. The division between the **Tutsis** and the **Hutus** has deep historic roots. In 1994, the centuries-old conflict between the two ethnic groups erupted into violence of unprecedented proportions and resulted in the senseless killing of hundreds of thousands of innocent people. The Hutus amassed large militias and took revenge on the Tutsis for years of oppression. Hutu militias rounded up and killed all Tutsis, moderates, and anyone not supporting the Hutu cause. The killing of hundreds of thousands of people continued from city to city throughout the countryside. Within a few months, the **genocide** is estimated to have caused the death of as many a million people.

The atrocities of the Rwandan genocide extended across the population. Men, women, and children were encouraged to kill their neighbors by hacking them to death with machetes. If they refused to comply, they themselves were threatened with death. Victims were herded into schools and churches, where they were massacred and the buildings would be burned to the ground. As people fled the region, the number of Tutsi refugees entering neighboring countries ballooned to more than a million.

Tutsi rebels finally gained strength, defeated the Hutu militias, and ended the slaughter. Fearing retribution for the more than one million Tutsis who had been massacred, more than a million defeated Hutus fled as refugees across the borders into Uganda, Burundi, the Congo, Uganda, and Tanzania. Refugee camps housing thousands of people each were hastily constructed without proper sanitation or supplies of drinking water. Diseases such as cholera and dysentery swept through the camps and killed thousands of refugees. The country of Rwanda was left divided and devastated.

The entire Central African region was devastated by the massive shifts of refugees and the brutal killing of so many people. The competition for the control of resources and the increase in military arms along the Zairian border became a major component of the civil war that plagued the Congo (Zaire) during the same period. Up to five million people died in the Congo because of warfare, disease, and starvation. The wars in the Congo involved military actions by the Tutsi from the Rwandan and Ugandan governments and Hutu militias.

The names of the **Democratic Republic of the Congo** (a former Belgian colony) and the **Republic of the Congo** (a former French colony) can be easily confused. At one point after they had become independent, both countries chose the name Republic of Congo. To keep them straight, they were commonly referred to by their respective capital cities—that is, Congo-Leopoldville (the larger eastern country, also known as Zaire) and Congo-Brazzaville (the smaller western country). The larger former Belgian colony has since become simply **The Congo**, or is referred to unofficially as the Belgion Congo, and the smaller former French colony has become simply **Congo.**

Two wars in The Congo resulted in the highest number of deaths since World War II. The **First Congo War** (1996 to 1997) occurred when President Joseph Mobutu was overthrown by militant forces led by rebel leader Laurent Kabila, who was a long-standing political opponent of Mobutu and was backed by Ugandan and Rwandan militant groups. Mobutu was eventually forced from office and fled the country. Kabila declared himself president and changed the official title of the country from Zaire to Democratic Republic of the Congo. This transition in political power caused a shift in the militant rebel groups, which created the conditions for the **Second Congo War** (1998 to 2003), which was even more brutal than the first. The assassination of President Kabila in 2001 permitted his adopted son **Joseph Kabila** to take power

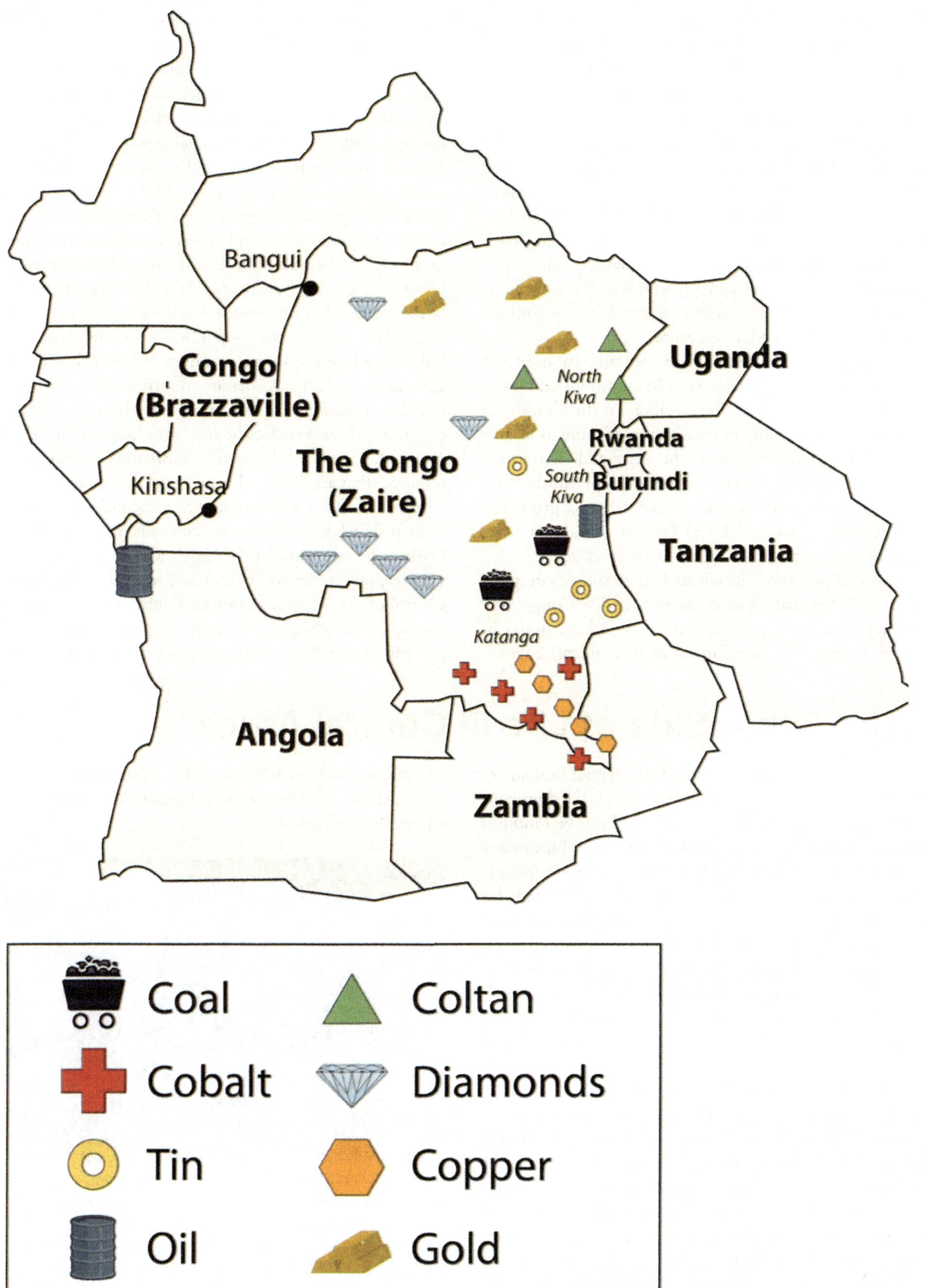

Figure 22. Minerals in The Congo. This map shows the locations of major minerals in the Congo. Other minerals can also be found in the region.

and run the country up to the present. The total number of deaths resulting from the civil wars in The Congo was estimated in 2008 at about 5.4 million—the majority by the wars and the remaining by disease or starvation.

Armed military conflicts in the Congo did not end with the Second Congo War. Conflicts continue in the eastern region. These armed confrontations are often referred to as the **Kivu Conflict** because they are taking place in the provinces of North and South Kivu in the Congo's eastern region along the borders with Rwanda and Burundi. It is estimated that nearly one million people were forced to flee their homes in The Congo in 2016, the largest number of internally displaced people of any country in the world that year. Such conflicts are not widely reported by news outlets in core areas such as the United States.

Understanding that most wars are fought over the control of resources is a major step in comprehending the conflicts in Central Africa. The conflicts in the Congo are likely to continue because of vast caches of mineral wealth in the country that have yet to be extracted. Economic pressure to control the extraction activities of saleable raw materials is often the driving force behind rebel groups in places such as the Congo. Local factions usually sell the raw materials for prices well below market value.

The sale of precious minerals such as cobalt, coltan, gold, and diamonds has helped fund the wars in the Congo. The country produces more cobalt ore than any other country in the world. Cobalt is a valued metal used in aircraft engines, medical implants, and high-performance batteries. The major mining operations for cobalt ore are located in the southeastern Katanga state, where there are also large reserves of copper.

Coltan is a mineral that tantalum comes from, which is highly prized for its use in capacitors for electronic circuits. Tantalum is found in most modern electronic devices in high demand worldwide, from video game systems to cell phones. Most of the coltan in the world comes from mines in the Congo in the eastern Kiva regions. The eastern region is also home to a high percentage of the world's industrial diamond reserves. Other minerals and ores are also found in abundance in the region. The sale of precious gems and rare minerals can bring huge profits, but the wealth seldom reaches the hands of those that labor in the mines in the extraction process.

Learning about the geopolitics of Central Africa is critical to developing an understanding of how colonial activity gave shape to Central African countries and why multinational corporations are now highly involved in creating the demand and markets for the resources found there. The core economic regions of the world require the raw materials and resources that are extracted from peripheral places such as the Congo to fuel their economic activities and to bring profits to their shareholders. The core economic players in the global markets are also some of the largest arms manufacturers that sell weapons to the local factions involved in the battle for control of the valuable resources. Globalization connects the core to the periphery. In resource-rich places such as the Congo, this relationship is only going become more interactive.

C. The Other Side of Life in Central Africa

Sports and entertainment seem to be a universal human activity. Football (known as soccer in America) is the dominant sport in the Central African states. Football can be found in even the poorest areas. The low-cost sport is accessible to most citizens of each country. Other traditional sports include wrestling, archery, and track-and-field events. In Chad, for example, a type of freestyle wrestling in which the wrestlers wear animal hides and cover themselves with dust to prepare for the match is widely practiced. Canoe racing can be seen in Cameroon. Board games such as mancala, a count-and-capture African stone game, are also common. Western-style sports such as basketball, baseball, and cricket are played in many of the urban areas. Olympic sports are practiced in countries as skill, ability, and funding provides.

A great variety of cultural traditions spring from the thousands of ethnic groups, distinct languages, and different religious beliefs found in Africa. Cuisine and drink vary as widely as the people's traditions and are important elements of any celebration or festivity. Every country has its own unique way of promoting and celebrating its cultural diversity. Even poor countries such as Chad possess a rich cultural heritage. Chad has opened its own Chad National Museum and the Chad Cultural Center. Rwanda, with its civil wars and ethnic divisions, still honors its holidays with ceremonies and celebrations. The people of Central Africa find time and energy to celebrate life.

Figure 23. Football in Burundi. A Sunday-afternoon football match takes place in Bujumbura, Burundi. The government has instituted a law that forbids any Burundians from participating in any activity that does not contribute to the communal good before 10 a.m. on a Saturday morning. Street cleaning, family events, and government tasks are examples of acceptable activities. Notice the livestock grazing on the edge of the field.

Figure 24. African Beer. Tusker is a popular African beer sold throughout the world.

Figure 25. African Coffee. Growing coffee is a major economic activity in Africa.

The cuisines of Central Africa can be as diverse as each local population, but they have retained much of their local tastes and flavors based on available ingredients. The slave trade introduced items that are now quite common, such as chili peppers, peanuts, and cassavas. Nevertheless, food preparation is based mainly on traditional methods. Plantains and cassavas are mainstays often served with spicy sauces and grilled meat. Tomatoes, onions, and spinach are common ingredients. Root crops of yams, sweet potatoes, and maniocs are widely served. Fruits and vegetables are popular. Specialty stews are choice entrees that may contain ginger, okra, chicken, and various spices. Bambara is a favorite sugary porridge with peanuts and rice that is common in Central Africa. Traditional meats of chicken, goat, or beef are universal, but wild game such as antelope, monkey, crocodile, or whatever is available is widespread. Fish is a main protein source served with other spicy dishes or dried to be eaten later.

D. The Role of Women

Women's issues remain a concern in Central Africa regarding social equality, economic development, and population growth. The treatment of women reflects the attitudes and traditions of a society. Attitudes and traditions vary with location and with the cultural heritage of an ethnic group indigenous to a given area. Most of the Central African countries are male dominated, a trend that can be witnessed in local activities or sports, in the construction of the government, or in the sex ratio of the business sector. With an average family size of five or more, few opportunities exist for women to break out of their traditional roles within their culture and receive a secondary education or enter a professional field of employment.

The Central African region has different types of marriages that vary from country to country. Some traditional societies still arrange their children's marriages, but this practice is becoming less common as tradition confronts more contemporary views. Marriages between two people with different ethnic backgrounds are becoming more widespread in the urban areas. Cities are places where people with diverse backgrounds come together to live and work, which can lead to an increase in mixed marriages. Cosmopolitan areas are also more likely to have connections with outside cultures through media and communication technology and might be more tolerant of diverse traditions.

One of the marriage traditions that exists in regions of Central Africa is **polygamy**, which is the practice of a man or a woman having more than one spouse. In Africa, as in most places in the world, it is far more likely for a man to have more than one spouse than for a woman to have more than one spouse. Polygamy is practiced in areas where traditional beliefs allow it. Approximately 30 to 40 percent of women in the Central African Republic are in polygamous unions. The law sanctions polygamy as long as spouses are aware of the arrangement and agree to it before they marry.

Women caring for large families do not always have the opportunity to earn an education or obtain the training required for employment opportunities or professional positions. Circumstances shaped by culture, tradition, or economic necessity are not necessarily supportive of women advancing in the private or public sectors. Even if the legal code of the country does not discriminate against women, traditional practices and the social status of men often keep women from making socioeconomic strides. Many societies are male dominated, and violence against women is still common and even accepted in parts of Central Africa.

One of the worst cases of violence and discrimination against women exists in the eastern regions of the Congo, where abuse against women, including sexual violence and rape, is listed by UN humanitarian officials as the worst in the world. The civil wars in the Congo have created a difficult climate for women, often with tragic results. Violence against women has become an accepted practice in some areas. During the Congo wars, women were often raped and taken as slaves to serve the soldiers. If and when the women were released from captivity, it was not uncommon for them to commit suicide. Improvements in health care, education, and child care are needed to improve living circumstances for women and may in turn positively affect population growth in the region.

Figure 26. Congo Woman Weaving.

Key Takeaways

1. Central Africa covers a large physical area about the size of the United States west of the Mississippi River. Most of the region is in the tropics and has a type A climate.

2. Central Africa is a relatively poor region economically. National wealth has been garnered traditionally from the export of food crops such as coffee or tea but more recently from mineral and oil extraction.

3. The civil war in Rwanda in 1994 was the result of conflict between the Tutsi and Hutu ethnic groups. The historic division was manipulated by colonial powers and erupted into brutal violence that led to the death of as many as a million innocent people; nearly as many refugees took shelter in neighboring countries.

4. The civil wars in the Congo are based on political control as well as control over valuable resources in the eastern states. The two main Congo wars resulted in an estimated 5.4 million deaths from either direct military conflict or the devastation that resulted through starvation and disease.

5. Central Africa resonates with cultural and ethnic diversity. Hundreds of ethnic groups live and work together with a high degree of interaction. Traditional cuisine and beverages such as coffee, tea, or beer can be found throughout the region.

6. Women's issues range from cosmopolitan in the urban areas, where mixed marriages are common, to harsh conditions in which violence and abuse is commonplace. Large family sizes and male-dominated traditions make it difficult for women to transition to positions of equality.

9.5 East Africa

A. Physical Geography

East Africa is a region that begins in Tanzania in the south and extends north through the great grasslands and scrub forest of the savannas of Kenya and Uganda and then across the highlands of Ethiopia, including the Great Rift Valley. The region also comprises the countries of Somalia, Djibouti, and Eritrea, which are located in the African Transition Zone between North Africa and Sub-Saharan Africa. Rwanda and Burundi are physically in East Africa but are covered in the lesson about Central Africa because of their border activities with the Congo. The world's second-largest lake by surface area is **Lake Victoria**, which borders Uganda, Tanzania, and Kenya. (Lake Superior, on the border between the United States and Canada, is considered the lake with the largest surface area.) Lake Victoria provides fish and fresh water for millions of people in the surrounding region. The **White Nile** starts at Lake Victoria and flows north to the city of Khartoum in Sudan, where it converges with the **Blue Nile** to become the **Nile River**. The source of the Blue Nile is **Lake Tana** in the highlands of Ethiopia.

Figure 27. East Africa.

The highest mountain in Africa, **Mt. Kilimanjaro** (19,340 feet), is located in Tanzania near the border with Kenya. The second highest peak, **Mt. Kenya** (17,058 feet), located just north of the country's capital of Nairobi, near the equator, is the source of Kenya's name. Both mountains are inactive volcanoes and have permanent snow at their peaks. They provide fresh water, which flows down their mountainsides, to the surrounding areas.

B. The Great Rift Valley

The **Great Rift Valley** provides evidence of a split in the African Plate, dividing it into two smaller tectonic plates: the Somalian Plate and the Nubian Plate. The Great Rift Valley in East Africa is divided into the **Western Rift** and the **Eastern Rift**. The Western Rift runs along the border with the Congo. A series of deepwater lakes run along its valley. On the western edge of the Western Rift are the highlands, which have a series of high-elevation mountain ranges, including the **Rwenzori Mountains**, the highest in the series. The **Virunga Mountains** on the Congo–Uganda border are home to endangered mountain gorillas. The Western Rift includes a series of deepwater lakes, such as Lake Tanganyika, Lake Edward, and Lake Albert. Lake Victoria is located between the Western Rift and the Eastern Rift.

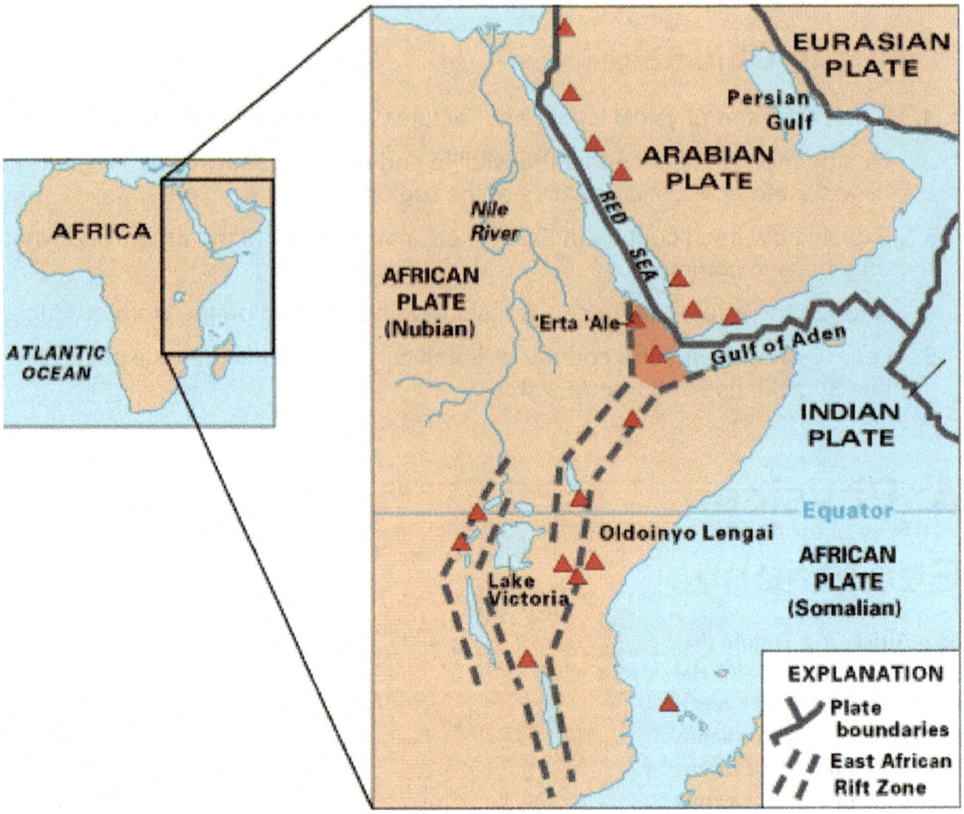

Figure 28. Great Rift Valley of East Africa. The African portion of the rift extends from Mozambique to the Red Sea. The valley is created by the movement of tectonic plates.

The Eastern Rift does not have deepwater lakes; rather, it is a wide valley or basin with shallow lakes that do not have outlets. The lakes have higher levels of sodium carbonate and mineral buildup because of a high rate of evaporation. The differences in water composition of the lakes along the Eastern Rift vary from freshwater to extremely alkaline. Alkaline water creates an ideal breeding ground for algae and other species of fish, such as tilapia, which thrive in this environment. Millions of birds feed off the abundant supply of algae and fish. Birds attract other wildlife, which in turn creates a unique set of environmental ecosystems. The eastern edge of the Eastern Rift is home to the inactive volcanic peaks of Mt. Kilimanjaro and Mt. Kenya. A number of other volcanic peaks are present in the Eastern Rift, such as Ol Doinyo Lengai, an active volcano.

Figure 29. The Great Rift Valley. Hiking into the Great Rift Valley region in Tanzania.

Figure 30. The Serengeti Plains in the Ngorongoro Crater, Tanzania.

Figure 31. Elephant Tusks in Mombassa, Kenya. Wildlife and game reserves are a large part of the tourism economy in places such as Kenya and Tanzania.

C. Serengeti and Game Reserves

The Great Rift Valley and the surrounding savannas in Kenya and Tanzania are home to some of the largest game reserves in Africa, with a broad variety of big game animals. One of these large regions is the vast **Serengeti Plain**, located in northern Tanzania and southern Kenya. The governments of Tanzania and Kenya maintain national parks, national game reserves, and wildlife sanctuaries in their countries, most notably in the Serengeti Plain. Legal protection for as much as 80 percent of the Serengeti has been provided. The protections restrict hunting and commercial agriculture and provide protection status for the wildlife. The word *Sereng eti* means "Endless Plains."

The Serengeti Plain is host to an extraordinary diversity of large mammals and fauna. The largest migration of land animals in the world occurs in the Serengeti. Every fall and spring, as many as two million wildebeests, antelope, and other grazing animals migrate from the northern hills to the southern plains in search of grass and food.

Dozens of protected areas throughout Eastern Africa have been established in an effort to protect and sustain the valuable ecosystems for the large animals that have found their habitat encroached upon elsewhere by the ever-expanding human population. Kenya has more than fifty-five nationally protected areas that serve as parks, reserves, or sanctuaries for wildlife. The Amboseli National Reserve and Mt. Kenya National Park are two of the more well-known areas. The "big five" game animals—elephants, rhinoceroses, lions, leopards, and buffalo—and all the other unique animals found in the same ecosystems, translate into economic income from tourists from around the world that wish to experience this type of environment. The national park systems in Uganda and Ethiopia have made provisions to provide more sanctuaries for wildlife in areas where the human population is growing and the political situation has not always been stable.

D. Kenya

During colonial times, the British considered the land area now called Kenya to be a Crown protectorate area. The coastal city of **Mombasa** has been an international shipping port for centuries and is now the busiest port in the region. Persian, Arab, Indian, and even Chinese ships made port in Mombasa during its earliest years to take part in the lucrative trade of slaves, ivory, and spices. Portugal sought early control of the trade center but eventually lost out to Britain. Arab and Middle Eastern shippers brought Islam to the region; Europeans brought Christianity. Hinduism and Sikhism from India found their way into the country with workers brought over by the British to help build a railroad to Uganda. Kenya gained independence in 1963 and worked throughout the latter part of the twentieth century to establish a stable democratic government.

Nairobi, Kenya's capital and primate city, has become a central core urban area that serves the greater East African region as an economic hub for development and globalization. The largest city in the region, Nairobi is an ever-expanding city that draws in people from rural areas seeking opportunities and advantages. It also has become a destination for international corporations planning to expand business ventures into Africa. Kenya has experienced economic growth and decline as market prices and agricultural production have fluctuated. The Kenyan government has been working with the International Monetary Fund (IMF) and the World Bank to support its economic reform initiatives and reduce waste and corruption in its fiscal processes. The countries of Uganda, Tanzania, and Kenya developed the **East African Commu-**

nity (EAC) as a trading bloc to support mutual development and economic partnerships.

Kenya has no one culture that identifies it. There are more than forty different ethnic groups in Kenya, each with its own unique cultural history and traditions. Of the many ethnic groups in Kenya, the **Maasai** have gained international attention and are often given wide exposure in tourism information. The Maasai are a small minority of Kenya's population but are known for wearing vivid attire and unique jewelry. Their historical lands have been the border region between Kenya and Tanzania. Cattle, a sign of wealth, have been at the center of Maasai traditions and culture and provide for their subsistence and livelihood. Tourism brings to the surface the diversity of cultures that coexist within Kenya's environmental attractions, and the country is working to enhance its international draw in the tourism marketplace.

E. Ethiopia

With more than eighty-five million inhabitants, Ethiopia has the second-largest population in Africa; Nigeria has the largest population. Ethiopia was never colonized by the Europeans in the scramble for Africa, but during World War II, it suffered a brief occupation by Italy (1936–41). From 1930 to 1974, **Emperor Haile Selassie** ruled the country until he was deposed in a military coup. After Selassie was deposed, the government shifted to a one-party Communist state. Successive years in Ethiopia were filled with massive uprisings, bloody coups, and devastating droughts, which brought about massive refugee problems and civil unrest. Famine in the 1980s caused the deaths of more than a million people. The Communist element in Ethiopia diminished when the Soviet Union collapsed in 1991. The country's first multiparty elections were held in 1995.

The region of Eritrea had been a part of a federation with Ethiopia. In 1993, **Eritrea** declared independence, sparking a boundary war with Ethiopia that eventually concluded in a peace treaty and independence in 2000. The final boundary is still disputed. The breaking off of Eritrea left Ethiopia a **landlocked** country with no port city.

The capital and largest city in Ethiopia is **Addis Ababa,** which is the center for various international organizations serving East Africa and Africa in general, such as the African Union and the United Nations (UN) Economic Commission for Africa. This city is the hub of activity for the country and for international aid for the region.

The Great Rift Valley bisects Ethiopia. Highlands dominate the northwest, and minor highlands exist southwest of the rift. The **Ethiopian Plateau** encompasses the Northwest Highlands and is home to Lake Tana, the source of the Blue Nile. Elevations on the Ethiopian Plateau average more than 5,000 feet, and the highest peak, Ras Dashan, reaches up to 14,928 feet. The climate includes sporadic rain in early spring. The typical rainy season extends from June to September, but the rest of the year is usually dry. The high elevations of the highlands cause a rain shadow effect in the deep valleys or basins on the dry side of the region. Eastern Ethiopia is arid, with desert-like conditions. The impact of overpopulation on the natural environment has been deforestation and higher rates of soil erosion; thus continued loss of animal species is inevitable. Fortunately, Ethiopia has established natural parks and game reserves to protect wildlife and big game.

Ethiopia has been inhabited by divergent kingdoms and civilizations, giving rise to a rich heritage and many cultural traditions—so much so that Ethiopia is home to eight UNESCO World Heritage Sites. More than 60 percent of the population claims Christianity as its belief system; about 30 percent of the population is Muslim. Many traditional religions prevail in rural areas. In contrast to other African countries, Christianity came to Ethiopia directly from the Middle East rather than from European colonizers or missionaries from Western countries. In Ethiopia, Christianity was structured into the **Ethiopian Orthodox Church**, a church that has endured through centuries.

Figure 32. Old Man in Harer, Ogaden Region of Eastern Ethiopia.

F. Rural-to-Urban Shift

The average family size (fertility rate) in East Africa is about 5.5, which is typical of the entire continent of Africa and translates into exploding population growth. In many areas of Africa in general and East Africa in particular, most of the population (as much as 80 percent) makes its living off the land in agricultural pursuits. Large families in rural areas create the conditions for the highest levels of **rural-to-urban shift** of any continent in the world. The large cities—with expanding business operations complete with communication and transportations systems that link up with global activities—are an attractive draw for people seeking greater employment opportunities. In East Africa, each of the three largest cities—Nairobi, Dar es Salaam, and Addis Ababa—is more populous than Chicago, the third-largest US city. In West Africa, the city of Lagos, Nigeria, is more populous than New York City and Chicago combined. These cities are all riding the worldwide wave of globalization and are core centers of economic activity for the business sector and corporate enterprises.

The other regions of Africa all have central cities that act as economic core areas and attract the multitudes from rural areas looking for employment and opportunities. International connections are indicative of local economic development, which is causing the urban areas to grow at exponential rates. Cities heavily affected by a high level of rural-to-urban shift often cannot build infrastructure fast enough to keep up with demand. Self-constructed slums and squatter settlements, which lack basic public services such as electricity, sewage disposal, running water, or transportation systems, circle the cities. All the large

Figure 33. Urban Slums of Nairobi. People walking to and from work along the railroad tracks in a poor neighborhood in Nairobi, Kenya.

Figure 34. Modern City of Nairobi. Since this photo was taken in 1980, the city center of Nairobi has expanded and grown ever larger to become a major core business hub for East Africa. The central business district at the center of this core locale has an enormous pull factor for individuals from rural areas seeking employment.

cities of Africa are expanding at unsustainable rates. Traffic congestion, trash buildup, higher crime rates, health problems, and air pollution are some of the common results.

Key Takeaways

1. The physical geography of East Africa is dominated by the Great Rift Valley, which extends through the middle of the region from north to south. Associated with the rift valleys are vast savannas such as the Serengeti Plain, large lakes, high mountains, and the highlands of Ethiopia.

2. East Africa has extensive habitat for large herds of big game animals, mountains for great apes, and enough prey for the big cats such as lions and cheetahs. Countries are establishing game preserves, national parks, and conservation areas to protect these animals and their natural habitats and to provide economic opportunities to gain wealth through tourism.

3. Most of the populations of countries in East Africa earn their livelihood through agricultural activities because the region has a high percentage of people living in rural areas. Large families fuel an ever-increasing rural-to-urban shift in the population, causing the cities to grow at an unprecedented rate.

4. Areas along the African Transition Zone have experienced a higher degree of political conflict and civil unrest. Somalia has the highest level of political division and fragmentation based on religion and ethnic backgrounds, which contributed to the secession of the two autonomous regions of Somaliland and Puntland. The central government is neither stable nor strong enough to govern the fragmented country.

5. Diversity in the region is remarkably high; there are hundreds of different indigenous ethnic groups, each with its own language and culture. Religious activity is split between Christianity, Islam, and traditional beliefs. Islam is more favored in the African Transition Zone and to the north, while Christianity and animism are more prominent to the south of the zone.

9.6 Southern Africa

Learning Objectives

1. Summarize the major physical features of the region and explain the role that natural resources play in the economic activities of the people.

2. Outline the differences between the mainland of Africa and Madagascar.

3. Determine how ethnicity has played a role in the political situation of each country.

4. Understand apartheid's effect on South Africa.

Southern Africa is the region of the African continent south of The Congo and Tanzania. The physical location is the large part of Africa to the south of the extensive Congo River basin. Southern Africa is home to a number of river systems; the **Zambezi River** is the most prominent. The Zambezi flows from the northwest corner of Zambia and western Angola all the way to the Indian Ocean on the coast of Mozambique. Along the way, the Zambezi River flows over the mighty **Victoria Falls** on the border between Zambia and Zimbabwe. Victoria Falls is the largest waterfall in the world based on selected criteria and is a major tourist attraction for the region.

Southern Africa includes both type B and type C climates. The Tropic of Capricorn runs straight through the middle of the region, indicating that the southern portion is outside the tropics. The **Kalahari Desert**, which lies mainly

Figure 35. Southern Africa and Madagascar.

Figure 36. Black-and-White Ruffed Lemur from Madagascar at the Dudley Zoo in England.

in Botswana, is an extensive desert region with an arid mixture of grasslands and sand. When there is adequate rainfall, the grasslands provide excellent grazing for wildlife. Precipitation varies from three to ten inches per year. The Kalahari is home to game reserves and national parks. Large areas of dry salt pans stretch over ancient lake beds. The salt pans fill with water after heavy rainfall but are dry the remainder of the year. The **Namib Desert**, found along the west coast of Namibia, receives little rainfall. Moderate type C climates are found south of the Kalahari Desert in South Africa, where conditions are suitable for a variety of agricultural activities, including fruit orchards and an expanding wine industry.

Madagascar is located to the east of the continent, in the Indian Ocean. Madagascar is the world's fourth-largest island and is similar in area to France. Surrounding Madagascar are the independent island states of the Seychelles, Comoros, and Mauritius. Madagascar is included in this section on Southern Africa but does not share its cultural geography or biodiversity. The early human inhabitants of Madagascar can trace their ancestry to the regions of Malaysia and Indonesia in Southeast Asia. People from the African mainland also joined the population. The whole island later came under the colonial domination of France but won its independence in 1960.

Madagascar's unique physical environment is home to many plants and animals found nowhere else in the world. At least thirty-three varieties of lemurs and many tropical bird species and other organisms are found only on Madagascar. It is an area of high biodiversity and is home to about five percent of all the animals and plants in the world. Tropical rain forests can be found on the eastern edge on

the windward side of the island. The western side of the island experiences a rain shadow effect because of the height of the central highlands, which reach as high as 9,435 feet. The western side of the island has a smaller population and receives less precipitation.

Since 1990, the eastern tropical rain forest has experienced a sharp decline because of extensive logging, slash-and-burn agriculture, mining operations, and drought. Population growth has placed a heavier demand on the environment, which in turn puts stress on the habitats of many of the unusual organisms that are unique to the island. Typical of many African nations, agriculture is Madagascar's main economic activity. About 80 percent of the twenty million people who live on the island earn their living off the land. Deforestation is occurring on all parts of the island and is more severe in areas where human habitation leads to a high demand for firewood used in cooking. In other parts of Africa, important environmental areas have been protected or transformed into national parks and wildlife preserves. Though protected areas do exist on Madagascar, efforts to protect the environment and the wildlife have been hampered by the lack of available funding and the population's high demand for natural resources.

The countries on the Southern African mainland share many of the demographic qualities of the rest of Africa: large family size, agrarian economies, multiple ethnic groups, rural populations, political instability, and a high rate of rural-to-urban shift. Southern Africa is set apart from other Sub-Saharan African regions because of its mineral resources, including copper, diamonds, gold, zinc, chromium, platinum, manganese, iron ore, and coal. Countries in Southern Africa are quite large in physical area, except three smaller landlocked states: Lesotho, Swaziland, and Malawi. The larger countries—South Africa, Botswana, Mozambique, Zimbabwe, Zambia, Namibia, and Angola—all have extensive mineral deposits.

The vast mineral resources make this one of the wealthiest regions of Africa with the greatest potential for economic growth. A physical band of mineral resources in Southern Africa stretches from the rich oil fields off the northwest coast of Angola, east through the diamond-mining region, and into the northern **Copper Belt** of Zambia. Diamond mining is found in parts of Botswana and along the coast of Namibia. Coal can be found in central Mozambique. Gold and diamonds are mined in South Africa. The countries that are able to conduct the necessary extractive processes are creating national wealth and increasing the standard of living for their people.

B. South Africa

Anchoring Sub-Saharan Africa to the south is the dominant country of South Africa. Its large land area and vast mineral resources support a population of about fifty million people. The Cape of Good Hope on the southern tip of the continent is a transition point from the Atlantic Ocean to the Indian Ocean. Its strategic location was important for the control of shipping during the early colonial era before the Suez Canal provided a shortcut between Europe and Asia. The European colonial era first brought Dutch explorers to the Cape of Good Hope, where they established the city of Cape Town as a stopover and resupply outpost on their way to Asia.

South Africa is home to many indigenous ethnic groups and is demonstrative of the diverse pattern of human geography. The country has a history of both ethnic diversity and ethnic division. Two of the largest African groups are the Xhosa and the Zulu. The European component of the ethnic mosaic was enhanced by colonial and neocolonial activities. After the arrival of the first Dutch ships, other Europeans followed and competed with the African groups for land and control. The discovery of first diamonds and then gold prompted Britain's involvement in South Africa. The **Boer Wars** (1880–81 and 1899–1902) were fought between the Dutch-based Boers and Britain for control of South Africa's mineral resources. South Africa became a British colony dominated by a white power structure. The **Boers** (later known as **Afrikaners**) spoke **Afrikaans** (a Dutch dialect that is now considered a separate language) and were prominent in the South African political system.

Segregation first developed as an informal separation of the racial groups but evolved into the legally institutionalized policy of **apartheid**, which separated people into black, white, and coloured (mixed race) racial categories. A fourth category was developed for people from Indian or Asian backgrounds. Apartheid eventually found its way into every aspect of South African culture. In the larger scale of society, access and separation was based on race. Each racial group had its own beaches, buses, hospitals, schools, universities, and so on. The legal system divided the population according to race, with the white minority receiving every advantage. There were extensive and detailed rules for every aspect of daily activity, including which public restroom or drinking fountain could be used, which color an individual's telephone could be, and which park bench a person could sit on. The government also sanctioned separate homelands for people from different ethnic groups. People were physically removed from their homes and transported to their respective new homelands based on their racial or ethnic background. The policy of apartheid not only divided the country at that time but set up racial barriers that will take generations to overcome.

The controversial policy of apartheid in South Africa achieved international attention. Many countries condemned it and implemented economic sanctions and trade

Figure 37. "Petty Apartheid" Beach Sign Written in Afrikaans and English.

restrictions against South Africa. Opposition grew within the country and erupted into violence and social unrest. As a result, the white-dominated government of South Africa began to dismantle the apartheid system in the 1990s. The ban on political opposition parties, such as the **African National Congress**, was lifted, and after twenty-seven years in prison, **Nelson Mandela** was released from prison, where he had been held for his resistance activities. The apartheid legislation was repealed, and a new era began. Mandela was the first African to be elected president of South Africa in the new multiracial elections of 1994. His presidency, which ended in 1999, set the stage for a multiracial society. The country is still working through the ramifications of all those years of racial separation. As can be expected, the transition has not been without difficulty.

South Africa has large, modern cities such as Cape Town, Johannesburg, and Durban, each about the size of the US city of Chicago or larger. The cities of Pretoria, East Rand, and Port Elizabeth are also major metropolitan areas and have more than one million people each. These urban centers all contribute to and support the extensive mining and agricultural activities that provide national wealth. As the country that exports more diamonds than any other in the world, South Africa has gained much national income from the extraction of mineral resources, which are being tapped by some of the largest mining operations on Earth.

Much of South Africa has a moderate climate and good soils, which combine to produce enormous quantities of agricultural products, both for domestic consumption and for export. Mining and agriculture have provided many opportunities for employment—opportunities that draw in migrants from neighboring countries with political unrest or poor

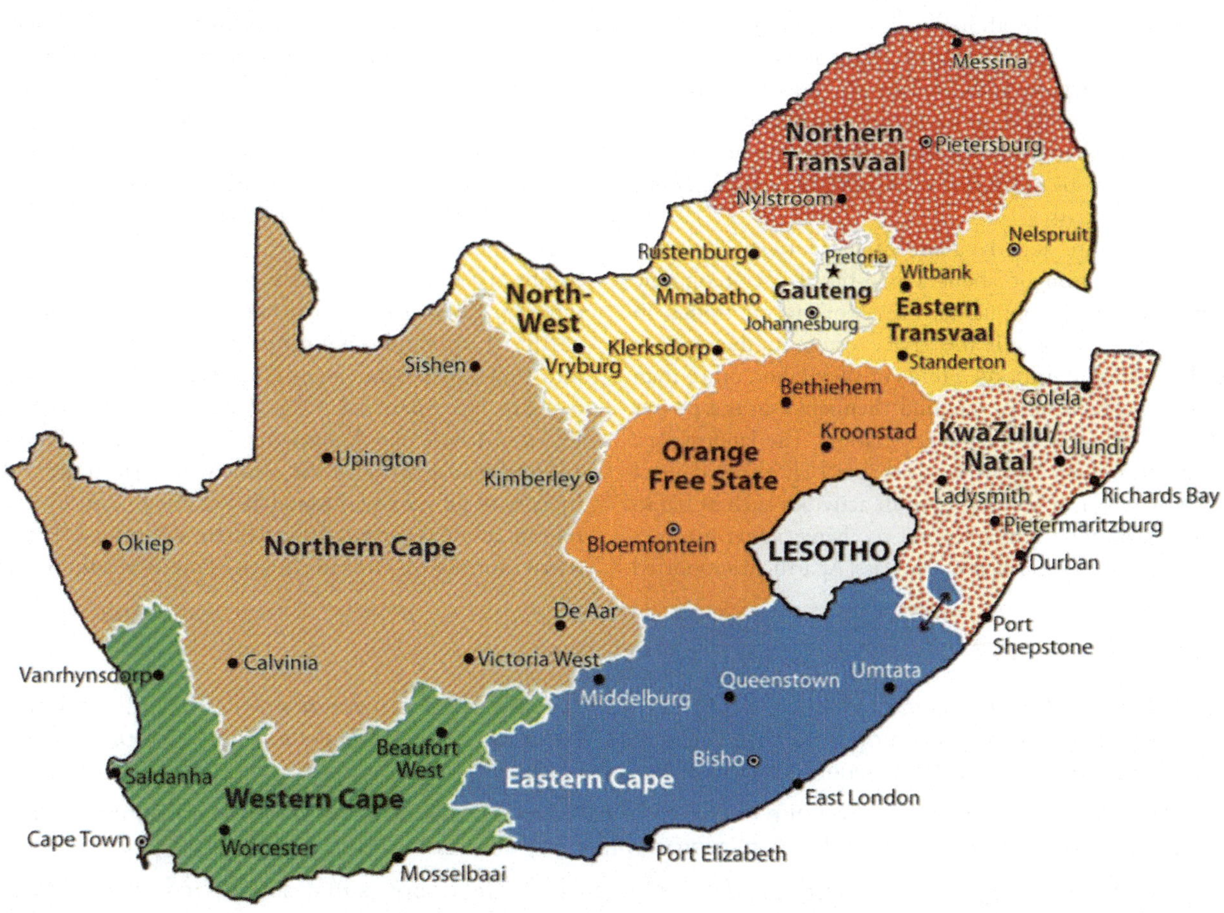

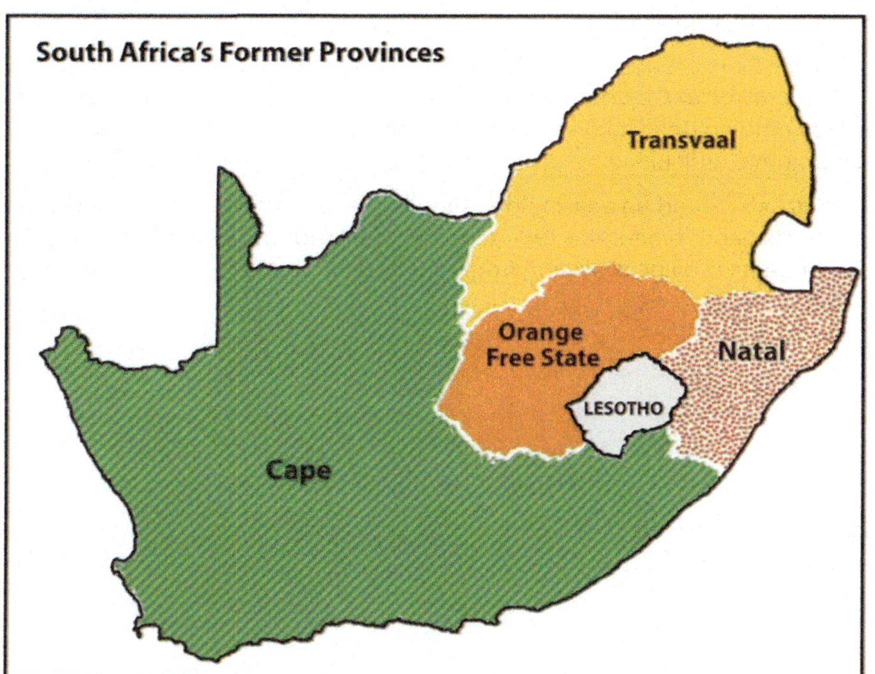

Figure 38. Provinces of South Africa after Reorganization in 1994. South Africa redesigned its internal provincial boundaries in 1994. The two large former provinces of Cape in the south and Transvaal in the north were divided up into seven smaller units. The province of Orange Free State was changed to Free State and the province of Natal was changed to Kwazulu-Natal; both kept their main boundaries intact.

economic conditions. These immigrants add to the cultural dynamics of an already ethnically diverse country.

South Africa's manufacturing sector is not well developed. The country depends on Europe, East Asia, and the United States, the three main core economic areas of the world, to provide industrial goods. There are few well-paying manufacturing jobs to provide for a growing or stable South African middle class.

The structure and dynamics of the current economic activities have brought about a two-tiered socioeconomic system. Most of the population may work in the mines, in agricultural activities, or in the service sector, but they are not directly benefiting from the profits of their labor, other than earning a wage. These people find themselves in the poorer working-class majority of the population. The landowners, mining corporation executives, and social elite that control the service sector or are employed in activities such as banking or the commodity markets are receiving higher incomes and have created a wealthier upper class. Apartheid supported this class division. The current free and open legal system has yet to bring about a change in the socioeconomic structure of the population. Millions of poor ethnic minorities find themselves in living conditions similar to their economically depressed neighbors in other parts of Africa, while the wealthier upper class has established a good standard of living similar to that of the core economic areas of the world.

South Africa redrew its internal provincial boundaries in 1994. The large former Cape Province in the south, which had been a former British center of power, was divided into three smaller provinces. The large Transvaal Province in the north, which had been a center for Afrikaners, was divided into four smaller provinces. Orange Free State changed its name to Free State and Natal Province changed its name to Kwazulu-Natal. The changes were made based on the size and political management of the country that had come with the transfer of power when Mandela became president. The European colonial pattern of settlement was adjusted to represent a more South African pattern of political administration.

Key Takeaways

1. Southern Africa is located along the Tropic of Capricorn and is host to various climate zones, including tropical, desert, and moderate type C climates. Extensive agricultural activity can be found south of the Tropic of Capricorn, where moderate climates prevail.

2. Madagascar is a separated from the mainland of Africa and provides a habitat for animals not found on the mainland. The mainland region is large with many different physical landscapes.

3. The mainland region of Southern Africa has extensive mineral deposits that provide resources for its countries to gain national wealth. Approaches to translating mineral deposits into wealth have varied among the countries.

4. Many ethnic groups can be found in Southern Africa, ranging from the San in the Kalahari to the Zulu groups to the south. Colonialism made its footprint on the region by creating borders and providing colonial names. The transition from colony to independent nation has caused political and economic difficulties.

5. South Africa had an apartheid system in place stemming from the colonial era. The apartheid system was dismantled after the 1994 election of President Nelson Mandela. South Africa went so far as to change its provincial system from four to nine provinces to more appropriately politically administer the country.

End-of-Chapter Summary

- Sub-Saharan Africa has many rapidly expanding core cities and large rural peripheral regions that are experiencing high rural-to-urban shift. Most countries are former European colonies that are growing their own economies as developing countries. The hundreds of tribal groups that live there with their own languages and cultures often clash with each other or their government.

- The informal sector is strong in this realm as people seek opportunities amid civil unrest and conflict. The weak formal sector in many countries does not always receive enough revenue from taxes or fees to operate the public sector and provide services to the people. The realm has a high potential for economic development in the extraction of minerals as well as tourism.

- West Africa has many countries with hundreds of separate ethnic groups with their own languages. The high diversity makes it difficult for social unity and political cohesiveness. Located on the African Transition Zone, there is often conflict between the Muslim north and the Christian south. Civil wars in recent years have devastated various countries. Nigeria is the most populous country in Africa with a growing diverse population. In spite of its diversity and issues, Sub-Saharan Africa retains a sense of vibrancy in the celebration of family and community.

- Central Africa covers a large region of tropical Africa, including the vast Congo River basin. Poor economic conditions have been partly a result of the many civil wars and conflicts in the region where the controlling party or military power has received most of the wealth, leaving the majority in need of infrastructure improvements and public services.

- The transition from colony to independent country has often resulted in short-lived unstable governments that were often replaced with leaders who took advantage of their power to place large amounts of public funds in their own foreign bank accounts. Such fraud and corruption have often left most of the people with fewer opportunities and little money left for public services.

- The civil war in Rwanda, which was based on ethnicity, led to as many as a million deaths. The civil war in the Democratic Republic of the Congo over the control of land and resources resulted in more than five million deaths, the most of any war since World War II. Hundreds of ethnic groups exist in an area that is being exploited for valuable mineral deposits.

- Large deposits of gold, diamonds, and rare minerals such as coltan are located in the Democratic Republic of the Congo and surrounding areas. The wealth that they can generate has fueled many armed conflicts by various groups that vie for control of these resources. Women have suffered serious setbacks in their rights and status in Central Africa because of these conflicts.

- East Africa is dominated by the Great Rift Valley, with its large lakes, high mountains, and vast savanna plains. The Serengeti and other game preserves are home to immense numbers of big game that are attractive for tourism. Most of the people work in agriculture. Diverse ethnic and religious groups exist here, and there is a high rural-to-urban shift in the population. Somalia is divided, with various major factions controlling parts of the country.

- Southern Africa, located on the Tropic of Capricorn, has moderate climates and extensive mineral deposits. The countries have a positive means to gain wealth from agriculture and minerals. There is a high diversity of tribal groups that have been impacted by colonial activity. South Africa had an apartheid system for years that finally broke down in the past few decades. Zimbabwe is an example of a country that has been devastated by dictatorial rule by one man for many years without democratic reforms.

- Madagascar is a large island off the east coast and has different flora and fauna from the mainland because of isolation by tectonic activity. The island of Madagascar was once connected to the African continent but drifted away with the activity of the tectonic plates in that area. The biodiversity on Madagascar is jeopardized by the high level of deforestation that is destroying the habitat for rare animal species.

Chapter 10

North Africa and Southwest Asia

Introduction

The realm of North Africa and Southwest Asia is large in physical area, but its regions share a number of common qualities. The physical area of this realm is divided into three regions: **North Africa**, **Southwest Asia**, and the countries of **Turkestan** (the geographic region of Central Asia). The countries in the North African region include the countries bordering the Mediterranean Sea and the Red Sea from Morocco to Sudan. The realm borders the Atlantic Ocean, the Sahara Desert, and the African Transition Zone. Egypt has territory in both Africa and Asia through its possession of the Sinai Peninsula. The second region, Southwest Asia, includes Turkey, Iran, the **Middle East**, and the **Arabian Peninsula**. The land on the eastern shores of the Mediterranean Sea is frequently referred to as the **Levant** and is often included as a part of the Middle East. Technically, the term *Middle East* only includes the five countries of Israel, Lebanon, Jordan, Syria, and Iraq, but in common practice *Middle East* refers to all of Southwest Asia. Central Asia, also referred to as Turkestan, includes the "stan" countries from Kazakhstan to Afghanistan in the region between China and the Caspian Sea. The suffix *stan*, meaning "land of," is a common suffix for country names in Central Asia. Afghanistan is the only country of Central Asia that was not officially a part of the former Soviet Union.

Figure 1. The Regions of North Africa, Southwest Asia, and Turkestan The African Transition Zone is the southern boundary of the realm.

10.1 Introducing the Realm

Learning Objectives

1. Understand three basic traits the countries of the realm all share.
2. Outline the two cultural hearths and explain why they developed where they did.
3. Describe how the people of this realm gain access to fresh water.
4. Understand how the events of the 2011 Arab Spring affected the realm.

The countries of the realm share three key dominant traits that influence all other human activities. The first key common trait relates to the climate of the region. Though various climate types can be found in this realm, it is the dry or arid type B climates (desert and steppe) that dominate. Other climate types include the type H highland climate (cold temperatures at the high elevations with moderate temperatures at the bases) of the mountains of the Maghreb, Iran, or Central Asia and the more moderate type C climates (especially Mediterranean climate) in the coastal regions bordering the sea. The Mediterranean climate along the coastal areas attracts human development and is home to many large port cities. The overall fact is that vast areas of each region are uninhabited desert. North Africa has the largest desert in the world—the Sahara—which borders the **Libyan Desert** and the **Nubian Desert**. About one third of the Arabian Peninsula is part of the **Empty Quarter** of the Rub' al Khali (Arabi-

an Desert). Kazakhstan, Uzbekistan, and Turkmenistan have vast regions of desert with few if any inhabitants. This aspect of the realm reveals the importance of water as a valuable natural resource. Most people in the realm are more dependent on the availability of water than on the availability of oil.

The second trait is **Islam**: most of the people in the realm are Muslims. The practice of Islam in day-to-day life takes different forms in the various divisions of the religion. The differences between the divisions have contributed to conflict or open warfare. Islam acts as more than just a religion. It also serves as a strong cultural force that has historically unified or divided people. The divisive nature of the religion has often resulted in serious political confrontations within the realm between groups of different Islamic ideologies. Concurrently, the religion of Islam is also a unifying force that brings Muslims with similar beliefs together with common bonds. Islam provides structure and consistency in daily life. The faith can provide comfort and a way of living. The holy cities of **Mecca** and **Medina** are located in Saudi Arabia. Other holy cities for other divisions of Islam include **Jerusalem** and the two cities holy to Shia Muslims: Karbala and Najaf in Iraq. Islam dominates the realm, but other religions are significant in various regions. Israel is a Jewish state, and Christianity is common in places from Lebanon to Egypt. There are also followers of the Baha'i faith, Zoroastrianism, and groups such as the Druze, just to name a few.

The third factor that all three regions of the realm share is the availability of significant natural resources. North Africa, Southwest Asia, and Turkestan all have significant reserves of oil, natural gas, and important minerals. Not every country has the same reserves and some of the countries have very few or none at all. However, it is the export of oil that has dominated the economic activity as it relates to the global community. This realm is a peripheral realm. The resource that the realm can offer to the core economic regions of the world is the energy to fuel their economies and maintain their high standard of living. Enormous economic profits from the sale of these resources have traditionally been held in the hands of the elite ruling leader or his clan and do not always filter down to most of the population. The control of and profits from natural resources have become the primary objectives of the countries; this fuels conflicts and armed military interventions in areas such as Iraq and Afghanistan.

A. Cultural Hearths

Availability and control of fresh water have typically resulted in the ability of humans to grow food crops and expand their cultural activities. Hunter-gatherer groups did not settle down in one area but were more nomadic because of their seasonal search for food. As humans developed the ability to grow crops and provide enough food in one place, they no longer needed to move. The earliest human settlements sprang up in what is the present-day Middle East. Early human settlements provide some indication of early urbanization patterns based on the availability or surplus of food. The shift to permanent settlements included the domestication of livestock and the production of grain crops. Fruits and vegetables were grown and harvested. The activities of this era created humanity's earliest version of the rural-to-urban shift. It is theorized that the ability to grow excess food provided the time and resources for urbanization and the establishment of organized communities, which often progressed into political entities or regional empires.

In the Middle East, two **cultural hearths** provide significant historical value to the concept of human development: **Mesopotamia** and the **Nile Valley** in **Egypt**. Both areas were settings for the growth of human civilization and are still being studied today. In Mesopotamia, a remarkable human civilization emerged along the banks of

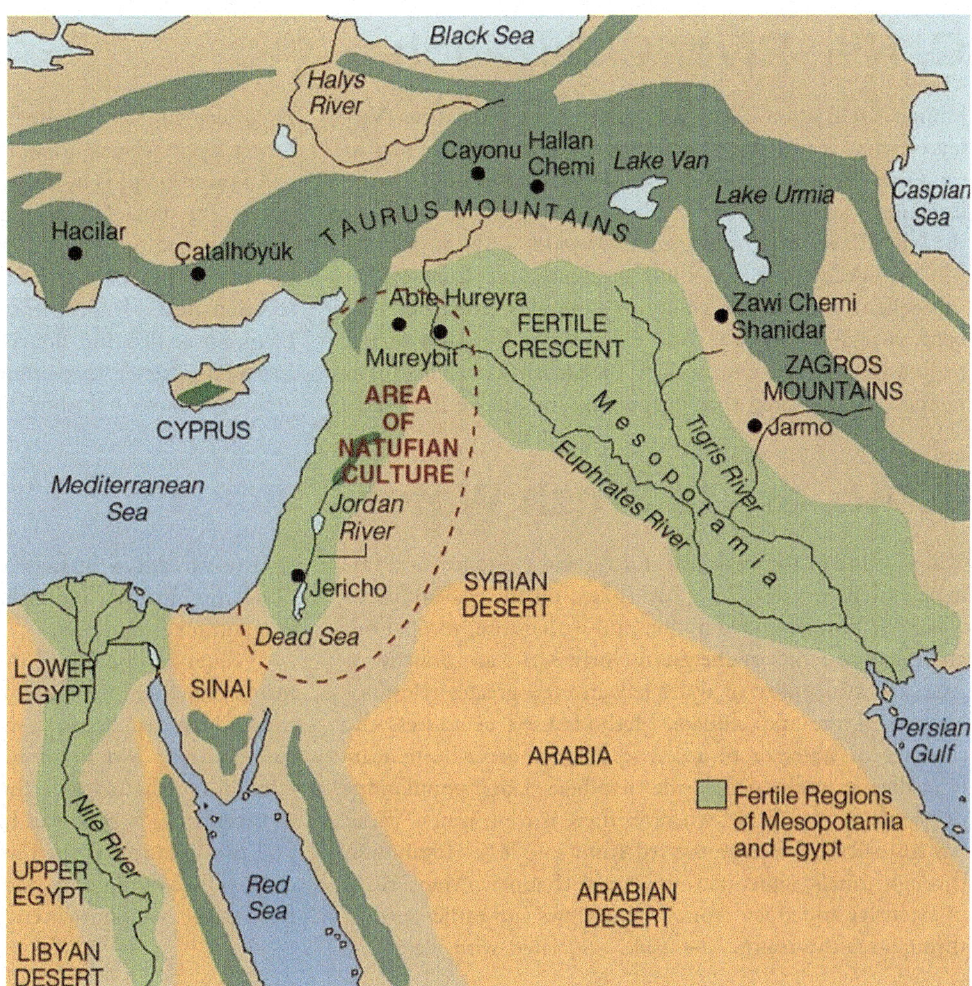

Figure 2. The two main cultural hearths in the realm: Mesopotamia and the Fertile Crescent in Asia and upper and lower Egypt in Africa.

the **Tigris** and **Euphrates Rivers** in what is present-day Iraq, Syria, and southern Turkey. The climate, soils, and availability of fresh water provided the ingredients for the growth of a human civilization that is held in high esteem because of its significant contributions to our human history.

B. Mesopotamia and the Fertile Crescent

Mesopotamia, meaning "land between rivers," is located between the Tigris and Euphrates Rivers. Neolithic pottery found there has been dated to before 7000 BCE. Humans in this area urbanized as early as 5000 BCE. People were settling in the Mesopotamia region, building magnificent cities, and developing their sense of human culture. Mesopotamia gave rise to a historical cradle of civilization that included the Assyrian, Babylonian, Sumerian, and Akkadian Empires, all established during the Bronze Age (about 3000 BCE or later). Famous cities such as Ur, Babylon, and Nineveh were located in the Mesopotamia region. The control of water and the ability to grow excess food contributed to their success. They developed extensive irrigation systems. Large grain storage units were necessary to provide the civic structure and to develop a military to protect and serve the city or empire. The human activity in this area extended around the region all the way to the Mediterranean Sea, which is where the term **Fertile Crescent** comes from.

Various ancient groups were well established on the eastern side of the Fertile Crescent along the Mediterranean coast. The cities of Tyre and Sidon were ports and access points for trade and commerce for groups like the Phoenicians who traded throughout the Mediterranean. Ancient cities such as Damascus and Jericho became established in the same region and were good examples of early human urbanization during the Bronze Age. These cities are two of the oldest continually inhabited cities in the world.

C. Nile River Civilization

Human civilization also emerged along the **Nile River Valley** of what is now Egypt. The pyramids and the Sphinx in the **Giza Plateau** just outside Cairo stand testimony to the human endeavors that took place here. Spring flooding of the **Nile River** brought nutrients and water to the land along the Nile Valley. The land could produce excess food, which subsequently led to the ability to support a structured, urbanized civilization. The Nile River is the lifeblood of the region. In the fifth century BCE, the ancient Greek historian Herodotus suggested that Egypt was "the gift of the Nile."

Egyptologists estimate the first dynasty ruled both Upper and Lower Egypt around 3100 BCE. Upper Egypt is in the south and Lower Egypt is in the north because the Nile River flows north. The terms "Upper" and "Lower" refer to elevation, or essentially "upriver" and "downriver." The ability of humans to harness the potential of the environment set the stage for technological advancements that continue to this day. The Egyptian civilization flourished for thousands of years and spawned a legacy that influenced their neighbors in the region, who benefited from their advancements.

D. Access to Fresh Water

Water is one of the necessities for human existence, and human settlements have long been based on the availability of water for human consumption and agriculture, navigation, and the production of energy. In North Africa and Southwest Asia, the availability of water has an even greater relevance because of the arid climate. Methods used to address the shortage of water or to access fresh water have been nearly as diverse as the people who live here. Large populations of people can be found wherever there is fresh water. Water has historically been transferred from source to destination through canals, aqueducts, or special channels. Many ruins of extensive aqueducts from Roman times and earlier remain throughout the realm. The issues associated with water use continue to affect the lives of the people of this realm. Rapid population growth and industrialization have intensified the demand for fresh water.

Water can be found in the desert regions in a range of forms. For example, there are oases, springs, or noted wells from which people can draw underground water that is close to the surface. Mountainous regions such as the Atlas Mountains in North Africa or the Elburz Mountains in Iran trap moisture, which produces higher quantities of precipitation. The precipitation is then available in the valleys to irrigate crops. Discovering or developing other methods of acquiring fresh water is a requirement in areas without mountains.

Figure 3. Roman Aqueduct near Caesarea.

E. Nile Water in Egypt

Egypt draws water from the Nile to irrigate fields for extensive food production. For thousands of years, floods of the Nile annually covered the land with fresh silt and water. This made the land productive, but the flooding often caused serious damage to human infrastructure. The building of the **Aswan High Dam** in the 1970s helped control the flooding of the Nile Valley. The river no longer flooded annually, and water had to be pumped onto the land. Over time, the constant and extensive use of this type of irrigation causes the small quantities of salt in the water to build up in the soil to serious levels, thereby reducing the land's productivity. This process, called **salinization**, is a common problem in arid climates. To rid the soil of the salts, fresh water is needed to flood the fields, dissolving the salt and then moving the salty water back off the fields. High salinization in the soil and the reduction in agricultural productivity is a growing concern for Egypt. Egypt's growing population places a high demand on the availability of food. More than half of the eighty million people in Egypt live in rural areas, and many of them make their living in agriculture, growing food that plays a critical role in the country's economic stability.

F. Water from the Tigris and Euphrates

The major source of water in the Fertile Crescent region comes from the Tigris and Euphrates Rivers. Both have their origins in Turkey and converge to form the **Shatt al-Arab** (literally, "River of the Arabs") waterway that flows into the Persian Gulf (which is called the Arabian Gulf by Arabs). The Euphrates is the longest river in Southwest Asia and flows through Syria from Turkey before entering Iraq. Turkey has developed large dams on both the Tigris and Euphrates for agricultural purposes and to generate hydroelectric power. As water is diverted for agriculture in Turkey there is less water flowing downstream for Syria or Iraq. Disputes over water resources continue to be a major concern in the **Tigris-Euphrates Basin. The Atatürk Dam** in Turkey is largest dam on the Euphrates, and it has a reservoir behind the dam that is large enough to hold the total annual discharge of the river. All three countries have dams on the Euphrates and both Turkey and Iraq have dams on the Tigris. The three countries signed a memorandum of understanding in 2009 to strengthen cooperation within the Tigris-Euphrates Basin. All three countries need the water for agriculture to produce food for a growing population. Agreements to share water have been difficult as a result of the Iraq War and the recent violence in Syria that have contributed to further political tension between the three countries.

G. Water Conservation in Israel

Israel has taken innovative steps to conserve water and use it efficiently. Drip irrigation mixed with fertilizers is called fertigation. Fertigation is used extensively in the area. Israel grows plantation crops such as bananas, which require large quantities of water. Banana groves are covered with material that allows sunlight to penetrate but reduces the amount of transpiration, which conserves water. Israel has worked to recycle water whenever possible. Gray water is water extracted from sewage that has been treated to be used in agriculture. Underground wells in the West Bank region provide water for a high percentage of people in both the Palestinian areas and Israel. The issue of control over the water is contentious at times. Just as the control of water may have been an important factor in the early Mesopotamian civilizations, it remains a point of political conflict in places such as Israel and the West Bank. The lack of fresh water and the heavier demand placed on water resources have caused countries that can afford it to desalinize seawater. Desalinization is used extensively in the oil-rich states of the Arabian Peninsula. Israel is implementing a similar plan to accommodate their increasing population and fresh water requirements.

Figure 4. Banana Grove in Israel near the Lebanese border. The grove is covered with material that allows sunlight to penetrate but helps reduce the loss of water through transpiration.

H. Diversion of Water in Turkestan

Fresh water is in short supply in many of the desert regions of Turkestan in Central Asia. Agricultural production has traditionally been dependent on water flowing in rivers and streams that originated with the precipitation from the mountains, but as humans have developed canals and irrigation systems, water from rivers has been diverted for agricultural use. Vast fields of cash crops such as cotton were developed during the Soviet era for economic reasons, and the result had devastating consequences for the Aral Sea, which depended on the water from these rivers for its survival and has now shrunk to a fraction of its former size. More than half the population of Central Asia depends on agriculture for their livelihood. The other half, of course, requires water and food for their existence.

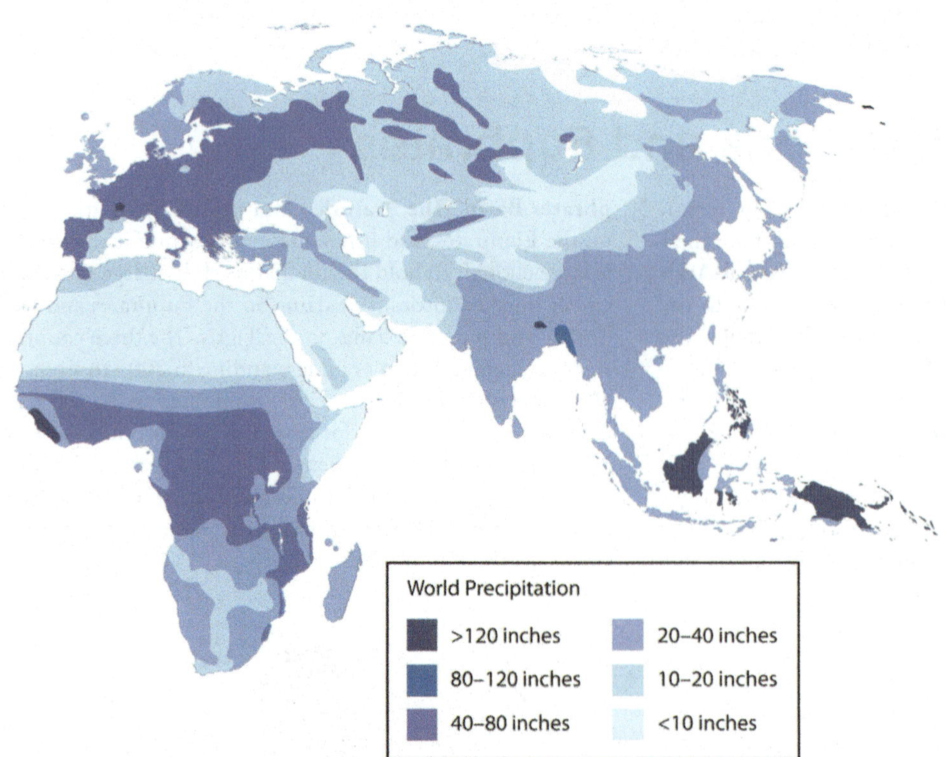

Figure 5. Precipitation patterns for North Africa and Southwest Asia.

I. Arab Spring of 2011

The year 2011 brought about important changes for the human geography of parts of this realm. The year ushered in a wave of human activity that awakened the power of the citizens to speak out against conditions in their country and actively protest against their governments. North Africa, the Middle East, and the Arabian Peninsula experienced the highest levels of protests and insurgency. Political leaders that had been in power for extended periods were challenged and removed from office. Democratic reforms were requested or demanded by citizens seeking more individual freedom and greater access to political power. Uprisings in some of the countries were internal; other countries received external support or intervention. Overall, demonstrations, protests, and outright revolution involved millions of people desiring improved living conditions and a better future for themselves and their families.

Protests emerged in North Africa in the beginning of 2011. Tunisia was the first country in which leadership felt the heat of civil resistance and open revolution. In January, the Tunisian president of more than twenty-three years was forced to flee to Saudi Arabia. In Egypt, millions of protesters demonstrated in the streets against political corruption and the lack of reforms. The revolution of Egypt's citizens was not an armed conflict, but it was an effective protest, because it eventually brought about the ouster of President **Hosni Mubarak**, who had been in office for almost thirty years. Demonstrations and protests continued against governments in Morocco and Algeria; the people voiced their concerns regarding issues such as high unemployment, poor living conditions, and government corruption. Libya's protests erupted into a full-scale armed revolution as antigovernment rebels took control of the city of Benghazi in an attempt to topple Muammar Gadhafi's forty-two years of authoritarian control of the government, oil revenues, and the people. The armed Libyan revolution was eventually successful in taking control of Tripoli and in removing **Muammar Gadhafi** and his family from power. The revolution in Libya was aided by North Atlantic Treaty Organization (NATO) air strikes and the implementation of a no-fly zone over the country.

The Middle East did not escape the Arab Spring of 2011. Protests in Jordan forced King Abdullah II to reorganize his government. The country experiencing the greatest impact was Syria. Major mass demonstrations and serious protests against the government were staged in a number of cities across the country. In Syria, the long-term leadership of an Alawite minority continues to run the government and control the military. The al-Assad family—a father and then his son—has ruled Syria since 1971. The Syrian government cracked down on the revolution with hardline measures aimed at subduing the protests and demonstrations. By September 2011, more than two thousand protesters had been killed in Syria, and many more were detained or tortured. Countless others fled to neighboring countries for their safety. The protesters in Syria wanted democratic reforms as well as the end of the al-Assad family reign.

The wave of change that swept over the realm in the Arab Spring of 2011 is an example of how centripetal and centrifugal cultural forces act on a state or region. The political landscape was altered or drastically changed in many countries. The impact of these changes will be realized in the years and decades to come.

Key Takeaways

1. The realm of North Africa and Southwest Asia extends from the Atlantic Ocean along the Moroccan coast to the western border of China. It includes the regions of North Africa, Southwest Asia, and Central Asia (often referred to as Turkestan).

2. Three basic features that dominate this realm include the arid type B climates, Islam as the predominant religion, and the export of petroleum and minerals to gain wealth. There are exceptions to all three features, but these three are found within most countries of the realm.

3. The two main cultural hearths in this realm are located along the rivers in Mesopotamia and in Egypt. Control of and access to water resources to grow excess food were the basis for the success of the empires that flourished in these two areas.

4. Fresh water is a valuable resource that is not always available in North Africa and Southwest Asia because of the climate and physical geography. Each region within the realm has developed its own methods to draw from or extract the valuable resource of fresh water.

5. The Arab Spring of 2011 was a massive wave of protests and demonstrations by citizens of the realm against their governments over such issues as poor living conditions, high unemployment, government corruption, and the lack of democratic reforms. Various leaders were removed from office and governments were pressed to reform their power structures to allow for more shared governance and reduced political corruption.

10.2 Muhammad and Islam

Learning Objectives

1. Summarize the early life of Muhammad and the origins of Islam.
2. Analyze the differences and similarities among the three main monotheistic religions.
3. Explain the process of spatial diffusion and the various forms it may represent.
4. Outline the main divisions of Islam and the approximate percentages of the followers of each division.
5. Explain how Islamic fundamentalism influences the debate between a religious state and a secular state

Located in the mountains of western Saudi Arabia, the city of Mecca (also spelled Makkah) began as an early trade center for the region and a hub for camel caravans trading throughout Southwest Asia and North Africa. Mecca is about forty-five miles from the Red Sea coast at an elevation of 531 feet. South of Mecca, the mountains reach more than 7,200 feet in elevation. According to Islamic tradition, the patriarch Abraham came to Mecca with his Egyptian wife Hagar and their son Ishmael more than two thousand years before the birth of the Prophet Muhammad (born 571 CE). When Hagar died, Abraham and Ishmael built the **Kaaba** (or Ka'ba), a rectangular shrine that included a special stone, in Mecca. The shrine was destined to become one of the holiest sites for nomadic groups in Arabia. Abraham later died in Palestine in what is now the country of Israel. Centuries after Abraham's death, the Kaaba and the rituals associated with it deteriorated and mixed with other local traditions.

A. The Prophet Muhammad

The traditional groups in the region of Arabia were **polytheistic** and worshiped their own gods. By the time of Muhammad, Mecca is said to have been a center of worship to more than 360 deities or gods; the greatest of these was Allah (meaning "the god"). Allah was known as the chief of the Meccan pantheon of gods and was worshiped from southern Syria to Arabia. Mecca was full of idols, temples, and worship sites. Tradition states that the god Allah was the only god without an idol; he would become the sole entity of Muhammad's new Islamic religion.

Muhammad, born in Mecca 571 years after the birth of Christ and about 100 years after the fall of the Roman Empire, was orphaned at an early age, and was employed in a camel caravan when he reached his teens. His travels introduced him to many people, places, and issues.

Islamic tradition states that the angel Gabriel appeared to Muhammad while he was meditating in a mountain cave in 610 CE, when he was about forty. Muhammad was given words from Allah, which he recited from memory to his followers. According to tradition, Muhammad was illiterate; his supporters wrote down his words, compiling them into the Koran (Qur'an), the holiest book of Islam. Muhammad was the founder of the new religion, which he called **Islam** (meaning "submission to Allah"). The term **Muslim** (meaning "one who submits") refers to a follower of Islam.

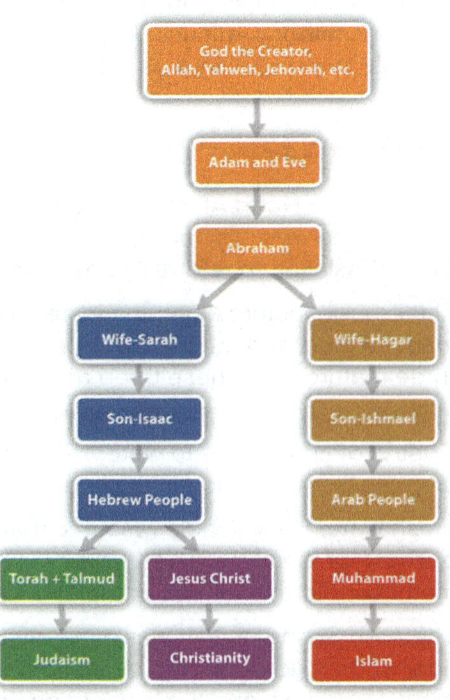

Figure 6. Traditional succession of the three main monotheistic religions of the Middle East: Judaism, Christianity, and Islam.

After Muhammad returned to Mecca and related his visions and Allah's words from the angel Gabriel, he began to speak out against the city's vices and many gods. He stated that there was only one god: Allah, the same creator god of Abraham. His message was not well received: in 622 CE the people of Mecca forced Muhammad out. He fled to the safety of the nearby city of Yathrib in a journey known as the Hegira. Muhammad later renamed the city **Medina** ("city of the prophet").

Launching out from Medina, Muhammad and those loyal to him defeated the army of Mecca and converted the city into Islam's holiest place. They destroyed all Mecca's idols and temples except the Kaaba. Muhammad's teaching united the many Arabian groups under one religion. Since the Koran was written in Arabic, Arabic became the official language of Islam. The Kaaba and the mosque built at Mecca became the center of the Islamic world and a destination for Muslim pilgrims. Islam brought a new identity, a faith in one god, and a set of values to the Arab world. Islam made sense in a world with many traditional beliefs and few unifying principles.

It is important to keep in mind that **monotheism** was not new: Christianity had been around for more than six hundred years. Judaism and Zoroastrianism in Persia had been around for centuries before Christianity. The principles of Islam and Muhammad's teachings are a continuation of Judaism and Christianity. All three traditions assert a faith in a divine creator, with important messages coming through prophets or holy messengers. All three religions acknowledge Abraham as a founding patriarch. Muslims believe that Moses and Jesus were prophets and that Muhammad was the greatest and final prophet. All three religions have stories about creation, Adam and Eve, the flood, and other stories that have been adapted to the traditions and characters of each religion.

Religion is a part of culture. The religions that emerged out of the Middle East absorbed many of the existing cultural traits, traditions, or habits of the people into their religious practices. Early Islam adapted many Arab cultural traits, styles of dress, foods, and the pilgrimage and folded them into its principles. Early Christianity and Judaism also adopted cultural traits, holidays, styles of dress, and cultural traditions.

B. Spatial Diffusion

The spread of Islam was accomplished through trade and conquest. Mecca was a center of trade. When camel caravans left Mecca, they carried Muhammad's teachings with them. Islam diffused from Mecca and spread throughout the Middle East and into Central Asia and North Africa. The geographic principle of **spatial diffusion** can be applied to any phenomenon, idea, disease, or concept that spreads through a population across space and through time. The spatial diffusion of Islam outward from Mecca was significant and predictable.

There are two main types of spatial diffusion: **expansion diffusion** and **relocation diffusion**. Expansion diffusion has two main subtypes: **contagious diffusion** and **hierarchical diffusion**. A religion can spread from individual to individual through contagious diffusion when a religion starts at one point and propagates or expands outward from person to person or place to place in a pattern similar to the spreading of a disease. Another way a religion can spread through expansion diffusion is hierarchically, when rulers of a region convert to the religion and decree it as the official religion

of their realm; the religion filters down the political chain of command and eventually reaches the masses. The second type of diffusion, relocation diffusion, takes place when the religion is carried by people to a new place as they migrate there, thus relocating the religion. When Islam jumped from the Middle East to Indonesia, it diffused through relocation. Relocation diffusion also occurred when Muslims migrated to the United States. As of 2010, Islam had attracted as many as 1.5 billion followers, second only to Christianity, which has about 2 billion followers.

Figure 7. Diffusion of Islam and the ten countries with the highest Muslim populations.

C. The Five Pillars of Islam

The basic tenets of the **Five Pillars of Islam** create the foundational structure of Islam. Prayer is an important part of the religion. A Muslim must offer prayers five times a day: before sunrise, at midday, at midafternoon, after sunset, and in the early evening. During prayer, Muslims face toward the compass direction of Mecca. Before clocks and time were well established, a mosque leader would climb a minaret (a tall tower next to the mosque, their place of public worship) and call the faithful to prayer at the required times of day. Muslims gather together for common prayer on Friday, which is a time to unite the community of believers. Mosques sprang up after Muhammad died, and they became the center of community activities in the Islamic world.

The Five Pillars of Islam can be translated as follows:

1. **Express the basic creed** (*Shahadah*). Profess that there is no god but Allah and his messenger and prophet is Muhammad.

2. **Perform the prayers** (*Salat*). Pray five times a day, facing Mecca.

3. **Pay alms or give to charity** (*Zakat*). Share what you have with people who are less fortunate.

4. **Fast** (*Sawm*). During the month of **Ramadan**, abstain from personal needs, drinking, and eating from dawn to dusk (as one's health permits).

5. **Make the pilgrimage to Mecca** (*Hajj*). Conduct at least one pilgrimage to the holy city of Mecca (if within one's capacity).

Figure 8. Faithful Muslims Praying toward Mecca in Umayyad, Damascus.

D. The Death of Muhammad

Muhammad died in 632 CE at the age of sixty-two. He never claimed to be a god or anything other than a mere mortal. His tomb is located in Medina, the City of the Prophet. No provision was made to continue Muhammad's work after he died. One division thought his successor should be a blood relative. This division led to the **Shia** (or **Shi'ite**) branch of Islam, which makes up about 15 percent of Muslims. Others felt that the successor should be a worthy follower and did not need to be a blood relative. This branch became known as Sunni, which makes up about 84 percent of Muslims. Various smaller branches of Islam also exist, including **Sufi**, which approaches the Islamic faith from a more mystical and spiritual perspective.

In understanding the Middle East, it is critical to understand the Sunni and Shia divisions of Islam. The Shia and Sunni divisions of Islam have sometimes had divergent beliefs, resulting in conflicts. In the early sixteenth century, the Persian Empire, which is now Iran, declared the Shia branch its official religion. Its surrounding neighbors were predominantly Sunni. This divergence is part of the basis for the current civil unrest in Iraq. The two divisions of Islam currently vie for political power and control in Iraq. The majority of the Arab population in Iraq, about 60 percent in 2010, follows the Shia division of Islam, but the leadership under Saddam Hussein until 2003 was Sunni. The Shia majority in Iraq who are Arab share their faith with the Shia majority in Iran who are ethnically Persian.

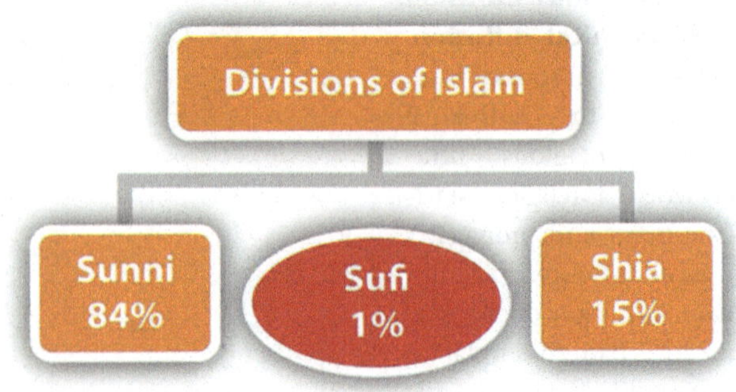

Figure 9. Three main divisions of Islam with approximate percentages.

E. Secular State versus Religious State

Islam has a code of law called the **Sharia** criminal code, which is similar to Old Testament law. The Sharia dictates capital punishment for certain crimes. For example, if a person is caught stealing, his or her arm would be severed. For more serious offenses, he or she would be beheaded or stoned to death. Some countries use the Sharia as the law of their country. Countries are called **religious states** (Islamic states in this case) when religious codes take precedence over civil law. States in which people democratically vote on civil law based on common agreement are called **secular states**. Whereas secular states attempt to separate religious issues and civil law, religious states attempt to combine the two. Iran is a good example of an Islamic religious state, and Turkey is a good example of a secular state. When the United States invaded Iraq in 2003 and Saddam Hussein was removed from leadership, the country entered a transitional period in which they had to decide if the country would develop into an Islamic state with the Sharia or move to a democratically elected government with civil law. The debate on these issues continually surfaces in many of the countries in North Africa and Southwest Asia whenever transition occurs.

The cultural forces of **democratic reforms** and **Islamic fundamentalism** have been pushing and pulling on the Islamic world. Democratic reformers push for a more open society with equality for women, social freedoms for the people, and democratically elected leaders in government. Islamic fundamentalists pull back toward a stricter following of Islamic teachings; they oppose what they consider the decadent and vulgar ways of Western society and wish to restrict the influence of liberal, nonreligious teaching. A rift between militant Islamic fundamentalists and moderate Islamic reformers is evident throughout the Muslim world. Militant leaders strive to uphold the Sharia criminal code as law. Moderate reformers work toward a civil law based on democratic consensus. This rift adds to the conflicts that have been occurring in this realm. Islamic fundamentalists push for a more traditional and conservative society and express opposition to the United States' intervention in the realm. The Muslim world will continue to confront such arguments over the future direction of Islam in a globalized economy.

Key Takeaways

1. Islam has its origins with Muhammad, who was born 571 years after the birth of Jesus, when Christianity was well established. Muhammad received his revelations through the angel Gabriel and passed them to his followers, who wrote down his words into what became the Koran.

2. Islam is the youngest of the three major monotheistic religions of the realm: Judaism, Christianity, and Islam. All three trace their origins back to the patriarch Abraham. Other monotheistic religions, such as the Baha'i faith and Zoroastrianism, are also evident in the Southwest Asia.

3. Spatial diffusion can be applied to any phenomenon, idea, disease, or concept that spreads through a population across space and through time. Islam has diffused through both expansion diffusion and relocation diffusion to become the second most followed religion in the world.

4. Since the death of Muhammad, Islam has divided into a number of different factions. The two most prevalent divisions of Islam are Sunni (followed by about 84 percent of Muslims) and Shia (followed by about 15 percent of Muslims). Other minor divisions of Islam, such as Sufi, also exist.

5. Religious states are structured around laws based on religious regulations that are usually determined by religious leaders. Secular states are structured around civil law, which is decided on by democratic consensus.

10.3 North Africa and the African Transition Zone

Learning Objectives

1. Summarize the historical geography of North Africa, identify the major physical features and the main cities, and understand who the people are and where most of the population lives in the region.

2. Understand the unique geographic qualities of the Maghreb and explain how this region is connected to Europe.

3. Outline the political issues in North Africa and understand the transitions and conflicts occurring in the governments of the region.

4. Describe the main qualities of the African Transition Zone and explain how the dynamics of this zone are affecting the country of Sudan.

North Africa's primary connection with the Middle East and Central Asia is that Islam diffused to North Africa from the Middle East and Central Asia. Today, it is a Muslim-dominated realm with Arabic as its primary language. Historically, the ethnicity of North Africa was predominantly **Berber** with the nomadic Tuareg and other local groups interspersed. When Islam diffused into North Africa, the Arab influence and culture were infused with it. Modern Egypt has become the cornerstone of the Arab world; more Arabs live in Cairo than in any other city on Earth. The three main areas of interest are the **Maghreb** of the northwest; the **Nile River** valley in the east; and the **African Transition Zone**, where the Sahara Desert transitions into the tropical type A climates of Central Africa's equatorial region.

Islam diffused through North Africa to the Berber people of the Maghreb and entered Europe across the **Strait of Gibraltar** to the **Iberian Peninsula**. The Arab-Berber alliance, called the Moors, invaded Spain in 711 CE. The Islamic influence thrived in Iberia and would have continued into mainland Europe if not stopped by Christian forces such as

Figure 10. North Africa and the Maghreb. The Maghreb traditionally includes Morocco, Algeria, and Tunisia, but Libya is also considered part of the Maghreb by many inhabitants of the region.

Charles Martel's army in the famous Battle of Tours. Islam was eventually pushed out of the Iberian Peninsula and held south of the Strait of Gibraltar. Islamic architecture and influence remain part of the heritage of Iberia.

The historical geography of North Africa is not complete without an understanding of the European influences that have dominated or controlled this region for centuries. The **Roman Empire** controlled much of the coastal area of the Mediterranean during its zenith. The Romans built ports, aqueducts, roads, and valuable infrastructure. After the fall of the Roman Empire, common bonds of religion and language were created for the people through the invasion of the Arabs, who introduced the Islamic faith. North Africa was later dominated by European colonialism. France controlled and colonized the region of the Barbary Coast along North Africa's western waterfront, including Algeria, Tunisia, and parts of Morocco. Italians colonized the region that is now Libya. The Barbary Coast of the Mediterranean was once a haven for pirates and a danger to shipping during the colonial era. Even the United States involved itself with wars against the pirates off the coast of the Berber states of North Africa during the early 1800s. After the fall of the **Ottoman Empire** at the end of World War I, Britain controlled Egypt and parts of the Sudan. The Spanish colonized parts of Morocco and Western Sahara. In due time, resistance movements were successful in defeating the colonial powers and declaring independence for all the countries of North Africa. However, the European influence remains through the region's dependence on trade and economic partnerships with Europe.

North Africa is separated from Sub-Saharan Africa by the African Transition Zone, a transitional area between Islamic-dominated North Africa and animist- and Christian-dominated Sub-Saharan Africa. It is also a transition between the Sahara Desert and the tropical type A climates of Africa's equatorial region. This is a zone subject to shifting boundaries. The region was once a major trade route between the Mali Empire of the west and the trade centers of Ethiopia in the east. Camel caravans have crossed this sector of Africa for centuries, and camel caravans from Mecca might have traveled across this zone. Many nomadic groups continue to herd their livestock across the region in search of grazing.

A. The Maghreb: "Isle of the West"

The **Maghreb** is a region extending from Morocco to Libya that is distinguished by the main ranges of the **Atlas Mountains**, which reach elevations of nearly thirteen thousand feet. The main Atlas range is often snow-covered at higher elevations. The *Maghreb*, which in Arabic means "Isle of the West," receives between ten and thirty inches of rainfall per year. This is substantially more rainfall than what is received in the Sahara Desert to the south. The Atlas Mountains extract precipitation from the air in the form of rain or snow, which allows fruits and vegetables to be grown in the fertile mountain valleys of the Maghreb. To the south of this region is the vast Sahara Desert with lower precipitation and warmer temperatures. Libya is actually outside the range of the Atlas Mountains but is associated with the Maghreb by most local inhabitants.

Aided by a moderate type C Mediterranean climate, the northern coastal region of the Maghreb and the mountain valleys are a center for agricultural production, including grapes, dates, oranges, olives, and other food products. Think about how geography affects population: Which climate type do most human groups gravitate toward? What conditions will you find when you combine this climate type and generous quantities of water and food? As you fit the pieces of the geographic puzzle together, you can understand why populations centralize in some places and not in others. The Maghreb is an attractive place for human habitation, but it borders on the inhospitable vastness of the Sahara Desert. Most of the Maghreb's residents live in cities along the Mediterranean coast. There are few people in the vast desert interior of these countries. The exceptions are groups such as the nomadic **Tuareg** that are found in the Sahara.

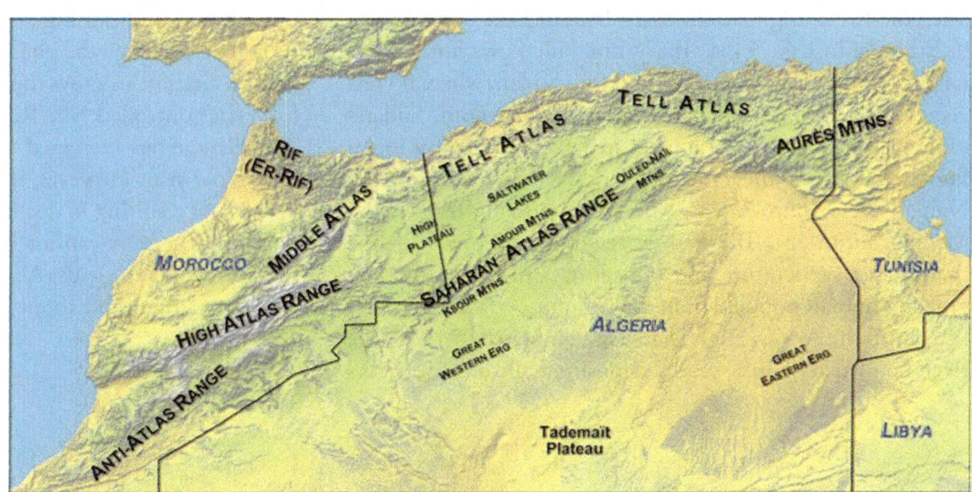

Figure 11. The main mountain ranges of the Maghreb.
The Atlas Mountains of the Maghreb extend to the east and west of the main ranges.

All the countries of the Maghreb have former connections to Europe. These ties have strengthened in recent years because of an increase in trade and the economic dependencies that have been created between Europe and the Mediterranean. North Africa can grow fruit and vegetable crops that are not as plentiful in the colder northern latitudes. In the last half of the twentieth century, an enormous amount of oil was discovered in the Maghreb, and Europe has a growing need for energy. The discovery of oil changed the trade equation: oil and natural gas revenues subsequently advanced past agricultural goods as the main export products. Oil and natural gas exports now make up 95 percent of the export income for Algeria and Libya.

Economic Geography of the Maghreb

Europe has small families with fewer young people to fill entry-level service jobs, and North Africa has a burgeoning population of young people seeking employment. Many people from North Africa speak the languages of their former colonial masters, and when they leave North Africa seeking employment, they find the transition to a European lifestyle relatively straightforward. Migration from the shores of North Africa to Europe is not difficult; the Strait of Gibraltar, for example, is only about nine miles across from Morocco to Spain.

European countries have attempted to implement measures to halt the tide of illegal immigration into their continent from North Africa but have not been successful. The need for cheap labor in European countries is a major economic factor in this equation. The **core-periphery spatial relationship** creates the **push-pull forces** of migration based on opportunities and advantages. Europe needs cheap labor and more energy, provides employment opportunities, and has an advantage in its higher standards of living: these forces attract immigration and pull people toward Europe. North Africa can supply labor and oil, has high levels of unemployment, and suffers from poor economic conditions: these factors push people to emigrate from North Africa to places where conditions are more attractive. Europe is the core economic region, and North Africa is the peripheral economic region. People usually shift from periphery to core in their migration patterns, and this is the case across the Mediterranean.

Figure 12. The Strait of Gibraltar. North Africa is separated from the Iberian Peninsula by the Strait of Gibraltar, which connects the Mediterranean Sea with the Atlantic Ocean. The distance from Morocco to Spain across the strait is about nine miles, making immigration to Europe from North Africa only a matter of a short boat ride.

B. The Nile River and Egypt

The Nile River originates in East Africa in Lake Victoria and in Ethiopia in Lake Tana. The White Nile flows north from Lake Victoria through Uganda and into Sudan, where it converges with the Blue Nile at the city of Khartoum, Sudan's capital. The Blue Nile originates in Lake Tana in Ethiopia. From Khartoum, the Nile River flows north through the Nubian Desert into Egypt, where it eventually reaches the Mediterranean Sea. The fresh water of the Nile is a lifeline that enables agriculture and transportation and supports a growing human population in the region.

Until the Aswan High Dam was completed in 1968, the river flooded its banks yearly, depositing silt and nutrients onto the soil and causing enormous damage to infrastructure. As far back as when the pharaohs ruled Egypt, the people used flood irrigation to grow their crops. Today, water is pumped from the controlled Nile River onto the fields to water crops. This change has increased the number of crops that can be grown per year. However, it has also caused a buildup of salt in the soil, resulting in declining soil quality. Without annual flooding, the salts cannot be dissolved away but remain in the soil, reducing yields. Almost a third of Egypt's population works in agriculture; about half the population is rural.

Figure 13. Nile River system. The white Nile originates in Lake Victoria, and the Blue Nile originates in Lake Tana. They converge at Khartoum.

Population Dynamics

Cairo, Egypt's capital, lies at the northern end of the Nile River. With a population of more than ten million, it is the largest North African city and home to more Arabs than any other city in the world. It is considered the cornerstone city of Arab culture. Cairo is so crowded that more than a million people live in its old cemetery, the City of the Dead. Cairo's residents, and the millions of people in Egypt, depend on the Nile River for their survival. About 95 percent of Egypt's population lives within fifteen miles of the Nile River. As the population has grown, urban expansion has encroached on the farmland of the Nile Valley. Egypt can no longer produce enough food for its people; about 15 percent of its food comes from other countries, mainly the United States.

Conflicts between democratic reforms and Islamic fundamentalism are evident in Egypt. The growing population of about eighty million in 2010 is a major concern. In Egypt's case, democratic reformers were able to promote a strong program of family planning and birth control to help reduce family size, which in 2008 was at 2.8 children per woman and declining. The government even created a popular Egyptian soap opera to promote the concept that it was appropriate in an Arab culture to use family planning and have a small family.

Television programming is popular in Egypt, and even reruns of old American shows such as *Bonanza* and *Dallas* are dubbed into Arabic and shown on Egyptian television. Egypt is a cultural mix with a strong heritage steeped in Arab history with a secular side that is open to the outside world. The cultural forces that create this paradox have not always been in unison. Egypt has a major connection to Western society because of tourism. The **Pyramids of Giza** and the **Great Sphinx** are major attractions that pull in millions of people per year from around the world. Tourism opens up Egypt to outside elements from various cultural backgrounds, most of which are secular.

Political Dynamics

The democratically elected government has had strong opposition from the **Islamic Brotherhood**, which advocates a more fundamentalist Islamic lifestyle and government structure. The democratic reformers that vie with the Islamic Brotherhood for political power support a more open and democratic civil government. These two elements are what drive Egyptian culture and society.

The political situation changed in Egypt with the Arab Spring of 2011. Student protests against government corruption and the lack of democratic reforms emerged with an intensity that gained the support of the Egyptian people and forced the Egyptian government to respond. Egyptian president Hosni Mubarak had been in power since 1981 after the assassination of the previous president, Anwar Sadat. President Mubarak was able to win every election for president that had been held since that time. Millions of Egyptians took to the streets in early 2011 in civil protests against the Mubarak government. Massive protests and demonstrations continued until February 11, when President Mubarak announced his resignation. The people and the government continue to search for progressive opportunities to address their issues. What started out as the Arab Spring turned into the Arab Year as all three long-term leaders in Tunisia, Libya, and Egypt were removed from positions of power.

Figure 14. Protesting in Egypt. On January 25, 2011, the "Day of Revolt" protests were held in Egypt. Tens of thousands of Egyptians went onto the streets to peaceably protest against the Mubarak government.

C. The African Transition Zone

Stretching across the widest part of Africa on the southern edge of the Sahara Desert is the **African Transition Zone**. Known as the **Sahel**, meaning "border or margin," this **steppe** zone is where the arid conditions of the desert north meet up with the moister region of the tropics. For thousands of years, the seasonal grazing lands of the Sahel have been home to nomadic groups herding their livestock across the zone and eking out a living held together by tradition and heritage. Changing climate conditions and overgrazing has enhanced the **desertification** process, and the region is slowly turning into desert. The Sahara Desert is shifting southward, altering the economic activities of the millions of people who live in its path. This desertification process has been occurring for centuries; it is not a new process. Human factors and climate change may be accelerating this process, but they did not create it.

Political stability is complicated to achieve in the African Transition Zone. The political borders established by European colonialism during the **Berlin Conference of 1884** remain basically intact and create barriers that hamper the nomadic groups from traveling through the Sahel in search of grazing land for their livestock. Political boundaries now restrict movement and keep people divided and separated into national identities. The African Transition Zone is also in transition from a rural, traditional agrarian culture to a society confronting the information age and modern technology. Camel caravans that once transported goods and materials across wide expanses of desert terrain are being replaced with motor vehicles and aircraft. The many traditional groups across this zone are adapting to the conditions of the modern world but work to retain their values and traditions. Today this region is unstable, with political and cultural conflicts between the local groups and governments. The recent conflicts in Sudan are examples of the instability.

Figure 15. Camel and Tuareg rider in the Southern Sahara Desert. The Tuareg are masters of the desert and camels. They often lead camel caravans on long trips through the desert.

D. Sudan: Slavery and Genocide

Prior to 2011 Sudan was the largest country in Africa, comparable in size to the entire United States east of the Mississippi River. The capital city of Khartoum lies where the Blue Nile River converges with the White Nile. Khartoum's government has a black Arab majority and follows Islam, complete with Sharia laws. The African Transition Zone crossed Sudan and separated the Arab-Muslim north from the mainly African-Christian south. There was civil war between the north and the south for decades. Before a peace agreement brokered in 2005, military soldiers from the north would raid the villages in the south, taking women and children as slaves. Though the Sudanese government denied the slave trade, thousands of Africans were owned by northern black Arabs in Sudan, and many still are. The world community has made little effort to intervene. The price for a slave in Sudan is about fifty US dollars.

The differences in religion, ethnicity, and culture always divided southern Sudan from the north. In January 2011, the southern region of Sudan voted on a referendum that would allow the south to break away and become an independent country called the Republic of South Sudan. The new **Republic of South Sudan** was formalized in July of 2011. Juba is designated as the capital with talk of moving it to the city of Ramciel in the center of the country in the future. The many clans and indigenous groups make it difficult for unity and cohesiveness in the new country. Armed groups in the various states continue to cause internal division, while at the same time boundary disputes continue to be worked out with North Sudan.

In 2003, various groups in **Darfur** complained that the Khartoum government was neglecting them. A militia group calling itself the Janjaweed was recruited by the local Arabs to counter the resistance in Darfur. The Janjaweed began an ethnic cleansing campaign that pushed into the Darfur region, burning villages, raping women, and killing anyone who opposed them. Refugees began to flee into the neighboring country of Chad.

Figure 16. Sudan, the region of Darfur and the republic of South Sudan. South Sudan elected to break away and become independent. The Darfur region has been experiencing genocide by Janjaweed militias backed by the Arab majority in northern Sudan.

In this particular case, the campaign was not based on religious divisions, because both sides were Muslim. This was an ethnic conflict in that the people of Darfur are of a traditional African background and the people of northern Sudan consider themselves Arab, even though they may have dark skin. Accurate numbers have been difficult to verify, but as of 2010 an estimated 300,000 people had died in this conflict. There were more than 2.7 million refugees, many of them in Chad. Just as the government of Sudan denied the slave trade, it denies that it supports the Janjaweed. The African Union provided a modest number of peacekeeping troops before the UN stepped in to provide security. It has been up to the world community and Sudan to take more action and provide more assistance. Food, water, and care for the refugees have taxed the region's aid and support system.

E. The African Union

Former Libyan leader Muammar Gadhafi was instrumental in the development of the **African Union** (**AU**) in the mid-1990s. The Sirte Declaration (titled after Gadhafi's hometown of Sirte in Libya) was issued by the Organization of African Unity, which outlined the need for the creation of the AU. The AU was launched in Durban, South Africa, on July 9, 2002. Fifty-three countries formed this intergovernmental organization. The focus of the AU is on the health, education, economic development, political stability, environmental sustainability, and general welfare of the people of Africa. The organization strives to integrate the socioeconomic and political stability of its members and promote a continent-wide effort for security and peace. The AU is working to create a proper political climate, one that helps its member states engage in the global economic marketplace by negotiating international issues and policies that affect Africa.

The dominating activities of colonialism and **neocolonialism** (**corporate colonialism**) are big concerns for the AU. The AU's objective is to bring more unity to the political and economic arena between the African countries to address the transition to a globalized world. It faces many challenges within its realm, including health care issues such as HIV/AIDS and malaria that have devastated much of Africa. The AU is working to bring political stability to countries such as Sudan and other countries experiencing civil unrest because of political turmoil or civil war, such as the Congo, Somalia, Sierra Leone, Ivory Coast, and Liberia. AU peacekeeping troops are assisting in this process. The legal issues regarding border disputes or territorial disputes such as that of Western Sahara are problems that the AU attempts to address.

In the global scale of economic and political **supranationalism**, the AU will be up against three main powerhouses: the European Union, the North American Free Trade Agreement, and the East Asian Community. Regions across the globe are working on trade associations to create economic networks to bring about greater cooperation and commerce between nations. The AU is one part of that network that represents a growing percentage of the world's population and the second-largest continent on Earth.

Key Takeaways

1. Three main physical features of North Africa are the Atlas Mountains, the Sahara Desert, and the Nile River. Most of North Africa's population lives along the Mediterranean coast or along the Nile River. The ethnic majority in the Maghreb are Berber, with Arabs dominating in Egypt.

2. The Maghreb centers on the Atlas Mountains, which traditionally has provided for a diversity of food production. Oil has been found in North Africa, the export of which has surpassed the export of food products.

3. Europe has had a strong influence on the region, ranging from the Roman Empire, to colonial activity, to becoming a destination for immigrants looking for employment and opportunities.

4. North Africa has experienced serious political conflicts. Political leaders in Tunisia, Libya, and Egypt who had been in power for decades were ousted in the Arab Spring of 2011 as people protested for economic and political reforms.

5. The African Transition Zone creates the southern boundary for North Africa. This zone serves as the transition between the arid type B climates and the tropical type A climates. It is also the transition between the dominance of Islam and the dominance of Christianity and animism.

6. The African Transition Zone cut through the center of Sudan and divided the country along religious and ethnic distinctions. Civil war has been waged in the south and in the Darfur region, which has split the country into separate regions. South Sudan became an independent country in 2011.

10.4 Israel and Its Neighbors

Learning Objectives

1. Summarize how the region of Palestine has evolved into the current Jewish State of Israel. Identify and locate the territories that have been annexed to Israel over the years.

2. Understand the division between the West Bank and the Gaza Strip and the Jewish State of Israel. Outline the complications of the one-state and two-state solutions to this division.

3. Describe the differences between the governments of Jordan and Syria.

4. Outline the political arrangements of the government leadership positions in Lebanon.

A. The State of Israel

At the center of the Middle East, on the shores of the Mediterranean in the Levant (the area bordering the eastern Mediterranean Sea), lies the country of Israel. Israel is bordered by Lebanon to the north, Syria and Jordan to the east, and Egypt to the south. Covering an area of only 8,522 square miles, Israel is smaller than the US state of Massachusetts. The coastal region, which has a moderate type C Mediterranean climate, receives more rainfall than the dry interior and the **Negev Desert** in the south, both of which have arid type B desert climates. The **Sea of Galilee**, also called Lake Kinneret or the Sea of Tiberias, is a major fresh water supply. The **Jordan River** flows from the Sea of Galilee to the **Dead Sea**. The Dead Sea is 1,300 feet below sea level, so it has no outlet. Over time, salts and minerals have built up, creating an environment that does not support fish or aquatic life. South of

the Negev Desert is the **Gulf of Aqaba**, which provides access to the Red Sea for both Israel and Jordan. Israel does not have substantial oil resources but has a potential for natural gas in offshore locations along the Mediterranean Sea.

Though most of the population in the Middle East is Islamic, there are exceptions, such as in Israel, which has a Jewish majority. Israel was established in 1948. Before that time, the country was called **Palestine**. The region went through a series of tumultuous transitions before it became the nation of Israel. Before 1948, most people in Palestine were called Palestinians and consisted primarily of Arab Muslims, Samaritans, Bedouins, and Jews. Most Jewish people were dispersed throughout the world, with the majority in Europe and the United States.

B. The Division of Palestine

Palestine was a part of the **Turkish Ottoman Empire** before the end of World War I. Britain defeated Turkish forces in 1917 and occupied Palestine for the remainder of the war. The British government was granted control of Palestine by the mandate of the Versailles Peace Conference in 1919 at the end of World War I. Britain supported the **Balfour Declaration of 1917**, which favored a Jewish homeland. The **British Mandate** included Palestine and **Transjordan**, the area east of the Jordan River, which was later renamed Jordan.

Between 1922 and 1947, during British control, most of the population of Palestine was ethnically Arab and followed Islam. In 1922, Jews made up less than 20 percent of the population. The Jewish settlements were mainly along the west coast and in the north. Jewish people from other countries—including Jews escaping German oppression in the 1930s—migrated to the Israeli settlements. Palestine was turned over to the control of the newly created United Nations (UN) in 1945 at the end of World War II.

The United Nations Special Committee on Palestine (UNSCOP) was created by the UN in 1947. To address the Palestine region, UNSCOP recommended in its **UN Partition Plan for Palestine** that Palestine be divided into an Arab state, a Jewish state, and an international territory that included Jerusalem. About 44 percent of the territory was allocated to the Palestinians, who consisted of about 67 percent of the population, which was mainly Arab. Approximately 56 percent of the territory was allocated to the minority Jewish population, who made up about 33 percent of the population. The city of Jerusalem was to remain under the administrative control of the UN as an international city. The Jewish State of Israel was officially recognized in 1948. The Palestinians, who were a majority of Israel's total population at the time and who owned about 90 percent of the land, denounced the agreement as unacceptable.

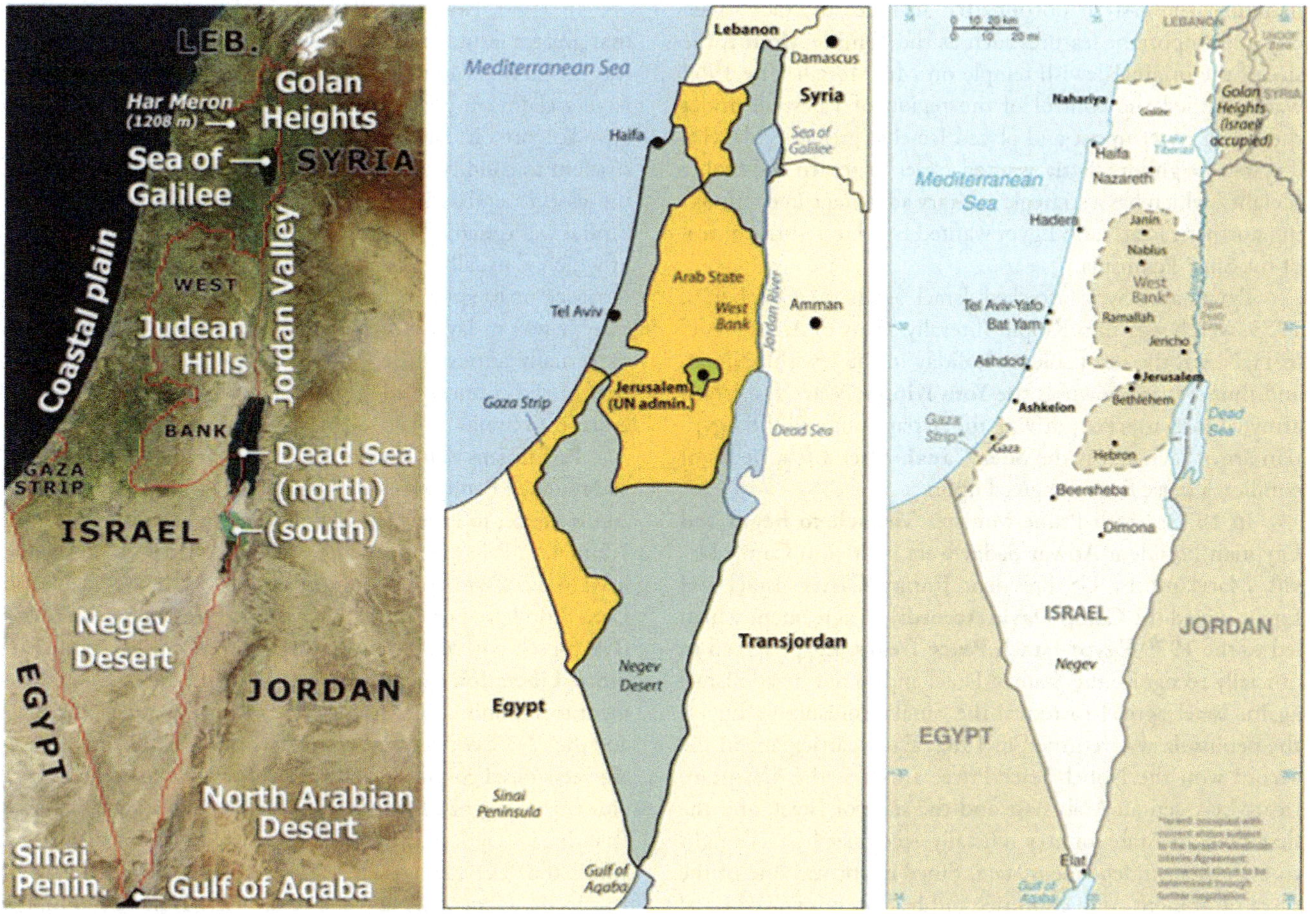

Figure 17. Satellite image of Palestine (left); 1948 UN division of Palestine into half Jewish state and half Arab state (center); Political Map of Israel in 2011 (right).

Israeli leaders accepted the UN plan and on May 14, 1948 declared Israel's independence within those boundaries. Palestine's Arab neighbors—Syria, Saudi Arabia, Lebanon, Iraq, and Egypt—sided with the Arab Palestinians and declared war on Israel, invading the very next day. The war did not end favorably for the Arabs. With support and aid from Britain and the US, Israel defeated the attacking Arab armies and took control of a larger portion of the land, including some of the land designated by the UN as a portion of the Arab state. Over 750,000 Palestinians living in the Jewish-controlled regions of Israel were forced out of their homes and into refugee camps. The result of the **1948 Arab-Israeli War** (also called the First **Arab-Israeli War**), was that Israel established itself as an independent country with a larger territory than called for in the UN Partition Plan.

By 1967, the Arab armies had regrouped and were willing to attack Israel again. The **1967 War** was short lived, lasting only six days, from June 5 through 10, and is hence often called the **Six Day War**. The Arab armies were devastated once again, and Israel gained even more territory. Israel took the **Sinai Peninsula** and the **Gaza Strip** from Egypt, the **Golan Heights** from Syria, and the **West Bank** from Jordan. The entire city of **Jerusalem** came under Israeli control,

Figure 18. The Western wall in Jerusalem. A remnant of the Jewish temple built by Herod the Great and destroyed in 70 CE, the Western Wall is the most holy place for the Jewish people. The Dome of the Rock mosque in the background is the third-holiest site for Muslims.

including the eastern portion that includes the old walled city and important features such as the **Temple Mount**, the site of the original Jewish temple on Mt. Moriah. The 1967 war solidified the control of the region of Palestine under the Israeli government and placed Israel at greater odds with its Arab neighbors. Syria wanted Israel to return the Golan Heights, which has a strategic military advantage in overlooking northern Israel, and Egypt wanted Israel to return control of the Sinai Peninsula.

Egypt and Syria attacked Israel again on October 6, 1973, which was Yom Kippur (literally, "Day of Atonement" in Hebrew), the most solemn holiday in the Jewish religion, and thus this is known as the **Yom Kippur War**. The Israeli army counterattacked, driving the Syrians out and the Egyptian army back across the **Suez Canal**. After a few weeks of conflict, a cease fire was agreed upon.

In 1978, Israeli Prime Minister **Menachem Begin** and Egyptian President **Anwar Sadat** were invited to Camp David, Maryland, by US President **Jimmy Carter**. Israel and Egypt signed the **Camp David Accords**, an agreement which led to the **1979 Egypt-Israeli Peace Treaty**. Egypt agreed to officially recognize the State of Israel and to not invade Israel again. Israel agreed to return the Sinai Peninsula to Egypt; the peninsula was returned in 1982. Each participant in the accord won the Nobel Peace Prize. This was the first peace treaty between an Arab state and the state of Israel, and the first time an Arab country officially recognized Israel's right to exist as an independent state. Since Egypt was one of the most prominent Arab countries and had been involved in all prior Arab-Israeli wars, this was monumental geopolitically.

In 1980, Israel passed the Jerusalem Law, which stated that greater Jerusalem was Israeli territory and that Jerusalem was the eternal capital of the State of Israel. The UN rejected Israel's claim on greater Jerusalem, and few if any countries have accepted it. Israel moved its capital from Tel Aviv to Jerusalem to solidify its claim on the city even though most of the world's embassies remain in Tel Aviv. The move of the capital was designed to create a **forward capital**, the purpose of which is usually either to lay claim to or protect a nation's territory or to spur the development of the country. In this case, it was to lay claim to valuable territory. In 1981, Israel formally annexed the Golan Heights, making it officially part of the country and signaling an unwillingness to ever r eturn it to Syria.

Palestinians were left with only the regions of the Gaza Strip and the West Bank, which is controlled by the Israeli government and is subject to Israel's national jurisdiction. As of 2010, about 1.5 million Palestinians lived in the Gaza Strip and 2.5 million lived in the West Bank. A number of cities in the West Bank and Gaza Strip have been turned over to the **Palestinian Authority** (**PA**) for self-governing. The PA was established between the **Palestine Liberation Organization** (**PLO**) and the Israeli government to administer internal security and civil matters. The PLO and the PA are two separate entities. The PLO is the internationally recognized governing body of the Palestinian people. It is legitimately recognized by the UN to represent the area known as Palestine in political matters. There are two main political parties within the PLO: Hamas and Fatah. The Hamas party is the strongest in the Gaza Strip, and the Fatah party is more prominent in the West Bank.

C. The Palestinians, Israel, and Possible Solutions

The future of the West Bank and Gaza Strip has been the focus of talks and negotiation for decades. There are various ways to approach this issue; a one-state solution and a two-state solution have been proposed. The **one-state solution** proposes the creation of a fully democratic state of Israel and the integration of all the people within its borders into one country. Integration of the Gaza Strip and the West Bank into the Jewish State of Israel is part of this plan; in oth-

Figure 19. A street in the West Bank City of Nablus. Cell phones are ubiquitous in Nablus.

Figure 20. Security wall between Israel and the West Bank.

er words, "Take the walls down and create one state." Many Palestinians support the one-state solution, but most of the Jewish population does not. Family size is much larger in the Palestinian side, so it would be only a matter of time before the Jewish population would be a minority population and would not have full political control with a democratic government. To have the Jewish State of Israel, the Jewish population needs to keep its status as the majority.

In a **two-state solution**, Palestinians would have their own nation-state, which would include the Gaza Strip and the West Bank. The rest of former Palestine would be included in the Jewish State of Israel. The two-state concept (Israel and a Palestinian state) has been proposed and supported by a number of foreign governments, including the United States. Implementation of a two-state solution is, of course, not without its own inherent problems. At the present, the West Bank and Gaza Strip are subjects under the Jewish State of Israel without full political or economic autonomy.

Parties to the negotiations have acknowledged that the most likely solution is to create a Palestinian state bordering Israel. However, it is not clear how to make this happen. Palestine is now divided between the Jewish State of Israel (with 7.3 million people) on one side and the Palestinians (with 4.0 million people) in the West Bank and the Gaza Strip on the other side. About 75 percent of Israel's population of 7.3 million people are Jewish, and about 25 percent are Arab. Travel between Israel and the Palestinian areas is heavily restricted and tightly controlled. A high concrete and barbed wire barrier separates the two sides for much of the border. The West Bank provides fresh water used on the Israeli side for agriculture and industrial processes. The industries also employ Palestinians and support them economically.

Jewish people from various parts of the world continue to migrate to Israel, and the Israeli government continues to build housing settlements to accommodate them. Since the West Bank is under Israeli jurisdiction, many of the new housing settlements have been built in the West Bank. The Palestinians who live there strongly oppose the settlements. In 1977, only about five thousand Jews lived in the West Bank settlements. As of 2010 there were more than two hundred thousand. The Palestinians argue that if they were to have their own nation-state, then the Jewish settlements would be in their country and would have to be either resettled or absorbed. Israel responds by indicating that the two-state solution is indefensible because the Jewish settlements in the West Bank cannot be protected if the West Bank is separated from Israel.

The issues in Israel are complicated. After a series of wars and considerable negotiations, the central problems remain: Jews and Palestinians both want the same land, both want Jerusalem to be their capital, and neither can find a compromise. Support for the Jewish State of Israel has primarily come from the US and from Jewish groups external to Israel. There are more Jews in the United States than there are in Israel,

and the US Jewish lobby is powerful. Israel has been the top recipient of US foreign aid for most of the years since 1948. Through charitable donations, US groups provide Israel additional billions of dollars annually. Foreign aid has given the Jewish population in Israel a standard of living that is higher than the standard of living of many European countries.

The problems between Israel and Palestinians are far from settled. The region has plenty of interconnected concerns. Israel has nuclear weapons, and Iran has worked at developing nuclear weapons. US involvement in the region has heightened tensions between Iran and Israel. Oil revenues

Figure 21. West Bank settlements and Palestinian-controlled Areas

are driving the economies of most of the Arab countries that support the Palestinians. Oil is an important export of the region, with the United States as a major market. The difficulties between Israel and the Palestinians continue to fuel the conflict between Islamic fundamentalists and Islamic reformers. Some Islamic groups have accepted Israel's status as a country and others have not. The Israel-Palestinian problem drives the geopolitics of the Middle East.

Figure 22. Jordan.

D. Jordan

North of the Arabian Peninsula are three Arab states that surround Israel: Jordan, Syria, and Lebanon. Each country possesses its own unique physical and cultural geography. The country of Jordan was created through the British Mandate after World War I, when Britain defeated the Turks in Palestine. The area east of the Jordan River became the modern country of Jordan in 1946. From 1953 to 1999, during the most volatile period of the region, the country was ruled by a pragmatic leader, King Hussein, who was able to skillfully negotiate his way through the difficult relationship with Israel and yet keep his country stable. When Palestine was divided by the UN to create the State of Israel, the region of Jordan received more than a million Palestinian refugees from the West Bank and Israel. Refugees make up a large portion of the more than six million people who live in Jordan today; about a half million refugees from the US war in Iraq are included in that total.

Jordan is not large in physical area. Natural resources such as oil and water are not abundant here, and the country often has to rely on international aid to support its economy. Inflation, poverty, and unemployment are basic issues. The government of Jordan is a constitutional monarchy. King Hussein's son 'Abdullah II took power after the king's death in 1999. Economic reforms were implemented by King 'Abdullah II to improve the long-term outlook of the country and raise the standard of living for his citizens. The king allowed municipal elections to be conducted, which allowed for 20 percent of the positions to be dedicated to women candidates. Parliamentary elections were held by a democratic vote.

Jordan has demonstrated how a country with few natural resources in a volatile region of the world can proceed down a progressive path despite difficult circumstances. Jordan has developed a positive trade relationship with Europe and the United States while at the same time working with its Arab neighbors to access oil and to maintain a civil state of affairs. Jordan is not without its challenges but has managed to confront each issue yet retain a sense of stability and nationalism.

Figure 23. King Abdullah II of Jordan visits US President Barack Obama in the White House in 2011. Jordan has had good political relations with the United States. King Abdullah II has worked to maintain a stable government in Jordan and maintain civil stability in spite of Jordan's lack of economic opportunities.

E. Syria

The strategically located country of Syria is at the center of the Middle East's geopolitical issues. Syria gained its independence from the French Mandate in 1946, the same year as Jordan. Syria has strived to work out and stabilize its political foundation. In a move to create greater Arab unity in the realm, Egypt and Syria joined forces and created the **United Arab Republic** in 1958. This geopolitical arrangement lasted until 1961, when the partnership was dissolved. Syria returned to its own republic. The Arab Socialist Baath Party gained strength, and in 1970 Hafiz al-Assad, of the Alawite minority (an offshoot branch of Shia Islam making up about 10 percent of the Syrian population), took over leadership in a coup that stabilized the political scene. It was during this era that the Golan Heights was lost to Israel in the Arab-Israeli War of 1967. This strategic geographical location is a point of contention in the peace negotiations between Syria and Israel.

Hafiz al-Assad served as the leader of Syria for twenty-nine years without having been democratically elected to the office by the people. His son Bashar took the reins of leadership after Hafiz died in 2000. The Alawite sect held power in Syria through the Assad family under military control. Syria has been accused of using its military power to influence conditions in Lebanon, where it brokered a peace deal in its civil war (1975–1990). Syria has also been accused of supporting the anti-Israel groups headquartered in Lebanon.

Figure 24. Female protesters in Douma, a suburb of Damascus, in 2011.

Syria is located in an ancient land with a long history of empires and peoples. The region of Syria was once part of the cradle of civilization that sprung up in Mesopotamia. Damascus claims to have been continually inhabited longer than any other capital city on Earth. The largest city and the center of industrial activity is Aleppo, which lies in the north of Syria. Syria's physical area is slightly larger than the US state of North Dakota. Overall, Syria's climate is characterized as an arid type B climate; some regions receive more rain than others. The western region, because it borders the Mediterranean Sea, is an area that receives more rainfall. The additional rainfall translates into extensive agricultural production. The northeast area of Syria is also productive agriculturally through water resources provided where the Euphrates River cuts through the country. Oil and natural gas have been the country's main export products. The petroleum reserves are being depleted, and few new fields are being developed. Eventually, the wealth generated by the sale of petroleum

Figure 25. Syria.

reserves, which are finite resources, is projected to diminish, even as the population continues to increase.

The Syrian government has exerted strict control over the economy. The country will face serious economic issues in the future. There is a high rate of unemployment. Because oil production has not been increasing, the government has been forced to take on additional national debt. The arid climate and the need to supplement agriculture production have placed additional pressure on precious fresh water supplies. The Euphrates River provides fresh water, but it originates in Turkey, where large dams restrict the flow. Water rights for the region are therefore an issue. One third of Syria's population is under the age of fifteen, which indicates a rapid population growth pattern that will tax future resources at an increasing rate. In 2010, Syria had about twenty-two million people. The country holds political significance; its strategic location between Iraq and Israel makes it is a vital player in any solution for lasting peace in the Middle East.

Syria experienced protests and demonstrations similar to those that swept through North Africa in the Arab Spring of 2011. Citizens expressed dissatisfaction with the government because of the lack of democratic reforms, high unemployment, and the loss of civil rights. Student protests escalated to massive citizen demonstrations that emerged in various Syrian cities in the spring of 2011. The government cracked down on protesters, killing some. After extensive demonstrations on March 15, the government arrested more than three thousand people. Hundreds were killed in violent clashes between the people and government security forces. The lack of democratic processes by President Bashar al-Assad's government continued to prompt protests and demonstrations in Syria, eventually resulting in armed conflict between protesters and the Syrian government and the start of civil war. As of 2017, the **Syrian Civil War** that began in 2011 continues.

Key Takeaways

1. The current Jewish State of Israel was recognized in 1948. Before this time, the region was called Palestine and the people who lived there were called Palestinians.

2. Victorious in war against their Arab neighbors, Israel acquired the Golan Heights, the West Bank, the Gaza Strip, and the city of Jerusalem. The West Bank and the Gaza Strip are considered Palestinian territory. Two plans have been proposed to address the division but have not been agreed upon.

3. Jordan is a constitutional monarchy led by King Abdullah II, who has worked to implement reforms to maintain a country that has few natural resources.

4. The government of Syria is led by Bashar al-Assad, a member of a minority ethnic group called the Alawites. Assad and his father have ruled Syria for more than forty years under a state of emergency. Civil war has raged since 2011, resulting in tens of thousands of deaths and creating millions of refugees.

10.5 The Arabian Peninsula

A. Overview

The Arabian Peninsula is a desert environment surrounded by saltwater bodies. The **Persian Gulf**, the **Arabian Sea**, and the **Red Sea** border the peninsula on three sides. Arid type B climates dominate the region. Saudi Arabia only receives an average of four inches of precipitation per year. The southern portions of the peninsula are some of hottest places on Earth. Summer temperatures can reach more than 120 °F. In the south is the **Rub' al-Khali (Empty Quarter)**, which is mainly desert and comprises about 25 percent of Saudi Arabia. It is extremely dry and virtually uninhabited, though oil discoveries have brought temporary settlements to the region. There are no natural lakes or major rivers on the peninsula. Agricultural activity is dependent on the availability of water by rainfall, underground aquifers, oases, or desalinization of seawater.

Most of the people living on the peninsula are Arabs, and most of the peninsula's countries are ruled by monarchs who rely on oil revenues to gain wealth. Minerals are mined in the mountains that dominate the peninsula's western and southern regions. The highest peaks reach more than twelve thousand feet in elevation in northern Yemen. Of the countries on the peninsula, Yemen has the fewest oil resources and has had the sole democratically elected government. Saudi Arabia dominates the region in size and in oil resources. Islam, the major religion, infiltrates all aspects of Arab culture.

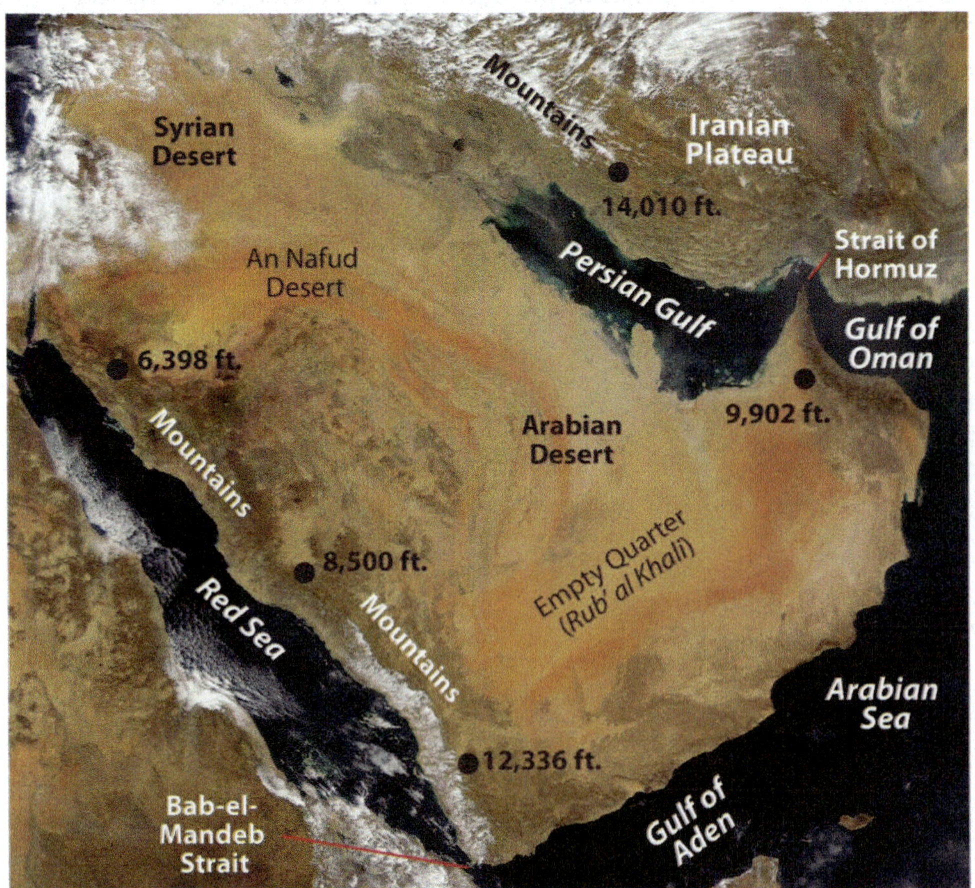

Figure 26. Satellite image from 2008 of the Arabian Peninsula illustrating the mountainous regions, the uninhabited Empty Quarter desert region, and the surrounding bodies of water

Saudi Arabia

The holy cities of **Medina** and **Mecca** are in Saudi Arabia, the birthplace of Islam. Islam first united the many traditional groups of Arabia with religion and then with the Arabic language. The region was further united after 1902, when Abdul Aziz Al-Sa'ud and his followers captured the city of Riyadh and brought it under the control of the **House of Sa'ud**. In 1933, the lands under the control of the king were renamed the **Kingdom of Saudi Arabia**. Saudi Arabia is an absolute monarchy. In 1938, US oil corporation Chevron found large quantities of oil in the region, which has sustained the royal family ever since. **Saudi Aramco** (originally the Arabian-American Oil Company, or Aramco for short) is the state-run oil corporation. Controlling about one-fifth of the world's known oil reserves, the Saudi royal family claims considerable power.

The Saudi royal family gave safe haven to thousands of Kuwaitis, including the emir and his family, during the First Persian Gulf War (1991). Saudi Arabia allowed US and Western military forces to use bases on its soil during Operation Desert Storm. Acquiescence to non-Muslims operating military bases on the same soil as the holy cities of Mecca and Medina gave extremist groups a reason to engage in terrorist activities. Out of the nineteen hijackers in the 9-11 attack in New York, sixteen were from Saudi Arabia. The Saudi government has been forced to step up its efforts against terrorism and domestic extremist groups.

The entire economy of Saudi Arabia is based on the export of oil, and more than 20 percent of the known oil reserves in the world are located in Saudi Arabia. The country is a key member of the **Organization of Petroleum Exporting**

Figure 27. Political map of the Arabian Peninsula.

Countries (**OPEC**) and has been the world's number one oil exporter. Millions of foreign workers in the petroleum industry make up a vital component of the country's economy.

A high rate of population growth has been outstripping economic growth in Saudi Arabia. In 2010, more than one-third of the population was younger than fifteen years old, and family size was about 3.8 children. The unemployment rate is high, and there is a shortage of job skills in the workforce. The government has been working to shift its focus away from a petroleum-based economy and increase other economic opportunities; it plans to heavily invest in the necessary infrastructure and education to diversify its economy.

Saudi Arabia has made several efforts to move forward and put the country more in line with globalization efforts that are modernizing the other Persian Gulf States. The World Trade Organization accepted Saudi Arabia as a member in 2005. In 2008, the king implemented the initiative for interfaith dialogue in an effort to address religious tolerance and acceptance. The first woman was appointed to the cabinet, and municipal councils held elections for its members.

Figure 28. Modern infrastructure illustrated on Medina road in Jeddah, Saudi Arabia.

Saudi Culture

The royal family and most of the people in Saudi Arabia are Sunni Muslims. The country has a strong fundamentalist Islamic tendency. The law of the state is strict and supports conservative Islamic ideals. The **Wahhabi** branch of Sunni Islam has a major influence on culture. Activities such as gambling, alcohol consumption, and the promotion of other religions are outlawed. Alcohol and pork products are forbidden in accordance with Islamic dietary laws. Movie theaters and other Western-style productions are prohibited but can be found in areas where workers from other countries live in private compounds. Though movie theaters are restricted, movies on DVDs are not prohibited and are widely available. The dress code in Saudi Arabia strictly follows the Islamic principles of modesty. The black abaya (an article of clothing that looks like a cloak or robe) or modest clothing is appropriate for women. Men often wear the traditional full-length shirt and a headcloth held in place by a cord.

In Saudi Arabia, human rights organizations, legal associations, trade unions, and political parties are banned. The country maintains a tight censorship of all local media. The press is only allowed to publish what the government permits it to report.

Communication with foreigners, satellite media, and Internet access are highly controlled. Those who speak out against the government can be arrested or imprisoned.

The Sharia is the basic criminal code in Saudi Arabia, along with whatever law is established by the king. A wide range of corporal and capital punishments—from long prison sentences to amputations (arm or foot), floggings, and beheadings—are proscribed for legal or religious offenses. Trials are most often held in secret without lawyers. Torture has been used to force confessions that are then used to convict the accused. Torture techniques—including the use of sticks, electric shocks, or flogging—can be applied to children and women as well as men. Executions are usually held in a public place every Friday.

Role of Women

Men hold the dominant roles in Saudi society. Under strict Islamic law, women do not have the same rights as men, so Saudi women do not have the opportunities that women in many Western countries have. For example, it is not customary for a woman to walk alone in public; traditionally, she must be accompanied by a family member so as to not be accused of moral offences or prostitution. Themutawa'een (religious police) have the authority to arrest people for such actions. The punishment could be as many as twenty-five days in prison and a flogging of as many as sixty lashes.

As of 2011, the following restrictions have been made on women:

- Women are not allowed to drive motor vehicles.

- Women must wear modest clothing such as the black abaya and cover their hair.

- Women can only choose certain college degrees. They cannot be engineers or lawyers, for example.

- Women cannot vote in political elections.

- Women cannot walk in a public spaces or travel without a male relative.

- Women are segregated from men in the workplace and in many formal spaces, even in homes.

- Women need written permission from a husband or father to travel abroad.

- Marriages can be arranged without the woman's consent, and women often lose everything in a divorce.

Saudi Arabia is a country steeped in tradition based on the heritage of its people. However, as the forces of globalization seep into the fabric of society, some of these traditions are evolving and changing to adapt to the times and to a more open society. Women are asserting themselves in the culture, and many long-standing traditions are starting to break down. Saudi Arabia is an example of how Islamic fundamentalism is being challenged by modernity and democratic principles.

Figure 29. Women in Saudi Arabia. Former first lady Laura Bush meets medical staff in Saudi Arabia. Note the women's attire.

Key Takeaways

1. The Arabian Peninsula is a desert region. The Rub' al-Khali (Empty Quarter) provides an example of desert extremes. The mountains along the western and southern edges receive the most rainfall. There are no rivers or major lakes on the peninsula. Water, a resource that is vital to human activity, is scarce throughout the region.

2. The export of oil and natural gas is what drives the economies of the region. Many of the states are working to diversify their economies with banking, free-trade zones, and even tourism.

3. Family size varies widely in the region, from more than 4.5 children per family in Yemen and 3.8 in Saudi Arabia, to around 2.5 in the progressive Gulf States of Qatar, the UAE, and Bahrain. Women's rights and opportunities have an inverse relationship with family size; that is, when women's rights and opportunities increase, family size usually decreases. In the Gulf States, the smaller countries in land area are more open to promoting women's rights and responsibilities in the public and private sectors.

Figure 30. The Tigris and Euphrates rivers and the Shatt al-Arab waterway between Iraq and Iran.

10.6 Iraq, Turkey, and Iran

Learning Objectives

1. Summarize Iraq's role in the Persian Gulf War and the Iraq War in 2003.
2. Understand how Iraq is divided ethnically and by the branches of Islam.
3. Explain why Turkey wants to be a member of the European Union (EU) and why it has not been accepted.
4. Outline Iran's physical geography and how it has used natural resources for economic gain.
5. Determine why young people might be dissatisfied with the policies of the Iranian government.

A. Iraq

Iraq lies in the **Fertile Crescent** between the **Tigris** and **Euphrates** Rivers, where the ancient civilizations of Mesopotamia were established. Ancient cities such as Nineveh, Ur, and Babylon were located here. Present-day Iraq and Kuwait were established out of the British Mandate territory gained following Britain's defeat of the Turkish **Ottoman Empire**. Britain established straight-line political boundaries between Iraq and Jordan, Syria, and Saudi Arabia. These types of boundaries are called **geometric boundaries** because they do not follow any physical feature. In 1961, when Britain withdrew from the region, the emir controlling the southern region bordering the Persian Gulf requested that Britain separate his oil-rich kingdom as an independent country. This country became Kuwait, and the rest of the region became Iraq. After a series of governments in Iraq, the Baath party came to power in 1968, paving the way for **Saddam Hussein** to gain power in 1979.

Iran-Iraq War (1980–88)

In 1980, a disagreement arose over the **Shatt al-Arab** waterway in the Persian Gulf on the border between Iraq and Iran, and the feud led to war between the two countries. The people of Iran are not Arabs; their ethnic background is Persian. Most Iranians are Shia Muslims. Saddam Hussein and his Baath party were ethnically Arabs and were Sunni Muslims. Ethnic and religious differences thus fueled the conflict. The war ended in 1988 without anyone declaring a victory. The **Iran-Iraq War** resulted in more than a million casualties and cost more than one hundred billion dollars. Before the war, the Iranian government had been taken over

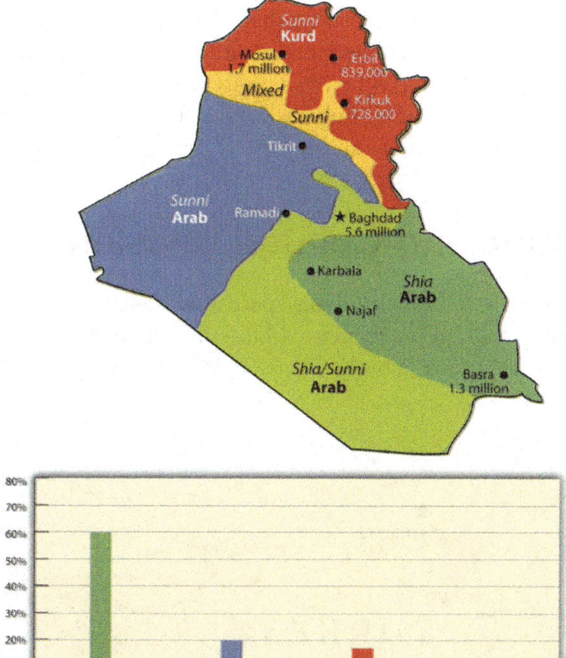

Figure 31. Iraq's Divisions of Islam as of 2008. The two ethnic divisions are Arab and Kurd. The two religious divisions of Islam are Shia and Sunni. Karbala and Najaf both have holy sites for Shia Muslims.

by Islamic fundamentalists who opposed the US intervention in the region; therefore, in the Iran-Iraq War, the United States supported Hussein and provided him with industrial supplies and materials.

The Persian Gulf War (1990–91)

After the Iran-Iraq War, Saddam Hussein looked to Kuwait to gain new oil wealth and expand access to the Persian Gulf. Hussein accused Kuwait of slant drilling oil wells along the Iraqi border and removing oil that was legally Iraq's. It was common knowledge that both sides engaged in this practice, but it was the excuse Hussein needed to invade Kuwait.

In 1990, the Iraqi military invaded and occupied Kuwait. Though the world community opposed this action, it was not until Hussein nationalized all the oil assets of the international oil corporations that resistance was organized. Under the leadership of US president George H. W. Bush, the UN organized a military coalition to remove Hussein from Kuwait. On January 16, 1991, **Operation Desert Storm**, which was a massive air campaign, began. After forty-five days of fighting, Iraq was overwhelmingly defeated and its military was ousted from Kuwait.

When it became evident that Hussein would lose Kuwait, his forces dynamited all the oil facilities and set all the oil wells in Kuwait on fire. His position was that if he could not have the oil, then nobody would. This was one of the worst environmental catastrophes regarding oil on record.

Figure 32. Operation Desert Storm. US fighter jets in Operation Desert Storm fly over burning oil well fires in Kuwait. The fires were set by retreating Iraqi forces.

Oil flowed into the Persian Gulf and covered the water's surface up to three feet thick. Most mammals, birds, and organisms living on the water's surface died. Oil flowed out onto the desert sand into large petroleum lakes. The air pollution caused by burning oil wells dimmed the sun and caused serious health problems.

Ethnic and Cultural Divisions

To keep Iraq from breaking apart after Operation Desert Storm, coalition forces allowed Hussein to remain in power. Ethnically and religiously, Iraq is divided into three primary groups that generally do not get along. Sunni Arabs dominate central Iraq in a region often referred to as the **Sunni Triangle**, which includes the three cities of Baghdad, Tikrit

Figure 33. Homelands of the Kurdish people, indicated by the shaded areas. Future Kurdistan would be the main portions in Iran, Iraq, and Turkey and a corner of Syria. The city of Diyarbakir in Turkey would be the Kurdish capital city.

(Hussein's hometown), and Ramadi. Sunnis were the most loyal to the Hussein government. Southeastern Iraq is dominated by Arabs who follow the Shia division of Islam, which is also followed by most of Iran's population. A group that is ethnically Kurdish and follows the Sunni division dominates northern Iraq. **Kurds** are not Arabs or Persians; rather, they originated from somewhere in northern Europe centuries ago with their own religion, language, and customs. Many have converted to Islam.

Hussein was a Sunni Muslim Arab, and when he was in power, he kept the Sunni Kurds and Shia Arabs in check. He used chemical weapons on the Kurds during the Iran-Iraq War. In 1988, he used chemical weapons on the Kurdish town of Halabja and killed about 10 percent of the eighty thousand who lived there. Thousands of Kurds died in other attacks, and thousands more continue to suffer serious health effects. After Operation Desert Storm, Hussein pushed the Kurds north until the UN and the United States restricted him at the thirty-sixth parallel, which became a security zone for the Kurds. The Arab Shia population in the south often clashed with Hussein's military in an attempt to gain more political power, and Hussein subjected them to similar harsh conditions and treatment.

The Kurds are the largest nation of people in the world without a country. About twenty-five million Kurds live in the Middle East, and most—about fourteen million—live in Turkey. About eight million Kurds live in Iran, about seven million live in Iraq, and the rest live in neighboring countries. At the 1945 conference of the UN, they petitioned to have their own country called **Kurdistan** carved out of Iran, Iraq, and Turkey, but they were denied. Israel was approved to become a state at the same UN conference.

The Iraq War (2003-11)

After the September 11, 2001, attack on the United States, there was a renewed interest in **weapons of mass destruction (WMD)**. Knowing that Hussein had used chemical weapons on the Kurds, the Iranians, and the Shia, there was a concern he would use them again. UN Weapons inspectors in Iraq never could confirm that Hussein retained WMD. They had been destroyed, moved out of Iraq, or hidden. US president **George W. Bush** decided to invade Iraq in 2003 to remove Hussein from power. In the invasion, Hussein's two sons were killed and Hussein was captured.

The US invasion of Iraq brought about the removal of the Baath Party from power and Iraq came under a military occupation by a multinational coalition. An Iraqi Interim Government was formed that assumed sovereignty in 2004. A new constitution was drafted and approved by vote of the Iraqi people. Elections were held and a new government was formed under the newly drafted constitution. Occupying troops continued to remain as the country strug-gled to adapt to the reforms. Insurgencies developed that brought about an increase in violence that peaked in about 2007. By 2010, the combat operations by occupying troops were ending and the country worked to sustain stable political conditions. Under an agreed upon mandate all combat troops withdrew by the end of 2011. A number of US troops remain in Iraq in an advisory capacity.

Figure 34. Soldiers from the 926th engineer brigade combat team and the army 432d civil affairs battalion patrol in Baghdad's district of Sadr City. The United States invaded Iraq in 2003 and removed Saddam Hussein from power.

Resources and Globalization

In geographic terms, the **Persian Gulf War** and the **Iraq War** were wars over resources—namely, oil. Wars have historically been fought over territory and resources. When the United States invaded in 2003, Hussein was contracting billions of dollars' worth of projects to oil companies in France, Russia, and China. Other support projects were contracted out to other European countries. When the United States invaded, the contracts with were summarily canceled, and British and US oil contractors took over the projects.

Iraq is an example of the second wave of globalization. Neocolonialism has been the dominant force in Iraq's economy since before the Persian Gulf War. Industrialization requires high energy demands; therefore, the industrialized countries of the world consume energy on a massive scale. Iraq is not a core economic country, but it holds vast petroleum reserves, making it vulnerable to exploitation by industrialized core countries. It is interesting to note that the Persian Gulf War, initiated in response to Iraq's invasion of Kuwait, can be traced directly to globalism. It was Britain that established the straight-line borders separating Kuwait and Iraq. The war over the control of Kuwait in 1991 was a war over the control of resources, just as the ongoing competition between Sunni Muslims and Shia Muslims, for example, is a competition for the control of political power or resources and not a competition related to ethnic principles.

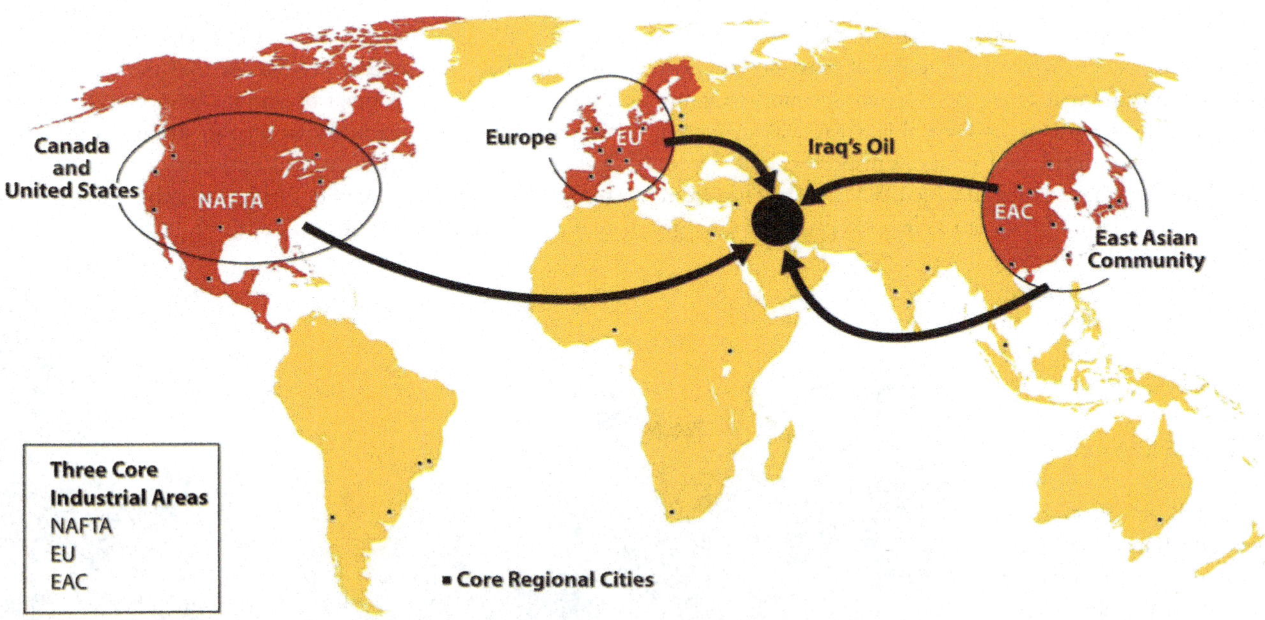

Figure 35. The three core economic areas of the world that consume high levels of energy. The East Asian Community (EAC) is not an official organization but a core economic area.

Figure 36. Turkey. The map shows the portion of Turkey in Europe to the west, the central Anatolian Plateau, the Euphrates and Tigris Rivers, and Mount Ararat on the Armenian border.

B. Turkey

Turkey is the only remaining country of the vast **Ottoman Empire**, which ruled the region for over six hundred years (1299–1923). When the empire was at its peak in the sixteenth and seventeenth centuries, it controlled parts of Europe, North Africa, the Middle East, and Arabia. Located on the **Bosporus**—the straits that connect the Black Sea with the Mediterranean Sea—the ancient city of **Constantinople** was the capital of the Ottoman Empire. Called Istanbul today, this city is the largest in Turkey, but it is not the current capital. Turkey moved its capital to **Ankara** on the **Anatolian Plateau**, which is centrally located in Turkey.

Turkey has a small portion of land on the western side of the Bosporus to claim its connection to Europe. Most of Turkey is a part of the Asian continent, and Turkey has often been referred to as Asia Minor. Mountains on Turkey's eastern border with Armenia include Mount Ararat, which is the highest peak in the country at 16,946 feet in elevation. Legend has it that Mount Ararat was the resting place of Noah's ark. The people of Turkey are neither Arab nor Persian; they are **Turkish** and speak the Turkish language. As much as 90 percent of the Turkish population is Sunni Muslim, which is similar to many of the other Muslim countries in the Middle East.

Turkey is a **secular state** with a democratically elected political leadership. To maintain its democracy, it has had to deal with Islamic fundamentalists, who often have supported a shift to an Islamic religious state. Turkey has also had to negotiate with its neighbors, Syria and Iraq, over water rights to the Tigris and Euphrates Rivers, which originate

Figure 37. A city park outside the Blue Mosque in Istanbul. The capital city of Turkey is Ankara, which is located in the center of the country. Istanbul remains the primate city of the country and is home to various world-class mosques.

in Turkey but flow through the other countries. As an upstream state, Turkey has built dams on these rivers, much to the dismay of its downstream neighbors, who want to use more of the water.

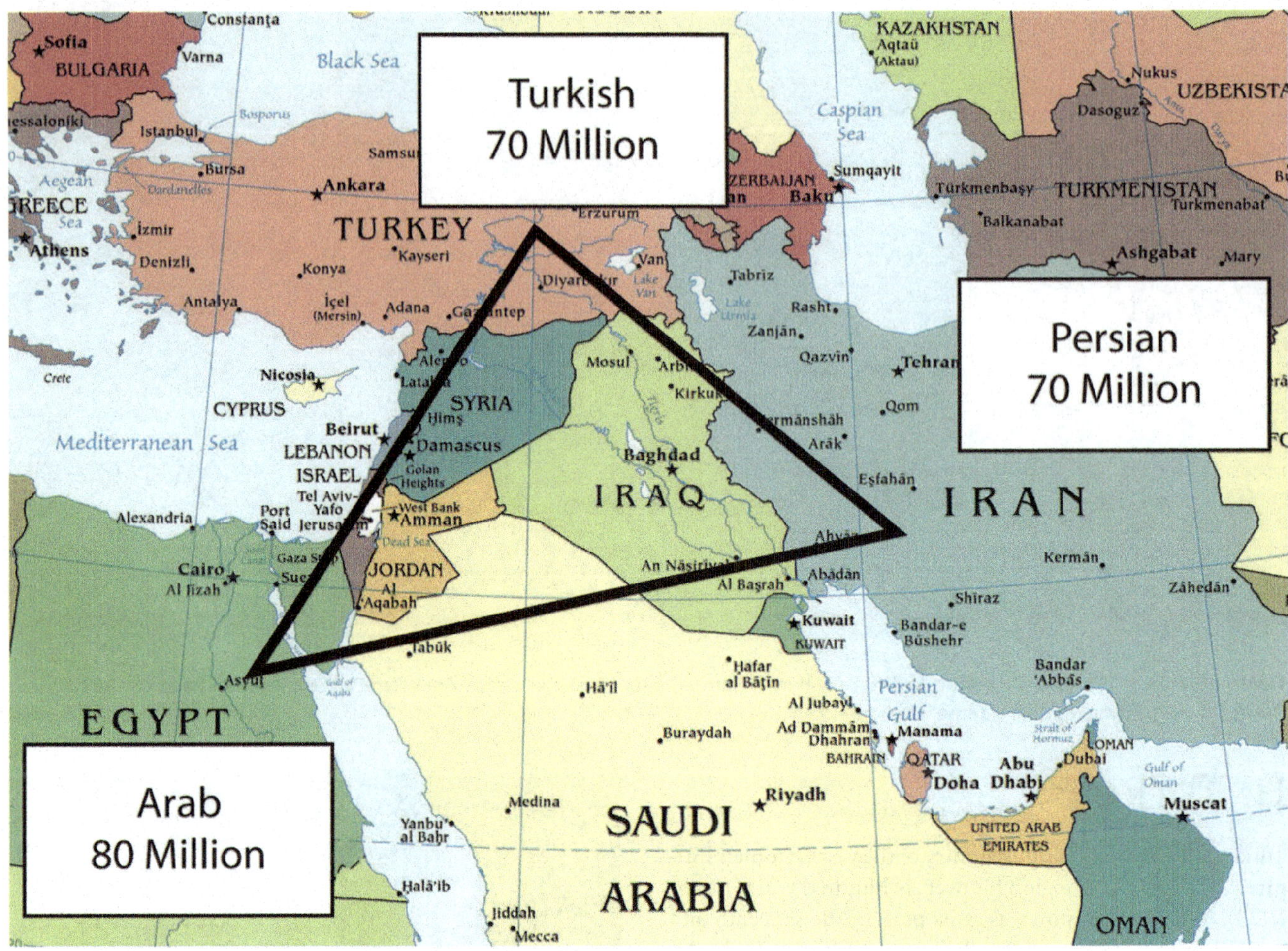

Figure 38. 2010 Population estimates for the Ethnic Triangle of the Middle East, with Egypt, Turkey, and Iran Anchoring Each Corner.

C. Iran

Iran covers a physical area larger than the US state of Alaska. It is a land of mountains and deserts: Iran's central and eastern regions are mainly desert with few inhabitants, and the northern and western regions of the country are mountainous. The **Elburz Mountains** in the north around the Caspian Sea reach as high as eighteen thousand feet in elevation near the capital city of **Tehran**. The **Zagros Mountains** run along the border with Iraq and the Persian Gulf for more than nine hundred miles and can reach elevations greater than fourteen thousand feet. Similar to the Atlas Mountains in the Maghreb, Iran's mountains trap moisture, allowing minor agricultural activities in the valleys. Most of Iran's population lives in cities along the mountain ranges. **Qanats**—systems of shafts or wells along mountain slopes—bring water from the mountains to the valleys for irrigation and domestic use.

Persian Empire to Islamic Republic

Iran was once the center of the **Persian Empire**, which has its origins as far back as 648 BCE, and the country was called Persia until about 1935. The **Ethnic Triangle of the Middle East** consists of Persians in Iran, Turks in Turkey, and Arabs in Egypt. Most of the seventy million people in Iran are Persian. Iran has a long history with the ancient Persian Empire and the various conquering armies that followed it. During the rise of Islam, Iran had major contributions to the arts, mathematics, literature, philosophy, and science. The highly advanced carpet-weaving traditions from centuries past are but one example of the advancements in design and the technical expertise of the people.

Since the **Iranian Revolution** of 1979, Iran has been an Islamic state with an ayatollah as the Supreme Leader. **An ayatollah** is a high-ranking Shia cleric that is an expert in the Islamic faith and the Sharia (Muslim code of law based on the Koran). Approximately 90 percent of Iran's population follows the Shia division of Islam.

Political Challenges

On the international front, Iran's leadership has indicated a drive to develop nuclear weapons and use them against Israel, which has caused concern in the global community. The government of Iran does not recognize the legitimacy of the nation of Israel. Iran is similar to Saudi Arabia in its restrictions of civil rights. A number of countries, including the United States, have placed trade sanctions against Iran regarding any materials associated with nuclear weapons or missiles. The US sanctions extend to an almost total trade embargo against Iran stemming from the 1979 revolution.

Open protests expressing a need for change have periodically erupted in the streets of Tehran. Ultimately, protesters are seeking personal freedoms and a more open society. The Arab Spring of 2011 was a phenomenon that spread across North Africa, the Arabian Peninsula, and the Middle East. Iran was not immune from the impact of the protests and demonstrations that occurred in their neighboring countries, but Iran was different. First of all, Iranians are not Arab but Persian in their ethnicity. Their history and heritage creates a distinct identity that separates them from the rest of the region. Many Iranian citizens want the same outcomes that the protesters and demonstrators wanted in the countries experiencing the Arab Spring revolutions. The difference is that Iranians have been demonstrating and protesting issues with their government in the years before 2011. Political tensions in Iran have been high since the 2009 elections and even earlier. During the 2009 election in Iran, students and other individuals used the Internet, Twitter, and cell phones to organize a massive protest against the current president and in support of opposition candidates. The demonstrations were called the **Twitter Revolution**.

Iran is at a crossroads in the conflict between conservative Islamic fundamentalists and Islamic reformers. The government of the Islamic state is controlled by Muslim clerics who tend to be more conservative in their rulings, but the young people are mainly on the side of the democratic reformers. Young people are becoming more familiar with Western culture. For example, the unofficial holiday of Valentine's Day has become extremely popular in Iran and is celebrated by a large sector of the population, mainly young people. In an effort to curb the influence of Western culture, on February 13, 2011, the government of Iran officially banned all symbols or activities associated with Valentine's Day. One claim was that the day was named after a Christian martyr and therefore was not supportive of Islam.

Figure 39. Main mountains and desert areas in Iran.

Economic Resources

Iran has abundant oil and natural gas reserves that are being exploited to form the base of their economy. Iran holds about 15 percent of the world's reserves of natural gas, which is second only to Russia. In 2010, the country was the fourth-largest oil exporter in the world and held about 10 percent of the world's known oil reserves. The UN has classified Iran's economy as semi-developed. The government has taken control of the oil and natural gas industry and implemented a type of central planning over many major businesses. Small-scale agriculture and village trading activities are not usually owned by the government. The Caspian Sea provides for fishing and has oil reserves under the seabed. Oil and gas revenues make up most of the state's income. However, fluctuations in commodity prices have resulted in a more volatile income stream, and Iran's manufacturing base has been increasing to support a more diversified economy.

Figure 40. Twitter revolution in Iran. During the 2009 election in Iran, students and others used the Internet, Twitter, and cell phones to organize a protest against the current president and in support of opposition candidates.

Key Takeaways

1. Iraq was ruled by Saddam Hussein from 1979 to 2003. Iraq invaded Kuwait in 1990, which prompted the First Persian Gulf War. Hussein's use of weapons of mass destruction provoked the US invasion in 2003, which eventually removed Hussein from power.

2. The majority Arab population in Iraq shares the country with the Kurds in the north. Sunni and Shia groups have divided the country and opposed each other since the US invasion of 2003.

3. Turkey has a relatively stable, democratically elected government. Turkey's physical geography provides for large supplies of food crops and hydroelectric energy.

4. Iran is a religious state that has often been at war or clashed with its Middle Eastern neighbors. Reformers and many young people in Iran would like to see the country become more democratic and more open to personal freedoms.

10.7 Central Asia and Afghanistan

Learning Objectives

1. Understand that Central Asia is a landlocked region that receives little rainfall and has to rely on water from major rivers flowing from the mountains in the east.

2. Summarize how Central Asia has been transitioning from a Soviet-dominated region to independent states and what has been occurring in the various states to adapt to the new economic environment.

3. Describe how the Aral Sea has been affected by the practices of water use in the region and the environmental consequences that have resulted from water use policies.

4. Explain the geopolitical history of Afghanistan and why this area has been so difficult to govern under a central government.

5. Learn why there is continual conflict in Afghanistan between Western military forces and local Taliban insurgents.

6. Understand the principle that globalization of the economy forces political units to compete over natural resources.

A. Central Asia (a.k.a. Turkestan)

Central Asia is a region in the Asian continent that extends from the mountains of western China to the shores of the Caspian Sea. Pakistan and Iran create the southern border of the region, and the vast expanse of Russia is to the north. Afghanistan is considered a part of the region even though it was never a formal part of the Soviet Union. Central Asia was located on what was known as the Silk Road between Europe and the Far East and has long been a crossroads for people, ideas, and trade.

Central Asia has an extremely varied geography, including high mountain passes through vast mountain ranges, such as the **Tian Shan**, **Hindu Kush**, and the **Pamirs**. The region is also home to the vast Kara Kum and Kyzyl Kum Deserts, which dominate the interior with extensive spans of sand and desolation. The expansive treeless, grassy **steppes** that surround the desert regions are considered an extension of the steppes of Eastern Europe. Some geographers think of the Eurasian Steppes as one single, homogeneous geographical zone.

Figure 41. Central Asia, formerly part of the Soviet Union. Afghanistan is also usually included as part of Central Asia, though it was never officially part of the Soviet Union.

Under the sand and prairie grasses lay the some of the most extensive untapped reserves of gas and oil on the planet. Natural resources are the main attraction of the region driving the economic forces that determine the development patterns of individual countries. Multinational corporations have vigorously stepped up their activity in the region.

The political systems are adjusting from the old Soviet Union's socialist policies to new democratic systems that are subject to high levels of authoritarian rule and corruption in business and politics.

The five countries of **Kazakhstan**, **Uzbekistan**, **Turkmenistan**, **Tajikistan**, and **Kyrgyzstan** were part of the former Soviet Union until its breakup in 1991. Today, with **Afghanistan**, they are independent countries that make up the region called Central Asia. The term *stan* means "land of," so, for example, Uzbekistan is the land of the Uzbeks. Central Asia is also referred to as Turkestan because of the Turkish influence in the region. The people of Turkey did not originate from the Middle East; they originated from northern Asia. They swept through Central Asia and dominated the region on their way to the Middle East. The Turkish language and heritage have had the most significant impact on the people of Central Asia. Turkmenistan's name is another reminder of the Turkish connection; it means "the land of the Turkmen."

Most of the groups of Central Asia were nomadic peoples who rode horses and herded livestock on the region's vast steppes. This way of life continued until the 1920s, when the Soviet Union forced many of the groups to abandon their lifestyle and settle on collective farms and in cities. Most of the people of Central Asia continue to identify culturally with their nomadic past. Central Asians who live in cities often demonstrate a mix of local and Russian culture in terms of dress and food because of the large influx of Russian populations in the region. More than six million Russians and Ukrainians were resettled into Central Asia during Soviet rule. Russian is often used as a lingua franca.

One of the primary ways in which people distinguish themselves culturally is through religious practices. Despite the area being part of the Soviet Union, where religious activities were discouraged, Islam was and still is the dominant religion. Most Central Asian Muslims are Sunnis.

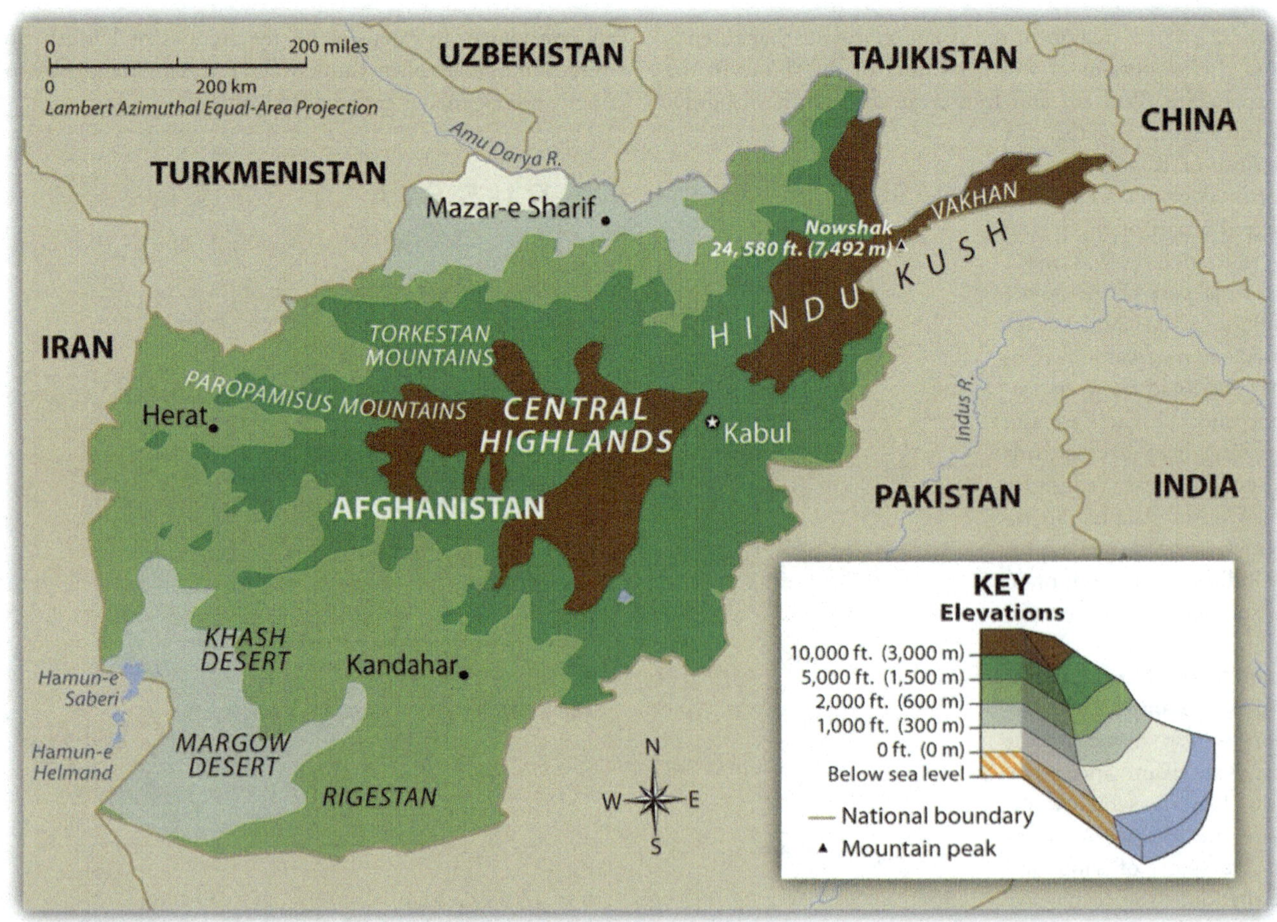

Figure 42. Afghanistan.
Kabul is the capital, and the southern city of Kandahar is the second-largest city.

B. Afghanistan

Present-day Afghanistan has been conquered by the likes of Genghis Khan, Alexander the Great, and the Mogul Empire and was a buffer zone for colonial feuds between Russia and British India. The high central mountain range of the Hindu Kush dominates the country and leaves a zone of well-watered fertile plains to the north and a dry desert region to the south. Afghanistan is a remote region without access to the sea and acts as a strategic link between the Middle East and the Far East.

The Soviet Invasion and the Taliban

In 1979, the Soviet Union took advantage of ongoing ethnic warfare in Afghanistan to inject itself into the country. The Soviets pushed in from the north and occupied much of Afghanistan until they completely withdrew in 1989. During the Soviet occupation, the United States supported anti-Communist resistance groups such as the **Mujahideen** with money, arms, and surface-to-air missiles. The missiles were instrumental in taking out Soviet aircraft and MiG fighters, which caused a critical shift in the balance of power in the war. One of the major connections between the Central Intelligence Agency (CIA) and Mujahideen was a Saudi national named **Osama bin Laden**. Support from the CIA through bin Laden to the Mujahideen was instrumental in defeating the Soviets.

The power vacuum left by the retreating Soviets allowed conflicts to reemerge between the many ethnic factions in Afghanistan. Dozens of languages are spoken in Afghanistan; the top two are Pashtu and Afghan Persian-Dari. There are also a dozen major ethnic groups; the top two are Pashtun and Tajik. The groups regularly fight among themselves, but they have also been known to form alliances. Rural areas are usually led by clan leaders who are not part of any official arm of a national government. Afghanistan is a place where forming any national unity or identity is not easy. The national government in the capital city of Kabul has little influence in the country's rural regions.

The Soviet invasion brought the internally warring factions together for a short period to focus on the Soviet threat. Chaos and anarchy thrived after the Soviet forces withdrew, but the Islamic fundamentalist group known as the **Taliban** came forward to fill the power vacuum. One objective of the Taliban was to use Islam as a unifying force to bring the country together. The problem with that concept was that there was much diversity in how Islam was practiced by the numerous local groups. Many of the factions in Afghanistan opposed the Taliban; one such group being the **Northern Alliance**, which was an association of groups located in the

Figure 43. Propaganda Poster in Afghanistan with Image of Osama bin Laden. Osama bin Laden was killed by US Navy Seals in 2011.

Figure 44. Operation Enduring Freedom in Afghanistan. Competing groups and the rough terrain make keeping the peace in Afghanistan difficult. The United States turned control over to NATO in 2006. In this photo, soldiers representing the 561st Military Police Company, Fort Campbell, Kentucky, are speaking with local Afghans about insurgent activity near Bagram, Afghanistan.

Figure 45. Voting in Afghanistan. After voting in Afghanistan, individuals dipped their fingers in ink to indicate they had voted and were not allowed to vote more than once. This photo indicates that women were also allowed to vote in Afghanistan in the 2005 elections for provincial councils and parliamentary positions.

northern portion of the country. The civil war between the Taliban and those that opposed them resulted in the deaths of more than fifty thousand people by 1996 when the Taliban emerged to take power in Kabul. The Taliban is a Sunni Muslim group that adheres to strict Islamic laws under the Wahhabi branch of the faith similar to that of Saudi Arabia. Under Taliban rule, women were removed from positions in hospitals, schools, and work environments and had to wear burkas (also spelled burqas) and be covered from head to toe, including a veil over their faces. Violators were either beaten or shot. The Taliban brought a sense of militant order to Kabul and the regions under their control. Various factions such as the Northern Alliance did not share the Taliban's strict Islamic views and continued to oppose their position in power.

Al-Qaeda and the US Invasion

After the war against the Soviet Union was over, the US role in Afghanistan diminished. The groups that the United States had supported continued to vie for power in local conflicts. Osama bin Laden remained in Afghanistan and established training camps for his version of an anti-Western resistance group called **al-Qaeda**. Just as he had opposed the Soviet Union, he opposed the United States, even though the United States had supported him against the Soviets. The Saudi government allowed the United States to establish military bases in Saudi Arabia during Operation Desert Storm in the Persian Gulf War, and this was one reason for bin Laden's opposition; he believed that non-Muslims should not be on the same ground as the Muslim holy sites of Mecca and Medina.

Figure 46. Vastness of Afghanistan. In this photo is the Lataband Road between Kabul and Surobi. The dry, treeless expanse of this region is home to Afghanistan's vast mineral wealth.

The 9-11 attack in New York City was traced back to al-Qaeda and bin Laden, who was residing in Afghanistan at the time. In a military action dubbed Operation Enduring Freedom, the United States invaded Afghanistan in 2001, removed the Taliban from power, and dismantled the al-Qaeda training camps. Eventually, in 2011, bin Laden was killed by a US Navy Seal at a compound in Pakistan, where bin Laden had been in hiding.

Resources and Globalization

In 2010, a US government report indicated that vast amounts of mineral wealth were discovered in Afghanistan by American geologists and Pentagon officials. Enormous deposits of iron, copper, gold, cobalt, and rare industrial minerals such as lithium are reported to be present in Afghanistan. Total reserves are unknown or have not been released but if extracted would result in trillions of dollars of economic gain for the country. Lithium is highly sought after and is used in the manufacturing of batteries, computers, and electronic devices. The report indicated that Afghanistan could become the world's premier mining country.

Discovery of vast resources helps place the war in Afghanistan in perspective with respect to global competition over the control of resources. Afghanistan does not have a long-standing tradition of mining. Agriculture has been the main focus of economic activity for the rural communities. A newfound potential for mineral wealth will change the future of Afghanistan. It will be interesting to watch how Afghanistan adapts to and benefits from the discovery of previously unknown resources.

Key Takeaways

1. Central Asia is a landlocked region that receives little rainfall. Two large desert regions are located at the region's core. Vast grasslands called steppes dominate the northern sector. High mountains to the east provide a border between Central Asia and China.

2. Central Asia (a.k.a. Turkestan) was dominated by the Soviet Union during the twentieth century. The transition to independence has challenged the region to adjust to changes in political and economic systems.

3. The demise of the Aral Sea is being caused by the diversion of water from the two rivers flowing into the sea. A once-thriving fishing industry has been destroyed and environmental damage has been catastrophic.

4. Armed conflicts in Afghanistan continue between Western military forces and the Taliban over the control of the country. Clan leaders are a main component of the political fabric of the country. The diversity and fragmentation of the country make it almost impossible to govern.

5. Afghanistan is one of the poorest countries in the world based on data on standards of living. However, the country has one of the largest deposits of valuable minerals and ores in the world waiting to be extracted. The deposits are the target of multinational corporations around the globe.

6. The following is a summary of trends in Central Asia:

- Shift from Soviet Union to independent states
- Rise in authoritarian governments and corruption
- Increase of cultural influence of Islamic institutions
- Global economic focus on extractive activities
- Increase in military activity over control of resources
- Decline in agriculture as an economic base
- Continued concern for environmental problems

End-of-Chapter Summary

- North Africa, Southwest Asia, and Central Asia (Turkestan) are included in a the realm of North Africa and Southwest Asia, which is dominated by the religion of Islam, arid type B climates, and the export of oil and natural gas. Large, expansive deserts make up most of the physical areas of all three regions of the realm. Water is of particular importance, as most of the population has traditionally made a living from agriculture. Division, warfare, and conflict have been constant elements. The Arab Spring of 2011 included a wave of mass citizen demonstrations in many Arab countries where the people demanded political and economic reform.

- The three main monotheistic religions of Judaism, Christianity, and Islam emerged from the Middle East and can be traced back to a common patriarch, Abraham. Most of the people in the realm are Muslims. The Five Pillars of Islam create a guide for Muslims to live by. Mecca, the holiest city for Muslims worldwide, is where the prophet Muhammad started the religion. The two main divisions of Islam are Sunni (84 percent) and Shia (15 percent). Other smaller divisions exist, such as Sufi.

- North Africa includes the countries bordering the Mediterranean Sea and Sudan on the Red Sea. Most of the people in North Africa are Muslims. The high-population areas are along the coast or along the Nile River. The Maghreb is included in the region with the Atlas Mountains in the west. The Atlas Mountains trap moisture, which provides precipitation for agriculture in the mountain valleys. The countries of North Africa have all witnessed political uprisings in which the people have demanded democratic reforms and increased civil rights. Egypt is the most populous Arab country and the cornerstone of the Arab world.

- The African Transition Zone is where the dry, arid conditions of the Sahara Desert meet the tropical conditions of Equatorial Africa. Islam is dominant in the north, and Christianity and tribal religions are dominant in the south. Sudan is one place where the two sides have clashed in warfare and division. Tribal Christian southern Sudan separated from the Arab Muslim north to create the independent country of South Sudan. Ethnic cleansing has been witnessed in the western part of Sudan in the region of Darfur.

- Palestine was divided in half in 1945 into an Arab state and a Jewish state. The Jewish state became the nation of Israel in 1948. Since that time, there has been a constant struggle between the mainly Arab Palestinians and the Jewish State of Israel. The West Bank and the Gaza Strip are Palestinian regions that have worked to be recognized with equal sovereignty. The solution to this division is debated; both sides have claims that are incompatible with each other. The neighboring Arab states of Jordan, Syria, and Lebanon have their own unique political situations.

- The Arabian Peninsula includes a number of states ruled by a single king, sultan, emir, or sheik, except Yemen, which has a democratically elected president. Saudi Arabia is the home of the two main cities of Islam, Mecca and Medina. Saudi Arabia is the largest country on the peninsula and has about 25 percent of all the known oil reserves in the world. Oil and natural gas are the main export products and the main source of income for the region.

- Most people in Turkey are of Turkish ethnicity, and most people in Iran are of Persian ethnicity. The country of Turkey is what remains from the once expansive Ottoman Empire. Iran is a country of mountains and deserts that was once the center of the Persian Empire. Turkey has a portion of its country in Europe, and its largest part includes the expansive Anatolian Plateau in Asia. Turkey is a democratic republic with elected public officials. Iran is a religious state in which the Islamic leadership controls who can be included in the government. Both countries have large populations of about seventy million each.

- Iraq was at the center of the Fertile Crescent of Mesopotamia, which spawned early human civilizations. Iraq was invaded by the United States in 2003 and has been working to establish a democratic government. The country is divided between a Shia majority and a Sunni minority. It is an Arab country with a large minority group of Kurds in the north. Iraq has enormous oil reserves that have not yet been developed for exploitation, which has attracted the attention of the world's energy-dependent core regions.

- Central Asia (Turkestan) is a landlocked region with mainly type B climates with cooler temperatures in the higher elevations of the eastern mountains. The prominently Muslim region has been transitioning from a Soviet-controlled domain to independent states. Some countries are making the transition with less difficulty than others. The Aral Sea has been one of the major environmental disasters in modern history. Globalization activities have targeted the region for its immense amount of natural resources—mainly oil, natural gas, and minerals.

- Afghanistan has been conquered by many empires and was a buffer state between British and Russian colonial efforts. The country has a dry, rugged terrain and a devastated economy. The people are poor with little infrastructure. The many tribal groups have made it difficult for a centralized government to be effective. The United States invaded Afghanistan in 2001 to remove al-Qaeda training camps run by Osama bin Laden. North Atlantic Treaty Organization (NATO) forces have taken over the mission in Afghanistan to defeat the Taliban extremists and support stability for the country. Afghanistan has enormous amounts of mineral resources that are in high demand by the world's core industrial regions.

Chapter 11

South Asia

Introduction

Of the world's seven continents, Asia is the largest. Its physical landscapes, political units, and ethnic groups are both wide-ranging and many. Besides Russia, Southwest Asia, and Central Asia, which are addressed in other chapters, Asian regions include South Asia, East Asia, and Southeast Asia.

The principal boundaries of South Asia are the Indian Ocean, the Himalayas, and Afghanistan. The **Arabian Sea** borders Pakistan and India to the west, and the **Bay of Bengal** borders India and Bangladesh to the east. The western boundary is the desert region where Pakistan shares a border with Iran.

The realm was the birthplace of two of the world's great religions, Hinduism and Buddhism, but there are also immense Muslim populations and large groups of followers of various other religions as well. Hinduism, Islam, and Buddhism are the top three religions of South Asia. While Pakistan and Iran are both Islamic republics, each represents a significant branch of that faith; Iran is predominantly Shia, and Pakistan is mostly Sunni. Religious differences are also evident on the eastern border of the realm, where Bangladesh and India share a border with Myanmar. Bangladesh is mainly a Muslim country, while most in India align themselves with Hinduism. In Myanmar, most follow Buddhist traditions. In addition, Sikhism is a major religion in the Punjab region, which is located on India's northern border with Pakistan.

The countries of South Asia include Sri Lanka, India, Bangladesh, Bhutan, Nepal, Pakistan, and the Maldives. The **Himalayas**, separating South Asia from East Asia along the border of China's autonomous region of Tibet, are the highest mountains in the world and the dominant physical feature of the northern rim of South Asia. Other countries that share the Himalayas include Nepal, Bhutan, India, and Pakistan. Farther north along the Himalayan range, the traditional region of **Kashmir** is divided between India, Pakistan, and China. On the opposite side of the Himalayas are two island countries off the coast of southern India. The first is Sri Lanka, a large tropical island off India's southeast coast, and the other is the Republic of Maldives, an **archipelago** (group of islands) off the southwest coast of India. The Maldives comprise almost 1,200 islands that barely rise above sea level; the highest elevation is merely seven feet, seven inches. Only about two hundred islands in the Maldives are inhabited.

The balancing of natural capital and population growth is and will remain a primary issue in the realm's future. South Asia is highly populated, with about one-and-a-half billion people representing a wide range of ethnic and cultural groups. The diverse population has been brought together into political units that have roots in the realm's colonial past, primarily under Great Britain. British colonialism had a significant impact on the realm; its long-term effects include political divisions and conflicts in places such as Kashmir and Sri Lanka.

Current globalizing forces are compelling South Asian countries to establish a trade network and institute economic policies among themselves. South Asia is not one of the three main economic core areas of the world; however, it is emerging to compete in the world marketplace. Economic advancements and global trade have catapulted the countries of South Asia onto the world stage.

Some would call India a part of the **semi-periphery**, which means it is not actually in the core or in the periphery but displays qualities of both. All the same, India remains the dominant country of South Asia and shares either a physical boundary or a marine boundary with all the other countries in the realm.

Figure 1. Political map of South Asia.

11.1 Introducing the Realm

Learning Objectives

1. Summarize the realm's physical geography. Identify each country's main features and physical attributes and locate the realm's main river systems.

2. Understand the dynamics of the monsoon and how it affects human activities.

3. Outline the early civilizations of South Asia and learn how they gave rise to the early human development patterns that have shaped the realm.

4. Describe how European colonialism impacted the realm.

5. Learn about the basic demographic trends the realm is experiencing. Understand how rapid population growth is a primary concern for the countries of South Asia.

A. The Physical Geography

The landmass of South Asia was formed by the Indian Plate colliding with the Eurasian Plate, giving rise to the highest mountain ranges in the world. Most of the South Asian landmass is formed from the land in the original Indian Plate. Pressure from tectonic action against the plates causes the Himalayas to rise in elevation by as much as one to five millimeters per year. Destructive earthquakes and tremors are frequent in this seismically active realm. The great size of the Himalayas has intensely influenced the beliefs and traditions of the people in the realm. Some of the mountains are considered sacred to certain religions that exist here.

The Himalayan Mountains dominate the physical landscape in the northern region of South Asia. **Mt. Everest** is the highest peak in the world, at 29,035 feet. Three key rivers cross South Asia, all originating from the Himalayas. The **Indus River**, which has been a center of human civilization for thousands of years, starts in Tibet and flows through the center of Pakistan. The **Ganges River** flows through northern India, creating a core region of the country. The **Brahmaputra River** flows through Tibet and then enters India from the east, where it meets up with the Ganges in Bangladesh to flow into the Bay of Bengal.

While the northern part of this region includes some of the highest elevations in the world, the Maldives in the south has some of the lowest elevations, some barely above sea level. The coastal regions in southern Bangladesh also have low elevations. When the summer wet **monsoon** arrives every year, there is heavy flooding and its effect on the infrastructure of the region is disastrous. The extensive **Thar Desert** in western India and parts of Pakistan, on the other hand, does not receive monsoon rains. In fact, much of southwest Pakistan—a region called **Baluchistan**—is dry, with desert conditions.

The mountains on the border between Pakistan and Afghanistan extend through Kashmir and then meet up with the high ranges of the Himalayas. The Himalayas create a natural barrier between India and China, with the kingdoms of Nepal and Bhutan acting as buffer states with Tibet. Farther south along the east and west coasts of India are shorter mountain ranges called ghats.

The **Western Ghats** reach as high as eight thousand feet, but average around three thousand feet. These ghats are home to an extensive range of biodiversity. The **Eastern Ghats** are not as high as the Western Ghats, but have similar physical qualities. The ghats provide a habitat for a wide range of animals and are also home to large coffee and tea estates. The **Deccan Plateau** lies between the Eastern and Western Ghats. The **Central Indian Plateau** and the **Chota-Nagpur Plateau** are located in the central parts of India, north of the two Ghat ranges. The monsoon rains ensure that an average of about fifty-two inches of rain per year falls on the Chota-Nagpur Plateau, which has a tiger reserve and is also a refuge for Asian elephants.

Figure 2. Trekking trail on the way to Mt. Everest in the Himalayas of Northern Nepal. Mt. Everest is the world's highest peak at 29,035 feet. The Himalayas are the highest mountain chain in the world and create a natural border between South Asia and China.

B. The Monsoon

A **monsoon** is a wind that blows steadily from on direction for part of the year, and then reverses direction for the other part of the year. This seasonal reversal of winds causes South Asia to have a dry period in the winter (the **dry monsoon**), when winds blow from land to sea, and a wet period in the summer (the **wet monsoon**), when the winds blow from water to land. The summer monsoon rains—usually falling heavily between June and September—feed the rivers and streams of South Asia and provide the water needed for agricultural production. In the summer, the continent heats up, with the Thar Desert fueling the system. The rising hot air creates a vacuum that pulls in warm moist air from the Bay of Bengal and the Indian Ocean. This action shifts moisture-laden clouds over the land, where the water is precipitated out in the form of rain.

The summer monsoon rains bring moisture to South Asia right up to the Himalayas. As moisture-laden clouds rise in elevation in the mountains, the water vapor condenses in the form of rain or snow and feeds the streams and basins that flow into the major rivers, such as the Brahmaputra, Ganges, and Indus. The Western Ghats creates a similar system in the south along the west coast of India. Some parts of Bangladesh and eastern India receive over 30 feet of rain during the wet monsoon season, and some areas experience severe flooding. The worst-hit places are along the coast of the Bay of Bengal, such as in Bangladesh. There is less danger of flooding in western India and Pakistan, because by the time the rain clouds have moved across India they have lost their moisture. Desert conditions are evident in the west, near the Pakistan border in the great Thar Desert. On average, less than ten inches of rain falls per year in this massive desert. On the northern rim of the region, the height of the Himalayas restricts the warm moist monsoon air from moving across the mountain range. The Himalayas act as a precipitation barrier and create a strong rain shadow effect for Tibet and Western China. The summer monsoon is responsible for much of the rainfall in South Asia.

By October, the system has run its course and the wet monsoon season is generally over. In the winter, the cold, dry air above the Asian continent blows from northeast toward the southwest, and the winter monsoon is characterized by cool, dry winds coming from the north. South Asia experiences a dry season during the winter months. A similar pattern of rainy summer season and dry winter season is found in other parts of the world, such as southern China and some of Southeast Asia.

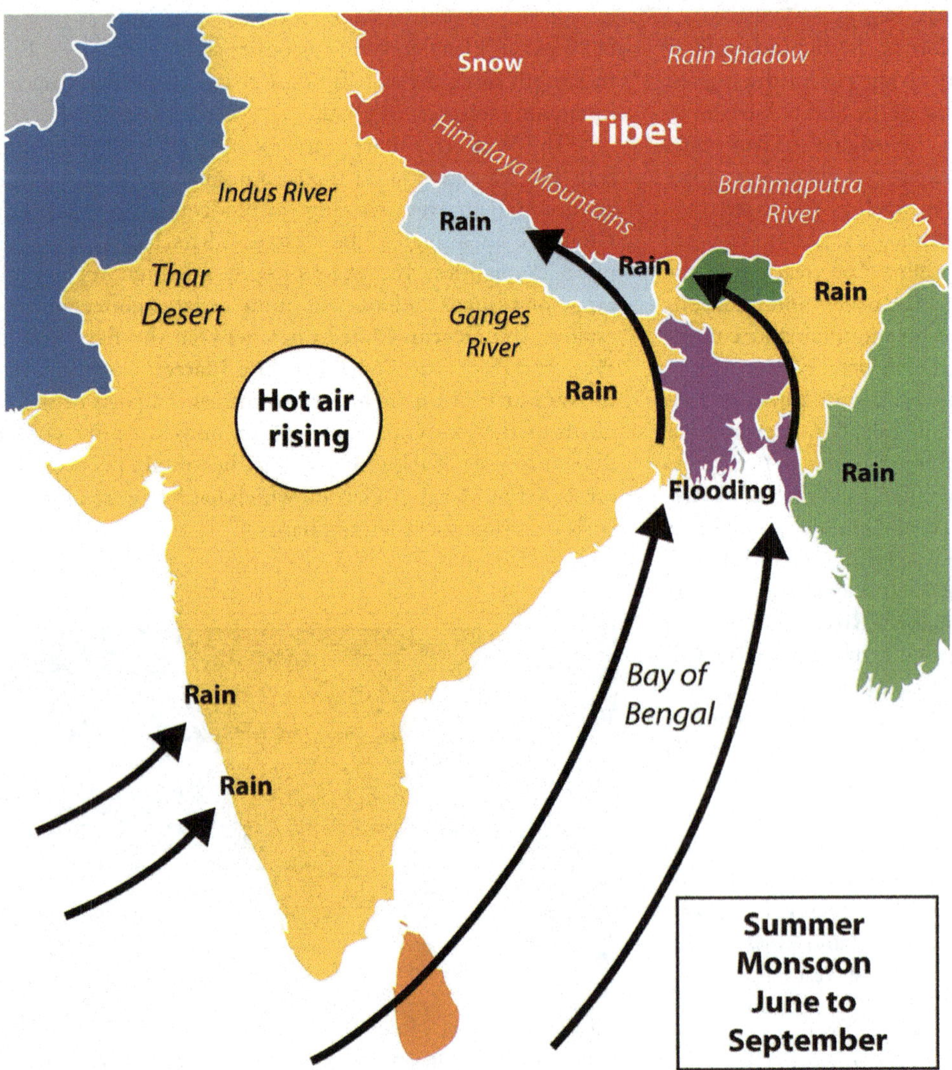

Figure 3. The wet monsoon system in South Asia.

C. Colonialism in South Asia

The force of colonialism was felt around the world, including in South Asia. South Asia provides an excellent example of colonialism's role in establishing most of the current political borders in the world. From the sixteenth century onward, ships from colonial Europe began to arrive in South Asia to conduct trade. The **British East India Company** was chartered in 1600 to trade in Asia and India. They traded in spices, silk, cotton, and other goods. Later, to take advantage of conflicts and bitter rivalries between kingdoms, European powers began to establish colonies. Britain controlled South Asia from 1857 to 1947.

The British withdrew from South Asia in 1947. Local resistance and the devastating effects of World War II meant the British Empire could not be controlled as it once was. Great Britain pulled away from empire building to focus on its own redevelopment. Upon the British withdrawal from India, Britain realized the immense cultural differences between the Muslims and Hindus and created political boundaries based on those differences. **West Pakistan** was a predominantly Muslim area carved out of western India, which was mostly Hindu; **East Pakistan** was a predominantly Muslim area carved from eastern India. However, the new borders separating Hindu and Muslim majorities ran through population groups, and some of the population now found itself to be on the wrong side of the border. The partition grew into a tragic civil war, as Hindus and Muslims struggled to migrate to their country of choice. More than one million people died in the civil war, and over 15 million refugees fled, with Muslims leaving India for Pakistan (either West or East), and Hindus leaving Pakistan for India. The **Sikhs**, who are indigenous to the **Punjab** region in the middle, also suffered greatly. Some people decided not to migrate, which explains why India has the largest Muslim population of any non-Muslim state.

Another civil war would erupt in 1971 between West Pakistan and East Pakistan. When the partitioning of South Asia occurred in 1947, they operated under the same government, despite having no common border and being over one thousand miles apart and populated by people with no ethnic or linguistic similarities. The civil war lasted about three months and resulted in the creation of the sovereign countries of Pakistan (formerly West Pakistan) and Bangladesh (formerly East Pakistan).

Language is probably one of the more pervasive ways Europeans affected South Asia. In modern-day India and Pakistan, English is the language of choice in secondary education (English-medium schools). It is often the language used by the government and military. Unlike many other Asian countries, much of the signage and advertising in Pakistan and India is in English, even in rural areas. Educated people switch back and forth, using English words or entire English sentences during conversation in their native tongue. Some scholars have termed this Hinglish or Urglish as the base languages of northern India and Pakistan are Hindi and Urdu, respectively.

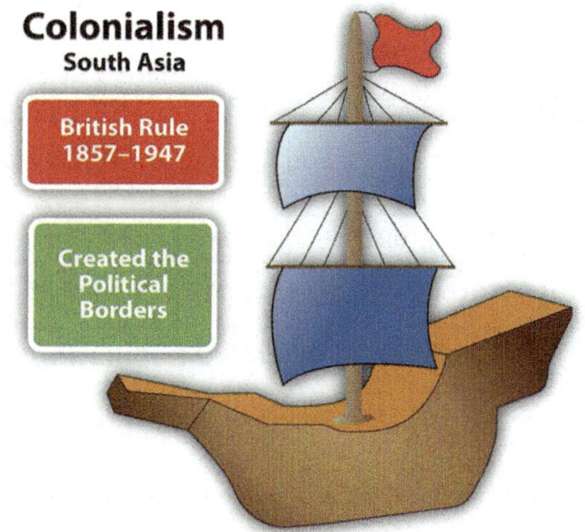

Figure 4. British colonialism in South Asia. British colonialism in South Asia began in 1857 and lasted until 1947.

D. Population in South Asia

South Asia has three of the ten most populous countries in the world. India is the second largest in the world and will likely surpass China within the next decade to become the world's largest. Pakistan and Bangladesh are numbers six and eight, respectively. Large populations are a product of a high fertility rate. The rural population of South Asia has traditionally had large families. Religious traditions do not necessarily support anything other than a high fertility rate. On the other hand, the least densely populated country in South Asia is the Kingdom of Bhutan. Bhutan has a population density of only fifty people per square mile. Bhutan is mountainous with little arable land. More than a third of the people in Bhutan live in an urban setting. Population growth for the realm is a serious concern. An increase in population requires additional natural resources, energy, and food production, all of which are in short supply in many areas.

South Asia's growing population has placed exceedingly high demands on agricultural production. The amount of area available for food production (known as **arable land**) compared to the total population may be a more helpful indicator of whether or not a country is overpopulated than simply using the average population density. For example, large

Rank	Country name	Population in millions	Total population density	Physiologic density	Fertility rate	Population growth rate (%)	Doubling time in years**	Percentage urban	Gross domestic product per capita ($)
1	China	1,336	361	2,405	1.54	0.49	143	47	7,600
2	India*	1,189	937	1,912	2.62	1.34	52	30	3,500
3	United States	313	84	468	2.06	0.96	73	82	47,200
4	Indonesia	245	331	3,013	2.25	1.07	65	44	4,200
5	Brazil	203	62	884	2.18	1.13	62	87	10,800
6	Pakistan*	187	604	2,414	3.17	1.57	45	36	2,500
7	Bangladesh*	158	2,852	5,186	2.60	1.57	45	28	1,700
8	Nigeria	155	435	1,319	4.73	1.94	36	50	2,500
9	Russia	138	21	301	1.42	−0.47		73	15,900
10	Japan	126	867	7,225	1.21	−0.28		67	34,000
11	Mexico	113	149	1,149	2.29	1.10	64	78	13,900
41	Nepal*	29	525	3,379	2.47	1.59	44	19	1,200
57	Sri Lanka*	21	862	6,001	2.2	0.93	75	14	5,000
165	Bhutan*	0.700	50	1,697	2.2	1.2	58	35	5,500
176	Maldives*	0.400	3,438	26,194	1.81	−0.15		40	6,900

*** Countries noted with an asterisk are part of South Asia**

**** Empty cell indicates a negative population doubling time.**

Table 1. Demographics of South Asia and the World's Most Populous Countries.

portions of Pakistan are deserts and mountains that do not provide arable land for food production. India has the Thar Desert and the northern mountains. Nepal has the Himalayas. The small country of the Maldives, with its many islands, has almost no arable land. The number of people per square mile of arable land, which is called the **physiologic density,** can be a better indicator of whether or not a country is overpopulated. Total population densities are high in South Asia, but the physiologic densities are even more astounding. In Bangladesh, for example, more than five thousand people depend on every square mile of arable land. In Sri Lanka the physiologic density reaches to more than 6,000 people per square mile, and in Pakistan it is more than 2,400. The data are averages, which indicate that the population density in the fertile river valleys and the agricultural lowlands might be even higher.

The population of South Asia is relatively young. In Pakistan about 35 percent of the population is under the age of fifteen, while about 30 percent of India's almost 1.2 billion people are under the age of fifteen. Many of these young people live in rural areas, as most of the people of South Asia work in agriculture and live a subsistence lifestyle. As the population increases, the cities are swelling due to the large influx of migrants arriving from rural areas. **Rural-to-urban migration** is extremely high in South Asia and will continue to fuel the expansion of the urban centers into some of the largest cities on the planet. The rural-to-urban shift that is occurring in South Asia also coincides with an increase in the region's interaction with the global economy.

Figure 5. Crowded street in New Delhi, India. Urban areas of South Asia are expanding rapidly.

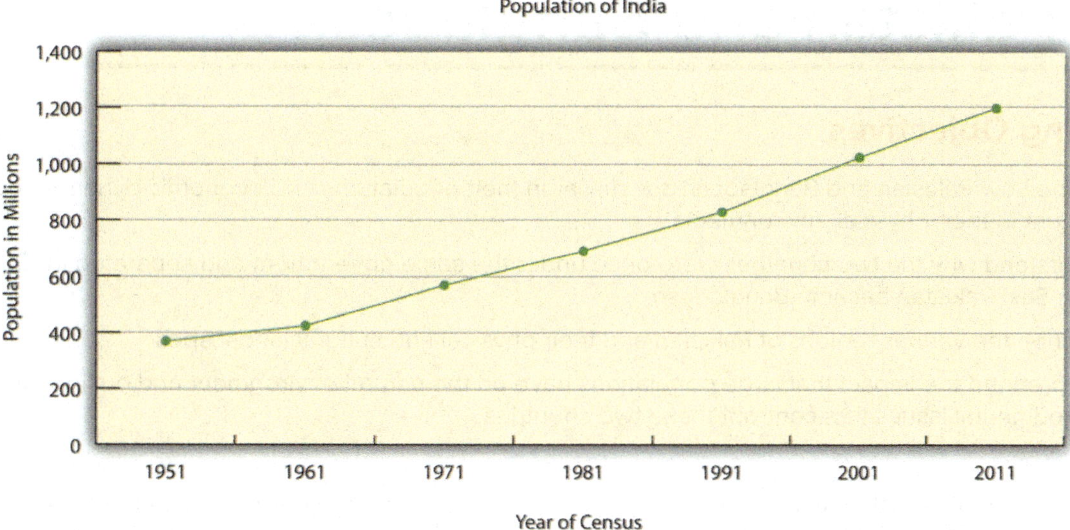

Figure 6. Population growth in India.

The South Asian countries are transitioning through the five stages of the demographic transition. The more rural agricultural regions are mostly still in stage 2. The realm experienced rapid population growth during the latter half of the twentieth century. As death rates declined and birth rates and family size remained high, the population swiftly increased. India, for example, grew from fewer than four hundred million in 1950 to more than one billion at the turn of the century. The more urbanized areas are transitioning into stage 3 and experiencing significant rural-to-urban shift. Large cities such as **Mumbai** (Bombay) have sectors that are in the latter stages of the demographic transition because of their urbanized work force and higher incomes. Family size is decreasing in the more urbanized areas and in the realm as a whole, and demographers predict that eventually the population will stabilize.

At the current rates of population growth, the population of South Asia will double in about fifty years. Doubling the population of Bangladesh would be the equivalent of having the entire 2011 population of the United States (more than 313 million people) all living within the borders of the US state of Wisconsin. The general rule of calculating doubling time for a population is to take the number seventy and divide it by the population growth rate. For Bangladesh, with a growth rate of 1.57%, the doubling time would be $70 \div 1.57 = 45$ years. The doubling time for a population can help determine the economic prospects of a country or region. South Asia is coming under an increased burden of population growth. If India continues at its current rate of population increase, it will double its population in fifty-two years, to approximately 2.4 billion. Because the region's rate of growth has been gradually in decline, this doubling time is unlikely. However, without continued attention to how the societies address family planning and birth control, South Asia will likely face serious resource shortages in the future.

Key Takeaways

1. All the South Asian countries border India by either a physical or a marine boundary. The Himalayas form a natural boundary between South Asia and East Asia (China). The realm is surrounded by deserts, the Indian Ocean, and the high Himalayan ranges.

2. The summer monsoon arrives in South Asia in late May or early June and subsides by early October. The rains that accompany the wet monsoon account for most of the rainfall for South Asia. Water is a primary resource, and the larger river systems are home to large populations.

3. The Indus River Valley was a location of early human civilization. The large empires of the realm gave way to European colonialism. The British dominated the realm for ninety years from 1857 to 1947 and established the main boundaries of the realm.

4. Population growth is a major concern for South Asia. The already enormous populations of South Asia continue to increase, challenging the economic systems and depleting natural resources at an unsustainable rate.

11.2 Pakistan and Bangladesh

Learning Objectives

1. Outline how Pakistan and Bangladesh are similar in their populations and economic dynamics but different in their physical environments.

2. Understand why the two countries were once under the same government and separated in 1971, when East Pakistan became Bangladesh.

3. Describe the various regions of Pakistan and their physical and cultural landscapes.

4. Comprehend the impact that large populations have on the natural environment and outline the main environmental issues that confront these two countries.

Since 1971, Pakistan and Bangladesh have existed as two separate countries physically divided by India. The two countries share a number of attributes. They both have Muslim majorities and high population densities. Their populations are youthful and mainly rural; agriculture is the main economic activity in each country. Rural-to-urban shift is a major trend affecting urban development. Infrastructure is lacking in many areas of each country. These similar factors indicate that both Pakistan and Bangladesh will face comparable challenges in providing for their large populations and protecting their natural environments.

The Muslim League was responsible for the formation of a united Pakistan, a predominantly Muslim state for South Asian Muslims. Pakistan was created from the former Indian territories of Sindh (Sind), North West Frontier Provinces, West Punjab, Baluchistan, and East Bengal. East Bengal, on the eastern side of India, was known as East Pakistan, while the remainder, separated by more than one thousand miles, was known as West Pakistan.

East and West Pakistan, administered by one government, became independent of their colonial master in 1947, when Britain was forced out. Pakistan (East and West) adopted its constitution in 1956 and became an Islamic republic. In 1970, a massive cyclone hit the coast of East Pakistan and the central government in West Pakistan responded weakly to the devastation. The Bengali populations were angered over the government's lack of consideration for them in response to the cyclone and in other matters. The Indo-Pakistan War changed the situation. In this war, East Pakistan, with the aid of the Indian military, challenged West Pakistan and declared independence to become Bangladesh in 1971. West Pakistan became the current country of Pakistan.

Figure 7. Provinces and Territories of Pakistan. The two core areas of Pakistan are the Punjab and the Indus River Valley.

Figure 8. Pakistan.

A. Pakistan

Much of Pakistan's land area consists of either deserts or mountains. The high Himalayan ranges border Pakistan to the north. The lack of rainfall in the western part of the country restricts agricultural production in the mountain valleys and near the river basins. The Indus River flows from the northern part of the Karakoram mountains and creates a large, fertile flood plain that comprises much of eastern Pakistan. Pakistan has traditionally been a land of farming. The **Indus River Valley** and the **Punjab** are the dominant core areas where most of the people live and where population densities are remarkably high.

Approximately 60 percent of the population lives in rural areas and makes a living in agriculture. Most of the people are economically quite poor by world standards. In spite of the rural nature of the population, the fertility rate has decreased from seven to four in recent decades. Nevertheless, the population has exploded from about 34 million in 1951 to just over 200 million as of 2016. About half of the population is under the age of twenty; 35 percent is under the age of fifteen. A lack of adequate medical care, an absence of family planning, and the low status of women have created an ever-increasing population, which will have dire consequences for the future of Pakistan. Service and infrastructure to address the needs of this youthful population are not available to the necessary degree. Schools and educational opportunities for children are rarely funded at the needed levels. As of 2010, only about 50 percent of Pakistan's population was literate.

The capital of Pakistan when it was under British colonialism was Karachi, a port city located on the Arabian Sea. To establish a presence in the north, near Kashmir, the capital was moved to **Islamabad** in 1960. This example of a **forward capital** was an expression of geopolitical assertiveness by Pakistan against India. It also indicates the importance of Islam in the country. Islamabad means, "City of Islam," and the city is the cultural focus of Islam in South Asia. The *lingua franca* of the country for the business sector and the social elite continues to be English, even though Urdu is considered the national language of Pakistan and is used as a lingua franca in many areas. More than sixty languages are spoken in the country. There are as many ethnic groups in Pakistan as there are languages. The three most prominent ethnic groups are Punjabis, Pashtuns, and Sindhis.

Regions of Pakistan

The three main physical geographic regions of Pakistan are the Indus River Basin, the Baluchistan Plateau, and the northern highlands. These physical regions are generally associated with the country's main political provinces. The four main provinces include the **Punjab, Baluchistan, Sindh,** and **Khyber Pakhtunkhwa (North West Frontier)**. To the north is the disputed region of Kashmir known as the **Northern Areas.** Each of these regions represents a different aspect of the country. The North West Frontier has a series of **Tribal Areas** bordering Afghanistan that have been traditionally under their own local control. Agents under Tribal Agencies have attempted to administer some type of structure and responsibility for the areas, with little success.

The Punjab

As explained previously, the Punjab is a core area of Pakistan, and has about 60 percent of Pakistan's population. The five rivers of the Punjab border India and provide the fresh water necessary to grow food to support a large population. Irrigation canals create a water management network that provides water throughout the region. The southern portion of the Punjab includes the arid conditions of the **Thar Desert**. The northern sector includes the foothills of the mountains and has cooler temperatures in the higher elevations. The Punjab is anchored by the cities of **Lahore**, Faisalabad, and Multan. Lahore is the cultural center of Pakistan and is home to the University of the Punjab and many magnificent mosques and palaces built during its early history. The Punjab is the most industrialized of all the provinces. Manufacturing has increased with industries producing everything from vehicles to electrical appliances to textiles. The industrialization of the Punjab is an indication of its skilled work force and the highest literacy rate in Pakistan, at about 80 percent.

Figure 9. Donkey cart on busy street in Lahore, Pakistan, in the Punjab. This is an example of traditional transportation mixing with modern technology Lahore is a large city with a wide range of methods of conducting business.

Figure 10. Man with his camel in the desert region of Baluchistan in Western Pakistan.

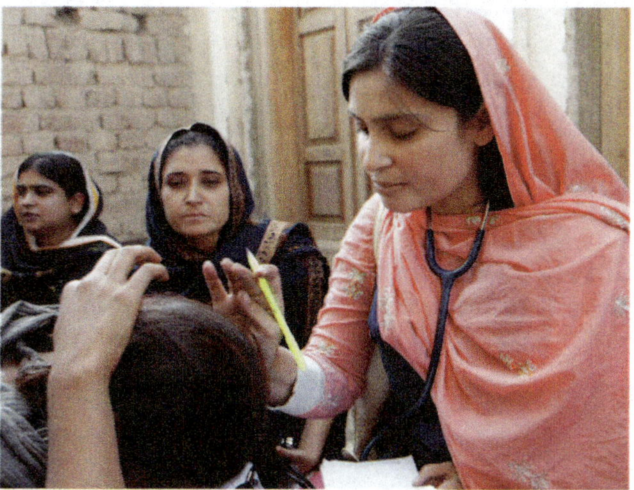

Figure 11. Female doctor examining patient from a mobile medical clinic in the Sindh Region of Pakistan.

Figure 12. Man firing AK-47 in the north west Frontier of Pakistan.

Baluchistan

Baluchistan encompasses a large portion of southwest Pakistan to the west of the Indus River. The region connects the Middle East and Iran with the rest of Asia. The landscape consists of barren terrain, sandy deserts, and rocky surfaces. Baluchistan covers about 44 percent of the entire country and is the largest political unit. The sparse population ekes a living out of the few mountain valleys where water can be found. Much of the coastal region is arid desert with sand dunes and large volcanic mountainous features.

The Sindh

The Sindh (Sind) region of the southeast is anchored by **Karachi**, Pakistan's largest city and major port. The Indus River is the border on the west and the Punjab region lies to the north. To the east of the Sindh is the border with India and the great Thar Desert. The Indus River Valley is a key food-growing area. Food crops consist of wheat and other small grains, with cotton as a major cash crop that helps support the textile industry of the region.

Rural-to-urban shift has pushed large numbers of Sindh residents into the city of Karachi to look for opportunities and employment. The central business district has a thriving business sector that anchors the southern part of the country. The city has a large port facility on the Arabian Sea. As an urban agglomeration of perhaps 20 million people or more, there are always problems with a lack of public services, law enforcement, or adequate infrastructure. The Sindh is the second-most populous region of Pakistan, after the Punjab.

Khyber Pakhtunkhwa (The North West Frontier)

The North West Frontier is a broad expanse of territory that extends from the northern edge of Baluchistan to the Northern Areas of the former Kingdom of Kashmir. Sandwiched between the tribal areas along the Afghanistan border and the well-watered lands of the Punjab, the Khyber Pakhtunkhwa Province is dominated by remote mountain ranges with fertile valleys. The famous **Khyber Pass,** a major **chokepoint** into Afghanistan, is located here. The frontier is a breeding ground for anti-Western culture and anti-American sentiments, mainly fueled by the US military activity in Afghanistan. A push for more fundamentalist Islamic law has been a major initiative of the local leaders. Support for education and modernization is minimal. The government of Pakistan has also stepped up its military actions in the region to counter the activities of the militant Islamic extremists.

The Tribal Areas

The North West Frontier borders the Tribal Areas, where clans and local leaders are standard parts of the sociopolitical structure. These remote areas have seldom been fully controlled by either the colonial governments (the British) or the current government of Pakistan. There are about seven main areas that fall under this description. Accountability for the areas has been difficult and even when the national government stepped in to exercise authority, there was serious resistance that halted any real established interaction. These remote areas are where groups such as al-Qaeda and the Taliban often find safe haven.

Northern Areas with Disputed Kashmir

Located in the high mountains of the north is the former **Kingdom of Kashmir,** a separate kingdom before the British divided South Asia. In 1947, when the British drew the boundary between India and Pakistan, the leader of Kashmir, the **maharajah,** chose not to be a part of either country but to remain independent. About 75 percent of the population in Kashmir was Muslim; the rest, including the maharajah, were mainly Hindu. This arrangement worked for a time, until the Muslim majority was encouraged by their fellow Muslims in Pakistan to join Pakistan. After a Muslim uprising, the maharajah asked the Indian military for assistance. India was more than pleased to oblige and saw it as an opportunity to oppose Pakistan one more time. Today **Kashmir** is divided, with Pakistan controlling the northern region, India controlling the southern region, and China controlling a portion of the eastern region. A cease-fire has been implemented, but outbreaks of fighting have occurred. The future of Kashmir is unclear. None of the countries involved wants to start a large-scale war, because they all have nuclear weapons.

The conflict in Kashmir is about strategic location and control of water rather than labor and resources. The Indus River flows through Kashmir from Tibet and into Pakistan. The control of this river system is critical to the survival of people living in northern Pakistan. If India were to place a dam on the river and divert the water to their side of the border, to the dry regions of the south, Pakistan could suffer a water shortage in the northern part of the country. Another aspect of the Kashmir conflict goes back to the division of Pakistan and India, which pitted Muslims against Hindus along the border region. The religious differences have come to the surface again in the conflict over the control of Kashmir. Extremist movements within Kashmir by the Muslim population have fueled the division between those who support Pakistan and those who support Hindu-dominated India.

Figure 13. The Highlands of the northern areas in Pakistan.

Figure 14. The issues with Kashmir. Pakistan controls the northern areas, India controls Jammu and Kashmir, and China controls the eastern portion, labeled Aksai Chin on this map. All three countries have nuclear weapons, and it seems apparent that none of them want to start a nuclear war.

Religion and Politics in Pakistan

Today most of the people living in Pakistan are Muslim. About 85 percent of the Muslim population in Pakistan is Sunni and about 15 percent is Shia, which is consistent with the percentages of the two Islamic divisions worldwide. Islam is considered the state religion of Pakistan. The state is a federal republic with a parliamentarian style of government. As an Islamic state following the Sharia laws of the Koran, it has been a challenge for Pakistan to try to balance instituting democratic reforms while staying true to fundamental Islamic teachings. Pakistan has held elections for government leaders, and the status of women has improved. Women have held many governmental and political positions, including prime minister. The military has been a foundation of power for those in charge. As a result of weak economic conditions throughout the country, it has been the military that has received primary attention and is the strongest institution within the government. Pakistan has demonstrated its nuclear weapons capability in recent years, which established it as a major player in regional affairs.

B. Bangladesh

Bangladesh is a low-lying country that is associated with the types of marshy environments found in tropical areas and river deltas. The region is extremely prone to flooding, particularly during the summer monsoon season because of the high amount of rainfall. One of the most important rivers of Bangladesh flows southward from the Himalayas through India and into Bangladesh. While in India, this river is known as the **Brahmaputra River,** but when it enters Bangladesh, it is known as the **Jamuna River.** It provides a major waterway for this region and empties into the **Bay of Bengal.**

Contributing to the immense flow of water through the country are the Ganges and the Meghna rivers, which join up with the Brahmaputra River near the sea. The **Ganges** flows through northern India and is a major source of fresh water for a large population before it reaches Bangladesh. The **Meghna** is a collection of tributaries within the boundaries of Bangladesh that flows out of the eastern part of the country. The Meghna is a deep river that can reach depths of over 1,600 feet with an average depth of more than one thousand feet. The hundreds of water channels throughout the relatively flat country provide for transportation routes for boats and ships that move goods and people from place to place. There are few bridges, so land travel is restricted when rainfall is heavy.

Figure 15. Bangladesh. Bangladesh has about the same area as the US state of Wisconsin. Bangladesh's population estimate in 2016 was 163 million; Wisconsin's was about 5.7 million.

Population and Globalization

Imagine a country the size of the US state of Wisconsin. Now imagine half of the entire population of the United States living within its borders. Welcome to Bangladesh. With an estimated population of about 163 million in 2016 and a land area of only about 57,000 square miles, it is one of the most densely populated countries on the planet. Most of the population in Bangladesh is rural, agriculturally grounded, and poor. The larger cities, such as the capital of **Dhaka**, have modern conveniences, complete with Internet cafes, shopping districts, and contemporary goods. The rural areas often suffer from a lack of adequate transportation, infrastructure, and public services. Poverty is common; income levels average the equivalent of a few US dollars per day. Remarkably, the culture remains vibrant and active, pursuing livelihoods that seek out every opportunity or advantage available to them.

There are many ethnic groups in Bangladesh, and many languages are spoken. The official and most widely used language in Bangladesh is **Bengali**,. Bengali is also the main language for the Indian state of West Bengal, which neighbors Bangladesh. English is used as the lingua franca among the middle and upper classes and in higher education. Many minor languages are spoken in Bangladesh and in the region as a whole. Most of the population, about 90 percent, is Muslim, with all but about 3 percent Sunni. There is a sizable minority, about 9 percent, which adheres to Hinduism, Buddhism, Christianity, or animism. The US State Department considers Bangladesh to be a moderate Islamic democratic country.

Figure 16. Street scene in Dhaka, the capital of Bangladesh.

Environmental Issues

The summer monsoons are both a blessing and a curse in Bangladesh. The blessing of the monsoon rains is that they bring fresh water to grow food. The northeast part of Bangladesh receives the highest amount of rainfall, averaging about eighteen feet per year, while the western part of the country averages only about four feet per year. Most of the rain falls during the summer monsoon season. Bangladesh can grow abundant food crops of rice and grain in the fertile deltas of the Ganges and Brahmaputra Rivers, rivers that ultimately empty into the Bay of Bengal. About 55 percent of the land area is arable and can be used for farming, but flooding causes serious damage to cropland by eroding soil and washing away seeds or crops. Every year, countless people die because of the flooding, which can cover as much as a third of the country. One of the worst flooding events in Bangladesh's history was experienced in 1998, when river flooding destroyed more than three hundred thousand homes and caused more than one thousand deaths, rendering more than thirty million people homeless.

Most parts of Bangladesh are less than forty feet above sea level. Storm surges from cyclones killed as many as one hundred fifty thousand people in 1991. In comparison, about two thousand people died when Hurricane Katrina hit New Orleans in 2006. The high death toll from flooding does not receive its due attention from Western news media.

Another environmental problem for Bangladesh is deforestation. Wood is traditionally used for cooking and construction. The needs of a larger population have caused widespread deforestation. Brick and cement have become alternative building materials, and cow dung has become a widely used cooking fuel even though it reduces the fertilizer base for agriculture. Even so, these adaptations have not halted the deforestation problem. The main remaining forests are located along the southern borders with India and Burma (Myanmar) and in the northeast sector.

Women and Banking in Bangladesh

Despite an overall languishing economy, economic success stories in this poor country do exist. The Grameen Bank has been working to empower women in Bangladesh for many years. The bank issues **microcredit** to people in the form of small loans. These loans do not require collateral. Loans are often issued to impoverished people based on the concept that many of them have abilities that are underutilized and can be transformed into income-earning activities. About 96 percent of these loans are to women, and the average loan is equivalent to about one hundred dollars. Women have proven to be more responsible than men in repaying loans and utilizing the money to earn wealth. The loan recovery rate in Bangladesh is higher than 98 percent. Microcredit has energized poor women to use their skills to make and market their products to earn a living. Millions of women have taken out such loans, totaling billions of dollars. This program has energized local women to succeed. It has been a model for programs in other developing countries.

Figure 17. Man working in a rice field in Bangladesh.

Key Takeaways

1. When Britain's colonialism ended in South Asia in 1947, the Muslim League was instrumental in creating the united Muslim state with East Pakistan and West Pakistan under one government. East Pakistan broke away and became the independent country of Bangladesh in 1971.

2. Both Pakistan and Bangladesh have large populations that are increasing rapidly. Both countries have agriculturally based economies. Rural-to-urban shift is occurring at an ever-increasing rate in both countries. Population growth places a heavy tax on natural resources and social services.

3. The political units within Pakistan include four main provinces. Tribal Areas border Afghanistan and are controlled by local leaders. The Northern Areas are disputed with India. Each of the provinces has its own unique physical and human landscapes.

4. Bangladesh is a low-lying country with the Brahmaputra River, Ganges River, and the Meghna River flowing into the Bay of Bengal. Flooding is a major environmental concern that has devastated the country on a regular basis.

11.3 India

Learning Objectives

1. Outline the basic activities of British colonialism that affected the realm.
2. Understand the basic qualities of the rural and urban characteristics of India.
3. Summarize the main economic activities and economic conditions in India.
4. Describe the differences between various geographic regions of India.

A. India and Colonialism

India is considered the world's largest democracy. As the historic geography and the development patterns of India are examined, the complexities of this Hindu state surface. European colonizers of South Asia included the Dutch, Portuguese, French, and, finally, the British. In search of raw materials, cheap labor, and expanding markets, Europeans used their advancements in technology to take over and dominate the regional industrial base. The **East India Company** was a base of British operations in South Asia and evolved to become the administrative government of the region by 1857. The British government created an administrative structure to govern South Asia. Their centralized government in India employed many Sikhs in positions of the administration to help rule over the largely Muslim and Hindu population. The English language was introduced as a lingua franca for the colonies.

In truth, colonialism did more than establish the current boundaries of South Asia. Besides bringing the region under one central government and providing a lingua franca, India's colonizers developed the main port cities of **Bombay, Calcutta,** and **Madras** (now called **Mumbai, Kolkata,** and

Chennai, respectively. The names of the port cities have been reverted to their original Hindi forms). The port cities were access points for connecting goods with markets between India and Europe. Mumbai became the largest city and the economic center of India. In 1912, to exploit the interior of India, the British moved their colonial capital from Kolkata, which was the port for the densely populated Ganges River basin, to **New Delhi**. Chennai was a port access to southern India and the core of the Dravidian ethnic south.

Britain exploited India by extending railroad lines from the three main port cities into the hinterlands, to transport materials from the interior back to the port for export. The **Indian Railroad** is one of the largest rail networks on Earth. The problem with colonial railroads was that they did not necessarily connect cities with other cities. The British colonizers connected rail lines between the hinterland and the ports for resource exploitation and export of commercial goods. Today, the same port cities act as focal points for the import/export activity of globalization and remain core industrial centers for South Asia. They are now well connected with the other cities of India.

B. The People of India

Contrasts in India are explicitly evident in the regional differences of its human geography. The north-south contrasts are apparent through the lingua franca and ethnic divisions. The main lingua franca in the north is Hindi. In the Dravidian-dominated south, the main lingua franca is English. The densely populated core region along the Ganges River, anchored on each end by Delhi/New Delhi and Kolkata, has traditionally been called the heartland of India. The south is anchored by the port city of Chennai and the large city of Bangalore. Chennai has been a traditional industrial center. The industrial infrastructure has shifted to more modern facilities in other cities, giving over to a "rustbelt" syndrome for portions of the Chennai re-

gion. India is a dynamic country, with shifts and changes constantly occurring. Any attempt to stereotype India into cultural regions would be problematic.

In 2016, India had more than 1.32 billion people, which is about one-sixth of the human population of the earth, and not far behind China's 1.38 billion. If current trends continue, India will overtake China as the most populous country in the world in less than ten years. An 80 percent majority follow Hindu beliefs. About 13 percent of the population is Muslim. That may not seem like a high percentage, but in this case it equates to over 170 million people. This is equivalent to all the Muslims who reside in the countries of Iraq, Saudi Arabia, Syria, and Egypt combined. India is the third-largest Muslim

Figure 18. The three main language families in India. Hindi is the official language of the government, and both Hindi and English are the lingua franca.

country in the world, after Indonesia and Pakistan, because of its large Muslim minority. India essentially has two lingua francas: English and Hindi, of which Hindi is the official language of the Indian government. India has twenty-eight states and fourteen recognized major languages. The languages of northern India are mainly based on the **Indo-European** language family. Languages used in the south are mainly from the **Dravidian** language family.

Urban versus Rural

Rural and urban life within the Indian Subcontinent varies according to wealth and opportunity. While concentrated in specific areas across the landscape, in general the population in rural areas is discontinuous and spread thinly. In urban areas, the populations are very concentrated with many times the population density found in rural areas. India has six world-class cities: Kolkata, Mumbai, Delhi, Chennai, Bangalore, and Hyderabad. There are many other large cities in India; in 2010, India had forty-three cities with more than a million people each.

India's interior is mainly composed of villages. In rural villages, much of the economy is based on subsistence strategies, primarily agriculture and small cottage industries. The lifestyle is focused on the agricultural cycles of soil preparation, sowing, and harvesting as well as tending animals, particularly water buffalo, cattle, goats, and sheep. About 67 percent of the population lives in rural areas and makes a living in agriculture. About 33 percent of the population—which is equal to 100 million more than the entire US population— is urbanized. India is rapidly urbanizing and industrializing. Changes in technology, however, tend to be slow in dispersing to the rural villages. More than half the villages in India do not have road access for motor vehicles. For residents of those villages, walking, animal carts, and trains are the main methods of transportation. Agricultural technology is primitive. Diffusion of new ideas, products, or methods can be slow. Modern communication technology is, however, helping connect these remote regions.

India's cities are dynamic places, with millions of people, cars, buses, and trucks all found in the streets. In many areas of urban centers, traffic may be stopped to await the movement of a sacred cow or a donkey or bullock cart loaded with merchandise. Indian cities are growing at an unsustainable rate. Overcrowded and congested, the main cities are modernizing and trying to keep up with global trends. Traditionally, family size was large. Large family size results in a swell of young people migrating to urban areas to seek greater opportunities and advantages. In modern times, family size has been reduced to about three children, an accomplishment that did not come easily because of the religious beliefs of most of India's people.

India is a country with considerable contrast between the wealthy urban elites and the poor rural villagers, many of whom move to the cities and live in slums and work for little pay. Low labor costs have enabled Indian cities to industrialize in many ways similar to Western cities, complete with computers, Internet services, and other modern communications services. India's growing middle class is a product of educational opportunities and technological advancements. This available skilled labor base has allowed India's industrial and information sectors to take advantage of economic opportunities in the global marketplace to grow and expand their activities. Development within India is augmented by **outsourcing** activities by American and European corporations to India. Service center jobs created by **business process outsourcing (BPO)** are in high demand by skilled Indian workers.

Figure 19. Farmer tilling a field with oxen in rural India.

C. India's Economic Situation

In the past decade, India has possessed the second fastest growing economy in the world; China is first. India's economy continues to rapidly expand and have a tremendous impact on the world economy. In spite of the size of the economy, India's population has a low average per capita income. Approximately one-fourth of the people living in India live in poverty; the World Bank classifies India as a low-income economy. India has followed a central economic model for most of its development since it declared independence. The central government has exerted strict control over private sector economic development, foreign trade, and foreign invest-

ment. Through various economic reforms since the 1990s, India is beginning to open up these markets by reducing government control on foreign investment and trade. Many publicly owned businesses are being privatized. Globalization efforts have been vigorous in India. There has been substantial growth in information services, health care, and the industrial sector.

The economy is extremely diverse and has focused on agriculture, handicrafts, textiles, manufacturing, some industry, and a vast number of services. A 60 percent majority of the population still earns its income directly from agriculture and agriculture-related services. Land holdings by individual farmers are small, often less than five acres. When combined with the inadequate use of modern farming technologies, small land holdings become inadequately productive and impractical. The wet monsoon is critical for the success of India's agricultural crops during any given season. Because the rainfall of many agricultural areas is tied to the monsoon rains of only a few months, a weak or delayed rainfall can have disastrous effects on the agricultural economy. Agricultural products include commercial crops such as coffee and spices (cardamom, pepper, chili peppers, turmeric, vanilla, cinnamon, and so on). Bamboo is an important part of the agricultural harvest as well. Rice and lentils provide an important basis for the local economy.

Over the last few decades, information technology and related services have transformed India's economy and society. In turn, India is transforming the world's information technologies in terms of production and service as well as the export of skilled workers in financial, computer hardware, software engineering, and software services. Manufacturing and industry are becoming a more important part of India's economy as it begins to expand. Manufacturing and industry account for almost one-third of the gross domestic product (GDP) and contribute jobs to almost one-fifth of the total workforce. Major economic sectors such as manufacturing, industry, biotechnology, telecommunications, aviation, shipbuilding, and retail are exhibiting strong growth rates.

A large number of educated young people who are fluent in English are changing India into a "back office" target for global outsourcing for customer services. These customer services focus on computer-related products but also in-

Figure 20. Mumbai (Bombay), the economic capital and largest city in India.

clude service-related industries and online sales companies. The level of outsourcing of information activity to India has been substantial. Any work that can be conducted over the Internet or telephone can be outsourced to anywhere in the world that has high-speed communication links. Countries that are attractive to BPO are countries where the English language is prominent, where employment costs are low, and where there is an adequate labor base of skilled or educated workers that can be trained in the services required. India has been the main destination for BPO activity from the United States. Firms with service work or computer programming are drawn to India because English is a lingua franca and India has an adequate skilled labor base to draw from.

Vehicle Manufacturing

Two examples of India's growing economic milieu are motor vehicle manufacturing and the movie industry. India's vehicle manufacturing base is expanding rapidly. Vehicle manufacturing companies from North America, Europe, and East Asia are all active in India, and India also has its own share of vehicle manufacturing companies. For example, Mumbai-based Tata Motors Ltd. is the country's foremost vehicle production corporation and it claims to be the second-largest commercial vehicle manufacturer in the world. Tata Motors is India's largest designer and manufacturer of commercial buses and trucks, and it also produces the most inexpensive car in the world, the Tata Nano. Tata Motors manufactures midsized and larger automobiles, too. The company has expanded operations to Spain, Thailand, South Korea, and the United Kingdom. The company is an example of an Indian-based international corporation that is a force in the global marketplace. In 2010, India was recognized as a major competitor with Thailand, South Korea, and Japan as the fourth main exporter of autos in Asia.

Figure 21. The Nano, made in India. The Nano is considered the world's most inexpensive car.

Indian Cinema

Cinema makes up a large portion of the entertainment sector in India. India's cinema industry is often referred to as **"Bollywood,"** a combination of Bombay and Hollywood. Technically, Bollywood only refers to the Hindi language segment of the Indian cinema industry that is based out of Bombay (Mumbai), but the title is sometimes misleadingly used to refer to the entire movie industry in India. Bollywood is the leading movie maker in India and has a world-class film production center. In the past few years, India has been producing as many as one thousand films annually. The highest annual output for the US film industry is only about two-thirds that of India. According to the Guinness Book of World Records, India's city of Hyderabad has the most extensive film production center in the world.

Figure 22. Bollywood filming in India. The film industry in India produces almost twice as many movies as the United States. Indian production scenes can be dramatic and expressive.

D. India: East and West

South Asia's physical geography—an overview of its physical features—was described at the beginning of the section. India makes up the largest physical area of the South Asia realm. Another way of looking at the physical and human landscapes of India is to study spatial characteristics. Additionally, the economic side of the equation can be illustrated by dividing India between east and west according to economic development patterns. To do this, on a map of India draw an imaginary line from the border with Nepal in the north, near Kanpur, to the Polk Strait border with Sri Lanka in the south. This division of India illustrates two sides of India's economic pattern: an economically progressive West India and an economically stagnant East India.

The progressive western side of India is anchored by **Mumbai** and its surrounding industrial community. Mumbai is the economic giant of India with the country's main financial markets, and has been a magnet for high-tech firms and manufacturing. Mumbai's port provides access to global

Figure 23. Dividing India. India can be divided either along north/south dimensions or along east/west dimensions.

markets and is solidly connected to international trade networks. Auto manufacturing, the film industry, and computer firms all have major centers in the large urban metropolitan areas of the west. Large industrial cities such as **Bangalore** and **Hyderabad** have established themselves as high-tech production centers, attracting international business in the computer industry and the information sector. Chemical processing has been ongoing in **Bhopal**, which is noted for an environmental disaster, a gas leak in 1984 that resulted in the deaths of as many as ten thousand people. The nation's capital is located in **New Delhi**, which borders the massive city of (Old) Delhi. The western half of India has been progressing along a pattern with a positive economic outlook that views the global community outside of India as a partner in its success.

The eastern half of India has not been as prosperous as the west in its economic growth. The renowned city of **Kolkata** has traditionally anchored the eastern sector, but its factories have deteriorated into rustbelt status with aging and outdated heavy industries. The high-intensity labor activities of textile and domestic goods manufacturing are not as economically viable as they were in the past. The stagnant economic scene in the east is signified by the low average income levels of many of the states in the eastern region. Neighboring Bangladesh offers little in support of economic growth, and Myanmar, another neighbor to the east, has its own set of problems and lacks support for East India. The eastern half of India does not have the strong partnerships with the global economy found in the west and thus relies more on internal resources for survival.

India: North and South

There are differences in the geographic patterns between the northern and southern halves of India as well as between the eastern and western halves—depending on the criteria used to compare them. Climate patterns, for example, are more diverse in the north, with a wide range of temperatures throughout the seasons. Winter temperatures in the mountainous north are cold and summer temperatures in the Thar Desert can be extremely high. Southern India has a more moderate range of temperatures throughout the year. The far north has high mountains. The south has only the low-lying Eastern and Western Ghats. The north has the extensive Ganges River basin. The south has different drainage networks based on the plateaus of the region.

Besides physical aspects, there are cultural differences between the north and south as well. India is a complex societal mix of many ethnic groups, languages, and traditions. The north is portrayed as a faster-paced society, with more edge and competitiveness. The south has been portrayed as more relaxed and less insistent. Indo-European languages are mainly spoken in the north and Dravidian languages are predominantly spoken in the south. Hindi is more commonly the lingua franca of the north, while English is more frequently the lingua franca of the south. People in the north are of Indo-Aryan descent, while the people in the south have a Dravidian heritage. Hinduism dominates all of India, but the north has a wider diversity of religions, such as Sikhism, Buddhism, and Islam, practiced by a large number of people. The south has a substantial Christian population along its west coast.

Key Takeaways

1. Colonialism had a tremendous impact on South Asia and its people. Colonial development patterns were implemented to control the people and to extract resources, not necessarily to benefit the realm.

2. India has a wide disparity between its poor rural areas with agricultural economies and its wealthier bustling cities with expanding business sectors.

3. Various urban centers of India have positioned themselves well to take advantage of the global economy and expand their manufacturing and industrial base. India is becoming a major manufacturing country for vehicles and high-tech industries.

4. There are noticeable economic differences between the more progressive Western India and the stagnant economic conditions of Eastern India. There is also a noticeable cultural difference between the North and the South in India in the categories of language, ethnicity, food, and society.

11.4 Major Religions of India and South Asia

Learning Objectives

1. Outline the basic religions of the realm. Name the largest minority religion.

2. Understand the basic structure and concepts of Hinduism, including the caste system.

3. Describe how Buddhism differs from Hinduism.

The realm of India and its surrounding countries is the native land for several ancient religions. The oldest world religions of India are Hinduism and Buddhism. Other important religions in the realm include Islam, Christianity, Sikhism, Zoroastrianism, Jainism, and the Baha'i faith. India is at times labeled a Hindu state, but the accuracy of the label is dubious. A more suitable way to describe India is to say that it is a secular country where approximately 80 percent of the population follows Hindu traditions. Islam is the second-most popular religion, practiced by about 13 percent of the population. Christianity is India's third-largest religion, practiced by about 3 percent of the population. Sikhism accounts for about 2 percent of the population of India. Buddhism and Jainism are two other minority religions that have their origins in South Asia. And finally, there are still Indians who practice animist religions, especially in remote areas.

Figure 24. Islamic Architecture in Hindu India. The Taj Mahal was constructed as a mausoleum for the wife of the ruler Shah Jahan in 1653 when the Muslim Mogul Empire controlled northern India. The Taj Mahal is located at Agra, India, and is a UNESCO World Heritage Site.

A. Hinduism

Hinduism is one of the world's oldest major religions still practiced. Its origins can be traced to ancient Vedic civilizations in India approximately three thousand years ago. The religion is found mainly in India, and it has the third-highest number of believers of religions in the world. Hinduism does not originate from a single teacher but from many traditions. The Hindu belief system consists of a number of schools of thought, with a wide variety of rituals and practices.

Predominantly, Hinduism follows the teachings of many gods or goddesses, frequently including a Supreme Being. While there are hundreds, if not thousands, of gods and goddesses, many are thought to represent different aspects of the same individual or Supreme Being. These individuals can be recognized by items that they are holding as well as by the vehicle or avatar that carries them. The three main deities and most widely venerated of the Hindu faith are Shiva the Destroyer, **Vishnu** the Preserver, and **Brahma** the Creator. There is a continuous cycle in which the original creation was accomplished by Brahma, Shiva destroys the universe, and Vishnu will recreate or preserve that universe from destruction. Different Hindu traditions have venerated each of the three main deities as the all-encompassing Supreme Being.

The **polytheistic** traditions of Hinduism consider a large number of deities or spiritual entities. Since there is no one creed or unified systems of beliefs, Hinduism has been referred to as more of a religious tradition than a religion. It has been said that Hinduism cannot be defined, but is instead experienced. This understanding allows a variety of beliefs to be included in the vast array of Hindu religious practices.

Hinduism is an extremely diverse religion, making it extremely difficult to define set doctrines that are accepted by all denominations. Within the wide spectrum of religious traditions, however, are general concepts that are common to Hindu beliefs. Hindus believe in **Dharma** (code of conduct or duty), **Samsara** (reincarnation/rebirth), **Karma** (personal actions and choices), and **Moksha** (salvation) by belief in God and through an individual path of faith. Reincarnation is a cycle of death and rebirth for a soul to transmigrate through until it reaches Moksha. Karma governs how the soul is reincarnated. Actions in this life determine the soul's life cycle for the next life. Positive and upright works will draw one closer to God and a rebirth through reincarnation into a life with a wider consciousness or higher caste level. Evil or bad actions take the soul farther from God and into a lower form of worldly life or caste level.

Pilgrimages are common in Hindu practice. Holy sites or temples are located throughout India and are regular destinations for the Hindu faithful. Pilgrimage is not required but is routinely conducted by many Hindus. Besides many holy temples, a variety of cities and other holy places are pilgrimage destinations for Hindus. **Varanasi**, one such city, is considered by many as the holiest city of Hinduism, although other cities also hold this distinction. Located on the **Ganges River,** Varanasi is home to a large number of temples and shrines. The most visited shrine in Varanasi is one in honor of a manifestation of Shiva. Hindu festivals are held in Varanasi throughout the year, many along the banks of the Ganges. Varanasi is also one of the holiest places in Buddhism; it is said to have been designated by Gautama Buddha as one of four prime pilgrimage sites.

More than one million Hindu pilgrims visit Varanasi annually. Mother Ganga, as the river is referred to in Hinduism, is considered holy by many Hindu followers. Devotees ritually bathe in the river or take "holy" water from it home to ill family members. Some Hindus believe that the water can cure illnesses. Others believe that bathing in the Ganges will wash away your sins. The nonspiritual truth is that the Ganges is a highly polluted waterway. The water is not considered safe for human consumption by most universal health standards.

Figure 25. Shiva Statue in Bangalore, India.

Figure 26. Ghat in Varanasi, India, where Hindu Faithful Access the Ganges River.

The Hindu Caste System

In the Hindu caste system, society is organized into separate groups or castes. Every person is born into an unchanging group or caste that remains his or her status for the rest of his or her life. All lifetime activities are conducted within one's own caste. The caste a person is born into is based on what they have done in a past life. The caste system has evolved differently in different parts of Asia. Each Hindu branch has its own levels of castes, and thousands of sublevels have been established over time. In Hinduism, the basic system originated around five main caste levels:

1. **Brahmin:** priests, teachers, and judges

2. **Kshatriya:** warrior, ruler, or landowner

3. **Vaishya:** merchants, artisans, and farmers

4. **Shudra:** workers and laborers

5. **Dalits (Untouchables or Harijan):** outcasts or tribal groups

The Dalits (Untouchables or Harijan) traditionally worked in jobs relating to "polluting activities," including anything unclean or dead. Dalits have been restricted from entering Hindu places of worship or drinking water from the same sources as members of higher castes. They often had to work at night and sleep during the day. In many areas, Dalits needed to take their shoes off while passing by upper-caste neighbors. Dalits could leave their Hindu caste by converting to Christianity, Buddhism, or Islam. The Indian government has implemented a positive affirmative action plan and provided the Dalits with representation in public offices and certain employment privileges. This policy has received harsh opposition by upper-caste groups. Technically, the caste system is illegal under current Indian law. Nevertheless, the opportunities that are available to the upper castes remain out of reach to many of the lowest caste. In some areas, education and industrialization have diminished the caste system's influence. In other areas, Hindu fundamentalists have pushed for a stronger Hindu-based social structure and opposed any reforms.

Traditional socioeconomic status tends to be more important in rural areas, where the caste system is more formally adhered to. If you live in a community of millions of people, caste affiliations tend not to be as important, but in a smaller, more rural community, these relationships and the status they hold can be very important, especially as many of the castes are associated with traditional village tasks, such as religious leaders, politicians, farmers, leather workers, or other activities.

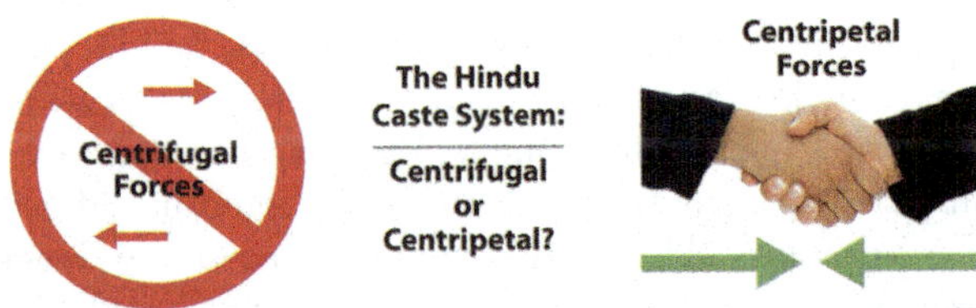

Figure 27. Centrifugal vs Centripetal forces (India). Is the Hindu caste system a centrifugal or centripetal cultural force for India?

B. Buddhism

Around 535 BCE in northern India, a prince by the name of **Siddhartha Gautama** broke from the local traditions that shaped Hinduism and taught religious salvation through meditation, the rejection of earthly desires, and reverence for all life forms. Siddhartha is recognized as the first **Buddha**. He taught that through many cycles of rebirth a person can attain enlightenment and no longer have a need for desire or selfish interests. Enlightenment is being free from suffering and is reaching a state of liberation often referred to as Nirvana. Buddhism is considered a "dharmic" faith that concerns following a path of duty for a proper life. According to Buddhism, life is dictated by karma, which connects our actions with future experiences. Buddhism spread across the Indian Subcontinent after the sixth century BCE and became the region's dominant religion within 1,500 years. However, since that time, the religion has diminished in the Indian Subcontinent, although it has seen some revival under the influence of Buddhist scholars. Buddhism predominates in the northernmost areas of India.

Buddhism is the majority religion in Bhutan, Sri Lanka, Tibet, and most of Southeast Asia. It was prominent in Chi-

Figure 28. Statue of Buddha at Bodh Gaya, India.

na, Mongolia, and North Korea before their governments adopted Communist ideology. Communist governments officially announced that their countries were nonreligious, although many people still followed religious systems. Various branches of Buddhism have developed, with many schools established within each branch. One feature common throughout all branches of the faith is that Buddhism does not have caste levels.

All branches of Buddhism teach nonviolence, honesty, selflessness, tolerance, and moral living. Buddhism holds to the **Four Noble Truths** and the **Eightfold Path (The Middle Way)** to enlightenment. Suffering is a standard component of humanity. Only through the Eightfold Path to enlightenment is freedom from suffering possible. Enlightenment comes through wisdom, ethical conduct, and meditation. Buddhism has become the world's fourth largest religion, with most of its followers in Asia.

The Four Noble Truths and the Eightfold Path

The Four Noble Truths

1. Suffering exists.
2. Suffering arises from attachment to desires.
3. Suffering ceases when desire ceases.
4. Freedom from suffering is possible by practicing the Eightfold Path.

The Middle Way or Eightfold Path

Attainable through wisdom

1. Right view
2. Right intention

Attainable through ethical conduct

3. Right speech
4. Right action
5. Right livelihood

Attainable through meditation

6. Right effort
7. Right mindfulness
8. Right concentration

Key Takeaways

1. Hinduism is one of the world's oldest still-practiced religions. There is no one specific path in the religion. Hinduism is more of a religious tradition based on core concepts than it is a formal religion.

2. The caste system is a Hindu practice of placing people in social layers with similar occupations, privileges, and status. The untouchables are the lowest caste.

3. Buddhism was created around 535 BCE from the traditions that shaped Hinduism by Siddhartha Gautama, who taught religious salvation through meditation, the rejection of earthly desires, and reverence for all life forms. There is no caste system in Buddhism, which has a number of branches that vary throughout Asia.

End-of-Chapter Summary

- The Himalayan Mountain ranges border South Asia to the north. Nepal is located along this border and is somewhat of a buffer state between India and China. Nepal has a high population growth rate. Most of its people work in agriculture. Deforestation is a major environmental concern and causes erosion of the landscape. Landlocked and poor, Nepal struggles to maintain a stable government and adequate public services.

- South Asia was colonized by Britain for ninety years. Colonialism brought a structured administration, a railroad system of transportation, and large port cities used for the export of goods from the interior. The political borders were established for South Asia by British colonizers, based on religious affiliation and economic advantages. The British elevated Sikhs from the Punjab to help rule over the Hindu and Muslim populations. English is widely used as a lingua franca.

- Conflicts continue in mountainous Kashmir and tropical Sri Lanka. Kashmir's remote territory in the northern part of the realm is divided between Pakistan, China, and India. All three countries have nuclear weapons.

- Port cities of South Asia are centers for international trade and development. There is a wide disparity between the rural poor and the affluent elites. India has been developing a strong economy based on a growing information sector, health care, and manufacturing. Motor vehicles and computer technologies are emerging in India and competing worldwide. Pakistan's economy struggles under the high population growth and Islamic extremism in the country.

- Pakistan and Bangladesh were once under the same government. Bangladesh was formerly East Pakistan. These Muslim countries have extremely high population densities and have agrarian economies. The Indus River flows through Pakistan and the two rivers of the Brahmaputra and the Ganges flow through Bangladesh. Monsoon flooding is a serious concern for Bangladesh; earthquakes have caused serious damage in Pakistan.

- Hindu and Buddhist traditions first developed in South Asia. India has the most of the world's Hindu followers. The concept of the caste system has created socioeconomic layers in the culture that are being tempered by high urbanization rates. Buddhism has a number of branches that can be geographically identified as eastern, northern, and southern. Bhutan and Sri Lanka have Buddhist majorities. South Asia is also home to Sikhism and Jainism. Islam is strong in South Asia: Pakistan is the world's second-largest Muslim country, India has the world's third-largest Muslim population, and Bangladesh is a Muslim country as well. South Asia is also home to a Christian minority in addition to various other minority religious groups.

Chapter 12

East Asia

Introduction

East Asia is a large expanse of territory with **China** as its largest country. The countries of **Mongolia, North and South Korea, and Japan** are China's neighbors. The island of Taiwan, off the eastern coast of China, has an independent government that has been separated from mainland China since shortly after World War II. On the southern coast of China is **Hong Kong**, a former British possession with one of the best ports in Asia. Under an agreement of autonomy, Hong Kong and its port were turned over to the Chinese government in 1997. Next door, to the west of Hong Kong, is the former Portuguese colony of Macau, which has also been returned to Chinese control. In western China is the autonomous region of Tibet, referred to by its Chinese name, **Xizang**. Tibet has been controlled by Communist China since 1949, shortly after the People's Republic of China (PRC) was declared a country. Lobbying attempts by the Dalai Lama and others for Tibetan independence have not been successful. The region of Tibet has recently become more integrated with the country of China because of the immigration of a large number of Chinese people to the Tibetan region.

Japan has emerged as the economic dragon of East Asia. Japanese people have a high standard of living, and the country has been an industrial and financial engine for the Pacific Rim. Up and coming economic tigers like Singapore, Hong Kong, Taiwan, and South Korea have also experienced strong economic growth and are strong competitors in the global economy. Balancing out the advances of the economic tigers and Japan is the extensive labor base of the Chinese people, which has catapulted the Chinese economy to its position as a major player in the global economy. Left behind in the region is North Korea, which has isolated itself behind an authoritarian dictatorship since World War II. A number of countries that were former enemies in World War II are now trading partners (e.g., China and Japan), as economic trade bridges cultural gaps with common goods and services. However, cultural and political differences between these countries remain.

East Asia is home to over one-fifth of the human population. The realm's location on the **Pacific Rim** provides access for interaction with the global economy. The location of Japan, South Korea, and Taiwan, just off the coast of mainland China, creates an industrial environment that has awakened the human entrepreneurial spirit of the realm. Manufacturing has fueled the high-tech engines of the Pacific Rim economies, which have recently taken advantage of the massive labor pool of the Chinese heartland. Across the Pacific from East Asia are the superpower of the United States and its NAFTA partners, countries that are both competing against and trading with the **East Asian Community (EAC)**. The Russian realm to the north of East Asia—especially its Pacific port of Vladivostok—continues to actively engage the East Asian nations.

12.1 Introducing the Realm

Learning Objectives

1. Outline the countries and territories that are included in East Asia. Describe the main physical features and climate types of each country.

2. Understand the relationship between physical geography and human populations in East Asia.

3. Summarize the main objectives in building the Three Gorges Dam.

4. Describe how colonialism impacted China. Outline the various countries and regions that were controlled by colonial interests.

5. Outline the three-way split in China after its revolution and where each of the three groups ended up.

A. Physical Geography

East Asia is surrounded by a series of mountain ranges in the west, Mongolia and Russia in the north, and Southeast Asia to the south. The Himalayas border Tibet and Nepal; the **Karakoram Ranges, Pamirs, and the Tian Shan Mountains** shadow Central Asia; and the Altay Mountains are next to Russia. The **Himalayan Mountains** are among the highest mountain ranges in the world, and Mt. Everest is the planet's highest peak. These high ranges create a rain shadow effect, generating the arid conditions of type B climates that dominate western China. The desert conditions of **western China** give rise to a large uninhabitable region in its center. Melting snow from the high elevations feeds many of the streams that transition into the major rivers that flow toward the east.

Figure 1. Six Countries of East Asia and Neighboring Countries. China is the largest country in East Asia in both physical size and population. Other countries of East Asia include Mongolia, North Korea, South Korea, Taiwan, and Japan. Hong Kong and Macau are former colonies that are part of mainland China.

Created by tectonic plate action, the many mountain ranges are also home to earthquakes and tremors that are devastating to human livelihood. The Indian tectonic plate is still pushing northward into the Eurasian plate, forcing the Himalayan ranges upward. With an average elevation of fifteen thousand feet, the Tibetan Plateau is the largest plateau region of the world. The plateau is sparsely populated and the only places with human habitation are the river valleys. **Lhasa** is the largest city of the sparsely populated region. Sometimes called "the Roof of the World," the **Tibetan Plateau** is a land of superlatives. Its landscape is generally rocky and barren.

The vast arid regions of western China extend into the **Gobi Desert** between Mongolia and China. Colder type D climates dominate the Mongolian steppe and northern China. The eastern coast of the Asian continent is home to islands and peninsulas, which include Taiwan and the countries of Japan and North and South Korea. North Korea's type D climates are similar to the northern tier of the United States, comparable to North Dakota. Taiwan is farther south, producing a warmer tropical type A climate. The mountainous islands of Japan have been formed as a result of tectonic plates and are prone to earthquakes. Since water moderates temperature, the coastal areas of East Asia have more moderate temperatures than the interior areas do. A type C climate is dominant in Japan, but the north has a colder type D climate. The densely populated fertile river valleys of central and southeastern China are matched by contrasting economic conditions. Rich alluvial soils and moderate temperatures create excellent farmland that provides enormous food production to fuel an ever-growing population.

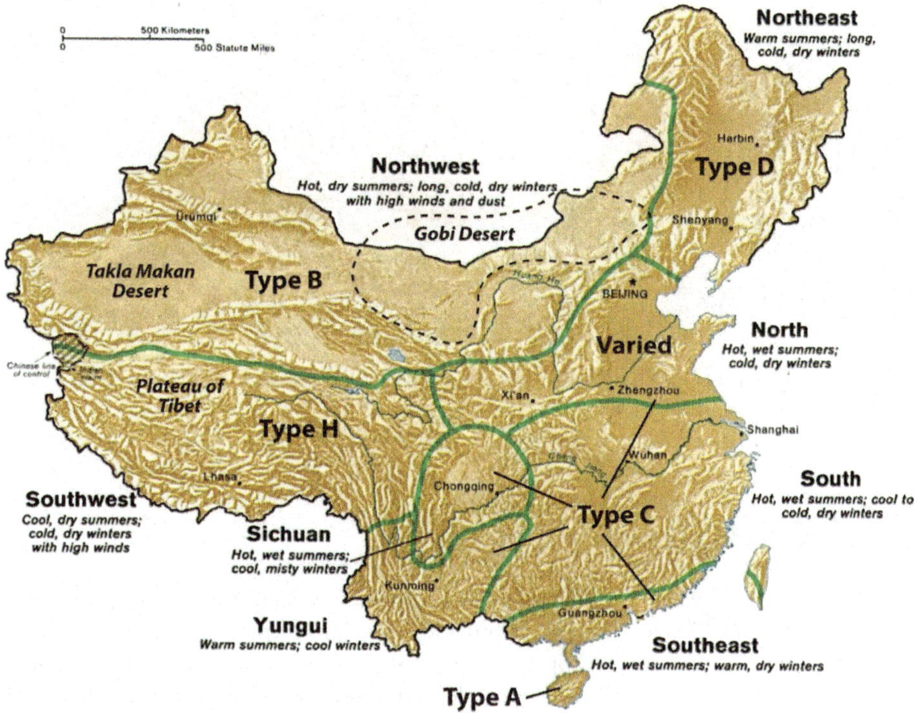

Most of China's population lives in its eastern region, called China Proper, with type C climates, fresh water, and good soils. China Proper has dense population clusters that correspond to the areas of type C climate that extend south from Shanghai to Hong Kong. Around the world, most humans have gravitated toward type C climates. These climates have produced fertile agricultural lands that provide an abundance of food for the enormous Chinese population. To the south the temperatures are warmer, with hot and humid summers and dry, warm winters. The climates of China Proper are conducive for human habitation, which has transformed the region into a highly populated human community. The North China Plain at the mouth of the **Yellow River** (*Huang He*) has rich farmland and is the most densely populated region in China.

Northwest of Beijing is **Inner Mongolia** and the Gobi Desert, a desert that extends into the independent country of Mongolia. Arid type B climates dominate the region all the way to the southern half of Mongolia. The northern half of Mongolia is colder with continental type D climates. Northeast China features China's great forests and excellent agricultural land. Many of China's abundant natural mineral resources are found in this area. Balancing mineral extraction with the preservation of agricultural land and timber resources is a perennial issue.

Lying north of the Great Wall and encompassing the autonomous region of Inner Mongolia is the vast **Mongolian steppe**, which includes broad flat grasslands that extend north into the highlands. North China includes the Yellow River basin as well as the municipalities of Beijing and Tianjin. Areas around parts of the Yellow River are superb agricultural lands, including vast areas of loess that have been terraced for cultivation. Loess is an extremely fine silt or windblown soil that is yellow in color in this region. Deciduous forests continue to exist in this region, despite aggressive clearcutting for agricultural purposes. The **Great Wall of China** rests atop hills in this region.

Most of western China is arid, with a type B climate. Western China has large regions like the **Takla Makan Desert** that are uninhabited and inhospitable because of hot summers and long cold winters exacerbated by the cold winds sweeping down from the north. In a local Uyghur language, the name *Takla Makan* means *"You will go in but you will not go out."* To the far west are the

Figure 2. China and its main climate regions.

high mountains bordering Central Asia that restrict travel and trade with the rest of the continent. Northwestern China is a mountainous region featuring glaciers, deserts, and basins.

The central portion of China Proper is subtropical. This large region includes the southern portion of the **Yangtze River** (Chang Jiang) and the cities of **Shanghai and Chongqing**. Alluvial soils give this area excellent agricultural land. Its climate is warm and humid in the sum-

mers with mild winters; monsoons create well-defined summer rainy seasons. Tropical China lies in the extreme south and includes **Hainan Island** and the small islands that neighbor it. Annual temperatures are higher here than in the subtropical region and rainfall amounts brought by the summer monsoons are at times very substantial. This area is characterized by low mountains and hills.

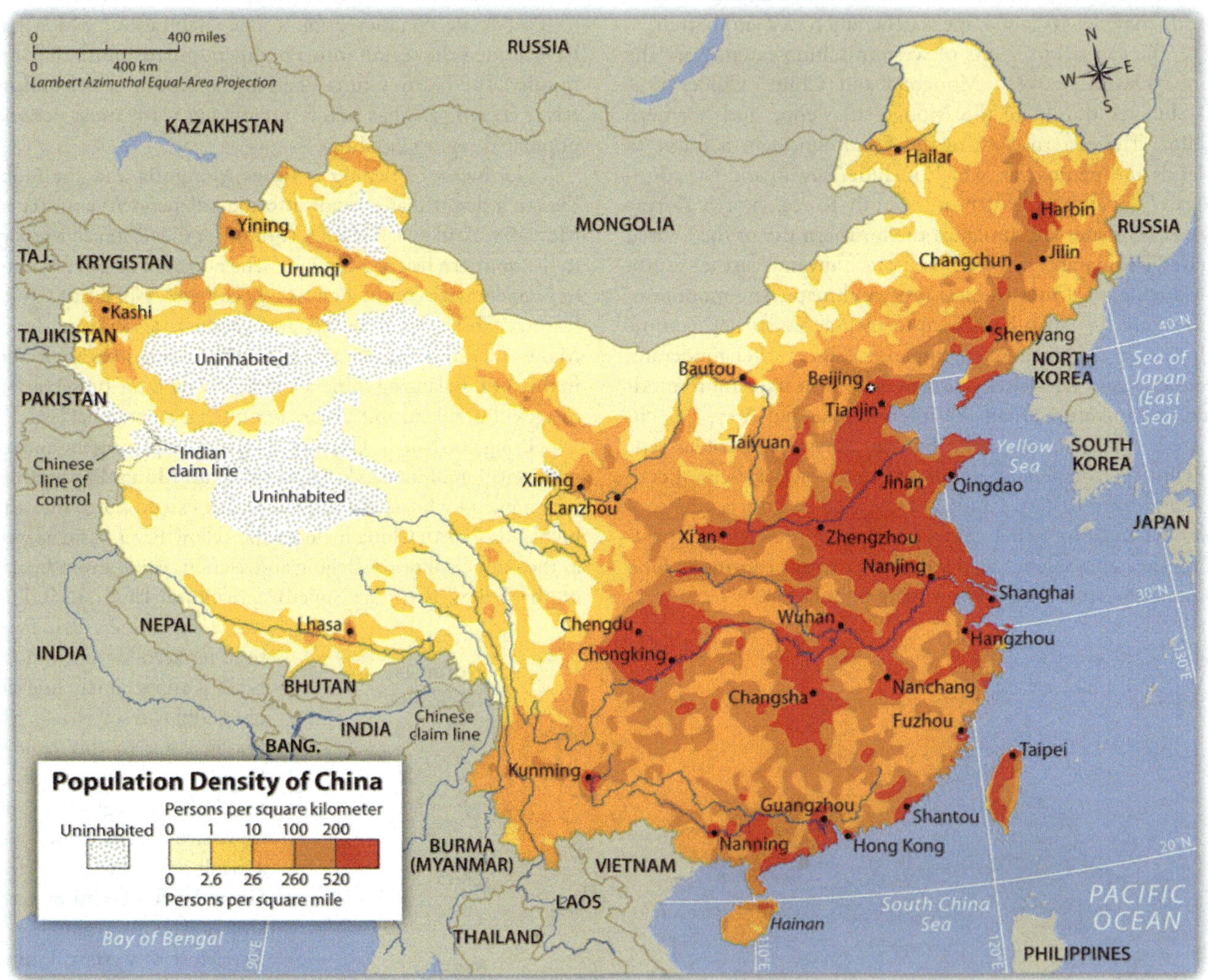

Figure 3. China and its main population regions. The region of eastern China that is favorable to large populations is called China Proper. River basins historically produce abundant food, which in turn leads to concentrated populations.

B. River Basins of China

There are two major river systems that provide fresh water to the vast agricultural regions of the central part of China Proper. **The Yellow River** (Huang He) is named after the light-colored silt that washes into the river. It flows from the Tibetan highlands through the North China Plain into

the Yellow Sea. Dams, canals, and irrigation projects along the river provide water for extensive agricultural operations. Crops of wheat, sorghum, corn, and soybeans are common with vegetables, fruit, and tobacco grown in smaller plots. **The North China** Plain has to grow enough food to feed its one thousand people per square mile average density. This plain does not usually produce a food surplus because of the high demand from the large population of the region. Beijing borders the North China Plain. Its nearest port, Tianjin, continues to expand and grow, creating an economic center of industrial activity that relies on the peripheral regions for food and raw materials. Cotton is an example of a key industrial crop grown here.

The Yangtze River (Chang Jiang) flows out of the Tibetan Plateau through the Sichuan Province, through the Three Gorges region and its lower basin into the East China Sea. Agricultural production along the river includes extensive rice and wheat farming. Large cities are located on this river, including **Wuhan and Chongqing. Nanjing and Shanghai** are situated near the delta on the coast. Shanghai is the largest city in China and is a growing metropolis. **The Three Gorges Dam on the Yangtze River** is the world's largest dam. It produces a large percentage of the electricity for central China. Oceangoing ships can travel up the Yangtze to Wuhan and,

utilizing locks in the Three Gorges Dam, these cargo vessels can travel all the way upriver to Chongqing. The Yangtze River is a valuable and vital transportation corridor for the transport of goods between periphery and core and between the different urban centers of activity. Sichuan is among the top five provinces in China in terms of population and is dependent on the Yangtze River system to provide for its needs and connect it with the rest of China.

Northeast China was formerly known as **Manchuria**, named after the Manchu ethnic group that had dominated the region in Chinese history. Two river basins create a favorable industrial climate for economic activity. The lower **Liao River Basin** and the **Songhua River Basin** cut through Northeast China. The cities of Harbin and Shenyang are industrial centers located on these rivers. This region is known as the **Northeast China Plain**. It has extensive farming activities located next to an industrial landscape of smokestacks, factories, and warehouses. Considerable mineral wealth and iron ore deposits in the region have augmented the industri-

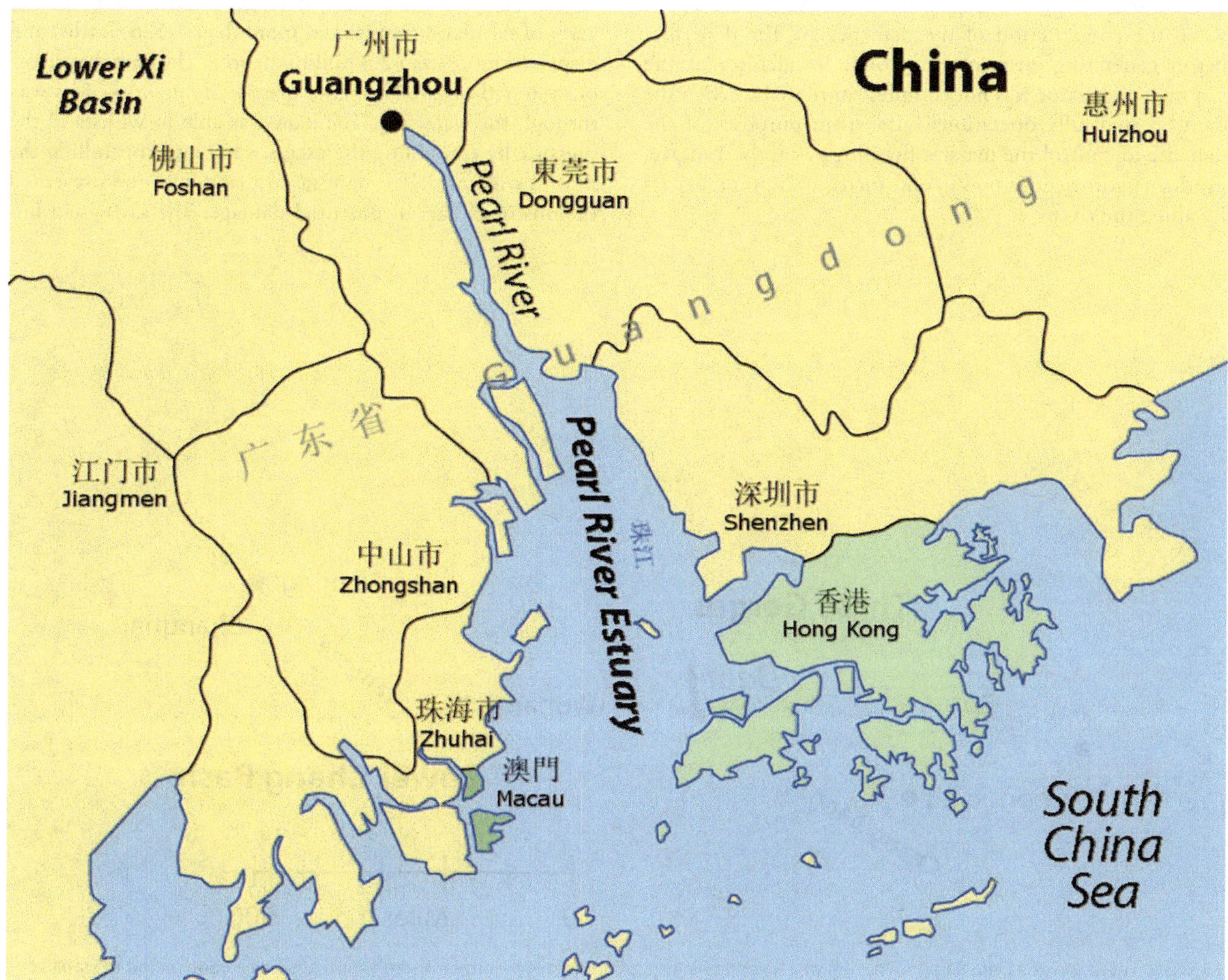

Figure 4. Xi-Pearl River System. An estuary is a wide area at the mouth of a river where it meets the sea. Hong Kong is located on the eastern side of the Pearl River Estuary, and the former Portuguese colony of Macau is located on the western side of the waterway.

al activities and have created serious environmental concerns because of excessive air and water pollution. In its zenith in the 1970s, this was China's main steel production area, but the region is being reduced to a rustbelt since many of China's manufacturing centers are now being developed in the southern regions of China Proper.

The southernmost region of China Proper is home to the Pearl River Basin, an important agricultural and commercial district. Though smaller in size than the Yangtze River Basin, major global urban centers are located on its estuary, where the mouth of the river flows into the South China Sea. The system includes the Xi River, **Pearl River**, and their tributaries. As the third-longest river system in China, these rivers process an enormous amount of water, and have the second-highest volume of water flow after the Yangtze. Guangzhou, Macau, and Hong Kong are the largest cities located here, alongside the rapidly expanding industrial center of Shenzhen. As mentioned earlier, Macau was a former Portuguese colony and Hong Kong was a former British colony. These urban areas are now hubs for international trade and global commerce.

C. Three Gorges Dam (The New China Dam)

The Three Gorges Dam on the Yangtze River is known in China as the New China Dam. Its hydroelectric production system is the largest on Earth, with an electrical generating capacity nearly four times that of Grand Coulee Dam on the Columbia River. The river system is the world's third longest, after the Nile and the Amazon. Plans and development for this project began in the decades before 1994, when the construction of the dam began. The dam first began generating electricity in 2003. Installation of the last main generator was not complete until 2012, when the dam became fully operational. The main purposes of the dam are to control the massive flooding along the Yangtze, produce hydroelectric power, and increase shipping capacity along the river.

Before construction of the dam, flooding along the Yangtze cost thousands of lives and billions of dollars in damage. In 1954, the river flooded, causing the deaths of more than thirty-three thousand people and displacing an additional eighteen million people. The giant city of Wuhan was flooded for three months. In 1998, a similar flood caused billions of dollars in damage, flooded thousands of acres of farmland, resulted in more than 1,526 deaths, and displaced more than 2.3 million people. The dam was rigorously tested in 2009, when a massive flood worked its way through the waterway. The dam was able to withstand the pressure by containing the excess water and controlling the flow downstream. The dam saved many lives and prevented billions of dollars in potential damage. The savings in hu-

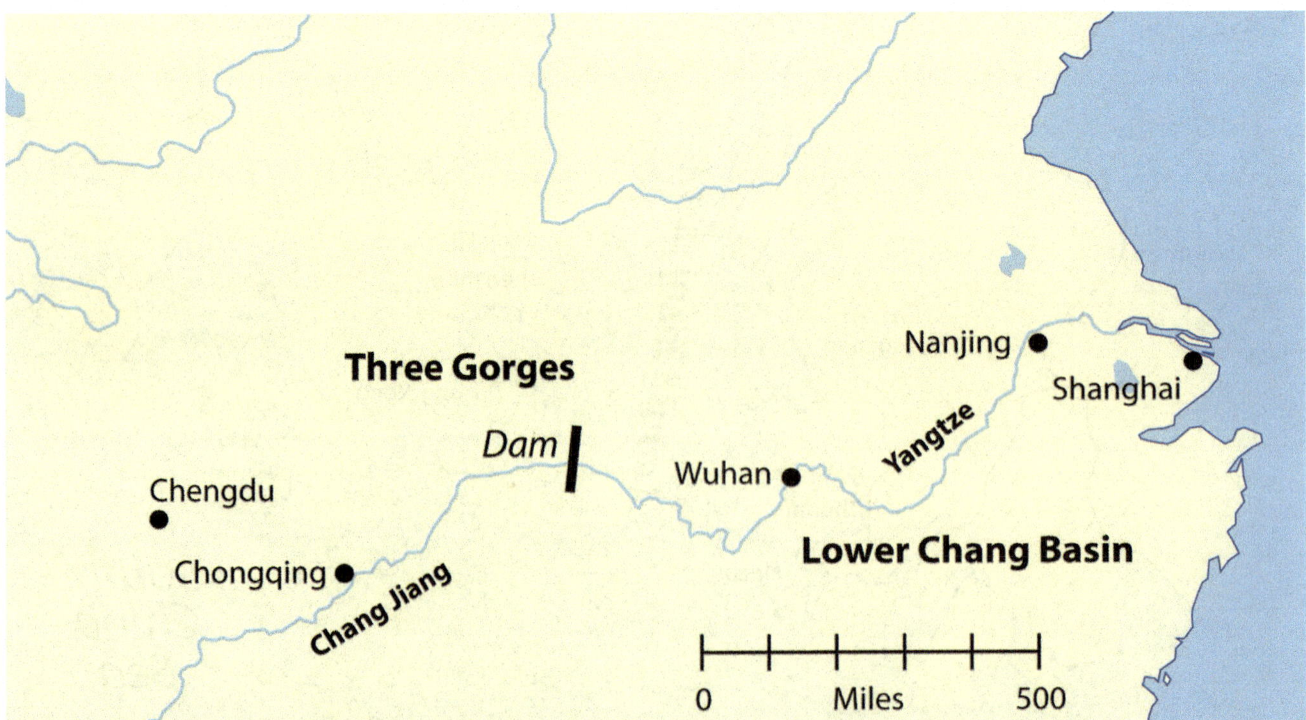

Figure 5. Three Gorges Dam Region. The Yangtze River flows through three deep gorges where a dam has been constructed to stabilize flooding, produce electricity, and support river transportation.

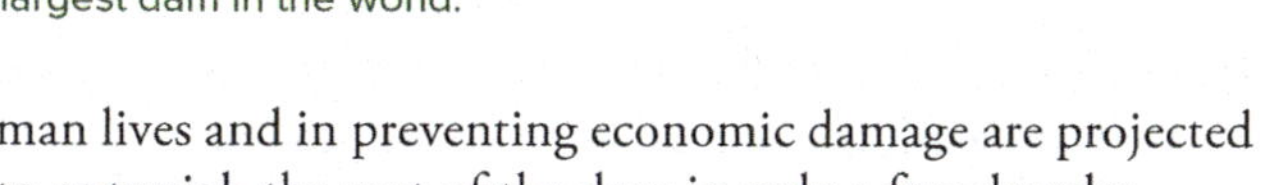

Figure 6. China's three Gorges Dam on the Yangtze River is the largest dam in the world.

Figure 7. Tourists on the Great Wall of China.

man lives and in preventing economic damage are projected to outweigh the cost of the dam in only a few decades.

The dam produces most of the electricity for the lower Yangtze Basin, including Shanghai, the largest city in China, and has reduced reliance on coal-fired power plants. This reduces the emission of millions of tons of carbon dioxide, sulfur, and nitrogen oxides into the atmosphere, which reduces air pollution and does not contribute to climate change. Heavy freight traffic on the Yangtze was the norm even before the dam was built; in fact, it has the highest rates of transport of any river.

The positive attributes of the Three Gorges Dam have contributed to the economic development of China. However, this project has also created its own problems and negative impacts on culture and the environment. By 2008, the number of people forced to relocate from the flooding of the reservoir had reached 1.24 million. Historic villages and hundreds of archaeological sites were flooded. Thousands of farmers had to be relocated to places with less productive soils. Compensation to the farmers for relocation was forfeited because

of corruption and fraud. Sadly, much of the scenic beauty of the river basin is now under water.

Animal species like the critically endangered Siberian Cranes, who had wintered in the former wetlands of the river, had to find habitat elsewhere. The endangered Yangtze River Dolphin has been doomed to extinction because of the dam and the amplified river activity. The dam restricts the flushing of water pollution and creates a massive potential for landslides along its banks, exacerbating the potential for the silting in of the reservoir and the clogging of the dam's turbines. The dam also sits on a fault zone and there is concern that the massive weight of the water in the reservoir could trigger earthquakes that may destroy the dam, with catastrophic consequences. Large development projects tend to have an enormous impact on the local people and environment. The building of the Three Gorges Dam has created controversy, with strong arguments on both sides of the issues. To further complicate the situation, other large dams are being proposed or are under construction along the same river.

D. Chinese Dynasties and Colonialism

The earliest Chinese dynasty dates to around 2200 BCE. It was located in the rich North China Plain. Organized as a political system, Chinese dynasties created the Chinese state, which provided for a continuous transfer of power, ideas, and culture from one generation to another. From 206 BCE to 220 CE, the **Han Dynasty** established the Chinese identity; Chinese people became known as People of Han or Han Chinese. The last dynasty, the **Qing (Manchu) Dynasty**, which ruled between 1644 and 1911, claimed control of a region including all of China, Mongolia, Southeast Asia, and Korea. Dynastic rule ended in China in 1911.

Europeans colonized the Americas, Africa, and South Asia, and it was only a matter of time before technology, larger ships, and the European invasion reached East Asia. European colonialism arrived in China during the Qing Dynasty.

China had been an industrialized state for centuries; long before the empires of Rome and Greece were at their peak, China's industrial cities flourished with clean drinking water, transportation, and technology. Paper, gunpowder, and printing were used in China centuries before they arrived in Europe. The Silk Road, which crossed the often dangerous elevations of high mountain passes, was the main link between China and Europe. European colonial powers met tough resistance in China. They were kept at bay for years.

Meanwhile, the Industrial Revolution in Europe, which cranked out mass-produced products at a cheap price, provided an advantage over Chinese production. British colonizers also exported opium, an addictive narcotic, from their colonies in South Asia to China to help break down Chinese culture. By importing tons of opium into China, the British were able to instigate social problems. The first **Opium Wars** of 1839–1842 ended with Britain gaining an upper hand and laying claim to most of central China. Other European powers also sought to gain a foothold in China. Portugal gained the port of Macau. Germany took control of the coastal region of the rich North China Plain. France carved off part of southern China and Southeast Asia. Russia came from the north to lay claim to the northern sections of China. Japan, which was just across the waterfront from China, took control of Korea and the island of Formosa (now called Taiwan). Claims on China increased as colonialism moved in to take control of the Chinese mainland.

Though European powers laid claim to parts of China, they often fought among themselves. China did not produce heavy military weapons as early as the Europeans did and therefore could not fend them off upon their invasion. Chinese culture, which had flourished for four thousand years, quickly eroded through outside intrusion. It was not until about 1900, when a rebellion against foreigners (known as the **Boxer Rebellion**) was organized by the Chinese people, that the conflict reached recognizable dimensions. The Qing Dynasty dissolved in 1911, which also signified an end to the advancements of European colonialism, even though European colonies remained in China.

Three-Way Split in China

European colonialism in China slowed after 1911, and World War I severely weakened European powers. The Japanese colonizers, on the other hand, continued to make advancements. Japan did not have far to travel to resupply troops and support its military. In China, a doctor by the name of Dr. Sun Yat-sen promoted an independent Chinese Republic, free from dynastic rule, Japan, or European colonial influence. Political parties of Nationalists and Communists also worked to establish the republic. **Dr. Sun Yat-sen** died in 1925. The Nationalists, under the leadership of **Chiang Kai-shek**, defeated the Communists and established a national government. Foreigners were evicted. The Communists were driven out of politics.

Nationalists, Communists, and Japan conducted a three-way war over the control of China. Japan's military took control of parts of Northeast China, known as Manchuria, and were making advancements on the eastern coast. Nationalists defeated the Communists for power and were pushing them into the mountains. The Chinese people were in support of the two parties working together to defeat the Japanese. **The Long March** of 1934 was a six-thousand-mile retreat by the Communists through rural China, pursued by Nationalist forces. The people of the countryside gave aid to the efforts of the Communists. The Chinese were primarily interested in the defeat of Japan, a country that was brutally killing massive numbers of China's people in their aggressive war.

In 1945, the defeat of Japan in World War II by the United States changed many things. Japan's admission of defeat prompted the end of Japanese control of territory in China, Taiwan, Korea, Southeast Asia, and the Pacific. By 1948, the Communists, who were becoming well organized, were defeating the Nationalists. Chiang Kai-shek gathered his people and what Chinese treasures he could and fled by boat to the island of Formosa (Taiwan), which in 1945 had just been freed from Japan. Taiwan was declared the official **Republic of China (ROC)**. The Communists took over the mainland government. In 1949, Communist leader Mao Zedong declared the establishment of the **People's Republic of China (PRC)** with its capital in Peking (Beijing). Japan was devastated by US bombing and defeated in World War II; its

infrastructure destroyed and its colonies lost, Japan had to begin the long process of rebuilding its country. Korea was finally liberated from the Chinese dynasties and Japanese colonialism but began to experience an internal political division. Political structures in the second half of the twentieth century in East Asia were vastly different from the political structures that had been in place when the century began.

Figure 8. Taiwan Currency with image of Dr. Sun Yat-sen. Both the Republic of China (ROC) and the People's Republic of China (PRC) consider Dr. Sun Yat-sen to be a famous Chinese historical figure.

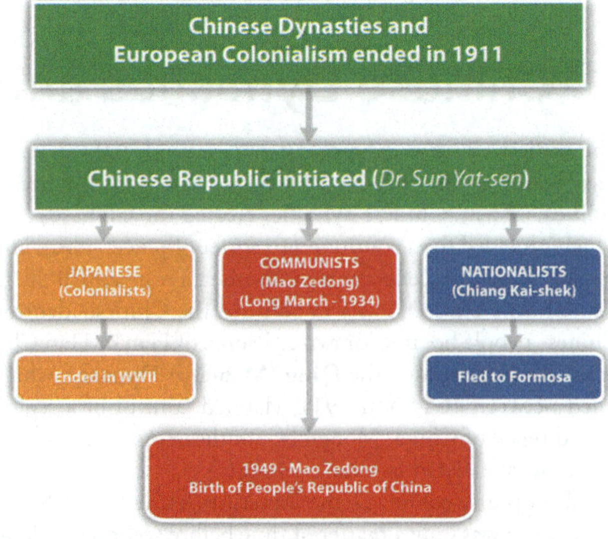

Figure 9. The Three-Way Split of China and the Emergence of Communist China in 1949.

Key Takeaways

1. China is the largest country in physical area and in population in East Asia. The realm is isolated by the high mountains in the west, which cause a major rain shadow and desert conditions in the western regions. Mongolia is the only landlocked country. The other countries and territories are located along the Pacific Rim.

2. Robust population growth has been supported by adequate food production in the major river valleys and coastal regions of East Asia. Coastal areas receive adequate precipitation and allow access to fishing for human activities.

3. The Three Gorges Dam was constructed on the Yangtze River to control flooding, generate electricity, and support shipping. The dam is the largest in the world. Downsides of the dam have included the relocation of human settlements, erosion, and other environmental concerns.

4. Colonialism infiltrated China and challenged its last dynasty. By 1911, both the dynasty and European colonialism were in demise. In a three-way battle for power in China, the Communists emerged in 1949 to take control. The Nationalists fled to Formosa to form their own government.

12.2 Emerging China

Learning Objectives

1. Summarize the main steps China has taken to transition from a strict Communist country established in 1949 to a more open society with a capitalist type of economy today.

2. Outline the location of the political units of China with the autonomous regions and special administrative units. Establish the connection between the autonomous regions and ethnicity.

3. Understand the population dynamics of China. Learn about the one-child-only policy and evaluate how it has affected Chinese society and culture.

4. Describe how China has shifted its economic policies and structure in the past few decades. Outline the changes in economic development with China's open door to Western trade.

5. Sketch out the historical geopolitical objectives of China and how they relate to China today.

A. The Emergence of Modern China

China is the world's largest Communist country. Isolated from Europeans and Central Asians by the Himalayas and other high mountain ranges, Chinese culture has endured for thousands of years. Rich in history, adventure, and intrigue, a tour through China would reveal a people with a deep, longstanding love of the land, traditions as old as recorded history, and a spirit of commerce and hard work that sustains them to this day. China is about the same land size as the United States, although technically it is slightly smaller than the United States in total area, depending on how land and water areas are calculated. China only has an eastern coast, whereas the United States has both an eastern and a western coast. If you recall the climate types and the relationship between climate and population, you can deduce the location of the heavily populated regions of China.

The form of Communism promoted by Mao Zedong was not the same as the type of Communism practiced in the Soviet Union. Various Communist experiments were forced upon the Chinese people, with disastrous results. For example, in 1958, the Great Leap Forward was announced. In this program, people were divided into communes, and peasant armies were to work the land while citizens were asked to donate their pots and pans to produce scrap metal and increase the country's industrial output. The goal was to improve production and increase efficiency. The opposite occurred, and millions of Chinese died of starvation during this era.

Another disaster began in 1966 and continued until Mao died in 1976. The **Great Proletarian Cultural Revolution** wreaked havoc on four thousand years of Chinese traditional culture in a purge of elitism and a drive toward total loyalty to the Communist Party. Armies of indoctrinated students were released into the countryside and the cities to report anyone opposing the party line. Schools were closed, universities were attacked, and intellectuals were killed. Anyone suspected of subversion might be tortured into signing a confession. Violence, anarchy, and economic disaster followed this onslaught of anti-Democratic terror. Estimates vary, but most sources indicate that about thirty million people lost their lives during the Mao Zedong era through purges, starvation, and conflict.

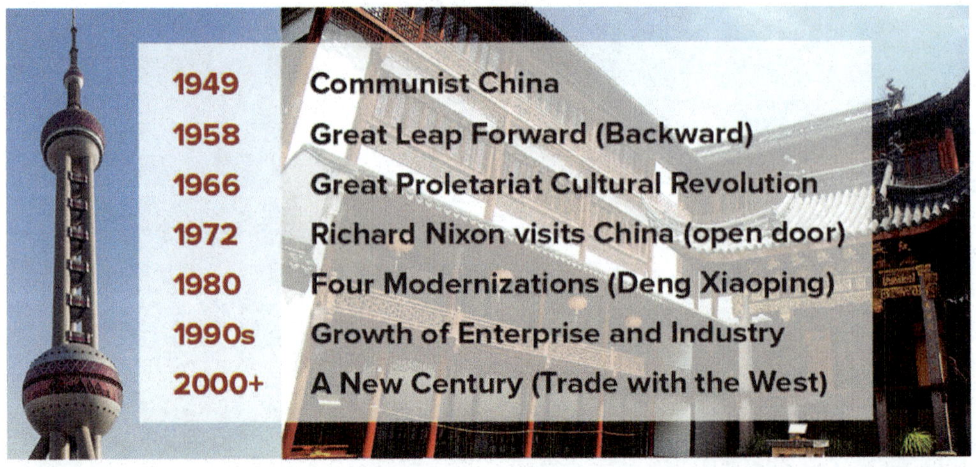

Figure 10. China's Communist timeline with photos of Shanghai Tower (left) and Pagoda.

During the early decades of Communism in China, all travel into or out of China was severely restricted by the so-called Bamboo Curtain. The United States had backed the nationalist movement during the Chinese Civil War and continued to support Chiang Kai-shek in Formosa (Taiwan). The United States did not recognize Communist China; the US embassy was in Taiwan, not in Beijing. As China was experiencing its disastrous experiments with Communism, the core economic areas of the world were advancing with commercial technology and high-tech electronics and sending rockets to the moon. China lagged behind in its industrial activities and became a country based on agriculture. A visit to China by US president Richard Nixon in 1972 signaled the opening of diplomatic relations with the United States and was also viewed as a Cold War move against the Soviet Union.

The Chinese Communist Party's approach when it took power was to institute a "planned economy." A planned economy, sometimes called a command economy, stands in marked contrast to a free market economy. In a planned economy, the government controls all aspects of the economy, including what goods and products should be produced, how much of each should be produced, how products will be sold or distributed and for what (if any) price, who should have jobs and what jobs they should have, how much people will be paid, and all other decisions related to the economy. In a planned economy, businesses are nationalized; that is, businesses are owned by the government rather than by any private entity. By contrast, in a free market economy, businesses are privately owned, and most decisions are driven by consumer and investor behavior.

The decade of the 1980s was a transition for China in that there was a shift of focus from China's Communist economy to a more market-oriented economy. The economic collapse of the Soviet Union in the early 1990s coincided with the opening of China to trade with the West. In 1992, China announced that it would transition to a socialist market economy, a hybrid of a Communist-planned economy and a market economy. A series of statements by China's political lead-

Figure 11. China and its Administrative Divisions, including Autonomous Regions, Municipalities, and Special Administrative Regions (SARs)

ers suggested that in order for China to enjoy a more mature form of socialism, greater national wealth was needed. They further indicated that socialism and poverty should not be considered synonymous and that the country was ready to turn its attention to increasing the wealth and quality of life of its citizens. During the next decade, China experienced an enormous growth in its economy. At the beginning of this century, China was ranked in the top five of the world's largest trading nations, joining ranks with the United States, Germany, Japan, and France. The sheer size of China's population contributes to the magnitude of its economy. This does not, however, mean that most of China's population has a high standard of living.

B. The People

Population

With four thousand years of culture to build on, China continues to press forward into the twenty-first century. In 2010, China had more people than any other country in the world, about 1.33 billion. Most of its people live in China Proper, in the eastern regions of the country. China Proper has the best agricultural lands in the country, the most fertile river basins, and the most moderate climates. For perspective, China has over one billion more people than the United States, with most of those people living in the southeastern portion of China. During Mao's time, there was little concern for population growth, but after his death China implemented measures to deal with its teeming population.

In 1978, China implemented the one-child-policy, limiting family size to one child. The policy allowed for exemptions under certain conditions. Couples living in rural areas and people of minority status are two examples of exempted conditions and could be permitted more than one child, especially if the first child was a girl. Peripheral administrative regions like Macau and Hong Kong were exempt. In 2013, China relaxed this policy somewhat, allowing couples to have two children if one parent was an only child. On January 1, 2016, the one-child policy officially ended, allowing for all married couples to have two children.

The one-child-only policy was implemented in an attempt to address environmental, economic, and social issues related to population growth. This policy helped reduce China's potential population by hundreds of millions of people, but the controversial policy was not easy to implement. There were growing concerns regarding the policy's negative impact on society. In response to the one-child-only policy, there were reports of female infanticide and a higher number of abortions. Some provinces in China have a severe shortage of women because of the policy; men in provinces where women are scarce may have to migrate to find a wife. In China there are more boys born than girls; the ratio averages more than 10 percent more boys than girls, with some provinces reaching more than 25 percent. This imbalance creates cultural issues that may have a negative impact on traditional society.

More than 50 percent of China's population lives in rural areas, meaning there is potential for high levels of rural-to-urban migration. The core industrial cities located in China Proper attract migrants in a periphery-to-core migration pattern. China hopes that this urbanization and industrial activity will also support their population control methods and fuel the industrial labor base. China's percent urban population increased at an unprecedented rate from 17 percent urban in 1978 to 47 percent in 2010. This rural-to-urban shift has been one of the largest in human history. It still continues; many workers shift from holding temporary employment in the cities to returning to their families in the countryside between jobs.

Figure 12. Young Girl and Her Grandfather in Southern China. The one-child-only policy eliminated brothers, sisters, cousins, aunts, and uncles in an extended family.

Language and Religion

Han Chinese are the largest ethnic group in China, making up about 90 percent of the country's population. The main language in China, Mandarin, is spoken in different ways in different parts of the country, particularly in the north and the south. The number of languages in China (over 290) roughly corresponds to the number of ethnic groups in the country. In Western China, where the percentage of Han Chinese is quite low, most of the population is Uyghur, a group that tends to be Muslim. There are also Kazahks, Kyrgyz, and Tajiks from Central Asia, who are also predominantly Muslim. In 2010, the precise number of Muslims in China was not known, but government reports indicated there were about twenty-one million. Other reports indicate there may be three times that many. Minority groups like the Uyghurs have often experienced discrimination by the Chinese government, which has taken measures to marginalize minority groups to keep them in check.

As a Communist country, there is no official religion in China, nor is any supported by the Chinese government. Before Communism took control, most of the people followed a type of Buddhism. Other belief systems included Taoism and the teachings of Confucius. Christianity does exist in China, but it is illegal to proselytize or recruit converts. Despite that, the Christian population in China has grown rapidly in recent decades. The exact number is unknown, but the *CIA World Factbook estimates 3 to 4 percent of the population to be Christian, or somewhere between thirty-nine and fifty-two million adherents.*

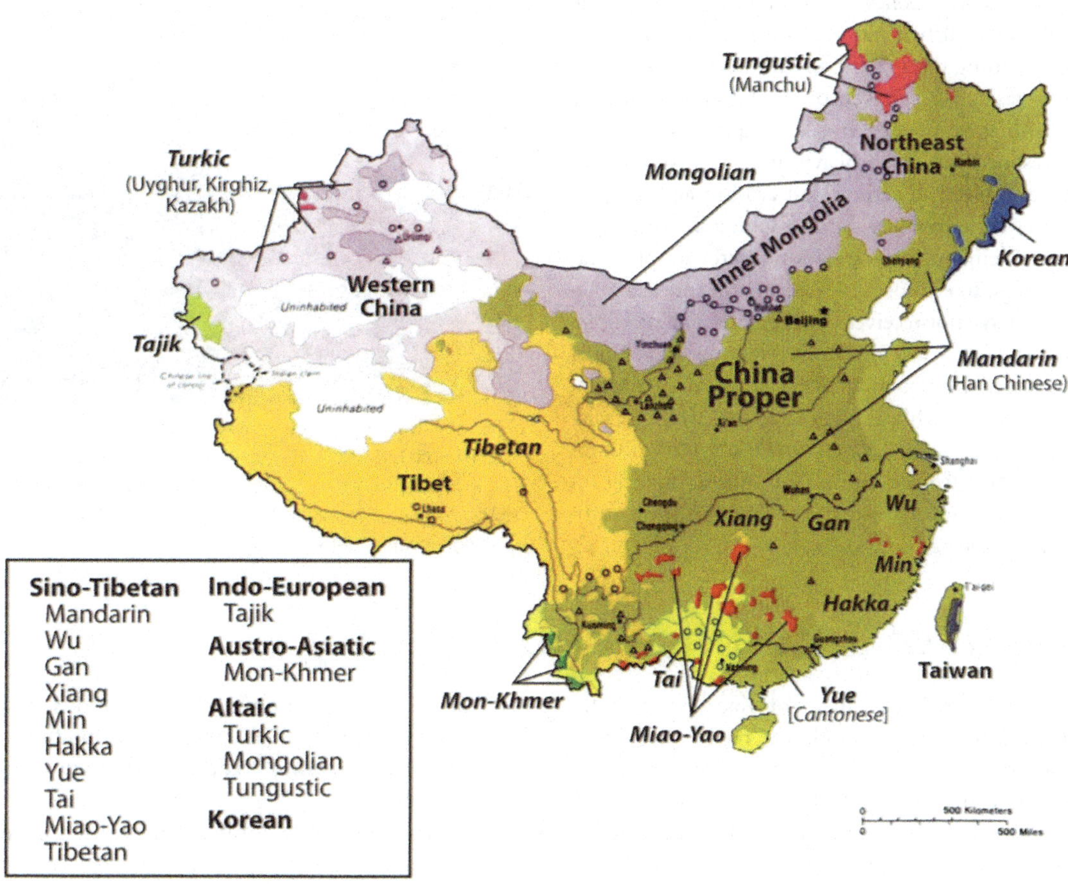

Figure 13. China and Its main Ethnolinguistic regions

C. Economic Summary

In the past few decades, China has shifted its economy from a closed system with a centrally planned, government-controlled market to one with more open trade and a flexible production structure. These economic reforms have allowed capitalistic tendencies to drive production, have promoted increased involvement in private enterprise, and have increased international investment in the Chinese economy. China phased out collective farms and has increased agricultural production; the approach to free enterprise and international trade and investment has become more open; and the Chinese economy has grown at a rapid rate. Open trade and interaction with the global community have allowed China to catch up with the rest of the world in terms of information and communication technology, and computer and Internet use in China has opened up many sectors to new opportunities and employment possibilities. China's policy of creating special economic zones (SEZs) increased urban and economic growth in coastal cities, fueling the strong rural-to-urban shift in the population.

The recent fast-paced growth of the Chinese economy has brought with it some negative consequences. The exploitation of resources and the heavy utilization of the environment have resulted in serious soil erosion and air pollution. The water table in many parts of China has decreased because of heavy demands on the nation's water supply. Arable land is being lost to erosion and inadequate land-use practices. Rural areas have not received consideration or resources equal to the coastal cities, so conditions remain poor for most rural people. Half of China's population earns the equivalent of a few dollars per day, while a fortunate few earn high salaries. Unemployment is at an elevated level for tens of millions of migrants who shift from location to location, looking for work. There is also an unfortunate degree of corruption within the government and state-run offices.

Compared to Western countries, China is an authoritarian state that does not allow labor unions, free speech, freedom of religion, or freedom of the press. There has been more openness in China's economic reforms and in travel, but other strict rules of the state remain. There is no minimum wage law for factory workers, who work long hours and do not receive benefits or sick leave. There are fewer safety requirements or government regulations for security. China is trying to have the best of both worlds: the efficiency of an authoritarian government and an efficient market-driven capitalist economy. Sustaining the largest standing army in the world, China has become a global superpower. The next great world conflict could be a cultural war between the United States and China that would involve economic, political, and human issues.

Figure 14. Beijing. Beijing, a metropolitan center of Chinese culture, is the capital of China. Freeway development has expanded in recent years as higher incomes have allowed for automobile purchases. As a result, traffic congestion and air pollution are part of China's current environment.

D. Growth of Enterprise and Industry

During the 1980s, following the death of Mao Zedong, China went through a transition period. The new leader, Deng Xiaoping, realized that in order for China to compete in the world market, its economy would have to be modernized. The challenge was to open up to the outside without the outside placing pressure on the Communist system inside of China. The so-called Bamboo Curtain, which referred to the restriction of movement of goods and people across Chinese borders, diminished. To attract business and tap into the global market economy, China established Special Economic Zones (SEZs) along the coast at strategic port cities. SEZs attracted international corporations who wanted to manufacture goods cheaply, while China's population

of 1.3 billion people provided an enormous labor pool and consumer market. China's modernization efforts paid off in the 1990s, when world trade increased and US trade with China exploded.

By the year 2000, China had profited greatly because of its expanded manufacturing capacity. The coastal cities and the SEZs had become core industrial centers, attracting enormous numbers of migrants—most of them poor agricultural workers—looking for work in the factories. Compared to other jobs, factory jobs are prized employment opportunities. Rural-to-urban shift has kicked in and China's urban growth is occurring at unprecedented rates. Chinese people are moving to the cities looking for work, just like people in many other areas of the world. The SEZs encourage multinational corporations to move their overseas operations to China and take advantage of the lower labor and production costs. China benefits from the new business opportunities and by the creation of jobs for its citizens. SEZs operate under the objectives of providing tax incentives for foreign companies, exporting market-driven manufactured products, and creating joint partnerships so that everyone benefits. In the past decade, four main cities have been designated as SEZs, along with the entire province of Hainan Island to the south. Many coastal cities were also designated as development areas for industrial expansion.

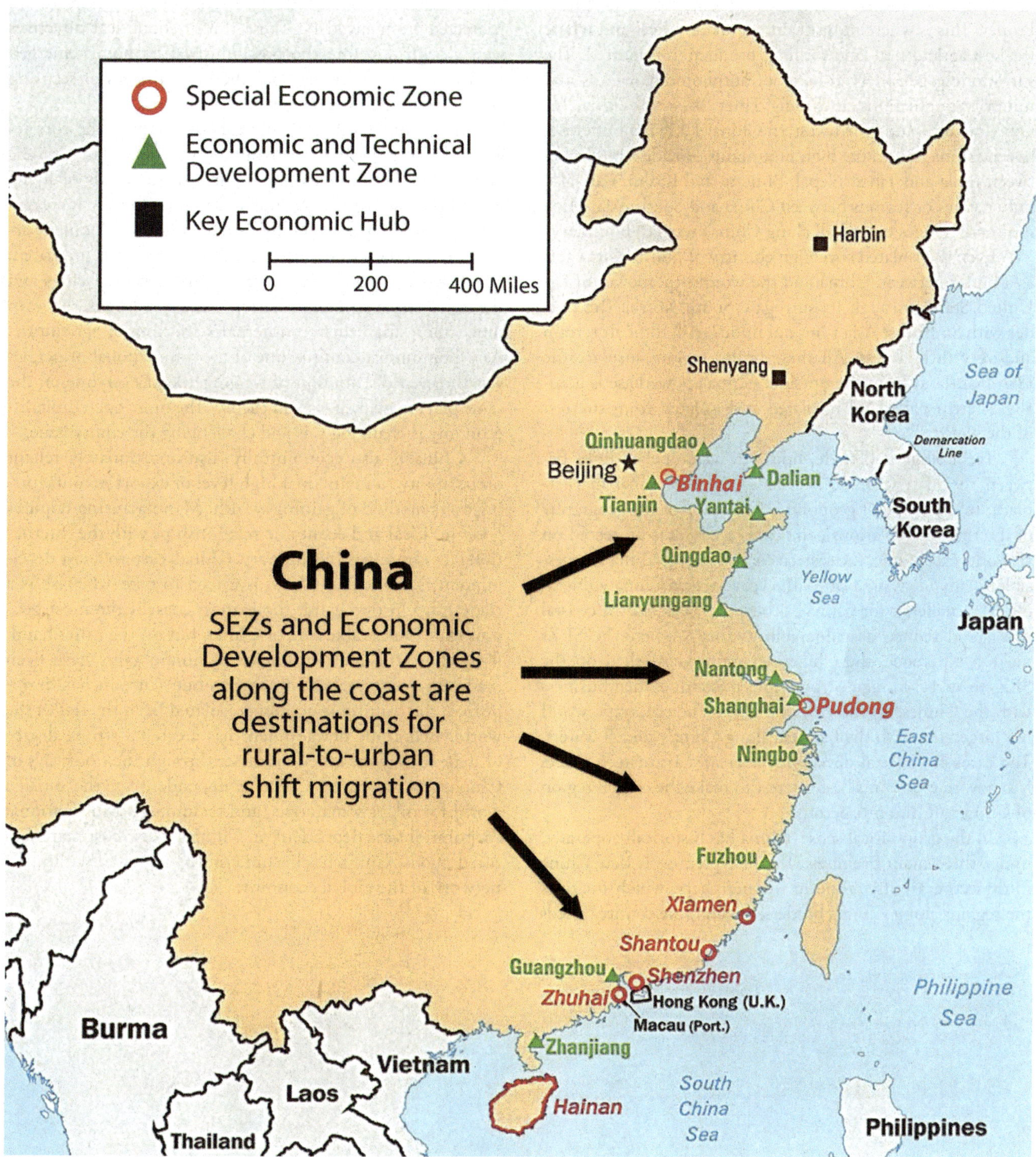

Figure 15. Coastal development and SEZs along the coast of China.

E. Geopolitics

The political environment of China cannot be separated from its geography. In terms of land area, China is a large country. It includes an immense peripheral region surrounding China Proper consisting of minority populations and buffer states. The core of China is the Han Chinese heartland of China Proper. This is where most of the population lives and where the best agricultural land for food production is located. The surrounding periphery consists of autonomous regions and minority populations, including Tibet, Western China, Inner Mongolia, and Northwestern China. Tibet is a buffer between China and India; high mountains provide a buffer between India and Tibet. Nepal, Bhutan, and Kashmir are also buffer states or regions between China and South Asia. High mountain ranges extend all along China's western boundary.

Even the isolated sovereign country of North Korea acts as a buffer between China and the world superpower of the United States through its surrogate South Korea. The border with Southeast Asia is mountainous with difficult terrain and very little access. All these buffer regions insulate the Han heartland from the outside. China's powerbase is actually a territorial island insulated and isolated from the rest of the world.

Technology and trade, however, have opened links between the Han Chinese heartland and the global marketplace. In the scope of geopolitics (the study of the geography of international relations), the core of China is protected on all fronts except one, the eastern coast. Coastal China is accessible to anyone with a ship. The coast makes China vulnerable. The challenge for China has been to neutralize the coastal threat and reduce its vulnerability. This is where the SEZs make a difference: the Chinese economy has relied on the SEZs to attract foreign corporations to conduct their business with the Chinese in these coastal cities. The economic world has forged ties with the Chinese along China's coastal waters. The extensive coastal development creates a situation where it is now in everyone's best interest to make the coastal region of China safe and prosperous.

In the geopolitical sense, China has historically operated under three main premises. The first premise is that China must secure a buffer zone in the periphery, which includes the regions along China's borders. Second, the country needs to continue to maintain unity within the Han Chinese majority that anchors the core region of China Proper. The third component of China's geopolitical strategy is to protect and secure its vulnerable coastal region. The country needs to find a way to marginalize outside influence and keep the heartland protected from invaders. These three geopolitical directives were designed before the postindustrial economy emerged with the information age, but the main themes of securing and protecting China remain.

China has additional political issues within the core region of China Proper. To maintain a unified Han Chinese powerbase, there should not be uneven standards of living within the core region. Manufacturing has rapidly increased in the coastal cities of the SEZs and other development areas. The standard of living in the coastal cities has improved, which has created a disparity between the coastal cities and the rural regions of the interior. Factory workers can earn much more than their counterparts working in agriculture. As a Communist country, one of the foundational principles was the even distribution of wealth. It kept everyone on the same playing field, at least in theory. The more open capitalist economy is changing this and challenging the equity issue.

China is also economically disproportionately reliant on its ability to maintain a high level of export manufacturing as its method of gaining wealth. Manufacturing requires good political and economic relationships with the international community. In summary, China's eastern coast development is a risk that China is forced to take. The risk is a trade-off. On one hand, the eastern coast is the most geopolitically vulnerable side of China, but on the other hand, it is where the most extensive economic gains have been made in recent years. China's Bamboo Curtain has disappeared; the country is no longer isolated from the rest of the world. Economic development has created a strong degree of dependency on other countries through the doorways of China's coastal cities. China is dependent on the outside world for oil, raw materials, and technology. Multinational corporations are dependent on China for low-cost manufactured goods. China has become more integrated within the network of the global economy.

Key Takeaways

1. China has embarked on a number of serious transitions since the Communist republic was declared in 1949. Since that time, the country has shifted from a command economy to a market economy and has opened up trade to the West. This has opened up more opportunities for the Chinese people as well.

2. China has a number of autonomous regions and municipalities. Many of these political units are also home to minority ethnic groups. Han Chinese make up about 90 percent of the people of China.

3. China's large population prompted the government to implement a one-child-only policy, which resulted in a decrease in population growth but has also caused an imbalance in the percentages of boys and girls being born in many of the provinces. That policy officially ended on January 1, 2016, and now married couples are allowed to have two children.

4. The creation of SEZs along the coast, and an open-trade policy, have rapidly developed China's manufacturing sector. This trend has attracted multinational corporations to move their operations to China to tap into the low labor and overhead costs and receive tax incentives from the Chinese government. The result has been an increase in rural-to-urban shift in China's population.

5. China's historical geopolitical situation was to secure a buffer around its heartland, unify the Han majority, and protect its venerable coastal region. Economic trade and interaction with the global economy has made China dependent on other countries for its continual economic success.

12.3 China's Periphery

Learning Objectives

1. Outline Hong Kong's progression from a British colony through the transformation into a special autonomous region of China. Describe why Hong Kong and Shenzhen have been important to the development of the Chinese economy.

2. Understand the development of and reasons for a government for Taiwan that is separate from the government of mainland China. Describe the current relationship between Taiwan and mainland China.

3. Explain the challenges that Tibet has experienced in becoming an autonomous region of China. Understand who the Dalai Lama is and describe his traditional position within Tibet.

4. Summarize the role Mongolia played during the Cold War and learn how that role is changing as Mongolia becomes more involved in the global economy.

5. Explain why the peripheral regions of Tibet, Hong Kong, and Taiwan are important to the Chinese government and the vitality of the country.

The heartland of China has always been China Proper. Surrounding China Proper are a number of autonomous regions, associated territories, and independent countries. These peripheral areas buffer China Proper from the rest of the world. They also provide China with either access to trade relationships or access to raw materials needed for development. Each peripheral political unit has its own unique physical and cultural characteristics and, in many cases, a minority population that is other than Han Chinese.

A. Hong Kong and Shenzhen

Hong Kong is a former British colony and includes a number of islands on the southern coast of China. Hong Kong includes an excellent port, Victoria—one of the best in Asia. Victoria's deepwater port is on the interior side of Hong Kong Island and is protected from the sea, allowing ship access to an extensive sheltered harbor area. To the north of Hong Kong Island is a peninsula called New Territories, which the British agreed to lease from China in 1898 for a period of ninety-nine years. Bordering the New Territories on the mainland is the rapidly growing industrial city of Shenzhen.

Hong Kong was a major entry point into China for Britain during the colonial era. The colony was a British outpost that created a doorway for British expansion into China. The port allowed for an early trade relationship to become established. Ships would import raw materials, which would be processed or manufactured into finished products. The products would then be shipped back out at a profit.

Manufacturing is traditionally the best means for countries to gain wealth, since it has the highest value-added profits. Hong Kong is a good example of this; manufacturing was a successful method of gaining wealth for Hong Kong. First, low-level goods such as clothing and textiles were produced

with Hong Kong's cheap labor and low costs. As business increased, higher-level technical goods were produced, such as radios and other electronic products. Incomes increased. Soon, Hong Kong was a business and banking center trading with all global markets. International trade attracted a strong financial network that enhanced the free market system in Hong Kong. Business in Hong Kong attracted attention in Beijing. Chinese leader Deng Xiaoping, who had begun China's open door policy in the 1980s, viewed Hong Kong as a major access point to establish trade and commerce with. In the 1990s, the special economic zone of Shenzhen became an important development area for Chinese industries and multinational corporations. China once again used Hong Kong as a major trade corridor so that about one-fourth of all the country's imports and exports were being shipped through the port of Hong Kong.

In 1997, the ninety-nine-year lease of the New Territories to Britain expired. Hong Kong became associated with mainland China as a special autonomous region, but remained capitalist and democratic in its operations. This opened the door for Taiwan and other trading partners to increase trade with China through Hong Kong. Shenzhen, a special economic zone (SEZ) across the border from Hong Kong in China, was ready to capitalize on its accessibility to the port and the enormous trade that had been established by Hong Kong. Shenzhen became one of the fastest growing cities in the world and has become a center of manufacturing and trade for the global economy. It grew from a moderately sized city of about three hundred fifty thousand in the early 1980s to a city of ten and a half million by 2010, and is still growing. Shenzhen has established a port of its own and is a magnet for international trade.

More than 95 percent of the seven million people living in Hong Kong are ethnically Chinese. The people have strong ties to mainland China but highly value their separate and independent economic and political status. Hong Kong is a major financial and banking center for Asia and has been working with the Chinese government to provide private banking services for Chinese citizens. The small land size of Hong Kong makes it a high-priced real estate destination. The cost of living is high, and space is at a premium and expensive. Nevertheless, Hong Kong attracts millions of visitors per year and has established itself as a tourism hub for people desiring to visit southern China. Tens of millions of tourists each year use Hong Kong as a base or stopover point to enter China's southern provinces. Hong Kong offers visitors immense shopping possibilities in a safe and modern environment that is attractive to people from all over the world. Cantonese is the official language, but English is widely spoken in Hong Kong because of Britain's influence and because of world trade relationships.

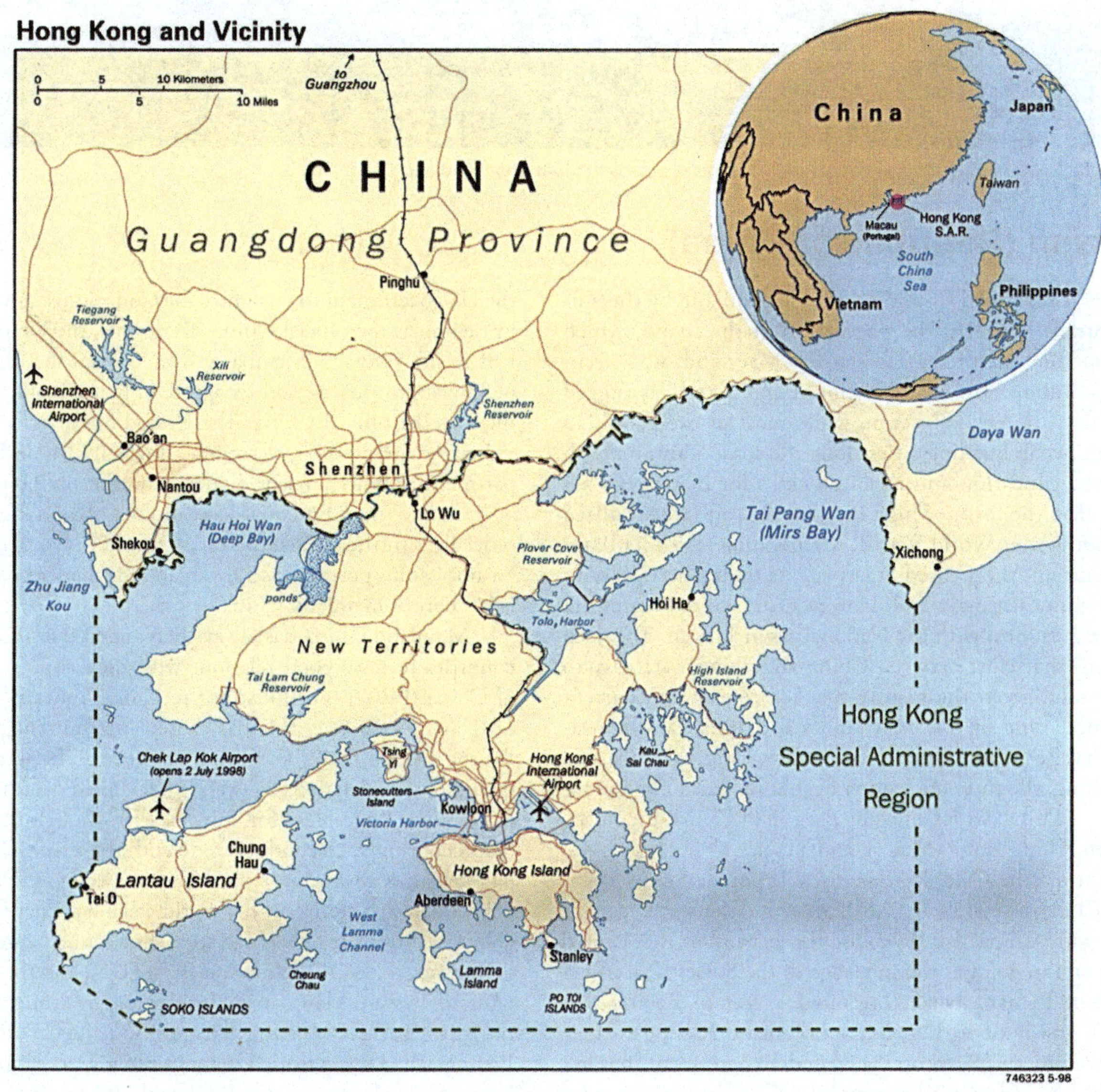

Figure 16. The Special Administrative Region (SAR) of Hong Kong, with Shenzhen across the Border from the New Territories.

Figure 17. Taiwan (ROC). Taiwan is located just across the Taiwan Strait from Mainland China (PRC).

Taiwan (Republic of China)

Taiwan is separated from the Chinese mainland by the Taiwan Strait. In 2010, the population of the island, which is about the size of the US state of Maryland, was twenty-two million. The island is mountainous and has rugged national parks in its interior, while most of the people in Taiwan live in bustling cities along the coast. Taiwan's push for good education and technical skills for its citizens has paid off in the form of high incomes and an industrialized economy. After World War II, when China raised its Bamboo Curtain, the United States cut its trade relations with the Communist mainland. However, the United States did trade a great deal with the Nationalists in Taiwan. The deep tensions that exist between China and Taiwan are easy to understand given the history that led to Taiwan's current existence. One of the flash points in China-Taiwan relations has been the "One China" policy. Originally asserted by Beijing, the principle contends that there is only one indivisible China, which encompasses all of China, including Taiwan, Hong Kong, Macau, and Tibet.

During the Cold War, the sizeable population of mainland China was attractive to US corporations that sought trade with China. In 1972, for the benefit of the United States and to counter the influence of the Soviet Union, US president Richard Nixon organized a visit to Beijing that opened the door to US-China diplomatic relations. There were conditions placed on the open door agreement that followed. One condition that mainland China imposed was for the US government to recognize the Communist government in Beijing as the official China. To do this, the United States had to move their US embassy from Taipei, the capital city of Taiwan, to Beijing and recognize only the Beijing government as the official China. The United States agreed to this policy. As a result, no government official in the US State Department, cabinet level or higher, can officially visit Taiwan; at the same time, no equal-level diplomat from Taiwan can officially visit the United States. This policy placed Taiwan in an ambiguous position: Is Taiwan an independent country or is it a part of mainland China?

Mainland China has always demanded that in order for countries to have good relations with the People's Republic of China (PRC), they must not recognize Taiwan (Republic of China, or ROC). This has caused intermittent problems for the United States and other countries. The termination of diplomatic relations between the United States and the ROC was a huge blow for Taiwan, even though the United States continued to conduct trade with the island. The United States has always supported Taiwan's economic independence. Taiwan's status on the world stage has fluctuated. For about twenty years after the Nationalists set up a government in Taiwan, the ROC held a seat in the UN, and the UN recognized Taiwan's claim as the legitimate government of China. In 1971, the UN changed its position and gave the seat to Beijing. The UN indicated it could not give a seat to Taiwan because Taiwan is not recognized as a legal country.

Figure 18. Map of Taiwan.

Taiwan's economy continues to use its skilled labor base. In years past, Taiwan's factories produced textiles and light clothing. As incomes and skills advanced, production shifted to electronics and high-end goods, which it has since been exporting around the world. Taiwan has achieved a high standard of living for its people. When trade opened up through Hong Kong with mainland China, many Taiwanese industries shifted production to the SEZs along China's coast. Labor in SEZs is less expensive, leading to a corresponding rise in profits. Unfortunately, this has resulted in higher unemployment levels in Taiwan.

In the past few years, Taiwan's focus has turned from secession to trade. In 2008, Taiwan elected a government that, though nationalist in its ideology, was committed to strengthening trade and investment with China. Taiwan entered the World Trade Organization (WTO) in 2001 using the name Chinese Taipei. That move started a cascade of actions that improved "cross-strait" trading conditions, and within just a few years, direct trade between China and Taiwan has become extensive. The question of Taiwan's political status is as yet unresolved and is clearly still volatile, but trade and investment between China and Taiwan are robust and growing. Taiwan has been able to independently compete in world-class sporting events, including the Olympics, under the name Chinese Taipei.

Figure 19. Street in Taiwan, ROC, illustrating modern businesses and use of scooters for transportation.

B. Autonomous Region of Tibet

Located in mountainous southwestern China, Tibet is classified as one of China's autonomous regions, a disputed political arrangement. It is debatable whether Tibet or any of the other Chinese regions are actually autonomous. The legal structures actually allow for very little self-governance, and most new initiatives require approval from the Chinese central government. Tibet had been an independent entity through much of its history and governed itself as a Buddhist theocracy. Its theocratic political system established the Dalai Lama as both the head of state and the religious leader of the Tibetan people.

Tibet came under the control of China during the Qing Dynasty. When imperialist rule ended in China in 1911, Tibet began to once again assert its independence. Chinese, British, and Tibetan officials met and came up with an agreement that partitioned Tibet into Inner Tibet, which would be controlled by China, and Outer Tibet, which would be independent. China later indicated its intention to control all of Tibet, a move greatly resented by the Tibetan people.

When Communists took control of China in 1949, the Dalai Lama was originally allowed control over domestic affairs while China would control all other governmental

Figure 20. Potala Palace in Lhasa, Tibet. Potala Palace in Lhasa, Tibet, rests at 12,100 feet above sea level. This is the former seat of government of the Dalai Lama in Tibet and is a UNESCO World Heritage Site.

Figure 21. The fourteenth Dalai Lama and Tibetan leader in exile.

functions. The Chinese government then took steps to greatly reduce the powers of the Dalai Lama and the Panchen Lama (spiritual leader). With the intention of spreading Communism to Tibet, China destroyed monasteries, religious symbols and icons, and other long-practiced ways of life were threatened. The Tibetans considered the Chinese political dominance a cruel and invading force. Monasteries were burned, monks were beaten or imprisoned, and Buddhism was brutally suppressed.

Tibetans revolted in 1956. Backed by the United States, the revolt continued even as China indicated it would suspend the transformation of Tibet. China brutally crushed the revolt in 1959, leaving tens of thousands of Tibetans dead or imprisoned. The Dalai Lama and thousands of Tibetans fled to Dharamsala, India, where they established a government in exile. At this point, China had a firm grip on Tibet, and in 1965 the reorganization of Tibet into a Chinese socialist region began in full force.

The population of Tibet is only about three million. They mainly live in the mountain valleys, below seven thousand feet. Lhasa is Tibet's main urban center as well as its capital. In the past couple of decades, China has softened its treatment of Tibetans and restored some of the monasteries. China has also moved thousands of ethnically Chinese people into Tibet to shift the ethnic balance of the population. A major rail line now directly connects Beijing with Lhasa, Tibet. Lhasa will soon have a majority Chinese population with cultural similarities that reflect more the attributes of China Proper than Tibet. The Tibetans who live there are being compromised simply by population dilution.

As indicated previously, China wants to maintain the mountainous, rural region—with little inhabitable land area and very few people—as a buffer state between India and the Han Chinese heartland of China Proper. Tibet is also becoming more important to the government of Beijing and the business people in Shenzhen and Shanghai. To understand why this is, consider the core-peripheral spatial relationship. What do the core and the periphery usually have to offer each other? Consider also how countries gain wealth. Tibet can play an important role in the larger picture of the Chinese economy. The mountains of Tibet potentially hold vast amounts of natural resources that could support the growing industrial economy in China, which is based on manufacturing goods for the global export markets. China will continue to see to it that Tibet becomes economically integrated with the rest of China in spite of the cultural differences that may exist.

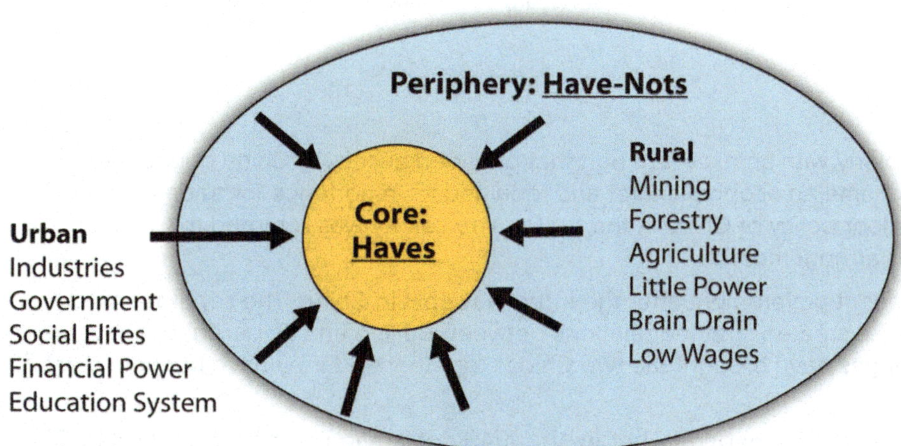

People usually shift from periphery to core.

Figure 22. Periphery: Have-Nots. The core-peripheral spatial relationship can be appropriately applied to Tibet and its relationship with China. Tibet is a peripheral region with raw materials that provide benefits to the core areas of China, which require the raw materials for industrial development

C. Mongolia

Mongolia shares a similar geography with much of Kazakhstan, which is the world's largest landlocked country; Mongolia is second-largest. Despite Mongolia's large land area (slightly smaller than the US state of Alaska or the country of Iran), its population is only about three million, and the country is the least densely populated country in the world. Mountains, high plains, and grass-covered steppe cover much of Mongolia, a country that receives only between four and ten inches of precipitation per year, most of it in the form of snow. The Gobi Desert to the south, extending from southern Mongolia into northern China, receives even less precipitation. Inner Mongolia is a sparsely inhabited autonomous region south of Mongolia that is governed by China.

Mongolia's modern capital city of Ulan Bator is home to about one-third of the people of Mongolia; it has the coldest average temperature of any world capital. Mongolia's history includes the great Mongol Empire of Genghis Khan, which was established in the thirteenth century. The Soviet Union used Mongolia as a buffer state with China. Mongolia's Communist party dominated politics until the Soviet Union collapsed in the 1990s. The current government has to contend with Mongolia's position between the two large economies of Russia and China.

Tibetan Buddhism is the dominant religion in Mongolia and is practiced by about 50 percent of the population. The Communist influence is evident in that approximately 40 percent of the population considers itself nonreligious.

Figure 23. Men wrestling on the rugged grasslands of Mongolia. Wrestling is a major sport of the country.

Most of the people are of Mongol ethnicity. Today, about 30 percent of the people are still seminomadic and migrate seasonally to accommodate good grazing for their livestock. A round yurt, which can be constructed at whatever location is selected for the season and disassembled for mobility, is the traditional dwelling. Mongolian culture and heritage revolve around a rural agrarian culture with an extensive reliance on the horse. Archery, wrestling, and equestrian events are some of the most popular sporting activities.

Mongolia's economy has traditionally centered on agriculture, but mining has grown in recent years to be a major economic sector. Mongolia has rich mineral resources of coal, molybdenum, copper, gold, tin, and tungsten. Being landlocked cuts Mongolia off from the global economy, but the large amounts of mineral reserves are in demand by core industrial areas for manufacturing and should boost the poor economic conditions that dominate Mongolia's economy. China has increased its business presence in Mongolia and has been drawing Mongolia's attention away from its former Soviet ally to become a major trading partner.

Key Takeaways

1. Hong Kong was a British colony with an excellent port that provided access to China during the colonial era. Hong Kong became an economic tiger and regained its importance for trade with the mainland during the open door policy of China in the 1980s. The colony was returned to China in 1997 under an autonomous arrangement.

2. Taiwan created an independent government after the three-way split in China. The separation from the mainland government has created strained relations between Beijing and Taipei. In recent years, trade and economic exchanges have brought the two Chinas together and they have become more dependent on each other.

3. Tibet established itself as a Buddhist theocracy before the 1949 declaration of the Communist PRC. The Dalai Lama ruled as both political and religious leader of Tibet until China took control. The Dalai Lama fled to India, where he set up a government in exile to continue to promote Tibetan culture and support Tibet.

4. Mongolia is an independent landlocked country with a small population and a large land area. It receives only modest amounts of rainfall. The country was a historic buffer state between the Communist countries of the Soviet Union and China. Mongolia is working to capitalize on its natural resources to engage the global economic situation.

12.4 Japan and Korea (North and South)

Learning Objectives

1. Summarize the physical geography of the Korean peninsula.

2. Understand how North Korea and South Korea developed independently.

3. Outline the political structure of North Korea.

4. Describe how South Korea developed a high standard of living and a prosperous economy.

5. Express how the concept of regional complementarity can apply to the Koreas.

A. Japan

Japan consists of islands that lie along the Pacific Rim east of China and across the Sea of Japan from the Korean Peninsula. Most of the archipelago, which has more than three thousand islands, is just north of 30° latitude. Four islands make up most of the country: Shikoku, Kyushu, Hokkaido, and Honshu. The islands have tall mountains originating from volcanic activity. Many of the volcanoes are active, including the famous Mt. Fuji, located just west of Tokyo, which is a widely photographed mountain because of its symmetrical volcanic cone. The physical area of all Japan's islands is equivalent to about the size of the US state of Montana. The mountainous islands of Japan extend into two climate zones. All but the northern region of Japan has a type C climate. The island of Hokkaido in northern Japan has a type D climate and receives enough snow for downhill skiing. Japan hosted the Winter Olympics there in 1998.

Cities like Mexico City, São Paulo, and Bombay, which have vast slum communities, each claim to be

the world's largest city but lack firm census data to verify the population. The largest metropolitan area in the world that can be verified is Tokyo, with a population of 37.8 million

Figure 24. Japan's four main islands and its core urban and industrial areas.

in 2016, just one million fewer than the entire US state of California and two million more than all of Canada. The Tokyo metropolitan area is located in an extensive agricultural region called the Kanto Plain and includes the conurbation of Tokyo, Yokohama, and Kawasaki. The second-largest urban area in Japan is located in the Kansai District, and includes the cities of Kobe, Osaka, and Kyoto. Japan is a mountainous country, and most of its large cities are located in low-lying areas along the coast. Most of the population (67 percent) lives in urban areas such as Japan's core area, an urbanized region from Tokyo to Nagasaki. The 2016 population of Japan was about 125 million, which is less than 40% of the

Figure 25. Expanse of the metropolitan area of Tokyo, with Mt. Fuji in the Background.

population of the United States. It is ironic that the world's largest metropolitan area is built on one of the worst places imaginable to build a city. Tokyo is located where three tectonic plates meet: the Eurasian Plate, the Philippine Plate, and the Pacific Plate. Earthquakes result when these plates shift, leading to possibly extensive damage and destruction. In 1923, a large earthquake struck the Tokyo area and killed an estimated 143,000 people. In 1995, an earthquake near Kobe killed about 5,500 people and injured another 26,000.

In March 2011, a magnitude 9.0 earthquake struck forty-three miles off the eastern coast of northern Japan. The earthquake itself caused extensive damage to the island of Honshu. A shockwave after the earthquake created a tsunami more than 130 feet high that crashed into the eastern coast of Japan causing enormous damage to infrastructure and loss of life. This was the strongest earthquake to hit Japan in recorded history. It resulted in more than 15,500 deaths and billions of dollars in damages. Nuclear power plants along the coast were hit hard by the tsunami, which knocked out their cooling systems and resulted in the meltdown of at least three reactors. The nuclear meltdowns created explosions that released a sizeable quantity of nuclear material into the atmosphere. This is considered the worst nuclear accident since the 1986 reactor meltdown at the Soviet plant at Chernobyl. The next big earthquake could happen at any time, since Japan is located in an active tectonic plate zone.

Development

The country of Japan is an interesting study in isolation geography and economic development. For centuries, shogun lords and samurai warriors ruled Japan, and Japan's society was highly organized and structured. Urban centers were well planned, and skilled artisans developed methods of creating high-grade metal products. While agriculture was always important, Japan always existed as a semi urban community because the mountainous terrain forced most of the population to live along the country's coasts. Without a large rural population to begin with, Japan never really experienced the rural-to-urban shift common in the rest of the world. Coastal fishing, always a prominent economic activity, remains so today. The capital city of Tokyo was formerly called Edo and was a major city even before the colonial era. Japan developed its own unique type of urbanized cultural heritage.

Encounters with European colonial ships prompted Japan to industrialize. For the most part, the Japanese kept the Europeans out and only traded with select ships that were allowed to approach the shores. The fact that European ships were there at all was enough to convince the Japanese to evaluate their position in the world. During the colonial era, Britain was the most avid colonizer with the largest and best-equipped navy on the high seas. Britain colonized parts of the Americas, the Middle East, Africa, South Asia, and Australia and was advancing in East Asia when they encountered the Japanese. Both Japan and

Great Britain are island countries. Japan reasoned that if Great Britain could become so powerful, they should also have the potential to become powerful. Japan encompassed a number of islands and had more land area than Britain, but Japan did not have the coal and iron ore reserves that Britain had.

Around 1868, a group of reformers worked to bring about a change in direction for Japan. Named after the emperor, the movement was called the Meiji Restoration (the

Figure 26. Japanese Colonialism. Japan was a major colonial power in the Asian Pacific Rim. Japan colonized parts of Russia, China, Taiwan, Philippines, Southeast Asia, and Korea.

return of enlightened rule). Japanese modernizers studied the British pattern of development. The Japanese reformers were advised by the British about how to organize their industries and how to lay out transportation and delivery systems. Today, the British influence in Japan is easily seen in that both Japanese and British drivers drive on the left side of the road. The modernizers realized that to compete on the high seas in the world community they had to move beyond samurai swords and wooden ships. Iron ore and coal would have to become the goods to fuel their own industrial revolution. Labor and resources were valuable elements of early industrialization in all areas of the world, including Asia.

Japan began to industrialize and build its economic and military power by first utilizing the few resources found in Japan. Since it was already semi urban and had an organized social order with skilled artisan traditions, the road to industrialization moved quickly. Japan needed raw materials and expanded to take over the island of Formosa (Taiwan) and the Korean Peninsula in the late nineteenth and early twentieth centuries. Japan was on its way to becoming a colonial power in its own right. They expanded to take control of the southern part of Sakhalin Island from Russia and part of Manchuria (Northeast China) from the Chinese. Japanese industries grew quickly as they put the local people they subjugated to work and extracted the raw materials they needed from their newly taken colonies.

The three-way split in China revealed that the Japanese colonizers were a major force in China even after the other European powers had halted their colonial activities. Japan's relative location as an independent island country provided both quick access to their neighbors and also protection from them. By World War II, Japan's economic and military power had expanded until they were dominating Asia's Pacific Rim community. The Japanese military believed they could invade the western coast of North America and eventually take control of the entire United States. Their attack on Pearl Harbor, Hawaii, in 1941, was meant to be only a beginning. History has recorded the outcome. The United States rallied its people and resources to fight against the Japanese in World War II. The Soviet Union also turned to fight the Japanese empire. The end came after atomic bombs, one each, were dropped by the United States on the Japanese cities of Hiroshima and Nagasaki. The terms of Japan's surrender in 1945 stipulated that Japan had to give up claims on the Russian islands, Korea, Taiwan, China, and all the other places they had previously controlled. Japan also lost the Kurile Islands—off its northern shores—to the Soviet Union. Japan offered Russia an enormous amount of cash for them, but the islands have never been returned. Japan also agreed not to have a military for offensive purposes. Japan was decimated during World War II, its infrastructure and economy destroyed.

Figure 27. Shinto shrine in Nagasaki. This photo, taken after the atomic bomb blast of World War II in 1945, illustrates all that was left standing of a Shinto shrine in Nagasaki.

Economic Growth

Since 1945, Japan has risen to become Asia's economic superpower and the economic center of one of the three core areas of the world. Japanese manufacturing has set a standard for global production. For example, think of all the automobiles that are Japanese products: Toyota, Honda, Mitsubishi, Subaru, Nissan, Isuzu, Mazda, and Suzuki. How many Russian, Brazilian, Chinese, or Indonesian autos are sold in the United States? The term "economic tiger" is used in Asia to indicate an entity with an aggressive and fast-growing economy. Japan has surpassed this benchmark and is called the economic dragon of Asia. The four economic tigers competing with Japan are Singapore, Hong Kong, Taiwan, and South Korea.

Japan's recovery in the last half of the century was remarkable. What was it that caused the Japanese people to not only recover so quickly but rise above the world's standard to excel in their economic endeavors? The answer could be related to the fact that Japan already had an industrialized and urbanized society before the war. The United States did help rebuild some of Japan's infrastructure that had been destroyed during World War II—things like ports and transportation systems to help bring aid and provide for humanitarian support. However, the Japanese people were able to not only recover from the devastation of World War II but rise to the level of an economic superpower to compete with the United States. Japan used internal organization and strong centripetal dynamics to create a highly functional and cohesive society that focused its drive and fortitude on creating a manufacturing sector that catapulted the country's economy from devastation to financial success.

The loss of colonies after World War II prompted Japan to look elsewhere for its raw materials. Extensive networks of trade were developed to provide the necessary materials for the rapidly growing manufacturing sector. The urbanized region around Tokyo, including the cities of Yokohama and Kawasaki, swelled to accommodate the growth of industry. Japan's coastal interior has a number of large cities and urban areas that rapidly increased with the growing manufacturing sector.

The basis for Japan's technological and economic development may be related to its geographic location and cultural development. Japan's pattern of economic development has similarities to the patterns of other Asian entities that have created thriving economic conditions. Hong Kong and Taiwan are both small islands with few natural resources, yet they have become economic tigers. Taiwan exports computer products to the United States, but Russia, with all its natural resources, millions of people, and large land area, does not. Explaining how or why countries develop at different rates can be complex, because there may be many interrelated explanations or reasons. One element might be centripetal cultural forces that hold a country together and drive it forward. The ability to manage labor and resources would be another part of the economic situation. The theory of how countries gain wealth may shed some light on this issue. Japan used manufacturing as a major means of gaining wealth from value-added profits. This is the same method the Asian economic tigers also developed their economies. Japan developed itself into a core economic country that took advantage of the peripheral countries for labor and resources during the colonial era. Japan took advantage of every opportunity that presented itself to become a world manufacturing center.

Modern Japan

Japan is a homogeneous society. In 2010, all but one percent of the population was ethnically Japanese. Japan is a good example of a nation-state, where people of a common heritage and aspirations hold to a unified government. This provides a strong centripetal force that unites the people under one culture and one language. However, religious allegiances in Japan vary, and there are a good number of people who indicate nonreligious ideologies. Shintoism and Buddhism are the main religious traditions. Shintoism includes a veneration of ancestors and the divine forces of nature. There is no one single written text for Shintoism; the religion is a loosely knit set of concepts based on morality, attitude, sensibility, and

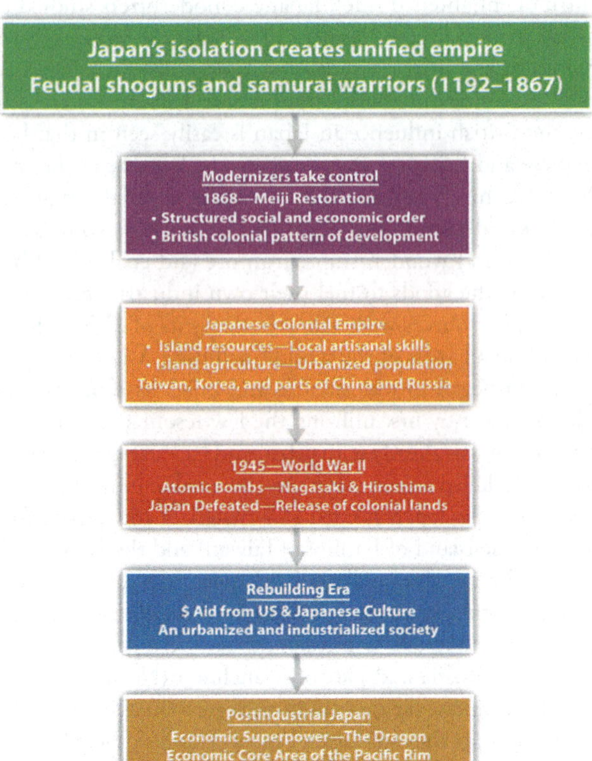

Figure 28. Timeline of Japanese Development.

right practice. It is possible for a Shinto priest to conduct a wedding for a couple, whereas the funeral (hopefully taking place much later) for either of the spouses might be conducted by a Buddhist priest. The Buddhism that is practiced in Japan is more meditative in nature than mystical.

Modernization has brought new opportunities and problems for the Japanese people. With a highly urban and industrial society, personal incomes in Japan are high and the fertility rate is quite low, at only about 1.2 children. The replacement level for a given population is about 2.1 children, which means Japan has a declining population. The population is aging and there are not enough young people to take entry-level jobs. The family has been at the center of Japanese tradition, and the elderly have been venerated and honored. As the country has become more economically developed, higher incomes for young people have prompted a shift toward convenience and consumer amenities. Consumer goods are available, but all the oil and about 60 percent of the food products have to be imported. Income levels are high, but the cost of living in places like Tokyo is also quite high.

Japan is facing a labor shortage for many low-level service jobs that had traditionally been held by young people entering the workforce. Japan is now in stage five of the demographic transition, and is at the start of a serious population decline. While Western Europe and the United States also have a declining percentage of young people, in those

Figure 29. Centripetal v Centripetal Forces (Japan). What cultural centripetal forces have been holding Japan together?

countries the declining youth cohort has been replaced by an elevated level of immigration, both legal and illegal. Culturally, this is not an attractive option for Japan, but they are forced to address the labor shortage in any way they can. Their economy will depend on the ability to acquire cheap labor and resources to compete in the global marketplace.

B. Korea

The Korean Peninsula juts out into the Pacific Ocean from northeastern Asia. The peninsula is bound by the Sea of Japan (the East Sea) and the Yellow Sea. North and South Korea share the peninsula. These countries have been separated by the Korean demilitarized zone (DMZ) since 1953. To the west and north is China and to the far north along the coast is Russia. Korea is separated from China by the Yellow Sea and by the Yalu and Tumen Rivers to the north, which form the actual border between North Korea and China. Japan is located just east of the Korean Peninsula across the Korea Strait. The capital city of North Korea is Pyongyang, and Seoul is the capital of South Korea.

Approximately 70 percent of the Korean Peninsula is mountainous, with relatively little arable land. The highest peak in North Korea rises more than nine thousand feet. The Korean Peninsula has four general areas:

1. Western Region with an extensive coastal plain, river basins, and small foothills

2. Eastern Region with high mountain ranges and a narrow coastal plain

3. Southeastern Basin

4. Southwestern Region of mountains and valleys

North and South Korea have very different, yet related, environmental issues, primarily related to the degree of industrialization. With a low level of industrialism, North Korea has severe issues of water pollution as well as deforestation and related soil erosion and degradation. Other issues in North Korea include inadequate supplies of clean drinking water and many waterborne diseases. South Korea has water pollution associated with sewage discharge and industrial effluent from its many industrial centers. Air pollution is severe at times in the larger cities, which also contributes to higher levels of acid rain.

For centuries, Korea was a unified kingdom that was often invaded by outsiders. After the fall of the Chinese Qing (Manchu) Dynasty, the Japanese took control of the Korean Peninsula (1910) and controlled it as a colony. Japanese colonial rule ended in 1945 when the United States defeated Japan, forcing it to give up its colonial possessions.

The conclusion of World War II was a critical period for the Korean Peninsula. The United States and the Soviet Union both fought against the Japanese in Korea. When the war was over, the Soviet Union took administrative control of the peninsula north of the thirty-eighth parallel, and the United States established administrative control over the area south of the thirty-eighth parallel. The United States and the

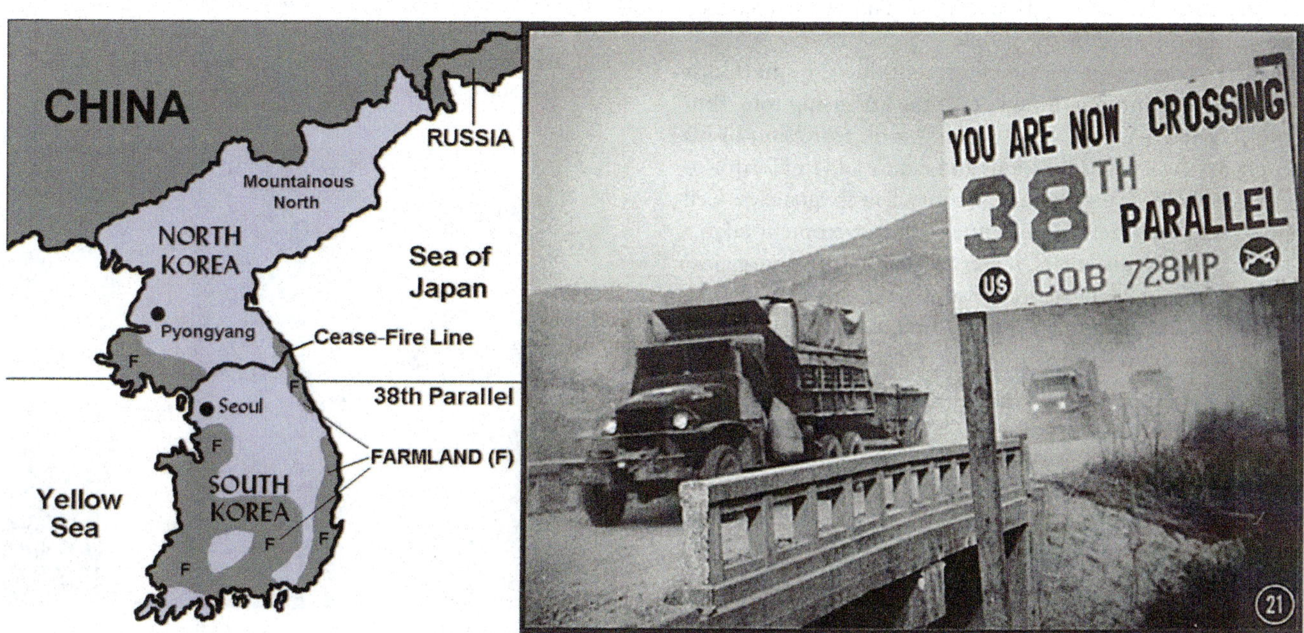

Figure 30. The Korean Peninsula between the North and the South. The thirty-eighth parallel cuts across the peninsula in the area of the DMZ farmland. The DMZ is the Cease-Fire Line between North and South Korea.

Soviet Union divided Korea approximately in half and eventually established governments in their respective regions that were sympathetic to each nation's own ideology. The Soviet Union administered the northern portion; the United States administered the southern region. Politics deeply affected each of the regions. Communism dominated North Korea and capitalism dominated South Korea. In 1950, with aid from China and the Soviet Union, the Communists from the north invaded the south. After bitter fighting, a peace agreement was reached in 1953 to officially divide the Korean Peninsula near the thirty-eighth parallel. Korea remains divided to this day. The United States has thousands of soldiers stationed along the Cease-Fire Line (or DMZ), which is the most heavily guarded border in the world.

North Korea

The government of North Korea (Democratic People's Republic of Korea, or DPRK) was formed with the leadership of Communist dictator Kim Il Sung, who shaped his country with a mix of Soviet and Chinese authoritarian rule. Having few personal freedoms, the people worked hard to rebuild a state. Using the threat of a US military invasion as a means of rallying his people, Kim Il Sung built up a military of more than one million soldiers, one of the largest in the world. People could not travel in or out of the country without strict regulation. North Korea existed without much notice until the 1990s, when things suddenly changed.

When the Soviet Union fell in 1991, North Korea lost a valued source of financial support and oil. Without fuel and funding, many of the factories closed. Unemployment rates rose significantly. Meanwhile, China adopted a more open posture and began to increase its level of trade with the West. The result was that China began to lose interest in propping up North Korea politically. North Korea was facing serious problems. Massive food shortages caused famine and starvation. Thousands of people died. Kim Il Sung died in 1994 and the country lost their "Great Leader." He had ruled the country since World War II and in the end was deified as a god to be worshiped by the people.

The government dictatorship continued as the Great Leader's son, Kim Jong Il, took over the leadership role. Kim Jong Il, known as the "Dear Leader," took repression of his people to new levels. Televisions and radios sold in North Korea can only receive government-controlled frequencies. Cell phones and the Internet are banned. The government takes a hard line against dissenters. If you are caught speaking against the state or taking any action to support dissent, you can be arrested, fined, placed in a prison camp, or executed. People are not allowed to leave or enter the country without government approval. Only a couple hundred tourists are allowed into the country per year and are closely escorted.

Kim Jong Il took hard measures to stay in power and to avoid yielding to the capitalist frenzy of corporate colonialism. He developed nuclear weapon capability and used his military weapons production as a bargaining chip against the United States and other world powers. After his death in late 2011, his youngest son, Kim Jong Un, became the supreme leader of the country.

North Korea has demonized the United States as the ultimate threat and has used state-funded propaganda to in-

Figure 31. Giant statue of North Korea's "Great Leader" Kim Il Sung This giant statue of North Korea's "Great Leader" Kim Il Sung stands in the capital city of Pyongyang and is called the Mansudae Grand Monument. This mammoth bronze statue was built in 1972 to celebrate the Great Leader's sixtieth birthday.

Figure 32. North Korean soldier looks across the DMZ in the Joint Security Area (left). The modern "63 Building" in Seoul, South Korea, stands as witness to the prosperity and advancements in the South Korean economy (right).

doctrinate its people. North Korea's government continually tells its people that the United States will invade at any minute and to be prepared for the worst. Propaganda has been used to create and enforce military, economic, and political policies for an ideology that supports the unification of all of Korea under Communist control.

North Korea is mainly mountainous; there is little quality farmland. While only 2 percent of the land is in permanent crops, about a third of the population works in agriculture. The best farmland in North Korea is located south of the capital city of Pyongyang. The capital is a restricted area where only the most loyal to the state can live. There is a serious shortage of goods and services. Electricity is not available on a dependable basis. Massive international food aid has sustained the people of North Korea. Outside food aid is accepted even though the country has a long-standing policy of self-reliance and self-sufficiency called **Juche.** Juche is designed to keep Korea from becoming dependent on the outside. The policy of Juche has been quite effective in isolating the people of North Korea and keeping dictators in power. The policy of Juche also holds back the wave of corporate capitalism that seeks to exploit labor and resources in global markets for economic profit.

One way that North Koreans are finding out about the outside world is through smuggled-in cell phones and VHS videotapes from South Korea. Popular VHS productions include South Korean soap operas, because of the common heritage, language, and ethnicity. The North Korean government has attempted to crack down on smuggled videotapes by going from home to home. Electricity is turned off before entering the home. Once the electricity is turned off, videotapes cannot be removed from the video player. The government agents then recover the videotape. Citizens found with smuggled videotapes are punished with steep fines or imprisonment. Desperate North Koreans have escaped across the border into northern China, where thousands of refugees have sought out better opportunities for their future. It is ironic to think that Communist China, with historically few human rights, would be a place where people would seek refuge, but China's economy is growing and North Korea's economy is stagnant, which creates strong push-pull forces on migration.

South Korea

In 1993, South Korea became a fully-fledged democracy with its first democratically elected president. Seoul, the capital city and home to almost ten million people, is located just south of the DMZ. The United States has many military installations in this area.

South Korea manufactures automobiles, electronic goods, and textiles that are sold around the world. South Korea is a democracy that has used state capitalism to develop its economy. After the World War II era, South Korea was ruled by a military government that implemented land reform and received external economic aid. Large agricultural estates were broken up and redistributed to the people. Agricultural production increased to meet the demands of the population. South Korea has much more agricultural production than North Korea and is therefore in providing food for the high population density. A once-modest manufacturing sector has grown to become a major production center for export trade.

The fifty million people who live in South Korea have a much higher standard of living than the residents of North Korea. Personal income in the north is barely equivalent to a few dollars per day, while personal income in the south is similar to that of Western countries. The economic growth of the south was a result of state-controlled capitalism, rather than free-enterprise capitalism. The state has controlled or owned most of the industrial operations and sold its products in the global marketplace. Giant corporations, which forced industrialization along the coastal region, have promoted South Korea as the world's leading shipbuilding nation. South Korean corporations include Daewoo, Samsung, Kia Motors, Hyundai, and the Orion Group. As an economic tiger, South Korea continues to reform its economic system to adapt to global economic conditions.

South Korea has announced plans to conduct a comprehensive overhaul of its energy and transportation networks. Government funding will augment efforts to create more green-based initiatives. Part of this effort will focus on lower energy dependency with environmentally friendly energy developments such as wind, solar, bike lanes, and new lighting technologies. High-speed rail service and increased capacity in electronic transmissions lines are planned as part of a next generation of energy-efficient technologies that will increase economic efficiency. These policies have been enacted to update South Korea's economy and create new products for manufacturing export.

Buddhism has been a prominent religion in Korea for centuries. The teachings of Confucius are also widely regarded. About 30 percent of the population claims Christianity as their religious background; this is the highest Christian percentage of any Asian country. About two-thirds of the Christians are Protestant and one-third are Catholic. Up to half of the population makes no claims or professions of faith in any organized religion.

Unification of North and South

What if North Korea and South Korea would become unified again? In geography, there is a concept of regional complementarity, which exists when two separate regions possess qualities that would work to complement each if unified into one unit. North and South Korea are the classic illustration of regional complementarity. The North is mountainous and has access to minerals, coal, iron ore, and nitrates (fertilizers) that are needed in the South for industrialization and food production. The South has the most farmland and can produce large harvests of rice and other food crops. The South has industrial technology and capital needed for development in the North. If these two countries were reunited, they could work well as an economic unit.

The harder question is how and under what circumstances the two Koreas could ever come to terms of unification. What about the thirty-five to forty thousand US soldiers along the DMZ? What type of government would the unified Korea have? There are many young people in South Korea that would like to see the US military leave Korea and the two sides united. The generation of soldiers that survived the Korean War in the 1950s understands the bitterness and difficulties caused by the division. This segment of the population is highly supportive of maintaining the presence of the US military on the border with North Korea. Unification is not likely to take place until this generation either passes away or comes to terms with unification. The brutal dictatorships of Kim Jong Il and his son Kim Jong Un have been a main barrier to unification. This is a political division, not technically a cultural division, even though the societies are quite different at the present time. The geography of this situation is similar to that of East and West Germany after World War II and during the Cold War. Korea may have different qualities from Germany, but unification may be possible under certain conditions, foremost being different leadership in the north.

Figure 33. Reunification Buddha in South Korea, Erected to Signify the Unity of the Korean People

Key Takeaways

1. Japan's mountainous archipelago has provided isolation and protection for the Japanese people to maintain their unique culture and heritage. The islands were created by tectonic plate action, which continues to cause earthquakes in the region.

2. Japan was a colonial power that colonized much of the Asian Pacific region. Its defeat in World War II devastated the country and liberated all its colonies. Since 1945, Japan has been developing one of the strongest economies in the world through manufacturing.

3. Japan is a highly industrialized country that is in stage 5 of the demographic transition. The country's small family size and declining population are a concern for economic growth in the future.

4. The Korean Peninsula is divided in two by the DMZ. North Korea is a Communist dictatorship with a low standard of living and absolutely no civil rights. South Korea is a capitalist democracy with a high standard of living and a free-market economy.

5. North Korea is more mountainous and lacks arable land to grow food. South Korea has more farmland and fewer mountains. If the two countries were to be united with a common economy, they would complement each other with a balance of natural resources.

End-of-Chapter Summary

- East Asia is anchored by Communist China, a large country that dominates the physical region. High mountains on its western borders have isolated China. The peripheral regions of Tibet, Western China, Mongolia, and Manchuria act as buffer states that protect the core Han Chinese heartland in China Proper, where most of the population lives.

- To boost foreign trade and increase its manufacturing sector, China has established special economic zones (SEZs) along its coast in hopes of attracting international businesses. Shenzhen is one such SEZ, which is located across the border from Hong Kong. Urbanization and industrialization have prompted a major rural-to-urban shift in the population.

- The world-class port and strategic geographic location of Hong Kong have played a large role in its development as an economic tiger. High-tech manufacturing with banking and financial services have catapulted the small area into a major market center for Asia. In 1998, Hong Kong was united with China, opening the door for massive trade and development with China and Shenzhen.

- Taiwan is a small island off the southeast coast of China. It has developed into an economic tiger with high-tech manufacturing and high incomes. The island has a separate government from mainland China and has a capitalist economy. There has been historic tension between Taiwan and Beijing ever since the Nationalists moved here in 1948. Economic trade between Taiwan and mainland China has brought the two sides closer together.

- Japan is made up of a number of large islands. After its defeat in World War II, Japan worked to recover and become a core manufacturing country and the second-largest economy in the world, although that title is now held by China. Japan has a homogeneous society in which about 99 percent of the population is Japanese. Family size is low and the population has begun to decline in numbers, causing an economic concern about the lack of entry-level workers.

- Korea is split between an authoritarian-controlled North Korea and a capitalist South Korea. North Korea is isolated with a weak economy. Its citizens are denied basic human rights. South Korea has developed into one of the economic tigers of Asia. South Korea has a strong manufacturing sector that produces high-tech electronics and information technology. Thousands of Korean and US soldiers remain stationed on the demilitarized zone (DMZ) between the two Koreas.

Chapter 13

Southeast Asia

Introduction

The region between China, India, Australia, and the Pacific Ocean is known as Southeast Asia. Southeast Asia includes countries with political boundaries creating many shapes and sizes. The political borders were created through a combination of factors, including natural features, traditional tribal distinctions, colonial claims, and political agreements. The realm also has the fourth-most populous country in the world, Indonesia. Southeast Asia is a region of peninsulas and islands. The only landlocked country is the rural and remote country of Laos, which borders China, Vietnam, and Thailand.

Southeast Asia can be divided into two geographic regions. The **mainland** portion, which is connected to India and China, extends south into what has been called the **Indochina Peninsula** or **Indochina**, a name given to the region by France. This mainland region consists of the countries of Vietnam, Laos, Cambodia, Thailand, and Myanmar (Burma). This region has been influenced historically by India and China. The islands or insular region to the south and east consist of nations surrounded by water. The countries in this region include Malaysia, Singapore, Brunei, Indonesia, East Timor, and the Philippines.

13.1 Introducing the Realm

Learning Objectives

1. Understand the geographical differences between the mainland region and the insular region.
2. Summarize how the region was colonized. Learn how colonial activities influence each country's cultural situation.
3. Realize how the physical geography has been influenced by tectonic activity.
4. Outline the main ethnic and religious affiliations of Southeast Asia and explain why they are so diverse.
5. Comprehend the impact and influence of the overseas Chinese in the region.

A. Physical Geography

The islands and the mainland of Southeast Asia include a wide array of physical and cultural landscapes. The entire realm is located in the tropics except the northernmost region of Burma (Myanmar), which extends north of the Tropic of Cancer. A tropical Type A climate dominates the region and rainfall is generally abundant. The tropical waters of the region help moderate the climate. Southeast Asia is located between the Indian Ocean on the west and the Pacific Ocean on the east. Bordering the many islands and peninsulas are various seas, bays, straits, and gulfs that help create the complex maritime

Figure 1. Southeast Asia: the mainland region and the insular region (the Islands).

boundaries of the realm. The **South China Sea** is a major body of water that acts as a separator between the mainland and the insular region. The thousands of islands that make up the various countries or lie along their coastal waters create a matrix of passageways and unique physical geography.

The three longest rivers of the realm, Mekong, Red, and Irrawaddy, are located on the mainland and have their headwaters in the high elevations of Himalayan ranges of China. The **Mekong River** makes its way from the high Himalayas in China and helps form the political borders of Laos and Thailand on its way through Cambodia to Vietnam where it creates a giant delta near Ho Chi Minh City (Saigon). The **Red River** flows out of China and through Hanoi to the Red River delta on the Gulf of Tonkin. The **Irrawaddy River** flows through the length of Burma providing for the core area of the country. Another major river of the mainland is the Chao Phraya of Thailand. With its many tributaries, the **Chao Phraya** creates a favorable core area that is home to the largest population of the country. Many other rivers can be found on both the mainland and the insular region. The rivers transport water and sediments from the interior to the coasts, often creating large deltas with rich soils that are major agricultural areas. Multiple crops of rice and food products can be grown in the fertile river valleys and deltas. The agricultural abundance is needed to support the ever-increasing populations of the realm.

Tectonic plate activity has been responsible for the existence of the many islands and has created the mountainous terrain of the various countries. High mountain ranges can have peaks that reach elevations of over fifteen thousand feet. The high-elevation ranges of New Guinea, which are along the equator, actually have glaciers, ice, and snow that remain year-round. The island of **Borneo**, in the center of the insular region, is actually a segment of ancient rock that has been pushed upward by tectonic forces to form a mountainous land mass. The mountains on Borneo have been worn down over time by erosion. Mountains and highlands stretch across the northern border of the realm along the borders with India and China. The interior nature of this border makes it less accessible. Similar dynamics can be found in the interior of the islands of the insular region, where the isolation and remoteness have helped create the environmental conditions for unique flora and fauna. In the highland areas the human cultural landscape can be diverse. Time and isolation have worked together to form the traditions and cultural ways that give local groups their identity and heritage.

Tectonic activity makes the region vulnerable to earthquakes and volcanic eruptions. The volcanic peak of **Mt. Pinatubo**, in the Philippines, erupted in 1991, spewing ash and smoke into the atmosphere and impacting much of the planet. An earthquake of 9.0 magnitude occurred off the coast of the Indonesian island of **Sumatra** in 2004 and caused widespread disaster throughout the wider region of the Indian Ocean. Over 230 thousand deaths were reported, mainly from flooding. A thirty-five-foot-high wall of water from the tsunami devastated many coastal areas from Thailand to India.

B. Impact of Colonialism

Southeast Asia has not escaped the impact of globalization, both colonial and corporate. As Europeans expanded their colonial activities, they made their way into Southeast Asia. Southeast Asia was heavily influenced by European colonialism. The only area of the region that was not colonized by the Europeans was Thailand, which was called Siam during the colonial era. It remained an independent kingdom throughout the colonial period and was a buffer state between French colonizers to the east and British colonizers to the west and south. The Japanese colonial empire controlled much of Southeast Asia before World War II.

Some of the countries and regions of Southeast Asia became known by their colonial connection. Indonesia was once referred to as the **Dutch East Indies**, which was influential in the labeling of the Caribbean as the West Indies. **French Indochina** is a term legitimized for historical references to the former French claims in Southeast Asia. Independence from the European powers and freedom from Japanese imperialism by the end of World War II provided a new identification for the various countries of the realm. Cultural and economic ties remain between many former colonies and their European counterparts.

Figure 2. South Asian colonialism. Southeast Asia was colonized by Europeans and later by Japan.

C. Cultural Introduction

Southeast Asia has a population of more than six hundred million people; more than half the population lives on the many islands of Indonesia and the Philippines. The island of **Java** in Indonesia is one of the most densely populated places on Earth. More than half of the two hundred forty-five million people who live in Indonesia live on the island of Java. The island of **Luzon** in the Philippines is also one of the more densely populated areas of the insular region. The Philippines has over one hundred million people, Vietnam has more than ninety million, and Thailand has about sixty-seven million. Local areas with high food-producing capacity are also high population centers, which would include deltas, river valleys, and fertile plains.

The **ethnic mosaic** of Southeast Asia is a result of the emergence of local differences between people that have evolved into identifiable cultural or ethnic groups. Though there are a multitude of specific ethnic groups, a number of the larger ones stand out with recognizable populations. On the mainland the Burmese, Thai, Khmer, and Vietnamese are the largest groups, coinciding with the physical countries from Burma to Vietnam. A similar situation can be found in the insular region. Many distinct groups can exist on the many islands of the region. The island of **New Guinea**, for example, has hundreds of local groups with their own languages and traditions. The large number of ethnic groups is dominated by Indonesians, Malays, and Filipinos, coinciding with the countries of Indonesia, Malaysia, and the Philippines. Each of these main groups has many subgroups that hold to their own cultural heritage in the areas where they exist. The many islands of Indonesia and the Philippines create the opportunity for diversity to continue to thrive, in spite of the globalization process that increased the interaction and communication opportunities between groups.

All major religions can be found here. Indonesia is home to the largest Muslim population in the world. The Philippine population is predominantly Christian, but there is a minority Muslim community. Most people in Malaysia follow Islam. About 95 percent of the people in Thailand and more than 60 percent of the people in Laos are Buddhist. Hinduism is present in the Indonesian island of Bali and in various other locations in the region. Animism and local religions can be found in rural and remote areas. Clearly, Southeast Asia is a mix of many ethnic groups, each with its own history, culture, and religious preference.

D. Overseas Chinese

Southeast Asia is also home to over thirty million overseas Chinese—ethnic Chinese who live outside of China. The Chinese exodus to the realm was the greatest during the last Chinese dynasties and during the colonial era. European colonial powers enhanced this migration pattern by leveraging the use of people with Chinese heritage in their governing over the local populations in the realm. Life has often been difficult for overseas Chinese. The Japanese occupation of the realm during World War II was a time of harsh discrimination against Chinese. Japanese occupation and colonialism diminished with the end of World War II. The overseas Chinese minority retained an economic advantage because of their former colonial status and their economic connections. Chinatowns emerged in many of the major cities of Southeast Asia. The discrimination against the Chinese, fueled by religious or socioeconomic differences, often continued after World War II by the local ethnic majorities. Nevertheless, overseas Chinese in Southeast Asia have been instrumental in promoting the global business arrangements that have established the Pacific Rim as a major player in the international economy.

Key Takeaways

1. Southeast Asia can be divided into two geographic regions: the mainland and the insular region. The mainland borders China and India and has extensive river systems. The insular region is made up of islands and peninsulas between Asia and Australia, often with mountainous interiors.

2. France and Britain colonized the mainland region of Southeast Asia. Burma was a British colony and the rest was under French colonial rule. The Japanese took control of the region briefly before World War II ended in 1945. Siam was the only area not colonized. Siam became the country of Thailand.

3. The physical geography of the mainland and the insular region is dominated by a tropical type A climate. Cooler temperatures may be found in the mountainous regions and more even temperatures ranges can be found along the coasts. Tectonic plate activity is responsible for the many earthquakes and volcanic eruptions that occur in the realm.

4. Southeast Asia is ethnically, religiously, and linguistically diverse. A number of major ethnic groups dominate in the mainland and insular region but are only examples of the multitude of smaller groups that exist in the realm. One minority group is the overseas Chinese, who immigrated to the realm during the colonial era.

13.2 The Mainland Countries

Learning Objectives

1. Summarize the main economic activities of each country.
2. Understand how Vietnam was divided by civil war and the impact the war had on the country.
3. Realize how the country of Laos is addressing its rural landlocked economic situation.
4. Describe the radical conditions that led to the creation of Democratic Kampuchea.
5. Outline the physical geography of Thailand and how this country has developed its economy.
6. Comprehend the conditions in Burma. Learn why the Burmese people would be opposing the government.

A. Vietnam

The elongated state of Vietnam is slightly larger than Italy and about three times the size of the US state of Kentucky. In 2010 it was estimated to have a population of about ninety million people. Sixty percent of the population is under age twenty-one. Vietnam has two main urban core areas: **Ho Chi Minh City (Saigon)** in the south and the capital, Hanoi, in the north. The middle region of Vietnam is narrow, with higher elevation. Each core area is located along a major river delta. The **Red River** delta is located east of Hanoi in the north, and the mighty **Mekong River** delta is located next to Saigon in the south. These river deltas deposit silt from upstream and provide excellent farmland for growing multiple crops of rice and food grains per year.

Vietnam has a tropical Type A climate with a long coastline. Fishing provides protein to balance out nutritional needs. More than 55 percent of the population works in agriculture. Family size has dropped dramatically because of population growth and a trend toward urbanization. Rural-to-urban shift has caused the two main urban core cities to grow rapidly. Saigon is the largest city in Vietnam and has a port that can accommodate oceangoing vessels. Hanoi, the capital, is not a port city and is located inland from the nearest port of Haiphong on the coast of the **Gulf of Tonkin**.

Political Geography

An understanding of Vietnam is not complete without understanding the changes in political control the country of Vietnam has experienced. Different Chinese dynasties controlled Vietnam at different times. When France colonized Vietnam, it imposed the French language as the *lingua franca* and Christianity as the main religion. Both changes met resistance, but the religious persecution of Buddhism by the French colonizers created harsh adversarial conditions within the culture. The French domination started in 1858. The Japanese replaced it in 1940; this lasted until the end of World War II. With the defeat of Japan in 1945, the French desired to regain control of Vietnam. The French aggressively pushed into the country, but met serious resistance and were finally defeated in 1954 with their loss at the battle of Dien Bien Phu.

In the mid-1950s, the Vietnamese began asserting their request for an independent country. The dynamics were similar to that of Korea. After 1954, Vietnam needed to establish a government for their independent country. They were not unified. The northern section rallied around Hanoi and was aligned with a Communist ideology. The southern region organized around Saigon and aligned itself with capitalism and democratic reforms.

During the Cold War, the United States opposed Communism wherever it emerged. Vietnam was one such case. Supporting South Vietnam against the Communists in the north started not long after the defeat of France. By 1960, US advisors were working to bolster South Vietnam's military power. After the assassination of John F. Kennedy, President Lyndon Johnson had to make a choice to either pull out of Vietnam or push the US military to fully engage the Communists in North Vietnam.

Not wishing Vietnam and its neighbors to "go Communist" through a **domino effect**—where if one country fell to Communism its neighbors would follow—President Johnson decided to escalate the war in Vietnam. By 1965, more than one half million US soldiers were on the ground in Vietnam. History has recorded the result. Just as Vietnam was divided by political and economic ideology, the

Figure 3. Southeast Asia and Vietnam. The two main cities of Vietnam are both located next to large rivers. The capital, Hanoi, in the north is on the Red River. Ho Chi Minh City (Saigon) to the south is next to the delta of the Mekong River.

Vietnam War also divided the US population. Protests were common on college campuses and public support for the war was often met with public opposition.

The US government, under President Richard Nixon, finally decided to pull all US troops out of Vietnam after a cease-fire was agreed upon in a Paris peace conference in 1973. More than fifty-seven thousand US soldiers had died in the Vietnam War. Two years later, in 1975, the North Vietnamese Communists invaded South Vietnam and took control of the entire country. Vietnam was unified under a Communist regime. More than two million people from South Vietnam escaped as refugees and fled to Hong Kong, the United States, or wherever they could go. Thousands were accepted by the United States, which caused ethnic rifts in US communities. The United States placed an embargo on Vietnam and refused to trade with them. The United States did not open diplomatic relations with Vietnam again until 1996. The Vietnam War devastated the infrastructure and economy of the country. Roads, bridges, and valuable distribution systems were destroyed. Vietnam could only turn to what it does best: growing rice and food for its people.

Modern Vietnam

For the past three decades, Vietnam has been recovering and slowly integrating itself with the outside world. Its population has doubled; most of the population was born after the Vietnam War. Their main goal is to seek out opportunities and advantages to provide for themselves and their families. Vietnam has been a rural agrarian society. The two main core cities, however, are now waking up to the outside world, and the outside world is discovering them. Looking for cheap labor and economic profits, economic tigers such as Taiwan are turning to Saigon to set up light manufacturing operations. People from the rural areas are migrating to the cities looking for employment. **Saigon** has more than 8.5 million people and has a **special economic zone (SEZ)** located nearby. Rural-to-urban shift is kicking in. After 1975, the city of Saigon was renamed **Ho Chi Minh City** after the victorious Communist leader, Ho Chi Minh. Many of the people who live there and who live in the United States still refer to it as Saigon.

The future of Vietnam may be similar to most of Southeast Asia as it balances out the strong adhesive forces of local culture and the demands of a competitive global economy. The growing population will add to the demand for resources and employment opportunities. Vietnam has been a relatively poor country but it still has been able to export rice and other agricultural products. In recent years, the Communist government has implemented a series of reforms moving toward a market economy, which has encouraged economic development and international trade.

Globalization has prompted a strong rural-to-urban shift within Vietnam. The rural countryside is still steeped in its agrarian heritage based on growing rice and food crops, but the urban centers have been energized by modern technology and outside economic interest. Vietnam has enormous growth potential. The country's urban centers are shifting from stage 2 of the index of economic development into stage 3, where the urbanization rates are the strongest. The rapid rise of the global economy that is connected to Vietnam's major cities has provided jobs and opportunities that are highly sought after by the growing population. The city streets are filled with a sea of motorbikes and bicycle traffic. Cars are becoming more plentiful. Saigon has been a major destination for the export textile industry and other industries seeking a cheap labor base. Cell phones and Internet services have connected a once-isolated country with the rest of the world.

Figure 4. Man hauling cut wood on a bicycle cart (pedicab) by the Perfume River near the city of Hue, Vietnam.

B. Laos

The geography of Laos centers on the **Mekong River** basin and rugged mountain terrain. Laos is the only landlocked country in southeast Asia, with no ports connecting it to the outside world. The mountains reach up to 9,242 feet. The Type A **tropical savanna** climate provides a rainy season and a dry season. The rains usually fall between May and November, followed by a dry season for the remainder of the year. The Mekong River flows through the land and provides fresh water, irrigation, and transportation. The country's capital and largest city, **Vientiane**, is located on the Mekong River. Laos is about the same size in area as the US state of Utah.

The **Lao Kingdom** coalesced in the 1500s and was eventually absorbed by the **Kingdom of Siam**, which thrived during the eighteenth and nineteenth centuries. France muscled in during the colonial era and created a French Indochina. Laos received independence from France in 1949. Laos is a rural country with about 80 percent of the population working in agriculture. Globalization has not yet had much impact in this country and infrastructure is less developed. Electricity is not available on a consistent

Figure 5. Woman in Attapeu Province, Laos, cooking midmorning snacks for schoolchildren. These snacks are a corn-soy blend shipped in from the United States and sponsored by the US McGovern-Dole Act. Food is provided in schools in rural Laos to encourage attendance and enrollment in school. Thousands of children benefit from this program.

basis and transportation systems are quite basic. There are no railroads and there are few paved roads. Clean water for human consumption is not always available. The economy is based on agriculture, with some outside investments in mining and natural resources.

Two-thirds of the people in Laos are Buddhists. Animist traditions and spirit worship have the next highest percentage of followers. Muslims and Christians make up a small percentage of the population. **Lao** make up the largest ethnic group and 70 percent of the population. Other ethnicities include the **Hmong** and mountain tribal groups, which are found in various remote regions of the country. The remoteness and rural heritage of the many tribal people have started to attract tourism. Tourism has increased in recent years partially due to the Chinese government allowing its citizens to travel outside their borders from China into Laos. Laos has two **UNESCO World Heritage Sites:** the historic town of Luang Prabang, and the southern site of Wat Phou (Vat Phu), which is an ancient Hindu temple complex.

Laos is a poor country. It has fewer employment opportunities for its citizens than most other developing countries. The one-party Communist political system of the Cold War has been decentralizing control and working to encourage entrepreneurial activities. Foreign investments are increasing in the areas of mining, hydroelectric production, and major construction projects. The World Bank and other agencies have supported efforts to improve infrastructure and provide opportunities for the people of Laos. China has partnered with the Laotian government to help build rail transport in the country. These efforts have assisted in reducing poverty and increasing the economic and physical health of the country.

C. Cambodia

A Notorious History

Cambodia is about the same size in area as the US state of Missouri. The population in 2008 was estimated at 14.5 million. The **Khmers** created the **Angkor Empire**, which reached its peak between the tenth and thirteenth centuries. Preceding the colonial period, the Angkor Empire entered into a long era of decline. France took control of the region in the latter part of the seventeenth century. Japan took control of the region before World War II and then relinquished it when they surrendered to end World War II. France regained control of Cambodia after the Japanese army was defeated. Cambodia finally received independence from France in 1953.

To understand Cambodia, one has to understand its recent history. This country has undergone some of the most extreme social transitions in modern times. The **Khmer Rouge**, under the leadership of **Pol Pot**, turned society upside down, giving the country a legacy that it will carry forward as integration continues into the world community.

Between 1969 and 1973, while the United States was fighting the Vietnam War, US forces bombed and briefly invaded Cambodia in an effort to disrupt the North Vietnamese military operations and oppose the Khmer Rouge. Millions of Cambodians were made refugees by the war, and many ended up in **Phnom Penh**. The number of casualties from the US bombing missions in Cambodia is unknown. The US war in Vietnam thus had spilled over into Laos and Cambodia and advanced the opportunities for the Khmer Rouge regime to gain power. Pol Pot's Communist forces of the Khmer Rouge finally captured Cambodia's capital of Phnom Penh in 1975. The Khmer Rouge evacuated all cities and towns and forced the people to move to the rural areas. The country's name was changed to **Democratic Kampuchea**. China's Great Cultural Revolution and the Great Leap Forward disaster were influential for Pol Pot's radical experiment. Since Vietnam was supported by the Soviet Union, the Khmer Rouge looked to China for arms and support.

Pol Pot was creating an agricultural model for a new country based on eleventh-century ideals. People in urban areas were forcibly marched off into the countryside for labor in agriculture. Anyone who resisted or even hinted at dissent was killed. All traces of Westernized ideas, technology, medical practice, religion, or books were destroyed. Thousands of people were systematically killed in an attempt to bring into being a rural agrarian utopian society. The thousands upon thousands who were systematically eliminated gave rise to the term **Killing Fields**, meaning fields where massive groups of people were forced to dig their own graves and then were

Figure 6. Skulls of the victims from the killing fields of Pol Pot and the Khmer Rouge in Cambodia from the 1970s.

killed. The mass killings were reminiscent of those carried out by Hitler, Stalin, and Mao. Pol Pot's regime also targeted ethnic minority groups. Muslims and Chinese suffered serious purges. Professional, educated people, such as doctors, lawyers, and teachers, were also targeted for execution. According to some reports, the very act of wearing eyeglasses was a death sentence as it was a symbol of intellectualism. In a country of eight million in 1970, more than two million people were executed or died as a result of Pol Pot's policies. The total number will never be known. Hundreds of thousands became refugees in neighboring countries.

By 1978, the Khmer Rouge was isolated in the countryside. Vietnamese forces controlled the urban areas. A decade of civil war and unrest followed. Paris peace talks, cease-fires, United Nations–sponsored elections and coalition governments have since helped provide political stability. Pol Pot died under unclear circumstances in 1998 while being held under house arrest. As of 1999, the Khmer Rouge elements that were still in existence had surrendered or were arrested. Many of the Khmer Rouge leaders were charged with crimes against humanity by United Nations–sponsored tribunals.

Modern Cambodia

Cambodia is working to become a democratic and open country with established trade relationships with global markets. The people have struggled to create a stable society that can rebound from their legacy of turmoil and conflict. The country's population is relatively young. More than half the population is under age twenty-five; one-third is under fifteen. The rural areas and the generations who remain there continue to lack the basic amenities of modern society. Education, electricity, and modern infrastructure are lacking. More than half the population works in agriculture. Since less than 25 percent of the population lives in cities, Cambodia is likely to experience a high rural-to-urban shift in its future.

People are returning to religious practices that were banned during the Pol Pot era. Buddhism is the dominant religion of about 95 percent of the population. Small percentages of the population also practice Christianity, Hinduism, Islam, or tribal beliefs. There are at least twenty distinct hill tribes that hold to their own traditions and cultural ways. The country has historically been self-sufficient with food, but the rapid population growth, political instability, and lack of infrastructure are challenging the future of the country.

Agriculture has been the main economic activity, though **textiles** (clothing manufacturing) have increased in recent years because of the low cost of labor combined with an abundant workforce. The international business sector has sought to exploit this opportunity, but multinational corporations are hesitant to invest in a country that suffers from political instability or a high level of corruption within the public and private sector.

Cambodia has been attempting to build a sustainable economy. The textile industry is the number one source of national wealth. Sweatshops and low-tech manufacturing have begun to take root in the expanding capital city of **Phnom Penh**. Tourism is another sector that has experienced rapid growth. Though nonexistent in earlier decades, tourism has taken off. Cambodian tourism provides travelers with an experience that is more pristine and less commercialized. Tourism has been rated as the second-largest sector of the economy. One of the main sites that attract many visitors is the extraordinary ancient site of **Angkor Wat** (Angkor means "city" and Wat "temple"). This site is one of the best-preserved showcases of Khmer architecture from its early empire years. Angkor Wat is being developed as a major tourist attraction. The twelfth-century complex was first a Hindu site dedicated to Vishnu, and then it was converted to a Buddhist site. Angkor Wat has become an international tourist destination. It is one of the largest temple complexes in existence in the world and is a UNESCO World Heritage Site. The city of Angkor has been estimated to have been the largest city in the world at its peak. As many as a thousand other temples and ancient structures have been recovered in the same area in recent years.

Cambodia has pressing environmental problems. The country has the notorious designation by the UN as the nation with the third-highest number of land mines on Earth. Since 1970, more than sixty thousand people have been killed, and many more injured or maimed because of unexploded land mines in rural areas. The growing population, attempting to recover from decades of devastation, has cut down the rainforest at one of the highest rates in the world. In 1970, rainforests covered about 70 percent of the country. Today there is only about three percent of the rainforest left. A rise in the need for resources, along with illegal timber activities, has devastated the forests, resulting in a high level of soil erosion and loss of habitat for indigenous species. The loss of natural resources is likely to hinder the country's economic growth.

Figure 7. A small section of the Angkor Wat temple complex in Northwest Cambodia.

Figure 8. Modern City of Bangkok with High-Rise Office Buildings and Business District.

D. Thailand

Thailand is larger than Laos and Cambodia combined but smaller than Burma. The physical regions that make up Thailand include the mountainous north, where peaks reach up to 8,415 feet; the large southeastern plateau bordering the Mekong River; and the mainly flat valley that dominates the center of the country. The southern part of the country includes the narrow isthmus that broadens out to create the **Malay Peninsula**. The tropical Type A climate has dry and rainy seasons similar to Cambodia. The weather pattern in the main part of Thailand, north of the Malay Peninsula, has three seasons. The main rainy season lasts between June and October, when the southwest monsoon arrives with heavy rain clouds from over the Indian Ocean. After the rainy season, the land cools off and starts to receive the northeast monsoon, which is a cool dry wind that blows from November to February. Considered the dry season, its characteristics are lower humidity and cooler temperatures. From March to May, the temperatures rise and the land heats up. Then the cycle starts over again with the introduction of the rainy season. The weather pattern in the southern part of Thailand in the Malay Peninsula receives more rain throughout the year, with two rainy seasons that peak from April through May and then again from October through December.

Thailand was formerly known as the **Kingdom of Siam**. In 1932, a constitutional monarchy was established after a bloodless revolution erupted in the country. The name was officially changed to Thailand in 1939. The ruling monarch remains head of state but a prime minister is head of the government. Siam was never colonized. The leaders of Siam played France and Britain against each other and remained independent of colonial domination.

About three-fourths of the population is ethnically **Thai**. There is a noticeable Chinese population and a small percentage of people who are ethnically **Malay**. There are various minority groups and hill tribes. The country's official language is Thai. Buddhism is adhered to by about 95 percent of the population. The ruling monarch is considered the defender of the Buddhist faith.

Thailand has an excellent record of economic growth and has been one of Southeast Asia's best performers in the past couple of decades. Thailand is developing its infrastructure and has established measures to attract foreign investments and support free-enterprise economic activities. The recent slowdown in the global economy and internal political problems have caused a sharp downturn in Thailand's economic growth. Nevertheless, Thailand remains a strong economic force and one of the best economies in the region. The positive indicators include a strong focus on infrastructure, industrial exports, and tourism.

Urbanization rates are increasing; at least one-third of the population lives in cities. Family size has fallen to lower than two children per family, while education rates have increased. The country has also tapped into its natural resources for export profits as the world's third-largest exporter of tin and the second-largest exporter of tungsten. Light manufacturing has taken off and become a major component of the economy, accounting for about 45 percent of the gross domestic product (GDP). The country is a major manufacturer of textiles, footwear, jewelry, auto parts, and electrical components. Thailand has been the major exporter of rice in the world and has a strong agricultural base.

Thailand is a newly industrialized country and has all its bases covered to build national wealth: a balance of agriculture, extractive activities, manufacturing, and postindustrial activities (tourism). Thailand is considered the third-largest manufacturer of motor vehicles in Asia, after Japan and Korea. Vehicle producers from the United States and Asia are manufacturing large numbers of cars and trucks in Thailand. Toyota dominates the market in both truck and auto production. Truck production is augmented by Mitsubishi, Nissan, Chevrolet, Ford, and Mazda. Honda, from Japan, and the Tata Motor Corporation, from India, are expanding their operations in Thailand. Thailand is in a good position to advance its economy and shift the whole country into the next stage of development to become a major participant in the global economic marketplace.

The tourism industry has grown immensely in Thailand over the past few decades. Green and lush tropical mountain landscapes, the exquisite architecture of ancient Buddhist temples, and beautiful golden beaches along warm tropical coastlines make for an excellent tourism market. Some of the best world-class tropical beach resorts are located along the sandy and sunny shores of Thailand. The country is open to outsiders and has welcomed tourism as part of its economic equation. The relatively stable country provides a safe and exciting tourism agenda that has a global clientele. The downside of the thriving tourism industry is the sex trade. Relaxed laws on sexual activity have made Thailand a destination for people from around the world seeking "sex tours" and erotic experiences. Not surprisingly, a sharp increase in the number of individuals infected with sexually transmitted diseases has been documented. Approximately one million people in Thailand tested HIV positive in the mid-1990s. The sex industry has been big business for Thailand and at the same time has created an unfortunate negative stereotype for the overall tourism situation. There is much more to the tourism industry in Thailand than the sex trade.

The country of Thailand has the potential to recover from the global economic downturn and once again claim its role as an economic tiger of Southeast Asia. If political stability serves to enhance economic investments, the country will continue to experience economic growth. The low population growth is a model for other countries in the region. Thailand provides a good example of the theory that as a country urbanizes and industrializes, family size will usually go down. Thailand is also moving forward in the index of economic development. It is in stage 3, where there is a strong rural-to-urban shift in the population. The capital city of Bangkok has stage 4 development patterns and is an economic core area for the country and the region. As large as New York City, Bangkok has developed into the political, cultural, and economic center of Southeast Asia. Often referred to as the "Venice of the East" because of its city canals, Bangkok has become a global city with a population of more than eight million people officially and more than fifteen million unofficially.

E. Myanmar or Burma

The Union of Myanmar (Union of Burma) is the official name for Burma. Since 1989, the military authorities in Burma have promoted the name Myanmar as a conventional name for their state. The US government and many other governments have not recognized or accepted the name change. Some groups within Burma do not accept the name because the translation of Myanmar is also the name of an ethnic minority in Burma. The use of the name Burma or Myanmar is split around the world and within the country.

Burma is the largest country on the Southeast Asian mainland in terms of physical area. It is about the same size in area as Texas and had a population in 2010 of about fifty-four million. The country has a central mass with a southern protrusion that borders Thailand toward the Malay Peninsula. The northern border area between India and China has high mountains that are part of the Himalayas, with towering peaks extending to 19,295 feet. The **Irrawaddy River** cuts through the center of the country from north to south, creating a delta in the largest city, **Rangoon** (Yangon). Most of the country's population lives along this river valley.

Figure 9. Burma (Myanmar): The Irrawaddy River and the north/south layout.

There are differences in the physical landscape between the north and south. The northernmost area is mountainous with evergreen forests. Cool temperatures are found in the north and warmer annual temperatures are found in the south. To the west of the Irrawaddy River and north of Mandalay the land cover is mainly deciduous forests. The eastern region from **Mandalay** to the Laos border is scrub forests and grasslands. This area is considered the dry zone, with an annual rainfall of about forty inches. The more tropical south and coastal areas can receive higher levels of precipitation. The area around the core city of Mandalay was a major focus of agricultural development before British colonialism. Dryland crops were most common. During the colonial era, the British looked to the rich farmlands of the southern Irrawaddy delta and emphasized Rangoon as the center of their exploitations. Wetland rice is a major crop of the southern Irrawaddy basin. The southwest and the southern protrusion are mainly tropical evergreen forests. There has been oil exploration along the coastal regions of the Bay of Bengal and along the **Andaman Sea**.

The country was colonized by the British and was once a part of Great Britain's empire in South Asia as a province of India. Burma was one of the most prosperous colonies of Britain until World War II, when the Japanese invaded and war devastated the region. Democratic rule existed from 1948 until 1962, when an authoritarian military dictatorship took over the country. A revolutionary council ruled the country between 1962 and 1974. This government nationalized most of the businesses, factories, and media outlets. The overall operating principle of the council was a concept called the **Burmese Way of Socialism**. This concept was based on central planning and Communist principles mixed with Buddhist beliefs.

Between 1974 and 1988, the sole political party of the country was the **Burma Socialist Program Party**, which was controlled by the same military general and his comrades who had been in control for decades. During this time, the rest of the world was advancing in technology and economic development and moving forward with advancements in health care and education. Burma remained an impoverished and isolated nation. A number of countries, including the United States, have trade restrictions with Burma. For decades, the authoritarian regime in Burma has been accused of serious human rights violations, which have largely been ignored by the outside world.

Protests against the military rule have always existed in Burma but been suppressed by the armed forces and the authoritarian government. In 1962, the government cracked down on demonstrations at Rangoon University, resulting in fifteen students being killed and many others in need of medical attention. The military government has taken serious action against any antigovernment protest activities. By 1988, the people of Burma were taking to the streets with widespread demonstrations and protests against the government over claims of oppression, mismanagement, and lack of

democratic reforms. A total crackdown on the people was implemented, with thousands of protesters killed. A new council led by a military general created the **State Law and Order Restoration Council** a year later. Martial law was imposed and even harsher policies were imposed on anyone opposing the government. This is when the name of Myanmar was first used for the country.

Burma has been placed in the same category as North Korea and Somalia in terms of authoritarian rule, lack of human rights, and stagnant economy. Economic conditions are poor. The military rulers have gained control of the main income-generating enterprises in the country, including the lucrative drug trade from the prime opium growing region of the northern **Golden Triangle**, where Burma borders Laos and Thailand. Precious gemstones such as rubies, sapphires, and jade are abundant in Burma. Rubies bring the highest incomes. Burma produces about 90 percent of the world's supply, with superb quality. The Valley of Rubies in the north is noted for quality gem production of both rubies and sapphires. Most of the gems are sold to buyers in Thailand. All the profits go to Burma's military rulers in the government, and since there is a high level of corruption and mismanagement within the government and business, the income from the gems produces limited economic development for the main population and discourages foreign investment in the country. Burma has become one of the poorest countries in Southeast Asia. China has emerged as the main trading partner with Burma and has been propping up the dictatorial military regime. China supplies the

Figure 10. Demonstrators marching to express discontent with the government of Rangoon (Yangon), 2007. The banner, written in Burmese, refers to a national movement to promote nonviolence. A Buddhist monk is in the foreground, and the Shwedagon Pagoda is in the background.

regime with arms, constructs many of the infrastructure projects, and supplies natural gas to the country.

Burma is ethnically diverse. Though it is difficult to verify, the government of Burma recognizes one hundred thirty-five distinct ethnic groups within its borders. It is estimated that there are over a hundred different ethnolinguistic groups in Burma. About 90 percent of the population is Buddhist. This high level of diversity can allow for strong centrifugal forces that are not generally conducive to unity and nationalism. The heavy emphasis on the national military is one of the only centripetal forces within the population, even though the military leadership is also looked at with distain by those desiring more openness and democratic conditions.

Key Takeaways

1. France and Britain colonized the mainland region of Southeast Asia. Burma was a British colony and the rest was under French colonial rule. The Japanese took control of the region briefly before World War II ended in 1945. Siam was the only area not colonized. Siam became the country of Thailand.

2. Vietnam was divided by a Communist north and a capitalist south during the Cold War. Vietnam is emerging from decades of isolation to provide the global economy with a large low-cost labor pool that has been attracting foreign investments by multinational corporations.

3. The rural and landlocked nation of Laos has strong Buddhist traditions and an agrarian society.

4. Cambodia was impacted by the Vietnam War and then by the devastation of Pol Pot's Khmer Rouge radical experiment in agrarian socialism, which killed as many as 2.5 million people. Recovery has been slow, but the textile industry and tourism have contributed to economic growth.

5. The Buddhist country of Thailand has been experiencing major economic development in recent decades and has established itself as a major economic power in the region. The modern capital city of Bangkok is a major center of manufacturing and cultural activities.

6. The people of Burma (Myanmar) continue to suffer under an authoritarian regime that offers few civil rights or democratic processes to its people. Poor, isolated, and militarily controlled, Burma has been at the center of many human rights violations in recent decades with little response from the international community.

13.3 The Insular Region (Islands of Southeast Asia)

Learning Objectives

1. Summarize the economic development of each of the countries in this section.
2. Understand that Malaysia is divided between the Malay Peninsula and the island of Borneo.
3. Outline how the structured island nation of Singapore became an economic tiger.
4. Describe the physical geography of Indonesia and the population dynamics of the island of Java.
5. Summarize the cultural characteristics of the Philippines. Learn why this country is a popular destination for business process outsourcing (BPO).

The **insular region** of Southeast Asia includes the countries of Malaysia, Singapore, Brunei, East Timor, Indonesia, and the Philippines. Of the Southeast Asian countries, East Timor most recently gained its independence, as was mentioned previously. In comparing these island nations, extensive diversity in all aspects will be found. There are major differences in cultural, economic, and political dynamics, and in the ethnic groups that make up the dominant majorities in each. There is also a high level of linguistic and religious diversity. The physical geography varies from island to island; some have high mountain relief and others are low-lying and relatively flat. Active tectonic plate action in the region causes earthquakes and volcanic activity, resulting in destruction of infrastructure and loss of life; both acutely impact human activities.

Economic forces continue to prompt the countries of Southeast Asia to enter into trade relationships that integrate them with global networks based on dependency and reliance. The old colonial powers may no longer control them politically but may affect them economically. The new dynamics of corporate colonialism, with their economic power located in the core economic regions, still seek to exploit the countries of Southeast Asia for their labor and resources. These Asian nations are working to develop their own economies and use their own labor and resources to gain national wealth and increase the standard of living for their people. Each country has to contend with globalization forces within the international network of economic relationships.

A. Malaysia

Malaysia is a country made up of various British colonies that came together as a federation and then became an independent country. Britain started establishing colonies in the region in the late 1700s. The two main areas include the western colonies on the Malay Peninsula and the eastern colonies on the island of Borneo. The western settlements were part of the Malay Peninsula, which included the colonies of

Pinang and **Singapore**. Eventually, the British took control of the eastern colonies of **Brunei, Sarawak,** and **Sabah** on the island of **Borneo**. In 1957, the western colonies on the mainland peninsula broke from their British colonizers and became an independent country called the **Federation of Malaya**. In 1963, the British Borneo colonies of Sarawak and Sabah joined the Federation of Malaya to form the current country, which is called **Malaysia**. In 1965, Singapore broke off from Malaysia and became an independent country. Brunei, which was still a British protectorate, became independent in 1984.

Malaysia has two main land areas separated by the South China Sea. The regions of Sarawak and Sabah,

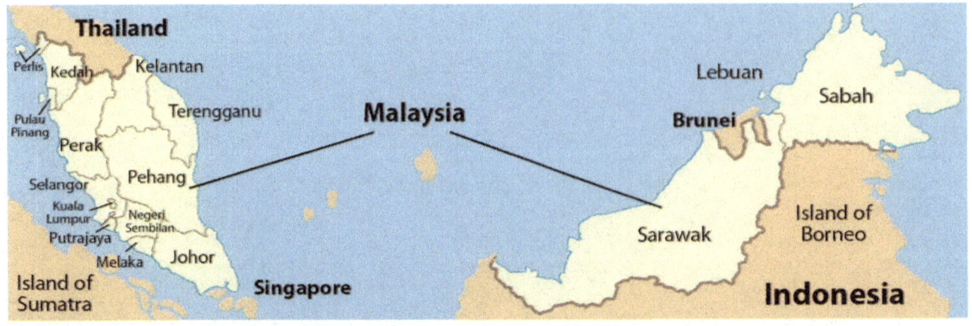

Figure 11. Provinces in east and west Malaysia.

on the island of Borneo, are called **East Malaysia**; the mainland on the Malay Peninsula is called **West Malaysia**. These regions have similar physical landscapes, which include coastal plains with nearby densely forested foothills and mountains. The highest mountains, rising 13,436 feet, are in East Malaysia on the island of Borneo. Located near the equator, Malaysia has a Type A **tropical rainforest** climate with abundant precipitation and warm temperatures year-round.

Diversity of Culture and Ethnicity in Malaysia

Malaysia's culture is diverse in that several major religions are practiced within its borders. Islam is considered the official religion and is supported by at least 60 percent of the population. About 20 percent of the people are Buddhists, 10 percent Christians, and 6 percent Hindu. The remaining percentages of the population include traditional Chinese religions and local tribal beliefs. In this Islamic country, there are

Figure 12. The tuaran road of Kota Kinabala City in Malaysia during a time of slow traffic. Note that the cars are driving on the same side of the road as in Great Britain, Malaysia's former colonizer.

concerns that Muslims get preferential treatment by government programs and policies. There are even special judicial legal courts for Muslims only to work out issues regarding marriage, custody, inheritance, or other conflicting Islamic issues regarding faith and obligation. This court only hears Islamic issues and no other legal matters. There have been movements by minority extremist groups that would like to see Malaysia shift toward a true Islamic state, complete with the Sharia Criminal Code as the law of the land. The movement, however, has been cracked down on by the government. Since the 9-11 incident in the United States, there has been more concern about extremist religious views.

People of Malay ethnic background make up more than 50 percent of the population. People of Chinese descent are the second-largest group at about 24 percent. An additional 11 percent of the population is made up of indigenous groups. During British colonialism, a number of people from South Asia were brought to Malaysia.

Malaysia's diverse ethnic and cultural mix often results in strong **centrifugal forces** that push and pull on the societal dynamics of the country. As a minority group, Chinese citizens of Malaysia have felt discrimination. Since the official language is Malay and the official religion is Islam, there have been concerns about discrimination against all minority groups. Working through the cultural and ethnic diversity has been a major challenge for the country.

Economic Development in Malaysia

Malaysia has rapidly advanced its economy in recent decades and is modernizing its infrastructure—roads, bridges, highways, and urban facilities. In the capital city, **Kuala Lumpur,** Malaysia built a modern central business district with a twin high-rise office building claimed to be the world's tallest at the time of construction. Before the global economic downturn that started in 2007, Malaysia had developed a fast-growing economy and was industrializing at a rapid rate. Malaysia has taken advantage of its location on a major shipping lane and has shifted to manufacturing as an important sector of its economy. The country has been a leader in the export of natural resources such as tin, rubber, and palm oil and has developed its agricultural and extractive sectors to gain income. The 1980s and 1990s were prosperous times for the country and it matured its manufacturing base from light textiles into electronics and heavy industries.

One aspect of the country that is looming on the horizon and may cause problems is the high population growth rate. In 2010, Malaysia's population was estimated at more than twenty-five million, with a doubling time of about forty years. Though the country is 70 percent urban, the fertility rate is still at about 3.0, which indicates an increasing population growth pattern. One-third of the population is under the age of fifteen. Malaysia is one case where the general principle that if a country urbanizes and industrializes the family size will go down has not taken place fast enough. The fertility rate has dropped from 5.0 to 3.0, but it needs to get below a rate of about 2.0 if the country is going to successfully stabilize its population growth. Unless the country addresses this population growth, the demand for resources might outstrip economic progress in the future.

B. Singapore

Under British colonial rule, the island of Singapore was included in the Malaysian federation. It broke away and became independent in 1965. It is a small island measuring about thirty miles long at its widest point. Singapore is about two hundred forty square miles in area. Singapore's most valuable resource is its **relative location**. Singapore is similar to Hong Kong in its development. With a good port, Singapore is a hub for ships sailing between Europe and China. It serves Southeast Asia as an **entrepôt**, or **break-of-bulk point,** where goods are offloaded from large ships and transported to smaller vessels for distribution to the Southeast Asian community.

Singapore has made good strategic utilization of its geographic location by serving as a distribution center for goods and materials processed in the region. Crude oil from Indonesia is unloaded and refined here. Raw materials are shipped in, manufactured into finished products, and then shipped out to global markets. Since Singapore is small, it has had to concentrate on manufacturing goods that provide for optimal profits. As an economic tiger, Singapore has transitioned through the same stages as Taiwan, South Korea, and Hong Kong to become an economic power in Southeast Asia.

The government of Singapore has entered into trade agreements with two of its neighbors to provide raw materials and cheap labor. A trade triangle has been established between Singapore, Malaysia, and Indonesia. Malaysia and Indonesia provide Singapore with raw materials and cheap labor; Singapore provides its neighbors with technical know-how and financial support. Everyone benefits. Singapore is an excellent example of the upper end of the economic spectrum in Southeast Asia. Countries like Laos or Vietnam would be at the opposite end, since they have a largely rural population based on agriculture that is just beginning to shift to the cities with industrialization. Singapore is already 100 percent urban with high incomes based on high-tech manufacturing and processing of raw

materials. Singapore is an economic hub for Southeast Asia, complete with global airline connections and is located on a major shipping lane. Singapore's world-class port is one of the busiest in Asia. The rest of Southeast Asia is somewhere in between these two ends of the spectrum as far as economic development is concerned.

Figure 13. Singapore modern architecture.

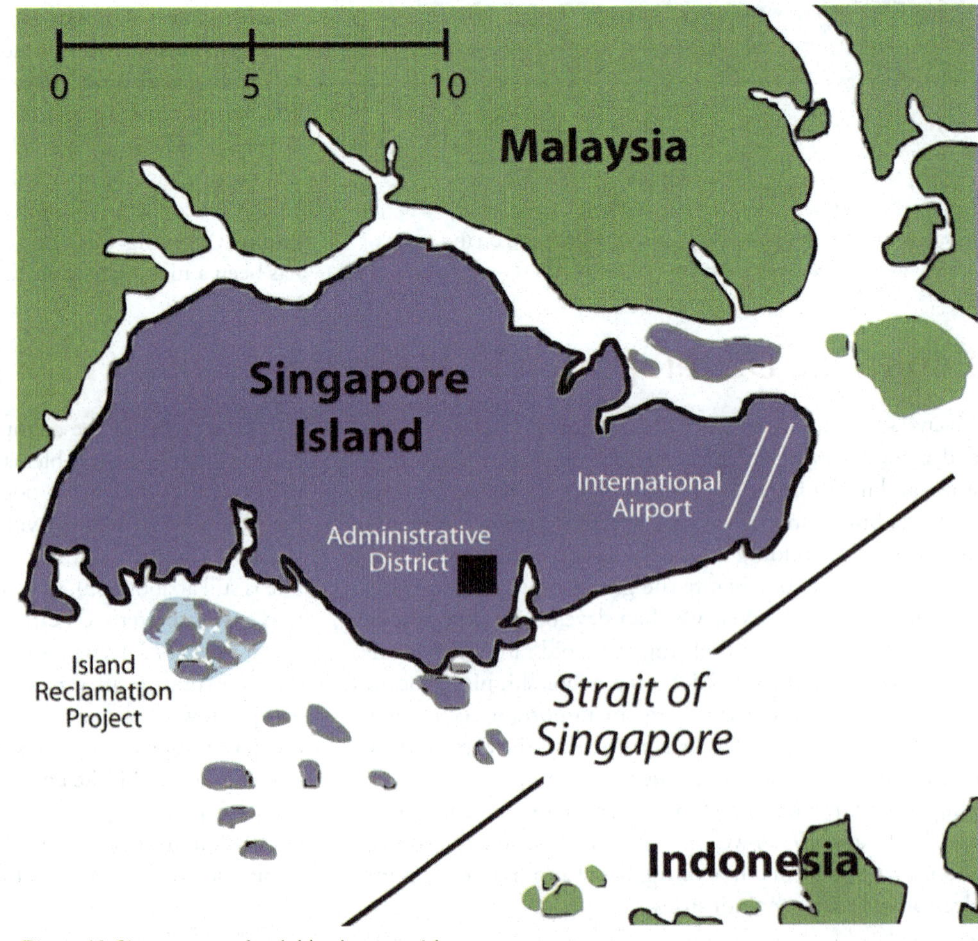

Figure 14. Singapore and neighboring countries.

Figure 15. Major islands and the thirty-three national provinces of Indonesia.

C. Indonesia

The country of Indonesia is the world's largest **archipelago** state, consisting of more than 17,500 islands, about one-third of which are inhabited. Indonesia is the sixteenth-largest country in the world by area. The combined area of all the islands and regions of Indonesia would equal about the size of the country of Mexico. The country shares land borders with the Borneo side of Malaysia, the western half of the island of Timor, and the western portion of the island of New Guinea, which is shared with the country of Papua New Guinea.

The country's location on both sides of the equator provides a tropical Type A climate, complete with a monsoon season. Average rainfall can vary from seventy to two hundred forty inches per year. The highest mountain is in **West Papua** and rises to about 16,024 feet. Indonesia is located on the Pacific Rim, where tectonic plate activity produces earthquakes and volcanic activity. The country is home to over one hundred fifty active volcanoes, including two of the most famous ones, **Krakatoa** and **Tambora**. Both had devastating eruptions in the past two centuries. One of the most violent volcanic explosions ever recorded in human history came from Krakatoa, which is located between the islands of **Java** and **Sumatra**. A series of eruptions in 1883 were heard as far away as the coast of Africa. Shockwaves reverberated around the globe seven times. Ash erupted into the atmosphere to a height of about fifty miles. The official death toll was 36,417, but estimates from local sources place it as high as 120,000. Global temperatures fell by about 2 °F, and weather patterns were disrupted for the next five years. Krakatoa remains active. Over the past few decades, the volcanic peak has been growing at the average rate of about five inches per week.

The tropical climate and the archipelago nature of the country provide for enormous **biodiversity** within the environment. Second only to Brazil in its biodiversity, Indonesia is host to an enormous number of unique plants and animals. The habitats of many of these creatures are being encroached upon by human activity. The remote islands have more of a chance of escaping habitat devastation and remaining intact, but agricultural and extractive economic activities have converted much of the natural environment into a cultural landscape that is not conducive to environmental sustainability.

In 2010, the estimated population of Indonesia was about 245 million. Indonesia has the fourth-largest population of any country in the world, after China, India, and the United States. Indonesia also has more Muslims than any other country in the world. More than half the population of Indonesia lives on **Java**, the island where **Jakarta**, the capital city, is located. Java is the most populous island in the world, and has a population density of more than 2,400 people per square mile. Java is the size in area of the US state of Louisiana. Java has 135 million people, whereas Louisiana has 4.5 million people. Jakarta is a world-class city that is larger than New York City and encompasses a large metropolitan area,

Figure 16. The skyline of Jakarta. Jakarta has a population of about ten million and is located on the island of Java. Java has more than 135 million people and has about the same physical area as the US state of Louisiana, which has about 4.5 million people. Java has more people than any other island in the world.

complete with many manufacturing centers, business complexes, and housing districts.

The many islands of Indonesia are home to a large number of diverse ethnic and religious groups that vary as widely as any Southeast Asian nation. There may be as many as three hundred different ethnic groups in Indonesia. Many of the ethnic groups are further divided by islands or distance. More than seven hundred separate languages and hundreds of additional dialects are spoken. The most prevalent language group in the country as a whole is Javanese, which is spoken by about 42 percent of the population. Javanese includes the official language of Indonesian, which is taught in schools and used in business and politics as the *lingua franca* of the country. Many people speak more than one language or even a number of languages to communicate throughout the country.

Islam diffused to Indonesia in the thirteenth century and by the sixteenth century had become the dominant religion. The Indonesian constitution allows for religious freedom, although more than 85 percent of the population follows Islam. There are at least four other religions that are officially recognized: Christianity (both Protestantism and Roman Catholicism), Hinduism, Buddhism, and Confucianism. Since Islam is followed by such a large percentage of the population, the other religions do not carry the same influence.

In spite of the diversity within the population, the country of Indonesia has established a substantial degree of **nationalism** as a **centripetal force** that holds the country together. There is a relatively high degree of stability in spite of the surface tensions and ethnic and religious conflicts that may erupt. An example of the social tensions is demonstrated in the case of Chinese citizens of Indonesia, who only make up about one percent of the population but impart a substantial influence over the privately owned business sector of the economy. This seemingly inequitable relationship has resulted in considerable resentment by other portions of the population, often with violent results. The many islands have

become natural divisions between cultural groups. It is not easy to create national unity with such a diverse population scattered throughout such a large archipelago.

Agriculture has been the historic base of the Indonesian economy. In 2010, it accounted for about 13 percent of the nation's gross domestic product (GDP). Agriculture is the largest employment sector—approximately 42 percent of the workforce. This equates to more than half of the population being rural. Many of the agricultural methods in rural areas are traditional; for example, farmers use water buffalo or oxen for tilling the land. The tropical climate and adequate rainfall provide for multiple crops of rice per year in many areas. Spices, coffee, tea, palm oil, and rubber are also produced in substantial quantities.

Industries are an important building block for how a country gains wealth. In the case of Indonesia, industry accounts for about 40 percent of its GDP and employs about 20 percent of its workforce. Major industries include oil, natural gas, mining, and textiles or clothing manufacturing. Indonesia's economy has been affected by global markets, but in 2005 still managed to run a trade surplus. Japan has been its main trading partner, and China has also been a major supplier of imported goods. Indonesia has been taking advantage of the trade triangle it has with its neighbors, Singapore and Malaysia, to increase its import and export trade activities.

The political background of Indonesia is similar in dynamics to many of its neighbors. Colonized by Europeans, Indonesia was previously called the **Dutch East Indies**, which explains why the islands of the Caribbean were called the **West Indies**. The Dutch colonized Indonesia in the early seventeenth century but had to relinquish possession of the archipelago to the Japanese in World War II. In 1945, after the Japanese surrendered, Indonesia declared its independence, which was finally granted in 1949 after much negotiation. Since 1999, Indonesia has conducted free parliamentary elections and is now considered the third-largest democracy, after India and the United States.

F. The Philippines

Located on the eastern side of the Southeast Asian community is the archipelago state of the Philippines. With more than 7,100 islands, many volcanic peaks, and an expanse of coastal waterways, the Philippines is home to more than ninety million people in a combined land area about the size of Arizona. The Philippines were a Spanish colony. The name is taken from Spain's sixteen-century King Philip II. Spain relinquished its claim on the Philippines to the United States in 1898 after its defeat in the Spanish-American War. The

people of the Philippines wanted independence at that time and fought a bitter war with the United States in which more than a million people died. The United States allowed the Philippines to become a commonwealth in 1935. The independence movement was placed on hold while the Japanese invaded and controlled the Philippines during World War II. After the war was over, the United States granted the Philippines their independence in 1946.

Environmental Forces

The islands of the Philippines are of volcanic origin. They are mainly mountainous and covered in tropical rainforest. The highest mountain, at 9,692 feet, is Mt. Apo, which is located on the southern island of **Mindanao**. The Philippines has a number of active volcanoes. The northern island of Luzon is home to the Taal Volcano, Mt. Pinatubo, and Mt. Mayon. The Pacific tectonic plate reaches the southern edge of the Philippine plate where it meets up with the Eurasian Plate. The juncture of tectonic plates creates a similar situation to that of Tokyo, which is at the opposite end of the Philippine plate.

Active seismic forces result in many earthquakes. As many as twenty earthquakes a day can be registered here, though many are too weak to be noticed. In 1990, an earthquake on the island of Luzon registered at a magnitude of 7.8 and killed more than 1,621 people, causing extensive damage.

Luzon's **Mt. Pinatubo** volcano has been active in recent years. Before 1991, the mountain attracted little attention, was heavily forested, and was home to tribal indigenous people. The volcano had a colossal eruption in 1991 that was recorded as the second largest in a century, after Alaska's 1912 Novarupta eruption. Mt. Pinatubo began giving signs of an eruption, which were heeded by the government. Thousands of people were evacuated from the area, which saved many lives. The eruption caused billions of dollars in damage. More than eight hundred people were killed, and more than two million were directly impacted. The eruption destroyed more than eight thousand homes and the overall effects of the volcano were felt around the world. The damage had a massive impact on the entire economy of the Philippines.

Figure 17. The three main regions of the Philippines.

Earthquakes and volcanoes are not the only serious natural concerns of the Philippine Islands; they are also directly in the center of the Western Pacific's major typhoon belt. As many as twenty typhoons occur yearly in the area of the islands, and as many as half hit the islands directly. The 1991 typhoon Thelma/Uring killed as many as eight thousand people. The 1911 typhoon dumped over forty-six inches of rain in a twenty-four-hour period. Flooding is usually the main problem with typhoons and is the number one killer related to typhoon deaths.

Political Geography

The Philippines can be divided into three main geopolitical regions: **Luzon**, **Visayas**, and **Mindanao**. The northern island of Luzon is home to the nation's **national capital region** with **Quezon**, the largest city, and **Manila**, the capital. Both cities are a part of **metropolitan Manila**, which has a population of more than twenty million. The northern island of Luzon is home to half the population of the country. The central Philippines consists of the **Visayas Island group**, including the islands between the Sulu Sea and the Philippine Sea. The southern region of the country is anchored by the large island of Mindanao.

Cultural Geography

The Philippines is a diverse country with hundreds of ethnic groups. Many tribal groups as well as a large number of immigrants from Asia, Spain, and the United States have made the Philippines home. Together with Spanish influence, mixed ethnic groups have been created. They are an example of the confluence of cultures that make up the country. The Philippines is the only country in Asia where Roman Catholicism predominates, other than recently independent East Timor.

Christians make up about 90 percent of the population. All but 10 percent identify themselves as Roman Catholic. A modest Muslim population is prominent in the southern island of Mindanao and neighboring islands. Islamic fundamentalism has increased the insurgency in the region, causing political and economic turmoil and conflict. People of Chinese heritage often follow Buddhism, Taoism, or Chinese folk religions. Various tribal groups still follow their cultural animist beliefs and have traditional shaman religious leaders.

The Philippines is home to more than one hundred eighty native languages and dialects. English and **Filipino** were declared the official languages of the Philippines in 1987. **Tagalog** is the main language spoken. Filipino is a version of Tagalog that is used in many of the urban areas. English and Tagalog are used in different parts of the country. The population growth rate is considerable. The Philippines now has a population of just over one hundred million, and 32 percent of its citizens are under the age of fifteen. The fertility rate is 2.8, which will continue to influence the economic situation of the country.

Figure 18. Ash plume from Mt. Pinatubo during 1991 eruption. The island of Luzon in the Philippines has a number of active volcanoes. Ash from Mt. Pinatubo caused so much damage that it resulted in the permanent closure of major US military bases in the Philippines.

Figure 19. Ferdinand Magellan brought Christianity to the Philippines and converted members of the Cebuano tribe to Catholicism in 1521.

Figure 20. Street scene in Manila with jeepney on the left. The greater metropolitan region of Manila has more than twenty million people The city of Manila itself is one of the most densely populated large cities in the world.

The Global Economy and Outsourcing

The modest level of political stability has caused the Philippines to become an attractive destination for global corporations who seek to **outsource** their information and technology service jobs. Any work that can be conducted over the Internet or telephone can be outsourced to anywhere in the world with high-speed communication links. Countries that are attractive to **business process outsourcing (BPO)** are countries where the English language is prominent, where employment costs are low, and where there is an adequate labor base of skilled or educated workers that can be trained in the services required. All three of these requirements are met by the labor force of the Philippines. The historical influence of the United States has provided a base of English language speakers. The country also has an adequate population base with the education or professional skills necessary to meet these demands. **Corporate colonialism** has the Philippines in its business focus and is finding a good source of available labor.

Key Takeaways

1. Malaysia was a former British colony made up of various regions from both the Malay Peninsula and the island of Borneo. Modern Malaysia has diverse cultural dynamics and is modernizing its economy to compete with the core economic areas of the world.

2. Singapore is an economic tiger that doesn't have natural resources but makes good use of an excellent location. High-tech manufacturing has been Singapore's main method of gaining wealth.

3. Indonesia is made up of thousands of islands and hundreds of ethnic groups. Indonesia is the fourth-most populous country in the world and has the world's largest Muslim population. More than half the population lives on the island of Java.

4. The Philippines has more than ninety million people on thousands of islands. The country was colonized by Spain and was then a possession of the United States before it gained independence. Roman Catholicism and the English language are common in the Philippines, both of which augment a large outsourcing industry.

End-of-Chapter Summary

- Southeast Asia consists of two main geographic regions: the mainland portion that borders China and the insular region that consists of islands or portions of them between Asia and Australia. The large island of Borneo is split between the three countries of Brunei, Malaysia, and Indonesia.

- The only region of Southeast Asia that was not colonized by European countries was the Kingdom of Siam, which is part of the current country of Thailand. This French-colonized region has been often referred to as French Indochina. Britain, the Netherlands, Portugal, and Spain were also primary colonizers of the realm.

- Southeast Asia is diverse in both its human and its physical landscapes. Tropical climates dominate the realm with mountains and coastal areas covering the main land surfaces. This realm has a high rate of seismic activity with many active volcanoes and is susceptible to earthquake activity.

- All the main world religions can be found here. Indonesia has the fourth-largest population in the world and is the most populous Muslim country in the world. East Timor and the Philippines are the only two predominantly Christian countries in Asia. Buddhism is the dominant religion of the mainland region. Both Malaysia and Singapore have sizeable Hindu minority groups.

- Economic activities vary in Southeast Asia, with Singapore being an economic tiger and Brunei being an oil-rich emirate. Thailand is becoming a major manufacturing center and the Philippines has been a destination for outsourced information jobs. Landlocked Laos and isolated Burma (Myanmar) have weak economies. Vietnam and Cambodia are recovering from political isolation.

Text Acknowledgment

Image Acknowledgments

Chapter 1

Figure 1.1 "Intel (Ronler Acres), Hillsboro, Oregon" by Sam Beebe, Ecotrust, is licensed under CC-BY 2.0 (https://www.flickr.com/photos/28585409@N04/3472141540/).

Figure 1.5 "45 Parallel, Half Way Between the Equator and North Pole, Interstate 84, Near Baker City, Oregon" by Ken Lund is licensed under CC-BY-SA 2.0 (https://www.flickr.com/photos/kenlund/265682075/in/photostream/).

Figure 1.6 "Graphic of the Four Seasons" by R. Berglee is licensed under CC-BY-NC-SA 3.0.

Figure 1.7 "Comparison information for State of Oregon" by Cierra Maher is licensed under CC-BY-SA 3.0 (Based on "Flag map of Oregon showing the obverse side" by DrRandomFactor, CC-BY-SA 3.0.).

Figure 1.8 "Glaciers in Glacier National Park, Montana" by R. Berglee is licensed under CC-BY-NC-SA 3.0.

Figure 1.10 "Coastal Belize, a type A climate example" by R. Berglee is licensed under CC-BY-NC-SA 3.0.

Figure 1.12 "Saguaro Cacti in Arizona, a type B climate example" by R. Berglee is licensed under CC-BY-NC-SA 3.0.

Figure 1.13 "Leaf Colors at Newfound Gap" by National Park Service is licensed under Public Domain (https://commons.wikimedia.org/wiki/File:Leaf-Colors-at-Newfound-Gap-NPS1.jpg).

Figure 1.14 "Snow covered neighborhood in Fargo, North Dakota" by Michael Rieger, FEMA is licensed under Public Domain (https://commons.wikimedia.org/wiki/File:FEMA_-_40243_-_Snow_covered_neighborhood_in_Fargo,_North_Dakota.jpg).

Figure 1.15 "Forest removal example" by R. Berglee is licensed under CC-BY-NC-SA 3.0.

Figure 1.16 "A man in Malawi carries firewood for cooking and heating purposes" by Bryce Sitter is licensed under CC-BY-NC-SA 3.0.

Figure 1.17 "Lumber Mill Processing Hardwood Timber" by R. Berglee is licensed under CC-BY-NC-SA 3.0.

Figure 1.18 "General Pattern of Tectonic Plates" by USGS is licensed under CC-BY-NC-SA 3.0.

Figure 1.20 "World human population density map" by daysleeperrr is licensed under

CC-0 1.0 (Public Domain) (https://commons.wikimedia.org/wiki/File:World_human_population_density_map.png).

Figure 1.21 "Population Pyramids" by US Census Bureau is licensed under CC-BY-NC-SA 3.0.

Figure 1.24 "Prevailing world religions map" by J Intela is licensed under CC-BY-NC-SA 3.0 (http://commons.wikimedia.org/wiki/File:Prevailing_world_religions_map.png.).

Figure 1.25 "Cultural Landscapes Representing the Urban Core Region of Los Angeles and the Peripheral Regions of Rural Montana" by R. Berglee is licensed under CC-BY-NC-SA 3.0.

Figure 1.27 "Walmart Store Exterior" by Walmart Stores is licensed under CC-BY-NC-SA 3.0 (http://www.flickr.com/photos/walmart-corporate/5266815680.).

Figure 1.28 "Three Core Economic Areas of the World: North American Free Trade Agreement (NAFTA); the European Union (EU); and the East Asian Community (EAC)." See above for licensing information.

Figure 1.30 "The three sectors of the economy" by Emily Evans is licensed under CC-BY-NC-SA 3.0.

Chapter 2

Figure 2.1 "Map of Europe" by University of Texas Libraries is licensed under CC-BY-NC-SA 3.0 (http://www.lib.utexas.edu/maps/europe/txu-oclc-247233313-europe_pol_2008.jpg).

Figure 2.4 "Eiger, Mönch und Jungfrau" by Steinmann is licensed under Public Domain (http://commons.wikimedia.org/wiki/File:Eiger,_M%C3%B6nch_und_Jungfrau.jpg).

Figure 2.5 "The Western Highlands meet the lowlands in central Scotland." by R. Berglee is licensed under CC-BY-NC-SA 3.0.

Figure 2.6 "Roman Empire 117AD" by Tataryn77 is licensed under CC0 1.0 Universal Public Domain (http://commons.wikimedia.org/wiki/File:Roman_Empire_117AD.jpg.).

Figure 2.12 "Which Countries Use the Euro?" by Evan Centanni is licensed under CC-BY-SA 3.0 (http://www.polgeonow.com/2014/08/map-which-countries-use-euro-plus-this.html).

Figure 2.15 "Stockholm Port" by Condor Patagónico is licensed under Public Domain (http://commons.wikimedia.org/wiki/File:Stockholm_Port.jpg).

Figure 2.16 "Via Toledo1" by Inviaggiocommons is licensed under Public Domain (http://commons.wikimedia.org/wiki/File:Via_Toledo1.jpg.).

Figure 2.18 "Ice bruxelles" by Clicgauche is licensed under Public Domain (https://commons.wikimedia.org/wiki/File:Ice_bruxelles.JPG).

Figure 2.19 "The Eiffel Tower" by Library of Congress is licensed under Public Domain.

Figure 2.20 "Reagan Berlin Wall" by White House Photographic Office is licensed under Public Domain (https://commons.wikimedia.org/wiki/File:ReaganBerlinWall.jpg).

Figure 2.21 "Swiss Cottage Cafe, No.33 The High Street, Ilfracombe" by Roger A Smith is licensed under CC-BY-SA 2.0 Generic (https://commons.wikimedia.org/wiki/File:Swiss_Cottage_Cafe,_No.33_The_High_Street,_Ilfracombe._-_geograph.org.uk_-_1267429.jpg).

Figure 2.22 "The Cliffs of Moher, Ireland" by Pixabay is licensed under CC0 Public Domain (https://pixabay.com/en/cliffs-of-moher-ireland-landscape-981873/).

Chapter 3

Figure 3.1 "Map of Russia" by CIA World Factbook is licensed under Public Domain.

Figure 3.4 "Map of Russian subjects, 2008-03-01" by Updated from map courtesy of Lokal Profil is licensed under CC-BY 2.5 Generic (http://commons.wikimedia.org/wiki/File:Map_of_Russian_subjects_recent.svg).

Figure 3.6 "October Revolution celebration 1983" by Thomas Hedden is licensed under Public Domain (http://commons.wikimedia.org/wiki/File:October_Revolution_celebration_1983.png).

Figure 3.7 "The first McDonald's restaurant in the former Soviet Union" by Mary Krueger is licensed under CC-BY-NC-SA 3.0.

Figure 3.8 "Red Square, Moscow, Russia 2" by Raul P is licensed under CC-BY-SA 3.0 Unported (https://commons.wikimedia.org/wiki/File:Red_Square,_Moscow,_Russia_2.jpg).

Figure 3.10 "Pyhän Vasilin katedraali" by Stoljaroff is licensed under Public Domain (http://commons.wikimedia.org/wiki/File:Pyh%C3%A4n_Vasilin_katedraali.jpg).

Figure 3.11 "Russia's Eastern Frontier, the Far East, and Siberia" by CIA World Factbook is licensed under Public Domain.

Figure 3.12 "Koryaksky volcano Petropavlovsk-Kamchatsky oct-2005" by Tatyana Rashidova is licensed under Public Domain (http://commons.wikimedia.org/wiki/File:Avacha_volcano_Petropavlovsk-Kamchatsky_oct-2005.jpg).

Figure 3.13 "Vladivostok tram" by Oxunhutch is licensed under CC-BY 3.0 (http://en.wikipedia.org/wiki/File:Vladivostok_tram.JPG).

Figure 3.14 "Chechnya and Caucasus" by Kbh3rd is licensed under CC-BY-SA 3.0 Unported (http://commons.wikimedia.org/wiki/File:Chechnya_and_Caucasus.png).

Figure 3.15 "Azerbaijanoil" by Indigoprime is licensed under CC-BY 2.0 Generic (http://commons.wikimedia.org/wiki/File:Azerbaijanoil.jpg).

Chapter 4

Figure 4.1 "The geographic center of North America located near Rugby, North Dakota" by R. Berglee is licensed under CC-BY-NC-SA 3.0.

Figure 4.2 "Physiographic regions of the US and Canada" by Cephas is licensed under CC-BY-SA 3.0 Unported (http://commons.wikimedia.org/wiki/File:Major_habitat_type_CAN_USA.svg.).

Figure 4.3 "European Influence in the Colonial United States" is licensed under CC-BY-NC-SA 3.0.

Figure 4.4 "U.S. Territorial Acquisitions" is licensed under CC-BY-NC-SA 3.0.

Figure 4.5 "U.S. Territorial Acquisitions" National Atlas of the United States is licensed under (http://commons.wikimedia.org/wiki/File:U.S._Territorial_Acquisitions.png).

Figure 4.6 "The Holy Trinity Serb Orthodox Church in Butte, Montana" by R. Berglee is licensed under CC-BY-NC-SA 3.0.

Figure 4.7 "Manufacturing Belt Turned Rust Belt" is licensed under CC-BY-NC-SA 3.0.

Figure 4.9 "Farm Resource Regions" by US Department of Agriculture is licensed under Public Domain (http://www.ers.usda.gov/publications/aib760).

Figure 4.10 "Freiheitsstatue NYC full" by US Government is licensed under Public Domain (http://commons.wikimedia.org/wiki/File:Freiheitsstatue_NYC_full.jpg).

Figure 4.11 "Hispanic Population in the United States and the US Sun Belt" by National Atlas of the United States is licensed under Public Domain (http://www.nationalatlas.gov/).

Figure 4.12 "Hispanic Population in the United States, 1970–2050" by Data from the US Census Bureau, 1970, 1980, 1990, and 2000 Decennial Censuses; Population Projections, July 1, 2010, to July 1, 2050. is licensed under Public Domain.

Figure 4.13 "La Frontera" by Omar Bárcena is licensed under CC-BY 2.0 Generic (http://www.flickr.com/photos/omaromar/316206516.).

Figure 4.14 "Civil War Division in the United States, 1861–1865" is licensed under CC-BY-NC-SA 3.0.

Figure 4.15 "Black/African American Population by County, 2000" by National Atlas of the United States is licensed under Public Domain (http://www.nationalatlas.gov/).

Figure 4.17 "Dehart's Bible and Tire, Eastern Kentucky" by R. Berglee is licensed under CC-BY-NC-SA 3.0.

Figure 4.18 "Map Canada political-geo" by E Pluribus Anthony is licensed under Public Domain (https://commons.wikimedia.org/wiki/File:Map_Canada_political-geo.png).

Figure 4.19 "Curling Canada Torino 2006" by Bjarte Hetland is licensed under CC-BY-SA 3.0 (http://commons.wikimedia.org/wiki/File:Curling_Canada_Torino_2006.jpg)

Figure 4.20 "Bilingualstopsign" by Steven Spell is licensed under Public Domain (http://commons.wikimedia.org/wiki/File:Bilingualstopsign.jpg).

Figure 4.21 "Untitled" by Marcin Wichary is licensed under CC-BY 2.0 Generic (http://www.flickr.com/photos/8399025@N07/2711707979).

Figure 4.22 "Evidence of Eastern European Immigration" by R. Berglee is licensed under CC-BY-NC-SA 3.0.

Figure 4.23 "Lufthansa cityline crj900lr d-acke takeoff london heathrow arp" by Adrian Pingstone is licensed under Public Domain (http://commons.wikimedia.org/wiki/File:Lufthansa_cityline_crj900lr_d-acke_takeoff_london_heathrow_arp.jpg).

Figure 4.24 "Main Regions of the United States and Canada" is licensed under CC-BY-NC-SA 3.0.

Figure 4.26 "Downtown Philadelphia's city hall" by R. Berglee is licensed under CC-BY-NC-SA 3.0.

Figure 4.27 "Long Lot Farms Typical in French Canada" by E. Ratajeski is licensed under CC-BY-NC-SA 3.0.

Figure 4.28 "The Dixie Grill restaurant in Moreland, Kentucky" by R. Berglee is licensed under CC-BY-NC-SA 3.0.

Figure 4.29 "Acres of Corn Harvested for Grain as a Percentage of Harvested Cropland Acreage, 2007" by US Department of Agriculture is licensed under Public Domain (http://www.agcensus.usda.gov/Publications/2007/Online_Highlights/Ag_Atlas_Maps/Crops_and_Plants/Field_Crops_Harvested/07-M165.asp).

Figure 4.30 "Satellite Image of Corn, Sorghum, and Wheat Fields in Southwestern Kansas" by NASA/GSFC/METI/ERSDAC/JAROS and U.S./Japan ASTER Science Team is licensed under Public Domain (http://earthobservatory.nasa.gov/Newsroom/NewImages/images.php3?img_id=17006).

Figure 4.31 "Typical Home in a Phoenix Suburb, Where Water Is a Valued Commodity" by R. Berglee is licensed under CC-BY-NC-SA 3.0.

Figure 4.32 "0161 Grand Canyon_Eagle Dancer 9/21/10" by Michael Quinn and Grand Canyon NPS is licensed under CC-BY 2.0 Generic (http://www.flickr.com/photos/grand_canyon_nps/5128920480).

Figure 4.33 "Rocky Mountains of Western Montana on US Highway 2" by R. Berglee is licensed under CC-BY-NC-SA 3.0.

Figure 4.34 "Urban Growth in Las Vegas, Nevada, from 1984 to 2009" by NASA is licensed under Public Domain (http://landsat.gsfc.nasa.gov/images/archive/e0018.html).

Figure 4.35 "The San Andreas Fault" by USGS is licensed under Public Domain (http://nationalatlas.gov/articles/geology/features/sanandreas.html).

Figure 4.36 "San Pedro Harbor" by Prayitno is licensed under CC-BY 2.0 Generic (https://www.flickr.com/photos/prayitnophotography/5124662348).

Figure 4.37 "Radar Station at Point Lay, Alaska" by US Air Force, TSgt Donald L. Wetterman is licensed under Public Domain (http://commons.wikimedia.org/wiki/File:Point_Lay_Alaska_DEW_Line.jpg)

Chapter 5

Figure 5.1 "Oceania (Reference Map) 2002" by University of Texas Libraries is licensed under CC-BY-NC-SA 3.0 (http://www.lib.utexas.edu/maps/australia/oceania_ref02.jpg.).

Figure 5.2 "Wallace's and Weber's Lines" updated from map courtesy of CIA World Factbook is licensed under Public Domain.

Figure 5.3 "Australian Colonialism" is licensed under CC-BY-NC-SA 3.0.

Figure 5.4 "Alice Springs" by Free Aussie Stock is licensed under CC-BY 3.0 Unported (http://freeaussiestock.com/free/Northern_Territory/alice_springs/slides/alice_springs_city_view.htm).

Figure 5.5 "Aerial view of Uluru (Ayers Rock), Located in the Interior of Australia near Alice Springs" by Huntster is licensed under Public Domain (http://commons.wikimedia.org/wiki/File:Uluru_%28Helicopter_view%29-crop.jpg).

Figure 5.6 "Australia states blank" by Golbez (updated) is licensed under CC-BY-SA 3.0 Unported (http://commons.wikimedia.org/wiki/File:Australia_states_blank.png).

Figure 5.7 "Sydney opera house and skyline" by Matthew Field is licensed under CC-BY 2.5 Generic (http://commons.wikimedia.org/wiki/File:Sydney_opera_house_and_skyline.jpg).

Figure 5.8 "Barossa Valley South Australia" by Louis Roving is licensed under CC-BY 2.0 Generic (http://commons.wikimedia.org/wiki/File:Barossa_Valley_South_Australia.jpg).

Figure 5.9 "Australia states blank" by Golbez (updated) is licensed under CC-BY-SA 3.0 Unported (http://commons.wikimedia.org/wiki/File:Australia_states_blank.png).

Figure 5.10 "Bondi Beach" by Dingy is licensed under Public Domain (http://commons.wikimedia.org/wiki/File:Bondi_beach_en_novembre.JPG).

Figure 5.11 "Southern Alps, New Zealand" by Remember is licensed under Public Domain (http://commons.wikimedia.org/wiki/File:Mountains_in_New_Zealand.jpg).

Figure 5.12 "New Zealand" by CIA World Factbook is licensed under Public Domain.

Figure 5.13 "Kiwi Bird" by Maungatautari Ecological Island Trust is licensed under Public Domain (https://commons.wikimedia.org/wiki/File:TeTuatahianui.jpg).

Figure 5.14 "New Zealand: Maori Culture 002" by Steve Evans is licensed under CC BY-NC 2.0 (tp://www.flickr.com/photos/babasteve/5397152967/in/photostream).

Chapter 6

Figure 6.1 "The Tropical Realm of the South Pacific with the Three Main Regions of Islands" is licensed under CC-BY-NC-SA 3.0.

Figure 6.2 "The Region of Melanesia" by is licensed under CC-BY-NC-SA 3.0.

Figure 6.3 "Malaitan Chief on the Solomon Islands" by Jim Lounsbury is licensed under Public Domain (http://en.wikipedia.org/wiki/File:Malaitan_Chief.jpg).

Figure 6.4 "Saint Joseph's Bay on the Isles of Pines, New Caledonia" by Bruno Menetrier is licensed under Public Domain (http://commons.wikimedia.org/wiki/File:Noum%C3%A9a_Ile_des_Pins_Saint_Joseph.JPG).

Figure 6.5 "American Flags in Guam" by Pixabay is licensed under CC0 Public Domain (https://pixabay.com/en/guam-sky-clouds-palms-palm-trees-83227/).

Figure 6.8 "Moorea Ferry in Papeete harbour" by Evil Monkey is licensed under CC-BY 2.5 Generic (http://commons.wikimedia.org/wiki/File:Moorea_Ferry_in_Papeete_harbour.JPG).

Figure 6.9 "Bora Bora" by tensaibuta is licensed under CC-BY 2.0 Generic (http://www.flickr.com/photos/97657657@N00/2092792187).

Figure 6.10 "The Southern Ocean and Antarctica" by British Antarctic Survey is licensed under Public Domain (http://www.photo.antarctica.ac.uk).

Figure 6.11 "Size Comparison of Antarctica and the United States" by NASA is licensed under Public Domain (http://lima.nasa.gov/antarctica).

Figure 6.12 "Mount Erebus" by Josh Landis (National Science Foundation) is licensed under Public Domain (https://commons.wikimedia.org/wiki/File:Mount_Erebus_craters,_Ross_Island,_Antarctica_(aerial_view,_18_December_2000).jpg).

Figure 6.13 "Emperor Penguins, Ross Sea, Antarctica" by Michael Van Woert (NOAA) is licensed under Public Domain (http://commons.wikimedia.org/wiki/File:Emperor_penguin.jpg).

Figure 6.14 "Antarctic Territorial Claims" by NASA is licensed under Public Domain (http://www.nasa.gov/audience/forstudents/5-8/features/what-is-antarctica-58.html).

Figure 6.15 "The Greenhouse Effect and Climate Change" is licensed under CC-BY-NC-SA 3.0.

Figure 6.16 "Ozone in the stratosphere" is licensed under CC-BY-NC-SA 3.0.

Figure 6.17 "One of the largest ozone holes" by NASA is licensed under Public Domain (http://commons.wikimedia.org/wiki/File:-160658main2_OZONE_large_350.png)

Figure 6.18 "AntarcticBedrock" by Cristellaria is licensed under CC-BY 3.0 Unported (http://commons.wikimedia.org/wiki/File:AntarcticBedrock.jpg).

Chapter 7

Figure 7.1. ""Political Central America" CIA World Factbook" by Central Intelligence Agency is licensed under Public Domain (https://commons.wikimedia.org/wiki/File:%22Political_Central_America%22_CIA_World_Factbook.jpg).

Figure 7.3. "Mayan Site of Uxmal in the Yucatán Region of Mexico" by R. Berglee is licensed under CC-BY-NC-SA 3.0.

Figure 7.4. "Catholic Cathedral across from a Plaza in the Yucatán City of Valladolid (left); Model of a Spanish Colonial Urban Pattern (right)" by R. Berglee is licensed under CC-BY-NC-SA 3.0.

Figure 7.6. "Major Volcanoes of Mexico" by Topinka, USGS/Cascades Volcano Observatory is licensed under Public Domain (http://vulcan.wr.usgs.gov/Volcanoes/Mexico/Maps/map_mexico_volcanoes.html).

Figure 7.8. "MexCityPolution" by USFirstGov is licensed under CC-BY 3.0 Unported (https://commons.wikimedia.org/wiki/File:MexCityPolution.JPG).

Figure 7.9. "Mayan Home in the Rural Village of Yachachen in the Yucatán Peninsula" by R. Berglee is licensed under CC-BY-NC-SA 3.0.

Figure 7.12. "Population Pyramids for Mexico in 1980 and 2010" Data courtesy of US Census Bureau International Programs, licensed under Public Domain.

Figure 7.18. "For the sovereignty of the people..." by laurizza is licensed under CC-BY 2.0 Generic (http://www.flickr.com/photos/ljel/5553854799.).

Figure 7.20. "US "Colonial" Influence in Cuba" by R. Berglee is licensed under CC-BY-NC-SA 3.0.

Figure 7.21. "Fasciculus:Fidel Castro" Photo on the left courtesy of Agência Brasil and Lucas; Photo on the right by R. Berglee, licensed under CC-BY 3.0 Brazil (http://la.wikipedia.org/wiki/Fasciculus:Fidel_Castro.jpg).

Figure 7.22. "Dump Truck Taxi" by R. Berglee is licensed under CC-BY-NC-SA 3.0.

Figure 7.23. "Viñales Valley in Western Cuba" by R. Berglee is licensed under CC-BY-NC-SA 3.0.

Figure 7.24. "US Government Building in San Juan, Puerto Rico, with Both US and Puerto Rican Flags" by Bobby Lemasters is licensed under CC-BY-NC-SA 3.0.

Chapter 8

Figure 8.2. "Peru" by Benedict Adam is licensed under CC-BY 2.0 Generic (http://www.flickr.com/photos/backpackerben/2791685045).

Figure 8.5. "La Compania de Jesus, Cusco" by James Preston is licensed under CC-BY 2.0 Generic (http://www.flickr.com/photos/jamespreston/1125218299).

Figure 8.7. "Salvador Carnaval Comanches 03" by Carnaval.com Studios is licensed under CC-BY 2.0 Generic (http://www.flickr.com/photos/aforum/4398649958.).

Figure 8.8. "Mother and Child" by Thomas Quine is licensed under CC-BY 2.0 Generic (http://www.flickr.com/photos/91994044@N00/93022902).

Figure 8.9. "Source of the Amazon River" by NASA is licensed under Public Domain (http://earthobservatory.nasa.gov/IOTD/view.php?id=7823).

Figure 8.10 "Venezuela" by CIA World Factbook is licensed under Public Domain (https://www.cia.gov/library/publications/the-world-factbook/geos/ve.html).

Figure 8.11. "El Salto Angel (completamente alucinante y magico)" by Inti is licensed under CC-BY 2.0 Generic (http://www.flickr.com/photos/inti/3102779830).

Figure 8.12. "Vista de Caracas and A tunnel and barrios surrounding it on the hill" Photo on the left by Cristóbal Alvarado Minic licensed under CC-BY 2.0 Generic (http://www.flickr.com/photos/ctam/4732562277); Photo on the right by Danila Medvedev. licensed under CC-BY 2.0 Generic (http://www.flickr.com/photos/danila/29987096).

Figure 8.14. "Ja-knikker op" Photo on the left by the DEA is licensed under Public Domain (http://www.justice.gov/dea/photos/cocaine/cocaine_bricks_scorpion_logo.jpg); Photo in the center by R. Berglee; Photo on the right by LennartBolks is licensed under Public Domain (http://commons.wikimedia.org/wiki/File:Ja-knikker_op.jpg).

Figure 8.15. "Machu Picchu 12" by Corey Spruit is licensed under CC-BY 2.0 Generic (http://www.flickr.com/photos/funkz/4034082685).

Figure 8.16. "Panoramic View - Quito, Ecuador - South America" by David Berkowitz is licensed under CC-BY 2.0 Generic (http://www.flickr.com/photos/davidberkowitz/4870874502).

Figure 8.17. "Urfolk i Bolivia" by Norsk Folkehjelp Norwegian People's Aid is licensed under CC-BY 2.0 Generic (http://www.flickr.com/photos/folkehjelp/4776227579).

Figure 8.18. "South America" by University of Texas Libraries is licensed under Public Domain (http://www.lib.utexas.edu/maps/americas/south_america_ref_2010.pdf).

Figure 8.19 "North Atlantic Ocean laea relief location map." by Robert Alvarez is licensed under CC-BY-SA 3.0 Unported (http://titanicdatabase.wikia.com/wiki/File:North_Atlantic_Ocean_laea_relief_location_map.jpg).

Figure 8.20. "Brazil States" by Darlan P. de Campos is licensed under CC-BY 3.0 Unported (http://commons.wikimedia.org/wiki/File:Brazil_States.svg).

Figure 8.21. "Rio de Janeiro" by Kirilos is licensed under CC-BY 2.0 Generic (http://www.flickr.com/photos/pedrokirilos/3640359022).

Figure 8.22. "Gusto 's favéla" by dany13 is licensed under CC-BY 2.0 Generic (https://www.flickr.com/photos/dany13/10410315896).

Figure 8.24. "Natural Vegetation of Brazil" by University of Texas Libraries is licensed under Public Domain.

Figure 8.25. "Slash and burn agriculture in the Amazon" by Matt Zimmerman is licensed under CC-BY 2.0 Generic (http://www.flickr.com/photos/16725630@N00/1524189000).

Figure 8.27. "Brasília" by Carla Salgueiro is licensed under CC-BY 2.0 Generic (http://www.flickr.com/photos/carlinha/4038849886).

Figure 8.28. "Vinícola Miolo | Miolo Vineyard" by Jeff Belmonte is licensed under CC-BY 2.0 Generic (http://www.flickr.com/photos/jeffbelmonte/87127297).

Figure 8.29. "The Three Main Regions of Chile (left); Argentina and Uruguay with the Regions of Argentina Outlined and Labeled (right)" by CIA World Factbook is licensed under Public Domain.

Figure 8.30. "Iguazu Falls" by Jeffrey Bary is licensed under CC-BY 2.0 Generic (http://www.flickr.com/photos/70118259@N00/2701569937).

Figure 8.31. "South America" by University of Texas Libraries is licensed under Public Domain (http://www.lib.utexas.edu/maps/americas/samerica_95.jpg).

Figure 8.32. "Palermo IMG 6897" by Jimmy Baikovicius is licensed under CC-BY-SA 2.0 Generic (http://commons.wikimedia.org/wiki/File:Buenos_Aires_-Argentina-_136.jpg.).

Figure 8.33. "Patagonia" by Josh and Erica Olson Silverstein is licensed under CC-BY 2.0 Generic (http://www.flickr.com/photos/sacire/5577982949).

Figure 8.34. "North and South America" by University of Texas Libraries is licensed under Public Domain (http://www.lib.utexas.edu/maps/americas/americas_pol96.jpg).

Chapter 9

Figure 9.1 "Africa" by University of Texas Libraries is licensed under Public Domain (http://www.lib.utexas.edu/maps/africa/txu-oclc-238859671-africa_pol_2008.jpg).

Figure 9.2 "BlankMap-Africa" by Andreas 06 is licensed under Public Domain (http://commons.wikimedia.org/wiki/File:Blank-Map-Africa.svg).

Figure 9.5 "BlankMap-Africa" by Andreas 06 is licensed under Public Domain (http://commons.wikimedia.org/wiki/File:Blank-Map-Africa.svg.).

Figure 9.6 "Freedom Square in Ghana" by Janet Gross is licensed under CC-BY-NC-SA 3.0.

Figure 9.7 "Market Kaolack and Zimbabwe Harare Eastgate Shopping Mall" Photo on the left by Radoslaw Botev is licensed under Attribution (see website) (http://commons.wikimedia.org/wiki/File:Market_Kaolack.jpg); Photo on the right by Gary Bembridge. Is licensed under CC-BY 2.0 Generic (http://www.flickr.com/photos/tipsfortravellers/557269907).

Figure 9.8 "Population Pyramids of African Countries" by Data Source; US Census Bureau International Programs is licensed under Public Domain.

Figure 9.9 "Africa ethnic groups 1996" by Library of Congress is licensed under Public Domain (https://en.wikipedia.org/wiki/Languages_of_Africa#/media/File:Africa_ethnic_groups_1996.jpg).

Figure 9.10 "BlankMap-Africa" by Andreas 06 is licensed under Public Domain (http://commons.wikimedia.org/wiki/File:Blank_Map-Africa.svg.).

Figure 9.11 "Rwandan Genocide Murambi skulls" by US Congress is licensed under Public Domain (http://commons.wikimedia.org/wiki/File:Rwandan_Genocide_Murambi_skulls.jpg).

Figure 9.12 "Map-of-HIV-Prevalance-in-Africa" is licensed under Public Domain (https://commons.wikimedia.org/wiki/File:Map-of-HIV-Prevalance-in-Africa.png).

Figure 9.13 "Africa" by University of Texas Libraries is licensed under Public Domain (http://www.lib.utexas.edu/maps/africa/txu-oclc-238859671-africa_pol_2008.jpg).

Figure 9.14 "Hands ondiamonds 350" by United States Agency for International Development is licensed under Public Domain (http://commons.wikimedia.org/wiki/File:Hands_ondiamonds_350.jpg).

Figure 9.15 "Linguistic Groups" by University of Texas Libraries is licensed under Public Domain (http://www.lib.utexas.edu/maps/africa/nigeria_linguistic_1979.jpg).

Figure 9.16 "Nigerian Drummers" and "The Harpist" by Melvin "Buddy" Baker are licensed under CC-BY 2.0 Generic (http://www.flickr.com/photos/58034970@N00/178631090) (http://www.flickr.com/photos/i_level_news/184542488).

Figure 9.17 "Africa" by University of Texas Libraries is licensed under Public Domain (http://www.lib.utexas.edu/maps/africa/txu-oclc-238859671-africa_pol_2008.jpg.).

Figure 9.18 "DSC_0615" by Dylan Walters is licensed under CC-BY 2.0 Generic (http://www.flickr.com/photos/50169083@N00/1204360972).

Figure 9.19 "Tayor Shop" by Mark Knobil is licensed under CC-BY 2.0 Generic (http://www.flickr.com/photos/36448457@N00/66825084.).

Figure 9.20 "Rwandan refugee camp in east Zaire" by CDC is licensed under Public Domain (http://commons.wikimedia.org/wiki/File:Rwandan_refugee_camp_in_east_Zaire.jpg).

Figure 9.21 "Congolese Chairman of the Joint Chiefs, General Kisempia" by Themalau is licensed under Public Domain (http://commons.wikimedia.org/wiki/File:Kisempia.jpg.).

Figure 9.22 "BlankMap-Africa" by Andreas 06 is licensed under Public Domain (http://commons.wikimedia.org/wiki/File:Blank-Map-Africa.svg.).

Figure 9.23 "Le Match" by Geordie Mott is licensed under CC-BY 2.0 Generic (http://commons.wikimedia.org/wiki/File:Le_Match.jpg).

Figure 9.24 "African Beer" by R. Berglee is licensed under CC-BY-NC-SA 3.0.

Figure 9.25 "African And Costa Rican Coffee" by Rex Roof is licensed under CC-BY 2.0 Generic (http://www.flickr.com/photos/rexroof/266242059).

Figure 9.26 "woman_weaves_basket" by di bo di is licensed under CC-BY 2.0 Generic (http://www.flickr.com/photos/bonettodiego/314254523).

Figure 9.27 "Africa" by University of Texas Libraries is licensed under Public Domain (http://www.lib.utexas.edu/maps/africa/txu-oclc-238859671-africa_pol_2008.jpg.).

Figure 9.28 "EAfrica" by USGS is licensed under Public Domain (https://en.wikipedia.org/wiki/East_African_Rift#/media/File:EAfrica.png).

Figure 9.29 "Tanzanie, Descente vers le lac Natron" by Guillaume Baviere is licensed under CC-BY 2.0 Generic (http://www.flickr.com/photos/84554176@N00/1969101840.).

Figure 9.30 "Zebras, Serengeti savana plains, Tanzania" by Gary is licensed under CC-BY 2.0 Generic (http://commons.wikimedia.org/wiki/File:Zebras,_Serengeti_savana_plains,_Tanzania.jpg).

Figure 9.31 "Tusks in City of Mombasa" by Sandro Senn is licensed under Public Domain (http://commons.wikimedia.org/wiki/File:Tusks_in_City_of_Mombasa.jpg).

Figure 9.32 "Anciano" by Gusjer is licensed under CC-BY 2.0 Generic (http://www.flickr.com/photos/gusjer/2436691167).

Figure 9.33 "Nairobi 2008" by ND Strupler is licensed under CC-BY 2.0 Generic (http://www.flickr.com/photos/strupler/2975363718).

Figure 9.34 "Nairobi, view from KICC" by Daryona is licensed under CC-BY-SA 3.0 Unported, 2.5 Generic, 2.0 Generic, 1.0 Generic (https://commons.wikimedia.org/wiki/File:Nairobi,_view_from_KICC.JPG).

Figure 9.35 "Africa" by University of Texas Libraries is licensed under Public Domain (http://www.lib.utexas.edu/maps/africa/txu-oclc-238859671-africa_pol_2008.jpg.).

Figure 9.36 "Black-and-White Ruffed Lemur, Mantadia, Madagascar" by Frank Vassen is licensed under CC-BY 2.0 Generic (http://www.flickr.com/photos/42244964@N03/4309657930).

Figure 9.37 "ApartheidSignEnglishAfrikaans" by El C is licensed under Public Domain (http://commons.wikimedia.org/wiki/File:ApartheidSignEnglishAfrikaans.jpg).

Chapter 10:

Figure 10.1 "Political Map of the World, 2011" by University of Texas Libraries is licensed under Public Domain (http://www.lib.utexas.edu/maps/world_maps/world_pol_2011.pdf).

Figure 10.2 "fertilecrescentnatufian" by PB_Hausarbeiten is licensed under CC-BY 2.0 Generic (https://www.flickr.com/photos/58696257@N03/5512116621).

Figure 10.3 "Roman Aqueduct Near Caesarea " by R. Berglee is licensed under CC-BY-NC-SA 3.0.

Figure 10.4 "Banana Grove in Israel Near the Lebanese Border" by R. Berglee is licensed under CC-BY-NC-SA 3.0.

Figure 10.7 "Mapping the Global Muslim Population: A Report on the Size and Distribution of the World's Muslim Population" by Pew Research Center is licensed under Attribution (http://pewforum.org/newassets/images/reports/Muslimpopulation/Muslimpopulation.pdf).

Figure 10.8 "Mosque" by Antonio Melina of Agência Brasil is licensed under CC-BY 3.0 Brazil (http://commons.wikimedia.org/wiki/File:Mosque.jpg).

Figure 10.10 "Political Map of the World, 2011" by University of Texas Libraries is licensed under Public Domain (http://www.lib.utexas.edu/maps/world_maps/world_pol_2011.pdf).

Figure 10.11 "Atlas-Mountains-Labeled-2" by Williamborg is licensed under Public Domain (http://commons.wikimedia.org/wiki/File:Atlas-Mountains-Labeled-2.jpg).

Figure 10.12 "The Strait of Gibraltar in 3D" by NASA SRTM Team is licensed under Public Domain (http://earthobservatory.nasa.gov/IOTD/view.php?id=3926.).

Figure 10.14 "IMG00064-20110125-1429" by Muhammad Ghafari is licensed under CC-BY 2.0 Generic (http://www.flickr.com/photos/70225554@N00/5390371651).

Figure 10.15 "Tuaregs with their dromedaries" by Marco Bellucci is licensed under CC-BY 2.0 Generic (http://www.flickr.com/photos/marcobellucci/34170104939).

Figure 10.16 "Sudan-CIA WFB Map" by CIA World Factbook is licensed under Public Domain (https://commons.wikimedia.org/wiki/File:Sudan-CIA_WFB_Map.png).

Figure 10.17 "Satellite Image of Palestine" (left) by Nasa is licensed under Public Domain (http://commons.wikimedia.org/wiki/File:Southeast_mediterranean_annotated_geography.jpg); "1948 UN Division of Palestine into Half Jewish State and Half Arab State" (center) by Kordas is licensed under Public Domain (http://commons.wikimedia.org/wiki/File:1947-UN-Partition-Plan-1949-Armistice-Comparison-es.svg); "Political Map of Israel in 2011" (right) " by CIA World Factbook is licensed under Public Domain

Figure 10.18 "The Western Wall in Jerusalem " by R. Berglee is licensed under CC-BY-NC-SA 3.0.

Figure 10.19 "A Street in the West Bank City of Nablus " by R. Berglee is licensed under CC-BY-NC-SA 3.0.

Figure 10.20 "Security Wall between Israel and the West Bank " by R. Berglee is licensed under CC-BY-NC-SA 3.0.

Figure 10.21 "West Bank Settlements and Palestinian-Controlled Areas " by CIA World Factbook is licensed under Public Domain.

Figure 10.22 "King 'Abdullah II of Jordan Visits US President Barack Obama in the White House in 2011 " by White House is licensed under Public Domain.

Figure 10.23 "Jordan " by CIA World Factbook is licensed under Public Domain.

Figure 10.24 "Syria Damascus Douma Protests 2011 - 05" by syriana2011 is licensed under CC-BY 2.0 Generic (http://www.flickr.com/photos/syriana2011/5650171577).

Figure 10.25 "Syria " by CIA World Factbook is licensed under Public Domain.

Figure 10.26 "Arabian Peninsula dust SeaWiFS-2" by NASA is licensed under Public Domain (http://commons.wikimedia.org/wiki/File:Arabian_Peninsula_dust_SeaWiFS-2.jpg).

Figure 10.27 "Political Map of the World, 2011" by University of Texas Libraries is licensed under Public Domain (http://www.lib.utexas.edu/maps/world_maps/world_pol_2011.pdf).

Figure 10.28 "Medina Road - Jeddah - Saudi Arabia" by Ammar Shakar is licensed under Public Domain (https://commons.wikimedia.org/wiki/File:Medina_Road_-_Jeddah_-_Saudi_Arabia.jpg).

Figure 10.29 "Women in Saudi Arabia " by White House is licensed under Public Domain.

Figure 10.30 "The Tigris and Euphrates Rivers and the Shatt al-Arab Waterway between Iraq and Iran " by CIA World Factbook is licensed under Public Domain.

Figure 10.32 "Kurdish Areas in the Middle East and the Soviet Union" by University of Texas Libraries is licensed under Public Domain (http://www.lib.utexas.edu/maps/middle_east_and_asia/kurdish_86.jpg).

Figure 10.33 "USAF F-16A F-15C F-15E Desert Storm edit2" by US Air Force is licensed under Public Domain (http://commons.wikimedia.org/wiki/File:USAF_F-16A_F-15C_F-15E_Desert_Storm_edit2.jpg).

Figure 10.34 "Soldiers from the 926th Engineer Brigade Combat Team and the Army 432d Civil Affairs Battalion Patrol in Baghdad's District of Sadr City" by US Department of Defense is licensed under Public Domain.

Figure 10.36"Turkey " by CIA World Factbook is licensed under Public Domain.

Figure 10.37 1Blue Mosque" by Jeremy Vandel is licensed under CC-BY-NC-ND 2.0 Generic (http://www.flickr.com/photos/jeremy_vandel/3742592396).

Figure 10.38 "Political Map of the World, 2011" by University of Texas Libraries is licensed under Public Domain (http://www.lib.utexas.edu/maps/world_maps/world_pol_2011.pdf).

Figure 10.39 "Iran" by University of Texas Libraries is licensed under Public Domain (http://www.lib.utexas.edu/maps/middle_east_and_asia/iran_rel_2001.jpg).

Figure 10.40 "3rd Day - The Green Protest Rally" by Hamed Saber is licensed under CC-BY 2.0 Generic (http://www.flickr.com/photos/44124425616@N01/3630995595).

Figure 10.41 "Commonwealth of Independent States - Central Asian States" by University of Texas Libraries is licensed under Public Domain (http://www.lib.utexas.edu/maps/commonwealth/central_asian_common_2002.jpg).

Figure 10.42 "Afghanistan" by CIA-UT Auction Map Library is licensed under Public Domain (http://www.lib.utexas.edu/maps/middle_east_and_asia/txu-oclc-310605662-afghanistan_rel_2008.jpg).

Figure 10.43 "Propaganda Poster in Afghanistan with Image of Osama bin Laden " by US Department of Defense is licensed under Public Domain.

Figure 10.44 "Operation Enduring Freedom in Afghanistan " \by US Department of Defense is licensed under Public Domain.

Figure 10.45 "Voting in Afghanistan " by US Department of Defense is licensed under Public Domain.

Figure 10.46 "Lataband Road hut" by Sven Dirks is licensed under CC-BY-SA 4.0 International, 3.0 Unported, 2.5 Generic, 2.0 Generic, 1.0 Generic (http://commons.wikimedia.org/wiki/File:Lataband_Road_hut.jpg).

Chapter 11

Figure 11.1 "Middle East and Asia" by University of Texas Libraries is licensed under Public Domain (http://www.lib.utexas.edu/maps/middle_east_and_asia/txu-oclc-247232986-asia_pol_2008.jpg).

Figure 11.2 "The Himalayas summed up in one picture!" by Steve Hicks is licensed under CC-BY 2.0 Generic (http://www.flickr.com/photos/shicks/2515990913).

Figure 11.3 "Asie" by historicair is licensed under Public Domain (http://commons.wikimedia.org/wiki/File:Asie.svg).

Figure 11.5 "Crowded Street in New Delhi, India" by John Haslam is licensed under CC-BY 2.0 Generic (http://www.flickr.com/photos/foxypar4/415375182).

Figure 11.6 "Population Growth in India" by CIA World Factbook is licensed under Public Domain.

Figure 11.7 "Pakistan" by CIA World Factbook is licensed under Public Domain.

Figure 11.8 "PAK AU T1" by Schajee is licensed under Public Domain (http://commons.wikimedia.org/wiki/File:PAK_AU_T1.svg).

Figure 11.9 "A guy and his donkey" by Guilhem Vellut is licensed under CC-BY 2.0 Generic (http://www.flickr.com/photos/o_0/10070267).

Figure 11.10 "CAMEL @ GADANI" by Kashif Muhammad Farooq is licensed under CC-BY 2.0 Generic (http://www.flickr.com/photos/kashiff/2592800506).

Figure 11.11 "A female doctor with the International Medical Corps examines a woman patient at a mobile health clinic in Pakistan" by UK Department for International Development is licensed under CC-BY 2.0 Generic (http://www.flickr.com/photos/dfid/5331065350).

Figure 11.12 "dcp_9062" by Kai Hendry is licensed under CC-BY 2.0 Generic (http://www.flickr.com/photos/hendry/73370895).

Figure 11.13 "The Highlands of the Northern Areas in Pakistan" by Tore Urnes is licensed under CC-BY 2.0 Generic (http://www.flickr.com/photos/urnes/2663083945).

Figure 11.14 "The Disputed Areas of Kashmir" by University of Texas Libraries is licensed under Public Domain (http://www.lib.utexas.edu/maps/middle_east_and_asia/kashmir_disputed_2002.jpg).

Figure 11.15 "Bangladesh" by CIA World Factbook is licensed under Public Domain.

Figure 11.16 "The streets of Dhaka" by Ben Sutherland is licensed under CC-BY 2.0 Generic (http://www.flickr.com/photos/ben-sutherland/2050055462).

Figure 11.17 "Rice Field" by US Agency for International Development is licensed under Public Domain (http://commons.wikimedia.org/wiki/File:Rice_Field.jpg).

Figure 11.18 "Asie" by historicair is licensed under Public Domain (http://commons.wikimedia.org/wiki/File:Asie.svg).

Figure 11.19 "Annadatha" by antkriz is licensed under CC-BY 2.0 Generic (http://www.flickr.com/photos/ananth/136310496).

Figure 11.20 "city father" by BOMBMAN is licensed under CC-BY 2.0 Generic (http://www.flickr.com/photos/ajay_g/1516457856).

Figure 11.21 "Tata Nano im Verkehrszentrum des Deutschen Museums" by High Contrast is licensed under CC-BY 3.0 Germany (http://commons.wikimedia.org/wiki/File:Tata_Nano_im_Verkehrszentrum _des_Deutschen _Museums.JPG).

Figure 11.22 "Bollywood Filming in India" by Pixabay is licensed under Public Domain (https://pixabay.com/en/india-bolly-wood-film-890959/).

Figure 11.23. "Dividing India" by CIA World Factbook is licensed under Public Domain.

Figure 11.24 "Taj Mahal" by particlem is licensed under CC-BY 2.0 Generic (http://www.flickr.com/photos/particlem/2466310898).

Figure 11.25 "Lord Shiva" by Andrea is licensed under CC-BY 2.0 Generic (http://www.flickr.com/photos/rivo/125806388).

Figure 11.26 "People washing at India ghats" by Rupert Taylor-Price is licensed under CC-BY 2.0 Generic (http://www.flickr.com/photos/rupertuk/534644162).

Figure 11.28 "large buddha, bodh gaya, india" by Man Bartlett is licensed under CC-BY 2.0 Generic (http://www.flickr.com/photos/manbartlett/3731927920).

Chapter 12

Figure 12.1 "Middle East and Asia" by University of Texas Libraries is licensed under Public Domain (http://www.lib.utexas.edu/maps/middle_east_and_asia/asia_east_pol_2004.jpg).

Figure 12.4 "Pearl River Delta Area" by Croquant is licensed under CC-BY 3.0 Unported (http://commons.wikimedia.org/wiki/File:Pearl_River_Delta_Area.png).

Figure 12.6 "ThreeGorgesDam-China2009" by Rehman is licensed under CC-BY 2.0 Generic (http://commons.wikimedia.org/wiki/File:ThreeGorgesDam-China2009.jpg).

Figure 12.7 "Tourists on the Great Wall of China" by Pixabay is licensed under Public Domain (https://pixabay.com/en/china-great-wall-great-wall-746576/).

Figure 12.10a "Oriental Pearl Tower in Shanghai" by Dmitry A. Mottl is licensed under CC-BY-SA 4.0 International (https://commons.wikimedia.org/wiki/File:Oriental_Pearl_Tower_in_Shanghai.jpg).

Figure 12.10b "Pagoda. Shanghai, China, East Asia" by Mstyslav Chernov is licensed under CC-BY-SA 3.0 Unported (https://commons.wikimedia.org/wiki/File:Pagoda._Shanghai,_China,_East_Asia.jpg).

Figure 12.12 "Young Girl and Her Grandfather in Southern China" by Joyce Minor is licensed under CC-BY 3.0 Unported.

Figure 12.13 "China: Ethnolinguistic Groups" by University of Texas Libraries is licensed under Public Domain (http://www.lib.utexas.edu/maps/middle_east_and_asia/china_ethnolinguistic_83.jpg).

Figure 12.14 "Beijing" by Jon Phillips is licensed under CC-BY 2.0 Generic (http://www.flickr.com/photos/jonphillips/4998737777).

Figure 12.15 "China: Special Economic Zones" by University of Texas Libraries is licensed under Public Domain (http://www.lib.utexas.edu/maps/middle_east_and_asia/china_econ96.jpg).

Figure 12.16 "Hong Kong and Vicinity" by University of Texas Libraries is licensed under Public Domain (http://www.lib.utexas.edu/maps/middle_east_and_asia/hong_kong_pol98.jpg).

Figure 12.17 "Aqua Luna, a Chinese Junk in Victoria Harbour (Hong Kong)" by Mk2010 is licensed under CC-BY-SA 4.0 International (https://commons.wikimedia.org/wiki/File:Aqua_Luna,_a_Chinese_Junk_in_Victoria_Harbour_(Hong_Kong).jpg).

Figure 12.18 "Map of Taiwan" by CIA World Factbook is licensed under Public Domain.

Figure 12.19 "Potala" by Ondřej Žváček is licensed under CC-BY 3.0 Unported (http://commons.wikimedia.org/wiki/File:Potala.jpg).

Figure 12.20 "Happiness. The Dalai Lama at Vancouver" by kris krüg is licensed under CC-BY-SA 2.0 Generic (https://commons.wikimedia.org/wiki/File:Happiness._The_Dalai_Lama_at_Vancouver.jpg?fastcci_from=3843330&c1=3843330&d1=15&s=200&a=list).

Figure 12.22 "Mongolian warriors" by hu:Burumbátor is licensed under Public Domain (http://commons.wikimedia.org/wiki/File:Mongolian_warriors.jpg).

Figure 12.23 "Japan" by CIA World Factbook is licensed under Public Domain (https://www.cia.gov/library/publications/the-world-factbook/geos/ja.html).

Figure 12.24 "Tokyo Tower view & Mt. Fuji" by paper balloon is licensed under CC-BY 2.0 Generic (http://www.flickr.com/photos/bonjourmeeshell/3201391144).

Figure 12.26 "Shinto shrine in Nagasaki" by US Marine Corps is licensed under Public Domain.

Figure 12.29"The Korean Peninsula between the North and the South" by US Information Agency is licensed under Public Domain (http://research.archives.gov/accesswebapp/faces/showDetail?file=Item_541822.xml&loc=108).

Figure 12.30 "Mansudae Grand Monument" by John Pavelka is licensed under CC-BY 2.0 Generic (http://www.flickr.com/photos/28705377@N04/4610364189).

Figure 12.31 "Korean DMZ" (left) by Expert Infantry is licensed under CC-BY 2.0 Generic (http://www.flickr.com/photos/expertinfantry/5439135543); "US Army Korea and 63 building" (right) by PhareannaH[berhabuk]. is licensed under CC-BY 2.0 Generic (http://www.flickr.com/photos/phareannah/2520374354).

Figure 12.32"Reunification Buddha in South Korea, Erected to Signify the Unity of the Korean People" by Jane Elizabeth Keeler is licensed under CC-BY 3.0 Unported.

Chapter 13

Figure 13.1 "Southeast Asia" by University of Texas Libraries is licensed under Public Domain (http://www.lib.utexas.edu/maps/middle_east_and_asia/southeast_asia_pol_2003.jpg).

Figure 13.3 "Bandovietnam-final-fill-scale" by NgaViet is licensed under Public Domain (http://en.wikipedia.org/wiki/File:Bandovietnam-final-fill-scale.svg).

Figure 13.5 "Vietnam_full_04" by Heiko Carstens is licensed under CC-BY 2.0 Generic (http://www.flickr.com/photos/hierundsonstwo/280323293).

Figure 13.6 "Attapeu" by Prince Roy is licensed under CC-BY 2.0 Generic (http://www.flickr.com/photos/princeroy/5457616535).

Figure 13.7 "Choeungek2" by Adam Carr is licensed under Public Domain (http://commons.wikimedia.org/wiki/File:Choeungek2.JPG).

Figure 13.8 "Buddhist monks in front of the Angkor Wat" by sam garza is licensed under CC-BY 2.0 Generic (https://commons.wikimedia.org/wiki/File:Buddhist_monks_in_front_of_the_Angkor_Wat.jpg).

Figure 13.9 "Bangkok Night Wikimedia Commons" by Benh LIEU SONG is licensed under CC-BY-SA 3.0 Unported (https://commons.wikimedia.org/wiki/File:Bangkok_Night_Wikimedia_Commons.jpg).

Figure 13.10 "Burma (Myanmar): The Irrawaddy River and the North/South Layout" by CIA World Factbook is licensed under Public Domain.

Figure 13.11 "2007 Myanmar protests 7" by racoles is licensed under CC-BY 2.0 Generic (http://commons.wikimedia.org/wiki/File:2007_Myanmar_protests_7.jpg).

Figure 13.13 "Busy Roads Of Kota Kinabalu City" by thienzieyung is licensed under CC-BY 2.0 Generic (http://www.flickr.com/photos/thienzieyung/4693359106).

Figure 13.15 "Singapore skyline from Elgin bridge (8195437887)" by Erwin Soo is licensed under CC-BY 2.0 Generic (https://commons.wikimedia.org/wiki/File:Singapore_skyline_from_Elgin_bridge_(8195437887).jpg).

Figure 13.16 "Indonesia provinces english" by Golbez is licensed under CC-BY-SA 3.0 Unported (http://commons.wikimedia.org/wiki/File:Indonesia_provinces_english.png).

Figure 13.17 "Jakarta skyline" by Netaholic13 is licensed under CC-BY 3.0 Unported (https://commons.wikimedia.org/wiki/File:Jakarta_skyline.jpg).

Figure 13.18 "Ash Plume from Mt. Pinatubo during 1991 Eruption" by USGS is licensed under Public Domain (http://pubs.usgs.gov/fs/1997/fs113-97).

Figure 13.19 "Island regions of the Philippines" by seav is licensed under CC-BY-SA 3.0 Unported (https://commons.wikimedia.org/wiki/File:Island_regions_of_the_Philippines.png).

Figure 13.20 "Magellans cross marker" by Editor999999 is licensed under CC-BY-SA 3.0 Unported (https://commons.wikimedia.org/wiki/File:Magellans_cross_marker.jpg).

Figure 13.21 "Manila_09440rt" by Stefan Munder is licensed under CC-BY 2.0 Generic (http://www.flickr.com/photos/insmu74/4314519466).

CPSIA information can be obtained
at www.ICGtesting.com
Printed in the USA
LVOW05s1921150118
562979LV00031B/1968/P